The Chemistry of Evolution
The Development of our Ecosystem

D1296756

The Chemistry of Evolution

The Development of our Ecosystem

Cover illustration

The cover diagram is the conical zone, with a cut-away sector, of ecological advance with time while the cross-section dimensions at right angles to the time axis are of advancing chemistry and energy used. The side figures show the extent of species inside chemotypes, see text, and the increasing absorption and degradation, use of energy with time.

The Chemistry of Evolution
The Development of our Ecosystem

R.J.P. Williams

Inorganic Chemistry Laboratory
University of Oxford, South Parks Road
Oxford, OX1 3QR, UK

and

J.J.R. Fraústo da Silva

Instituto Superior Técnico
Technical University of Lisbon
Lisbon, Portugal

ELSEVIER

Amsterdam – Boston – Heidelberg – London – New York – Oxford – Paris
San Diego – San Francisco – Singapore – Sydney – Tokyo

ELSEVIER B.V.	ELSEVIER Inc.	ELSEVIER Ltd	ELSEVIER Ltd
Radarweg 29	525 B Street, Suite 1900	The Boulevard, Langford Lane	84 Theobalds Road
P.O. Box 211, 1000 AE Amsterdam	San Diego, CA 92101-4495	Kidlington, Oxford OX5 1GB	London WC1X 8RR
The Netherlands	USA	UK	UK

First edition 2006

Library of Congress Cataloging in Publication Data
A catalog record is available from the Library of Congress.

British Library Cataloguing in Publication Data
A catalogue record is available from the British Library.

ISBN-13: 978-0-444-52155-2
ISBN-10: 0-444-52155-0

∞ The paper used in this publication meets the requirements of ANSI/NISO Z39.48-1992 (Permanence of Paper). Printed in The Netherlands.

Working together to grow libraries in developing countries

www.elsevier.com | www.bookaid.org | www.sabre.org

ELSEVIER   BOOK AID International   Sabre Foundation

Acknowledgements

Our special thanks go to Mrs. Susie Compton who has managed the whole manuscript and the authors.

R.J.P. Williams acknowledges the continued support of the Inorganic Chemistry Laboratory, Oxford University and in his career of Wadham College, Oxford and of The Royal Society.

J.J.R. Fraústo da Silva acknowledges the Foundation for Science and Technology, Ministry for Science and Technology Portugal, for general support of research activities. Special thanks are due to Dr. Marina Fraústo da Silva and Eng. José Nascimento for the reproduction and redrawing of most figures of the book, and to Mrs. Teresa Maria Carreiras da Silva and Miss Cristina Sequeira da Silva for proficient secretarial assistance and concomitant considerable patience.

Preface

Traditional attitudes to biological evolution were based on the examination of morphological and behavioural features of organisms. They led to the classification of "species" by scientists such as Linnaeus and later to the analysis of the relationship between species by Wallace and especially Charles Darwin. It is therefore of interest to note some of Darwin's remarks which anticipate more recent developments. In "The Origin of Species (by means of natural selection)" Darwin presented his view that evolution of living organisms is a slow incremental process of natural selection among randomly occurring variations in descendants. In his opinion, the diversity of organisms, living and extinct, was the product of blind chance and struggle. However, he also wrote that "...natural selection depends on there being places in the polity of nature which can be better occupied by some of the inhabitants of the country undergoing modification of some kind. The existence of such places *will often depend on the physical changes*, which are generally very slow, but the action of natural selection will probably still oftener depend on some of the inhabitants becoming slowly modified, *the mutual interaction* of many of the other inhabitants being thus disturbed". Therefore... "the greatest amount of life can be supported by great diversification of structure". (See the Introduction by J.W. Burrow to the Origin of Species, Penguin Books, 1968.) Note that there is no mention *of chemical change in the environment nor in life with time and there is no analysis of the sources and deployment of energy*. Since Darwin's days, this reductionist, organism centred, approach has changed considerably. Once the concept of "genes" was established the emphasis of the discussions on evolution shifted to a new dominant description – natural selection among randomly mutated pools of genes. The connection of these changes with changes of physical and biological organism surroundings was observed in line with Darwinian thinking. However, the effect of the surrounds of organisms on evolution was not deemed to be causative.

In the last half of the 20th century a different perspective on organism evolution emerged from ecological studies which have led to the more general concept of the "ecosystem", involving not just the changes in biota, but also of the environment now treated interactively in general thermodynamic terms. It is becoming evident that the study of evolution of life must be centred on such systems rather than on individual organisms or species and their habitats. Fluxes of materials and energy became the new focus and management of the whole ecosystem was seen to require synergism, positive or negative, among organisms. The new approaches are still consistent with Darwin's principle of the "natural selection of species", but the emphasis has changed; it is the ecosystem that has evolved. However, the absence of chemical detail in systems treatments and the success and appeal of the limited chemistry within genetics and molecular biology have kept the two separate.

The weakness of this new approach is then that it pays no attention to the chemical components of the environment or of their fluxes employed by organisms. Such components are not just derived from "organic" elements – carbon, hydrogen, oxygen, nitrogen, and some sulfur and phosphorus and their compounds – but include many *essential* metals and some other non-metals (and their ions) with which they interact and without which life would not exist. The need, stressed in this book, is to take into account the major features of the life's physical chemistry involving these essential elements, some 20, used free and in combination, by organisms, and thereby providing a detailed treatment of the ecosystem approach. To stress the nature of our approach we classify organisms not by morphology or genes, but by their chemical elements, their concentrations and those of the compounds they form, their energetics, the space they occupy and their organisation, all in flowing systems, that is by thermodynamic variables. We call these different groupings of organisms, chemotypes. The sequence in evolution is then seen to be directional in detailed physical–chemical differences between organisms in the order of their appearance: prokaryotes (anaerobic then aerobic), unicellular eukaryotes, multicellular eukaryotes, animals with brains and human beings, see cover 1. They differ in use of oxidised chemicals in particular, of energy capture, of size and shape, and of complexity of organisation. There are within each major chemotype sub-groups which are not yet well analysed. We show that the sequence is a natural directed consequence of the interaction between the energised organisms, and the environment because the environment changed in an inevitable way before organisms, which just adapted to the changes. Species are still seen as arising within chemotypes by Darwinian selection.

In order to show that the whole evolution of the ecosystem is in fact directional through the required physical/chemical chemistry of living and environmental processes, we have to describe first the known systematic changing oxidation of geo-chemicals of the surface of Earth over the 5 billion years of its existence, Chapter 1. The background of all this chemistry is the ability of the chemical elements to form compounds either in stable equilibrium or in kinetically long-lasting states (Chapter 2). The latter are largely organic compounds unavoidably energised by the sun and they, with a complement of concentrated inorganic elements, gave rise to life. This energisation of chemicals leads to unavoidable reactions of synthesis and decay so that the chemistry is within cycles enforced by the degradation of light to heat, that is the production of thermal entropy (Chapter 3). In Chapter 4, we give a general description of the basic special components, selected by energy and survival criteria which have come together through these energised cycles of available elements to engender life. They are a consequence of optimal energy flux. It will be seen that since life had to reduce environmental chemicals (CO and CO_2) to make such chemicals it therefore increased the oxidation of the environment. It is the combination of an increasing uptake and degradation of energy (with a corresponding increase in thermal entropy generation) together with an unavoidable utilisation of more oxidised environmental chemicals (produced through the activity of organisms) that caused evolution of the ecosystem in the direction we observe. These cycles strain to be element neutral recycling all material while degrading energy, producing no pollution except heat. The sequence is described in Chapters 5–10 following that of the order of chemotypes listed above.

Our discussion indicates that, in the light of this clearly directional evolution, a re-evaluation of the role and functioning of the genetic machinery (not just of the coded molecules, DNA, RNA, proteins) is necessary. How does chance mutation lead to directional change when DNA is both conservative and changes of its sequence are undeniably linked only to chance mutation? There is growing evidence of occurrence of the so-called "epigenetic" effects of various kinds, which can change the present views on how not only inherited but also environmentally directed acquisitions may be transmitted to the offspring. An added factor is that complexity of later organisms also makes it necessary for an efficient total system to rely on cooperation between later and more primitive, earlier, organisms including distribution of genes. Cooperation not competition has led to overall ecological fitness.

In this ecological system of organisms and the environment *one species* has developed a remarkable brain of such power that all evolution now depends upon it, namely *Homo sapiens* or mankind. Mankind is cognitive and has become a special chemotype able to handle all elements (90 no longer 20), all forms of energy in much of space and in a highly sophisticated organisation. Owing to its activity, organisation which started from being just inside organisms, linked to genetic change, has passed into the environment to create 'abiotic' novel forms, and can even adjust genes themselves, using brains. Although in an extreme form, this development is in line with the general evolution of chemotypes as they became increasingly interactive with the environment, this activity is not element neutral and produces pollution. From this position of strength mankind is now dominant and can affect the whole ecosystem, which includes itself, very quickly. The situation is made more difficult however by the development of the individual in this species, which relies on an isolated brain not genes for decision making. Use of scientific knowledge has increased the independence of the individual so that there is no longer overall communal control. The resultant conditions of the present ecosystem with a strong element of human self-interest are examined in the last chapter. Sooner or later, mankind has to see that it is a part of the ecosystem and cannot afford such a selfish individual or even a selfish communal lifestyle. Mankind must be educated to be able to manage and sustain a biological- and environmental-friendly ecosystem which has been inherited, otherwise selfish human activity could prove self-destructive though evolution will continue. This education, which we hope this book will provide, will need to generate a different political will in the management of our science-based society. It will require scientists not only to teach the nature of ecological systems, but to become political since only they have the necessary knowledge to advise. Mankind has now the power to maintain, perhaps to advance, its inheritance but that will require careful limitation of exploitation of the ecosystem.

We stress that in this book we follow certain earlier views on evolution while we remove any religious undertones. For example the essence of Descartes' thinking is as follows: If in the beginning the world had been given only the form of chaos (we replace chaos by the Big Bang) and provided the laws of nature were then established all material things could in the course of time come to be just as we see them. In this approach and in our book there are no acts of creation other than the Big Bang and there is no need for intelligent design, except for the laws of nature.

R.J.P. WILLIAMS
J.J.R. FRAÚSTO DA SILVA
Oxford and Lisbon, June 2005

Contents

Contents

CHAPTER 1

The Evolution of Earth–The Geochemical Partner of the Global Ecosystem (5 Billion Years of History)

1.1. Introduction

This book concerns the Earth's evolving ecology, where ecology is defined by the study of the relationship between organisms and their physical and chemical surroundings. We begin with an account of geological chemistry starting from the Big Bang, which created the first light chemical elements, especially hydrogen and helium, from which in order stars, e.g. the Sun, and planets, including the Earth, were formed. At the same time, this event generated the very uneven distribution of energy, including that of the Sun, which is the major source of the energy for biological chemistry on Earth. A main intention of the book is then to show that when our planetary system formed, less than 5 billion years ago, the chemical element content of the Earth and the forms in which the elements occurred, described in Chapter 2, imposed gross restrictions on whatever chemistry, biological or other, could develop on its surface before mankind appeared. These restrictions are mainly due to the availability of chemicals in the atmosphere and surface waters, which have access to the Sun's energy. The understanding of the possibilities of this chemistry has come with the development of studies of elements and compounds largely in the last 200 years (Chapter 2). Given the background of geological chemistry and our more general understanding of what are called inorganic and organic chemistries we can proceed to consider ways in which chemical *systems* such as we

1

see in primitive life could have been energised. In Chapters 3 and 4 we introduce the global ecological system for it is not just biological chemistry that must be examined when we consider such evolution; we must remember that the geological chemistry of Earth's surface changed when life developed waste, mainly oxygen gas. The new energised surface chemicals then back-reacted with the biological system and forced it to change. The whole of the geosphere, hydrosphere, atmosphere and biosphere is seen to be one large chemical system evolving in a strictly restrained chemical fashion (Chapters 5–9). Throughout these chapters we stress strongly the role of the chemical elements, rather than particular types of compound, e.g. DNA or proteins, in evolution since in this way we knit together the chemistry of the Earth's mineral surface and life. In Chapter 10, we will consider the influence of mankind on the ecosystem and then in Chapter 11, we shall bring all the previous developments together. In order to present this ecological system in a manageable way we describe separately in this chapter the initial state of Earth and then how oxygen, no matter what was its source (we know it was due to organisms), changed Earth's surface chemistry in the atmosphere and waters. This allows us to describe the more striking evolution of organisms in Chapters 5–9, the principles of the chemistry and biochemistry having been described in Chapters 2–4, with reference to its dependence upon the environment at different times, see cover picture.

1.2.   The Formation of the Atomic Elements: Abundances

Whatever the nature of the ecosystem, it has to be based on the chemical elements on Earth. Our first task then is to describe what they are and where they are to be found. We begin with an outline of their formation (see P.A. Cox in *Further Reading*). The chemical elements were formed in giant stars by successive thermonuclear fusion followed by other secondary processes – neutron and proton capture. The amounts of the different elements that were synthesised were controlled by the rates of reactions of the nuclei starting from hydrogen and helium, the lightest of the elements, which originated from the Big Bang some 13.7×10^9 years ago. These two elements were and have remained of very high abundance in the universe (a total of about 99%) showing that subsequent nuclear reactions used only a small fraction of them and have not been completed. The reactions in giant stars, followed by the explosion of these stars when the state of supernovae was reached, gave rise to heavier elements up to the radioactive and unstable nuclei, which finish at the nuclear stability limit, the element uranium. Scientists have investigated many of these nuclear reactions and have created even heavier radioactive elements in very small amounts (see Chapter 10). The explosive reactions in the giant stars produced a well-defined pattern of abundances of elements of intermediate size between helium and uranium, shown in Figure 1.1. The elements lithium, beryllium and boron did not accumulate in large amounts due to the kinetics of nuclear reactions in the giant stars, so that the next most abundant elements in the universe after hydrogen and helium are carbon, nitrogen and oxygen. Abundance then

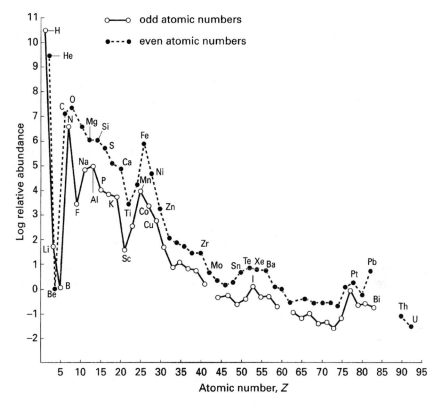

FIG. 1.1.   Relative abundances of the 'unchangeable' elements in the universe (based on log (abundance of Si) = 6). Filled circles (•), even atomic numbers; open circles (○), odd atomic numbers. ('Unchangeable' refers here to atoms of elements. Thus we ignore, for the purposes of this book, any transmutation of elements.)

decreases somewhat irregularly until above atomic weight 50, i.e. around atomic number 25. The appearance around this atomic weight of a few elements in high abundance, such as iron and nickel, is due to the special stability of their nuclei. After iron, abundance decreases to very low levels at atomic weight of uranium. Note that elements of odd atomic number are less common, hence there is more iron and nickel than there is cobalt, and more carbon and oxygen than nitrogen. We believe *all these facts to be important in the evolution of Earth and life.* The abundance of all the elements is now well understood in terms of nuclear reaction theory of processes in giant stars (see P.A. Cox in *Further Reading*).

It is believed that our small star, the Sun, was formed by gravitational attraction of the dust of the exploded giant stars giving the distribution of elements shown in Figure 1.1 at a very high temperature. Probably, a nearby interaction with another passing star generated a great plume of solar nebula extending billions of miles away from the Sun's central hot mass. Cooling of this plume led to the formation of local concentrations of elements, some of which condensed in a spatial sequence to give rise to basic planets and meteorites in orbit around the Sun. The concentration of gases was not altogether of uniform element composition before condensation due to the plume's formation under the gravitational attraction of the Sun. The heavier metallic elements, therefore, tended to stay closer to the Sun to some degree and helped to give rise to metal-rich planets, Earth, Mars and Venus, while further away from the Sun more gaseous, more largely non-metal planets, such as Saturn and Jupiter, formed.[*] As the planets nearest to the Sun are small and are heated by the Sun they also lost most of their hydrogen and helium. In the case of our star, the Sun, the planetary system is well analysed, but as yet, although several planets have been identified outside the solar system, we do not know of any parallel planetary system elsewhere in the universe.

Now, apart from the planets, many meteorites were formed, moving in quite different orbits and of quite different chemical composition. In particular, the so-called C-1 meteorites composed of carbonaceous chondrites have a composition of elements much closer to that of the Sun. It is proposed (see for example Harder and also Robert in *Further Reading*) that many of these meteorites collided with very early Earth and became incorporated in it, so that eventually some 15% of Earth came from this material (see Section 1.11). Other planets such as Mars and the Moon could have had similar histories, but the remote planets and Venus are very different.

1.3. Earth's Physical Nature: Temperature and Pressure

The planets nearest the Sun have a high-temperature surface while those further away have a low temperature. The temperature depends on the closeness to the Sun, but it also depends on the chemical composition and zone structures of the individual planets and their sizes. In this respect Earth is a somewhat peculiar planet, we do not know whether it is unique or not in that its core has remained very hot, mainly due to gravitic compression and radioactive decay of some unstable isotopes, and loss of core heat has been restricted by a poorly conducting mainly oxide mantle. This heat still contributes very considerably to the overall temperature of the Earth's surface. The hot core, some of it solid, is composed of metals, mainly iron, while the mantle is largely of molten oxidic rocks until the thin surface of solid rocks of many different compositions, such as silicates, sulfides and carbonates, occurs. This is usually called the crust, below the oceans, and forms the continents of today. Water and the atmosphere are reached in further outward succession. We shall describe the relevant chemistry in more detail later; here, we are concerned first with the temperature gradient from the interior to the surface (Figure 1.2). The Earth's surface, i.e. the crust, the sea and the atmosphere, is of

---

*There is much debate concerning the state of the original plume before planets formed so that there is no generally accepted theory of their formation. Turbulence in the plume could have led to a rather even distribution of elements in all planets.

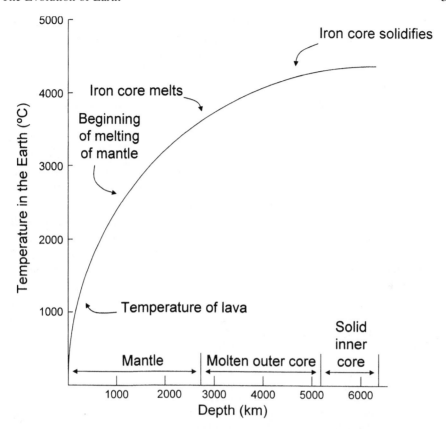

FIG. 1.2. A geotherm curve showing how temperature increases with depth from the surface to the centre of the Earth. Temperatures at Earth's centre reach about 4300°C. (Adapted from Press and Siever, see Further Reading.)

special interest in this book as it has shown the greatest evolution and here life appeared (see below), while, as stated, the almost invariant, precise size and composition of the core and mantle have helped to generate a particular surface temperature. Another contribution to the temperature is clearly from the Sun's radiation, but it can only be understood once the chemical nature of the Earth's atmosphere is described.

Most probably the Earth's atmosphere was initially mainly composed of the light gaseous non-metal nitrogen, free and as both ammonia and hydrogen cyanide, and carbon oxides, some hydrogen, methane and helium. At first water was possibly present as a gas. Note that no free molecular oxygen was present, becoming a major contributor only some 1–2 billion years ago. Because the gases were formed from light elements they tended to escape from the gravity field of the Earth, so that over long historical periods there has been considerable change in composition (Table 1.1). This is part of the evolution of the whole geological component of the

TABLE 1.1

SUMMARY OF DATA ON THE PROBABLE CHEMICAL COMPOSITION OF
THE ATMOSPHERE DURING STAGES I, II AND III[a]

Stage I (early Earth)	Stage II ($\sim 2 \times 10^9$ years ago)	Stage III (today)
Major components ($p > 10^{-2}$ atm)		
CO_2 (10 bar)		
N_2 (1 bar)	N_2	N_2 1 bar
CH_4		O_2 1 bar
CO		
Minor components ($10^{-2} > p > 10^{-6}$ atm)		
H_2 (?)	O_2 (?)	Argon
H_2O	H_2O	H_2O
H_2S	CO_2	CO_2 (10^{-3} bar)
NH_3	Argon	
Argon	(CO?)	
NO(?)	(NO?)	
Trace components ($p < 10^{-6}$ atm)		
He	Ne	Ne
Ne	He	He
	CH_4	CH_4
	NH_3 (?)	CO
	SO_2	NO
O_2 (10^{-13} bar)	H_2S (?)	

[a] We are able to give a good account of stage III and a good estimate of stage I,
but the evolutionary period, stage II, is hard to describe with any accuracy, see (?).

ecosystem. Not very much of the free light non-metals is left on the Earth but there
is much bound hydrogen, oxygen and carbon in condensed forms, and with nitrogen in organic compounds. The later production of gases from activity on the
solid/liquid surfaces of Earth has changed the composition further (Table 1.1).

We now return to the second contribution to the surface temperature – radiation
from the Sun. Many of the above gases in the atmosphere screen both the incoming Sun's UV light to Earth and outgoing (heat) radiation from Earth to some
degree (see Table 3.7), so that this heating, with that from the core, form the combined non-biological sources of energy which control the surface temperature. The
Sun's radiant energy as well as the atmosphere have altered with time; the Sun has
increased by 30% in luminosity over 5×10^9 years, while the CO_2 cover has
decreased more than 10-fold, and considerable amounts of methane may have been
introduced by early life (see Kasting in Further Reading). The curious and fortuitous fact (known as the "faint young Sun paradox") is that the combination of the
changes of the atmosphere, of CO_2 and probably CH_4 especially, and of the Sun
have been compensatory in total energy capture by the surface, so that over the
whole period of existence of the cool planet, Earth, the surface temperature has

been fixed within narrow limits of 300 ± 30 K. This has allowed the presence of condensed water for the whole period with perhaps minor fluctuations of lower temperatures due to local glaciations and at "snow-ball Earth" times (see Press and Siever and also Stanley in Further Reading). The atmospheric pressure has also steadily fallen from some 10 times as high an initial value to that of today. These factors are critical since they allowed the evolution of geochemical and biological chemistry at all the prevailing temperatures over 4×10^9 years.

We explain the loss of CO_2 as follows.

Much of the loss of CO_2 from the atmosphere is due to the run-off of calcium from the land that has reacted with the HCO_3^- in the sea to give sedimentary carbonates. The formation of much of the calcium carbonate will have arisen from initial precipitation of it on the surface of organisms but it is not then a biological product, while some of it is biogenic from shells. Deep burial of the calcite has transformed much of it to dolomite, which has reappeared even as mountain ranges. A parallel erosion is probably the cause of silica deposits. Another part of the CO_2 loss (and much nitrogen) is due to biological uptake into organic chemicals. Some of this carbon is not part of any rapid cycle but has become locked into "waste" as the carbon of methane, oil and coal. Today we know that carbon cycles exist without much loss between organisms and the environment including minerals which re-dissolve. It is clearly not correct to say that biological activity created the present CO_2 levels since much is due to water cycling generating calcium (and silicic acid) in the sea.

1.4.   Earth's Atmosphere and Its Composition

As mentioned, the pressure of gases on Earth's surface has probably fallen at least by a factor of 10-fold due to loss of gases and to other reactions to form solids, e.g. carbonates. (For a recent discussion of the early atmosphere see Chyba in Further Reading.) The pressure possibly settled close to its present value about 1 billion years ago. There has also been a dramatic change in composition of the atmosphere as described in Table 1.1. Comparison with other planets is of interest. The atmospheres of Venus, a very hot planet, and of Jupiter, a very cold planet, are compared in Table 1.2. The question arises as to the probability of existence not just of planetary systems but of planets of the particular atmospheric composition and temperature elsewhere in the universe, which could give rise to the peculiar initial and long-term suitable temperature/composition conditions for life as we know it, especially since life may only exist in the presence of *liquid water*. A second chemical observation of great interest here is that *the hydrides* other than water that were probably present in adequate amounts and which were required to initiate life notably CH_4, NH_3, HCN, H_2O, H_2S, H_2Se, contain all the elements to be found later in coded amino acids. Of these elements only carbon was also present as an oxide, CO and CO_2. No other element except the acidic halogens have chemically stable hydrides in H_2O. In the case of Earth the kinetics of its formation and early history are seen to be quite fortuitous (Table 1.2); this is particularly true of the

TABLE 1.2

Composition of Planetary Atmosphere (Today) – Molar% in Volume

	Planets				
Elements	Venus	Earth	Mars	Jupiter	Saturn
Hydrogen (H_2)	0.001	0.00005	—	89.8	96.3
H_2O	0.006	0–4	0.03	0.0005	0.0002
Carbon	96.5 (CO_2)	0.035 (CO_2)	95.3 (CO_2)	0.3 (CH_4)	0.45 (CH_4)
Nitrogen	3.5 (N_2)	78.1 (N_2)	2.7 (N_2)	0.26 (NH_3)	0.02 (NH_3)
Oxygen(O_2)	< 0.00003	20.9	0.13	—	—
Helium	0.0012	0.0005	—	10.2	3.25
Neon	0.0007	0.0018	0.00025	0.001	—
Argon	0.007	0.93	1.6	—	—
Temperature at surface (K)	735	288	223	165	134

Source: R.P. Wayne, *Chemistry of Atmosphere*. Oxford University Press, Oxford, 2002 and references therein.

surface, as we shall see, so that it is not possible yet to give an estimate of the likelihood of a very similar planet with this peculiar history in the universe. We stress that there are many factors that must be present to give the required fixed steady-state temperature around 300 K and hence water on the surface of a planet for almost 4.5×10^9 years and still others that have assisted or allowed the chemistry of life to appear and be maintained (see Section 1.8). (An interesting uncertainty is the quantity of NO present quite early in time. If NO was present very early in the history of Earth through the reaction of N_2 and H_2O activated by lightning, then it could have given rise to early redox chemistry, see Chapter 6.)

1.5.   The Initial Formation of Minerals

Given this description of the physical condition of the Earth as at first formed it is possible to turn to its chemical composition in more detail (Table 1.3). Immediately, we must consider different condensed mineral zones or compartments that became isolated from one another (Fig. 1.3). These zones formed slowly from the hot gaseous elements of the Earth. The earliest steps, chemical reactions, followed the equilibrium affinity of elements for one another (see Williams and Fraústo da Silva, 1999, in Further Reading). The major first reaction was therefore that between oxygen and the elements carbon, magnesium, silicon, iron, aluminium and phosphorus. Given the abundances of these elements, the reactions *removed virtually all the oxygen* as oxides leaving among metals an excess of mainly iron, some nickel, sodium, potassium and calcium, as well as many rarer heavy metals up to around Zn and Cu in the periodic table. Most other metal

TABLE 1.3

MAJOR COMPOUNDS OF THE EARTH

Non-metal element	Compound with other elements
Oxygen	Si, Mg, Fe, C, H, P, Al, Mn,
Nitrogen	H, C
Sulfur	H, Fe, Co, Ni, Cu, Zn, heavy group 13–16 metals
Hydrogen	O, C, S, (N)

Note: It is sometimes convenient to describe the element distribution between sulfides and oxides as chalcophiles (occurring in the Earth's crust as sulfides) and lithophiles (predominating as oxides and halides in the Earth's crust) (see Fig. 1.5). This geochemical classification includes also the siderophiles (remaining as metals or alloys, especially in the Earth's core) and the atmophiles (which occurs largely in volatile form in the atmosphere and dissolved in the oceans).
Note that the total amount of metals exceeded that of light non-metals so that Earth was very reducing and water could not have existed in an overall equilibrated system.

elements were present in only insignificant amounts. The major non-metals remaining were nitrogen, sulfur, and chlorine. Cooling below 2500 K led to condensation of the Mg, Fe and Al metal oxides, much with silicon, SiO_2, giving silicates, and to an excess of common heavy metals, iron and nickel, now in the metallic state (see Figure 1.3). Two immiscible liquids separated to form eventually the dense iron/nickel metallic core and the less dense molten mantle of oxides and silicates. Further cooling to 1500 K allowed residual metals such as copper and zinc to form sulfides while the outer surface of the mantle, the crust, solidified. Undoubtedly the upthrust now and then of molten material (see Section 1.11), led to a multitude of lesser chemical reactions, perhaps giving some iron sulfides, for example, on the surface. The relative heats (ease) of formation of oxides and sulfides are shown in Table 1.4. Oxides would be clearly thermodynamically favoured, provided oxygen was available, which was not the case for some elements (see above). These facts are of extreme consequence for the understanding of the nature of the geological/biological ecosystem on the surface of what is a very particular planet, and they have a huge influence on biological evolution.

Note that the excess of metal elements made the whole including the surface in a somewhat reduced state except for carbon as CO and CO_2. So far the process described is one of slow change as the temperature decreased, with reactions closely following affinities of the elements for one another. However, relative affinities change with temperature decrease and as temperature decreased further, the possibility of reactions to restore equilibrium appropriate to all relevant affinity orders was prevented by barriers to reactions, both physical and chemical at the lower temperatures. Thus, Earth developed a huge energy store beneath the cool surface and this is an important part of its later ecosystem.

Some compounds of major elements condensed much later, such as the chlorides of sodium and potassium and possibly calcium, and some excess magnesium,

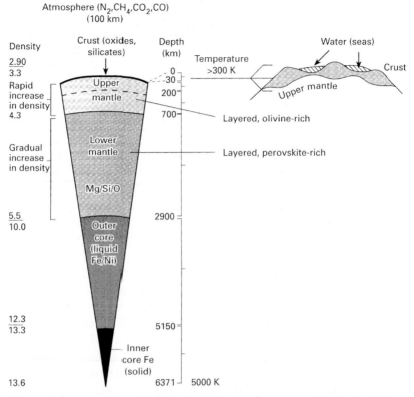

FIG. 1.3.   A schematic section through the Earth. Note that there is a huge source of energy in unbalanced chemical and physical (temperature) gradients of the Earth. This out-of-balance is due to the route of formation of the planet, i.e., it is a kinetically controlled situation due to gravity and cooling. Some dramatic consequences of this physiochemical out-of-balance are earthquakes and volcanic eruptions and (possibly) the beginnings of life itself. The out-of-balance shown in this figure has been increased later by living systems via the production of dioxygen (O_2) and carbon (coal and oil) deposits. In this last series of steps sunlight energy is trapped in the crust, atmosphere and waters in chemicals. Note that if we represent the radius of the Earth by a 15 cm radius sphere, the average thickness of the sea is about 0.2 mm and that of the atmosphere to 100 km is 2.0 mm.

which formed salts of one kind or another. Of course, nitrogen remained as N_2 and carbon remained largely as carbonates and CO_2, CO and some CH_4 in the atmosphere. The fate of many of these elements was decided subsequently by the formation of water on the surface of the Earth and then by living systems. Carbonates formed relatively late; note especially the formation of those of Ca/Mg, some due to later biological activity.

   The above general picture is of an early mineral and gaseous Earth formed by a sequence of events controlled not just by equilibrium chemistry but particularly by physical compartmentalisation due to condensation as the temperature decreased. There is no other way of explaining the central core of metals which,

TABLE 1.4

HEATS OF FORMATION OF OXIDES AND SULFIDES AT 25°C (KCALS)

	ΔH_O oxide/per O	ΔH_S Sulfide/per S	$\Delta H_O - \Delta H_S$ difference (O–S)
Mg	−143.8	−83.0	−60.8
Al	−130 (Al$_2$O$_3$)	−40.5	−89.5
Ca	−151.9	−115.5	−36.4
Mn	−92.0	−48.8	−43.3
Fe	−63.7	−22.7 (FeS$_2$−21.5)	−41.0
Co	−57.2	−20.2	−37.0
Ni	−57.8	−18.6	−39.2
Cu	−37.1	−11.6	−25.5
Zn	−83.2	−48.5	−34.7
Cd	−60.9	−34.5	−26.4
C	−47.0(CO$_2$) −26.4(CO)	+13.8 (CS$_2$)	−60.8
Si	−102.2(SiO$_2$)	−17.4 (SiS$_2$)	−84.8
P	−72.0(P$_4$O$_{10}$)	—	—
N	+21.6(NO)	+31.8 (NS)	−10.2
Na	−99.4(Na$_2$O)	−89.2 (Na$_2$S)	−10.2
K	−86.4(K$_2$O)	−100.0 (K$_2$S)	+13.6

Note: Heat of formation is the heat of reaction at 25°C, e.g.

M (solid) + ½ O$_2$ (gas) → MO (solid)

and the corresponding reaction for sulfur. Elements with strong element M lattices appear to have small ΔH.
Note the sequences Mg > Mn > Fe > Co (Ni) > Cu < Zn of MO and MS which also shows the preference for oxygen over sulfur (O–S).

after consuming all the oxygen, should have removed all the sulfur. Hydrogen sulfide and water could not have been present then or later had these metals been exposed to them. Every planet that forms will therefore be physically and chemically different according to its size, its age and the element content of its local gas, all dependent on its distance from its parent star, and to some degree on any physical barriers to reactions that may arise linked to rates of cooling, and on later collisions. The distribution of elements due to thermodynamic and kinetic factors is, therefore, open to a huge variety of possibilities. Other features are the rotation on Earth's own axis and its orbit around the Sun, both critical for its small temperature variations and fixed local average temperatures with time. We describe some additional physical features introduced after the initial cooling in Section 1.7. Of course, outbursts of the inner heavier layers into the outer layers and the impacts of meteorites still make some changes to the whole surface, as does weathering (see Section 1.11). The actual estimated element composition of the Earth's crust is shown in Figure 1.4.

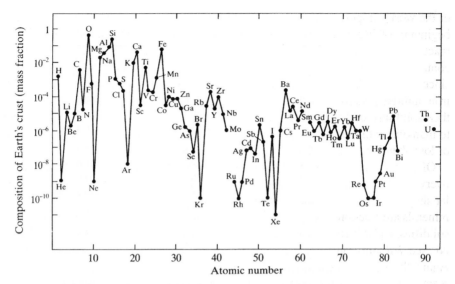

FIG. 1.4. Estimated composition of the whole Earth, showing the mass fraction of each element on a logarithmic scale. (Reproduced with permission from P.A. Cox (1989), see Further Reading.)

1.6.   The Reforming of Solids from Melts: Minority Solids

The major solids of Earth's mantle and crust are based on oxygen in the form of mixed oxides of various kinds (see Table 1.3). The melting and reforming of the solids that was and is continuous in the mantle could have ensured that thermodynamic equilibrium was approximately established among many of them.

It is clear however that there have been many secondary reactions between various solids on the surface, and, during the melting and resolidification, crystal forces have allowed local deposits of minor compounds to occur on the surface. Any inspection of the Earth shows that its rocks are different in different places and were formed at different times (see Figure 1.6). Well-known examples are the granite of Scotland and the Canadian Shield, the dolomites of northern Italy, and sandstones in all parts of the world. There are also strange outcrops of iron oxides in the Iron Knob in Australia, and large local deposits of ores such as molybdenum sulfide in Arizona, U.S.A. The appearance of these local deposits is often due to fractionation from melts but is sometimes due to secondary processing such as weathering and biological action. For example, it may be that all phosphate deposits are of biological origin. Of course, given the scale of the mantle many local minor minerals are due to sudden upwelling and recrystallisation of molten magma. We also draw attention to some particular sulfides such as those which are sulfide-deficient, e.g. NiS, those which have sulfide excess, FeS_2 (pyrite), and those which are mixed sulfides, Fe/Ni/S. As we shall show in later chapters all these compounds could have had great significance in prebiotic chemistry and occur as ores locally in many parts of the world. Importantly, all evidence shows that sulfur played a major role in the beginning of life and then and subsequently metal sulfides and their subsequent reactions have had a major influence on evolution. (Note that the crust of the Earth contains less than 1% sulfur, but it is not certain that this amount was present at first since certain meteorites that impacted Earth had considerable amount of especially iron sulfides.) Many of these minerals on the surface are not in equilibrium with their solid surrounds. Local formation of particular minerals is also of extreme value to mankind in his search for easily accessible elements (Figure 1.5), since many are of low abundance on the Earth (Figure 1.4; see also Chapter 10).

Of the major solids formed from melts, many, but not all, at equilibrium, the overwhelming influence is of cooperative interaction between ionic units of similar shape and size as we see in crystals. Trace elements apart from forming isolated minerals are fractioned in bulk oxides, for example, in particular orders as the melt solidifies, and this reduces the relative availability of some elements such as Cr and Ni (see Williams, and Williams and Fraústo da Silva (1999) in Further Reading). Again the interaction of selective molten minerals and water creates extremely reactive environments and such environments still exist, especially in the deep sea "black smokers" (hydrothermal vents), around which particular mixed minerals form, which could also have been involved in prebiotic chemistry and are still involved in the peculiarities of life in these "smokers". In Figure 1.6 we summarise

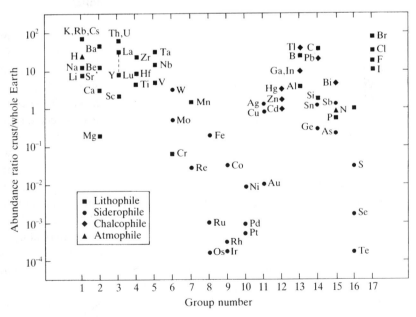

FIG. 1.5.   Comparison of crustal with whole-Earth abundances of the elements. Elements high up in this plot are enriched in the crust relative to the Earth as a whole; ones low down are depleted and concentrated in the inner parts of the Earth. (Reproduced with permission from P.A. Cox (1989), see Further Reading.)

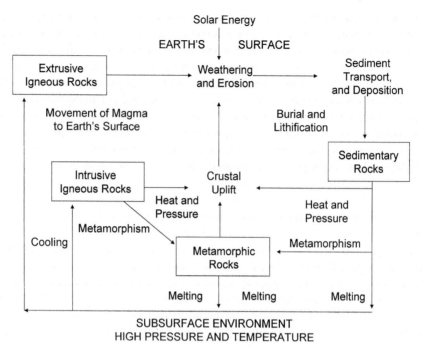

Fɪɢ. 1.6.   The rock cycle. Rocks are weathered to form sediment, which is then buried. After deep burial, the rocks undergo metamorphosis or melting, or both. Later they are deformed and uplifted into mountain chains, only to be weathered again and recycled. Some injection of rock from the upper mantle is irreversible, that is, non-cyclic. (Adapted from M.J. Pidwirny, www.geog.ouc.bc.ca/phys.geog/)

the geological processes that create and modify the Earth's rocks and changed their location. Even deep in the Earth there is constant flow with mixing which affects the chemical composition of the surface. Their interaction with the water layer is described in Section 1.8.

1.7.   The Settling Down of Earth's Physical Nature

We have noted that Earth formed as a metal-heavy planet and at a high temperature. At this temperature molten iron, the major dense metal, settled as a liquid core at least in part which, due to the motion of the Earth, produced a large magnetic field along an axis. This field protects the Earth from incoming ionised particles from the Sun. Without this protection it is doubtful if life could exist on the Earth's top surface. A second feature is the presence of the Moon. Well within the first few hundred million years of the Earth's existence after the mantle had formed it is thought that it was struck by an object not far from the size of the Moon, which caused a massive ejection of material from the Earth's surface into orbit around the Earth. The coalescence of this material made our present-day Moon. The circulation of the Moon around the Earth stabilises its rotational axis, preventing wobbling. The Earth and the Moon suffered further multiple collisions with external objects, which not only may have brought some water but could have brought many carbon compounds at least as large as amino acids. The intense bombardment completed about 4×10^9 years ago, Earth having existed for one-tenth of the time to the present day. For a summary of these physical changes see Table 1.5. Note that only a few features are of high probability, while others have a probability that is not easy to estimate since they involve unpredictable accidents. How singular is Earth in the universe?

1.8. The Initial Formation of the Sea and Its Contents

As stated above, the fact that there is any liquid water on the surface of the Earth is due to the maintained temperature and the isolation of the water from the metal core. The metals, essentially Fe and Ni, of the core would react according to the reaction

$$M + H_2O \quad \rightarrow \quad MO + H_2\uparrow$$

if contact were made. It is the thick oxide mantle of mainly Mg_2SiO_4, which protects the water from the metals. But where did the water come from initially? Clearly, in the above initial process of oxide formation it could not have existed. Once the Earth's atmosphere was cooled enough, below 100°C, it could, of course,

TABLE 1.5

PHYSICAL EFFECTS ON EARTH'S NATURE[a]

Physical feature	Effect
Gravitational settling of planetary material	Formation of metal (iron)-rich Earth 4.5×10^9 years ago
Sinking of molten iron to form a liquid hot core	Creation of magnetic field to protect Earth from Sun's particle emission
Creation of Moon by collision and loss of material	Resulted in a fixed rotational axis (no wobble) and an ever lengthening day from 5 to 24 h
Cooling of Earth	Creation of layers, mantle and crust
Bombardment of Earth by meteorites	Re-melting of mantle. Possibly introduction of water to give an acidic sea
Further bombardment of Earth by meteorites	Introduction of carbon compounds such as amino-acids?
Second cooling of Earth	4.0×10^9 years ago surface formed
Volcanic activity	Gave rise to a CO_2/H_2S-rich environment
Deep fissures in ocean bed with local chemistry	Protected environment for start of life?

[a] Many of the above observations depend upon the analysis of isotopes of H, C, N, O and S in minerals and fossils.

have trapped water from any passing object or by Earth's passage through space containing water. As mentioned above (Section 1.2), one well-supported hypothesis is that, after Earth was partly formed, certain meteorites, C-1, carbonaceous chrondrites, collided with it. These meteorites formed earlier than Earth and moved in very different orbits. Their composition was (and is) much closer to that of the Sun (see Zinner, and also Yin in Further Reading), indicating that some of the initial gas cloud that spread from the Sun cooled rapidly and trapped oxygen with some hydrogen but not large amounts of heavier elements. Some C-1 meteorites contained large amounts of water (~20% by weight) and considerable amounts of carbon compounds (5% as carbon by weight). Provided the accretion of such meteorites by collision amounted finally to some 15–20% of Earth, then the presence of water on the surface can be explained. Another possibility is that there was water release from, for example, hydroxylated silicates in the early mantle on rise of temperature by a process known as *degassing* and subsequent cooling. The explanations of the observations are not based on overall equilibria among the elements and hence the observed water content could be very different for a different planetary system. In passing note that the seas are of immense volume, oceans are often 4000 m deep and have trenches going down 11,000 m, but this represents only a very thin layer of the Earth's surface; compare the Earth's radius in Figure 1.3 (6371 km). The same applies to the atmosphere to the limit of the thermosphere (129 km).

Once the water layer had formed it would dissolve all chlorides and much of the non-metal oxides such as CO_2, P_2O_5 and any SO_2, introducing some carbonate, phosphate and perhaps sulfite into the sea. Note that only very small quantities of most elements other than Na, K, Cl, Mg, Ca, P and C were present with H and O in the early sea (see below). We expect the content of the sea would have equilibrated rapidly on the time scale of millions of years with all the minerals exposed to it (see Chapter 2), but note also that deeper layers of minerals cannot equilibrate (see Section 1.11).

Given this curious progression of chemical reactions and condensations to form first the core, mantle and crust and then the aqueous layer by accretion of meteorites and/or by degassing due probably to some melting reactions and mixing, Earth has formed as a body that is difficult to explain other than in outline. It is not a stable equilibrium body, except probably for the contents of the sea interacting with the surface minerals and atmosphere but a frozen and, on our time scale, permanently compartmentalised body. On the covering surface, which will interest us most, there may well have been a very large variety of solid-reduced oxides and sulfides (Table 1.6), lying below a complete covering of water and no land. As stated, the atmosphere was initially composed of ~10 atm of CO/CO_2 and N_2, with small amounts of hydrogen, H_2S, CH_4 and helium. It is in these very thin surface layers (atmosphere, water and the top of the mantle), which have evolved mostly at close to equilibrium that life emerged and it is this combination of a geo- and a bio-sphere in one ecosystem which we wish to explore as it evolved. We turn, therefore, to a more detailed analysis of the sea and the atmosphere and then to the known progression of oxidation of the surface generally, not asking yet why it became oxidised. Note that we have on Earth a huge still-cooling non-equilibrium but almost fixed underlying core and mantle, providing heat to the

TABLE 1.6

MINERALS ON EARTH'S SURFACE

Oxides	Sulfides
Mg, Si, Fe	Co, Ni, Fe, Cu, Zn
Al, Ti, B	Cd
V, Cr, Mn, W	Mo
H, C	As, Sb, Bi
	Sn, Pb

Note: The water (sea) contains much of the elements Na, K, Cl, Br, I and (F) with relatively small amounts (%-wise) of Mg, Ca, CO_2, Si, S and P. The approximate mineral content in % by mass is SiO_2, 55; Al_2O_3, 17; CaO, 10; FeO, 10; MgO, 5; Na_2O, K_2O, TiO_2, 3.

surface which was also open to radiation from the Sun. Together they forced the whole surface chemical system into evolving energised steady states (see Chapter 3). There is an inevitability about this process, probably including life, which we shall examine and which has a very long but eventually limited span of existence for it is open to risk.

1.9.   Detailed Composition of the Original Sea: Availability

We shall treat the chemicals of the sea to a large degree as an *equilibrium* in solution of *available* elements from the solid surface and the atmosphere at any one historical time. The elements will be seen to be in particular oxidation states and complex forms and differently hydrolysed except for nitrogen, remaining as N_2, and later oxygen as O_2. In the initial presence of H_2S and ferrous iron and water the reduction potential would be around –0.2 V against the H^+/H_2 potential of –0.42 V at pH 7. The initial equilibrium concentration of elements in solution at this pH is readily shown to be as in Table 1.7. Some values are estimates since their values have changed greatly from their initial condition due to life's processes, e.g. phosphate.

We see that the total element abundance on the continental crust of Earth today (see Figures 1.4 and 1.5), is poorly reflected in the availability of the elements in the sea. Two major reactions affected the availability of the non-metals and the metals apart from abundances; both concern solubility of salts:

(1) *Hydrolysis*. Precipitation of hydroxides at pH 7 restricted the availability of all trivalent and higher valent metal ions such as Al^{3+}, Ti^{4+}, Ga^{3+}, but the absence of hydrolysis allowed the availability of all monovalent ions at concentrations exceeding 10^{-1} M, and most divalent ions up to some 10^{-5} M while some are more soluble, e.g. Mg^{2+} and Ca^{2+} hydroxides. Hydrolysis gave also anions, such as $B(OH)_4^-$, HCO_3^-, HPO_4^{2-}, and later SO_4^{2-}, to be present

TABLE 1.7

AVAILABLE FREE CONCENTRATIONS IN THE SEA AS THEY CHANGED WITH TIME

Metal ion	Original conditions (M)[a]	Aerobic conditions (M)[a]
Na^+	$>10^{-1}$	$>10^{-1}$
K^+	$\sim 10^{-2}$	$\sim 10^{-2}$
Mg^{2+}	$\sim 10^{-2}$	$\sim 10^{-2}$
Ca^{2+}	$\sim 10^{-3}$	$\sim 10^{-3}$
Mn^{2+}	$\sim 10^{-6}$	$\sim 10^{-8}$
Fe	$\sim 10^{-7}$ (Fe^{II})	$\sim 10^{-19}$ (Fe^{III})
Co^{2+}	$\sim 10^{-9}$	$\sim (10^{-9})$
Ni^{2+}	$<10^{-9}$	$<10^{-9}$
Cu	$<10^{-20}$ (very low), Cu^{I}	$<10^{-10}$, Cu^{II}
Zn^{2+}	$<10^{-12}$ (low)	$\sim 10^{-8}$
Mo	$\sim 10^{-9}$ (MoS_4^{2-}, $Mo(OH)_6$)	10^{-8} (MoO_4^{2-})
W	WS_4^{2-}	10^{-9} (WO_4^{2-})
H^+	pH low (5.5?)	pH 8.5 (sea)
H_2S	10^{-2}	Low [SO_4^{2-} (10^{-2} molar)]
HPO_4^{2-}	$<10^{-3}$	$<10^{-3}$ molar
O_2^{b}	$<10^{-6}$ atm	$\sim 10^{-1}$ atm (21%)
CO_2^{b}	>10 atm	10^{-2} atm
N_2^{b}	$>?10$ atm (NH_3?)	~ 1 atm (78%)

[a] Except where other units are specified.
[b] NB Changes in the air, ? indicates uncertainty.

with F^-, Cl^-, Br^- and I^- and many organic anions and cations (Figure 1.7). Some of these oxyanions limited the presence of metal ions, e.g. the carbonates of Mg^{2+} and Ca^{2+}, to around 10^{-2}–10^{-3} M. Most of the products of hydrolysis formed in the absence of oxygen, but hydrolysis dominated availability even further in an oxidising solution when sulfide was removed, as we shall see, particularly in the case of Fe^{3+}. Metals such as V, Mo and W in high oxidation states hydrolyse to give *soluble* anions, e.g. VO_4^{3-}, MoO_4^{2-}.

(2) *Hydrogen sulfide.* It is known that H_2S was available in the early atmosphere and sea. The sulfides like the oxides of elements such as Na, K, Mg and Ca, if they ever formed, dissolve readily in water and would have given rise to H_2S in acid solution or to HS^- at about pH 7. As stated, the ions of these metals have been available in the sea in quite high concentration for billions of years, all of Earth's time. In the series of transition metal sulfides *insolubility increases very rapidly at the neutral pH of the sea*, e.g. $Mn^{2+} < Fe^{2+} < Co^{2+}, \leqslant Ni^{2+} < Cu^{2+}$ (Cu^+) $> Zn^{2+}$ (Figure 1.8), so that 4×10^9 years ago, Fe^{2+} was available at about 10^{-6} M, but the last four ions were only available at or below 10^{-10} M and copper as Cu^{2+} or Cu^+ was virtually absent. This series, the Irving–Williams series of increasing interaction between a metal ion and a non-metal ion, here S^{2-}, relative to water, is also general to the binding to all non-metal centres in compounds at equilibrium in water (see

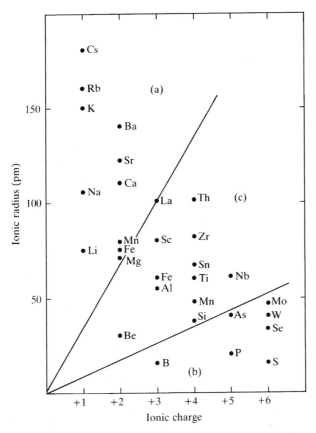

FIG. 1.7.   Ionic radius r, and charge z, of common forms of elements in water. The solid lines divide (a) elements with Z/r <0.03 pm^{-1}, which form soluble hydrated cations such as Ca^{2+}; (b) ones with Z/r >0.12 pm^{-1}, soluble as oxyanions such as SO_4^{2-}; and (c) those of intermediate Z/r, which form oxides or hydroxides insoluble around neutral pH. (Reproduced with permission from P.A. Cox (1989), see Further Reading.)

Section 2.17). We shall find that it dominates much of life's organic chemistry too. There is also the rather unusual insolubility of VS_4. We need to explain the presence of some V, Co, Ni and Zn in early life, but note that their sulfides are relatively soluble in somewhat acid solutions, pH < 7, which may have prevailed at least locally some 4×10^9 years ago. Molybdenum and heavy metals of Groups 13 to 16 are also insoluble as sulfides but tungsten does not precipitate so easily and is known to be more available than molybdenum in sulfide-rich seas.

These reactions are strict (thermodynamic) geochemical limitations on which elements could be used to initiate life, but their influence changed with time as oxidising agents replaced reducing agents (see Section 1.12), and this affected life's

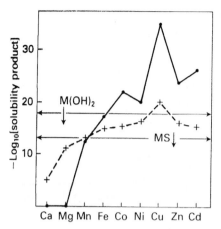

FIG. 1.8.   The solubility products of hydroxides (- - - -) and sulfides (—) which have been of extreme importance in evolution. The horizontal line indicates that solubility product that gives a precipitate at pH 7.0 when [M] = 10^{-4} M and [H_2S] = 10^{-3} M or [OH^-] = 10^{-7} M.

evolution greatly. Observe, however, that there has always been a hidden large reserve of sulfides near the surface under the sea, seen in smokers and volcanoes, and these could become more involved with the ecosystem at any time. Note also the absence from the sea of appreciable amounts of Li, Be and B and of the many heavy elements due to both low abundance and/or low solubility.

A feature of the sea, which occupies almost three-quarters of Earth's surface and probably covered it all at first, is that the water in it is easily vaporised especially by the heating of the Sun. This water vapour returns to the sea, and the mechanism will be discussed in Chapter 3. Here, we observe that much of its return is via clouds, rain and rivers on land. This activity over many billions of years has caused the erosion of the land and brought finely divided mixtures of rocks to form sediments and later still, after processing, sedimentary rocks (see Figure 1.6). The sediments and their combination with chemicals from organisms produced the soils and clays of the lowlands everywhere and have been a major source of materials for organisms from their aqueous surroundings. We stress here that this cycling of water on the land from and to the sea is an essential feature of life's evolution and is then a part of the total geological/biological system. The articles under the heading "Ecology in the Underworld" (see Further Reading) give an impression of the vital importance of soils today.

The second feature of the surface layer is the atmosphere, which we have described already under the discussion of Earth's temperature, but we need to add a note on its flow as it is deeply involved with water circulation. The flow of water vapour due to atmospheric flow, winds, is not independent of the flows of rivers and currents in the sea, e.g. the Gulf Stream. Of course, both are coupled to temperature and its gradients from North to South. In fact, there is intense linked

control between all these factors associated with heating of the fluid environments
and Earth's structures and motions.

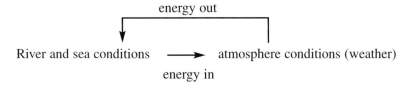

We shall see later how such linking affects many flow systems where elements or
compounds, here water, circulate. In fact, this is a way of reaching a *controlled
cyclic steady state*, a central thermodynamic objective in the Earth's ecosystem
evolution, but we must be aware that the cycles of one element or compound, here
water, are not independent of the changes in cycles of others. These considerations
are fundamental for the appreciation of ecosystems (see Chapter 3).

1.10.   Geological Periods – Chemical and Fossil Records

If we are to describe evolution of the geological/biological ecosystem then we
must have a way of estimating the time of appearance of various physical and chemical objects, including organisms. The best method we have of dating depends on
studies of minerals. The so-called geological periods (Table 1.8), have to be obtained

TABLE 1.8

GEOLOGICAL PERIODS

Era	Period	Date interval (10^6 years before present)
Cenozoic	Quartenary	1.6–present
	Tertiary	65–1.6
Mesozoic	Cretaceous	144–65
	Jurassic	208–144
	Triassic	245–208
Paleozoic	Permian	286–245
	Carboniferous	360–286
	Devonian	408–360
	Silurian	438–408
	Ordovician	505–438
	Cambrian	551–505
	(Pre-cambrian)	(700–551)
Proterozoic (eon)		2,500–551
Archean(eon)		3,800–2,500
Hadean (eon)		4,600–3,800

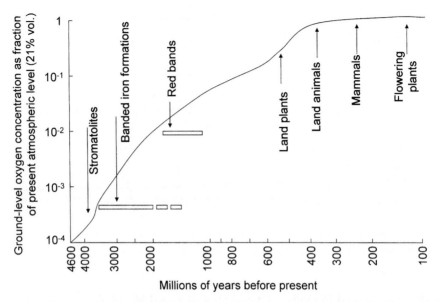

FIG. 1.9.   Cumulative history of O_2 levels through geologic time as recorded by minerals and fossil dating. Most of the oxygen is now present as Fe_2O_3 (some 65%). Molecular oxygen in the atmosphere represents 21% in volume (see Fig. 1.12), but only some 4% of the total oxygen on Earth.

from appropriate modes of the age determination of rocks based today on such studies as the uranium/lead ratios and other radioisotope contents. This dating is very useful in the discussion of the history of the Earth's geological/biological ecosystem. We are, of course, concerned to uncover changes in chemical composition during these periods. A very clear case is that of banded iron formations (BIFs), which are due to the deposition of ferric hydroxide/oxide, carbonate and silicate intercalated in cherts, and can be shown to have taken place first between 3.5 and 2.5 billion years ago (Figure 1.9). This would appear to relate to the reaction of oxygen with the reduced ferrous salts of the sea and has been called "rusting of Earth". Oxygen in the air only began to increase seriously around 2.0 billion years ago and the more continuous iron oxide (hematite) "red-bands" (not BIFs) then appeared. The earlier banded appearance of iron oxides (less oxidised, with higher Fe/O ratios) may well have been due to alternating periods of Fe^{2+} release from minerals and Fe^{3+} precipitation as the Fe^{2+} (and H_2S) removed O_2. The possible initial cause of O_2 generation was the action of UV light on the ferrous iron in water, but the yield is small and it is clear that the rise of oxygen later was due to biological photosynthesis, resisted by its removal by ferrous iron and sulfide, which prevented O_2 levels from rising greatly until around 2×10^9 years ago. In any case we know that the later red bands of iron give a time date for the major rise of oxidising power to above +0.0 V at about 2.0×10^9 years ago. There is a related dating of the appearance of sulfate, uncovered by the observation of different sulfur isotope ratios in minerals with time. The evidence is that sulfate found in barium sulfate (baryta) first appeared at around

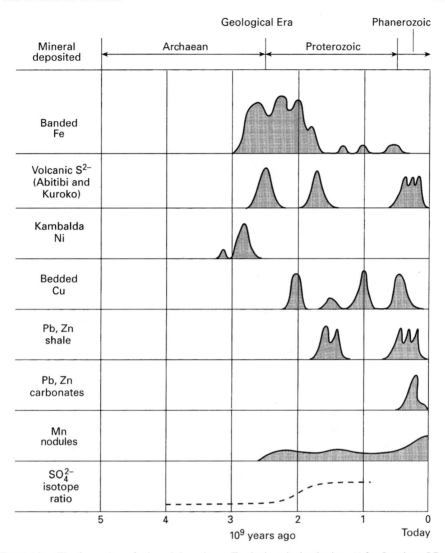

FIG. 1.10.   The formation of mineral deposits on Earth since its beginning. (After Lambert, I.B., Beukes, N.J., Klein, C. and Veizer, J. (1992). In J.W. Schopf and C. Klein (eds.) *The Proterozoic Biosphere*, Cambridge University Press, Cambridge, pp. 59–62.)

3.5×10^9 years ago but was only extensively formed by 2×10^9 years ago (Figure 1.10). Isotope analysis indicates that the sulfide/sulfate *biological* metabolism began well before 2.5×10^9 years ago. A further dating can be estimated by using carbon isotopes, which suggests that life on Earth began 3.8×10^9 years ago. Some other data of the occurrence of deposits of minerals are given in Figure 1.10.

Further evidence of the ages has led to the idea that land first appeared at around 4.0×10^9 years ago as separate masses of volcanic rock above sea water, which increased progressively into one unit and then broke up into continents (Figure 1.11).

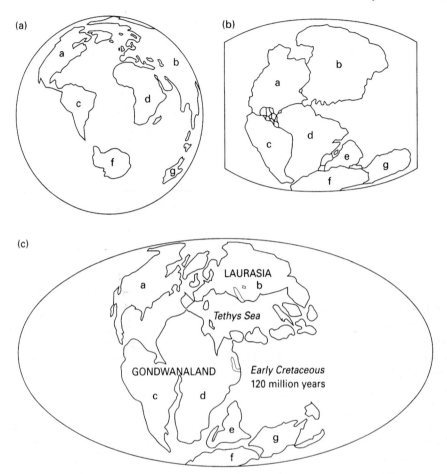

FIG. 1.11.   The development of the large, major continents: (a) today; (b) > 200 × 10⁶ years ago; (c) 120 × 10⁶ years ago. Fragments of the original supercontinent are the continents we know today: a, North America; b, Asia; c, South America; d, Africa; e, India; f, Antarctica and g, Australia.

Subsequently, they have floated on the molten magma below (see Figure 1.6), and have caused changes in contour structure through collisions. These have continued to cause considerable earthquake and volcanic activity which introduce reducing materials to the surface (see Figure 1.13).

The appearance of land and its later erosion was important for evolution of life since it gave new possibilities for adaptation and diversification of organisms. Again exact dating is not possible, but prokaryote invasion of land, possibly by symbiotic early cyanobacteria with early fungi, may have begun very quickly. The larger eukaryotes could only invade land much later and multicellular plants had to develop on land before such animals. Vascular plants evolved on the land some 450–400 million years ago and animals such as non-marine arthropods and amphibians precursors about 400–350 million years ago (Figure 1.9 and Table 1.9). Note that

TABLE 1.9

TIMES OF ORIGIN OF MAJOR GROUPS OF ORGANISMS[a]

Organisms	Approx. time of origin: years ago
Prokaryotes	$(4.0–3.5) \times 10^9$
Aerobic prokaryotes	$3.0–2.5 \times 10^9$
Single cell eukaryotes	$(2.5–2.0) \times 10^9$
Multi-cell eukaryotes	1.0×10^9
Modern plants + animals	500×10^6
Animals with nervous systems	$(250) \times 10^6$ (?)
Mankind	1×10^5

[a] Note that in this book we are concerned with groups of organisms which use chemical elements differently in different spatial divisions and organisation and not with species or conventional divisions of organisms (see Chapter 4). (?) indicates uncertainty.

none of this life might have developed on land if the ozone layer (from O_2) had not evolved about 1–2 billion years ago. All of the early development of life may well have had to be below the top surface of the sea since UV light does not penetrate deeply into water. Nearer the surface it would need some kind of protection, including that from remains of the same or other dead organisms and inorganic salts (carbonates), before the ozone layer formed. This may have been the case of the formation of the early stromatolites.

Life itself presents a way of dating since organism activity fractionates isotopes and their fossils can be subjected to isotope analysis. The fossil records lead to a hypothesis for the increase of oxygen in the atmosphere, which reached 10–20% of present day level by around 1.0×10^9 years ago (Figure 1.12), and this is the time, within a few hundred million years, when multicellular plants and animals first evolved in the sea and then on land (Figures 1.9 and 1.12). The best fossil records are of multicellular animal life with hard structures of $CaCO_3$ dating from some 0.5×10^9 years ago. By this time, the oxygen content of the atmosphere and the inorganic chemistry of the sea were probably not too far away from that of today, but there have been well-documented oscillations, for example of O_2 and CO_2 (see Chapter 8), and of the relative Mg/Ca concentrations in the sea, which had considerable effects on the composition of carbonate deposits (see Dickson in Further Reading). The dating of fossils is consistent with the dating of the minerals in which they have been found.

There is a chance that isotope studies of a great variety of elements will reveal much more of the history of Earth and the evolution of organisms since today mass spectroscopy allows the detection of the relative distribution of isotope concentrations for both light and heavy elements (see Zhu *et al.* and also Arnold *et al.* in Further Reading). The data from shells or minerals derived from such places as the White Cliffs of Dover show that there have also been many fluctuations in temperature, including ice-ages and "snow-ball" Earth periods (see Press and Siever, and Hoffmann and Schrag in Further Reading). However, in this book, we shall only deal with the

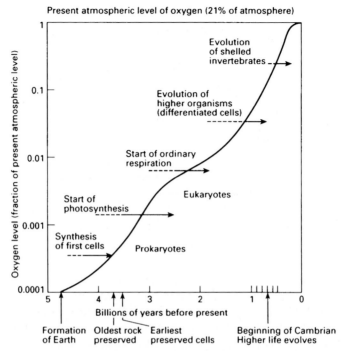

Fɪɢ. 1.12. One hypothesis of the evolution of oxygen in the atmosphere in relation to the origin of life and the evolution of higher organisms. (From *Earth* (4th edn) by Press and Siever.
Copyright 1986 W.H. Freeman and Company, with permission.)

gross development of the whole ecosystem as if it has been a continuous process of
rising oxygenation.

1.11. Fissures in the Surface and Impacts of Meteorites

Where the splittings between continents occurred (see Figure 1.11), there developed fissures in the mantle and outpourings of molten minerals still occur, e.g. the
"black smokers" (see Section 11.6), and trenches of the deep ocean (Figure 1.13),
where elements may be of very different availability, e.g. W > Mo (see Table 11.3).
Some of these fissures are 20 km wide and 10,000 km long and the massive circulation of the sea is associated with them. They may have a very steep temperature
gradient from above 100°C. Elsewhere the mantle has flowed down into a melt and
has then re-emerged in modified mineral form to the surface. All of these activities
have affected life, stressing the importance of examining the whole of geolo-
gical/biological chemistry as one ecosystem.

As stated, there have also been some late impacts of meteorites, which may
well have caused profound fluctuations in the development of biological species.

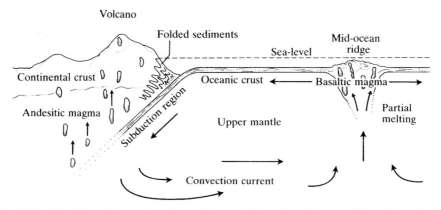

FIG. 1.13.   The flow of heat and molten rocks to and from the mid-ocean ridges. (Reproduced with permission from P.A. Cox (1989), see Further Reading.)

Some have left behind them traces of elements rare on Earth, such as iridium. The detection of traces of iridium from a large meteorite has led to a proposal that major extinction of species on Earth, namely the dinosaurs, could be due to the meteorite impact, causing chemical, light and temperature fluctuations, i.e. global climate changes, though this proposal is not universally accepted.

1.12.   The Geochemical Effects of Oxygen

We now wish to extend the analysis to the progressive change of the surface of the Earth, especially the sea, *as it must have taken place if, on the surface, equilibria dominated as molecular oxygen slowly came into the atmosphere* (see Figure 1.12), (see Kerr (2005) in Further Reading). We shall see in Chapter 3 that we can determine oxidative power on a quantitative scale but only at equilibrium. The time scale of this oxidation can only be given an approximate outline since the rate of oxygen production increased and the quantity of reduced materials of different kinds changed slowly. Only a small amount of the Earth is on the surface and hence equilibrates with atmospheric (O_2) zones, so that only surface chemicals need be considered, especially in the sea. We shall concern ourselves, therefore, mostly with this aqueous solution chemistry since it is largely associated with changes in the whole ecosystem. In this chapter *we are not concerned with organic compounds*, e.g., those of biological origin. Note that horizontal and vertical sea currents do cause mixing. In the absence of kinetic effects we can say that exposed substances of similar oxidability would change oxidation state at the same time in any equilibrium situation. However, some elements are not open to oxidative change and their combinations with other elements have remained closely fixed. Typical are ions of Na and K as Na^+ and K^+, and silicon and phosphorus as silicate and phosphate; chlorine remained largely as Cl^- and boron largely as borate. Some elements which have themselves remained the same, such as Mg^{2+} and Ca^{2+}, have found in part a limited number of new oxidised partners, so that the combined form in which they are found and used has changed. Let us look at the series of changes in time from the most easily oxidised considering their capacity and availability. Remember all changes were exceedingly slow due to the slow production of oxygen and its buffering by the original reduced contents of the sea.

The most readily oxidised common elements among non-metals were hydrogen, nitrogen, carbon and sulfur (Figures 1.14 and 1.15). As stated, silicon and phosphorus were in non-oxidisable states and halides are more difficult to oxidise. Among trace non-metals, selenide is slightly harder to oxidise than sulfur but is close to the potential for oxidation of iodide. Figure 1.14 summarises these observations. We consider, therefore, that there would have been progressive availability changes making it difficult for organisms to obtain the non-metals in the sequence H_2, then carbon (CH_4) and nitrogen (NH_3) before sulfur (H_2S) and selenium (H_2Se). The states of these elements gradually became more oxidised but this process is as yet incomplete, e.g. the oxidation of nitrogen to nitrate.

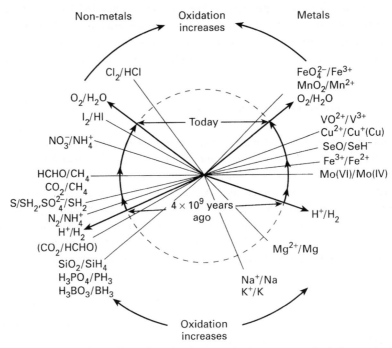

FIG. 1.14.   A scheme of the effect of the rise in atmospheric oxygen on the balance of oxidation states of several elements in the environment. The lines are drawn as oxidation strengths of elements at pH 7.0 against an H^+/H_2 oxidising power of -0.42 V at this pH. The initial balance of each element some 4×10^9 years ago is close to the H^+/H_2 line, hence many are reduced, while the state today is close to the O_2/H_2O line, whence many are oxidised. Thus in time, elements have moved from reduced states to oxidised states seen in their balances, which can be calculated, see Table 6.1. We are only concerned here with inorganic and not with bio-organic chemistry.

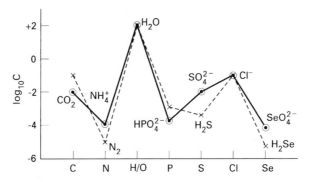

FIG. 1.15.   The estimated non-metal element content of the primitive (X) and the modern (⊙) sea.

Turning to metals, the sequence of oxidation is more difficult to state clearly since it depended on the removal (oxidation) of sulfide to sulfate solution, which gave the corresponding metal ions to the sea coincidentally. The sequence of increased availability of metal ions is then roughly in the order of the solubility of the sulfides and hydroxides given in Section 1.9 and Figure 1.8. Hence, manganese is little affected by oxygen levels though later some of it would be precipitated as MnO_2. Ferrous iron, originally present in large quantities as its sulfide and in silicates is not very insoluble, was oxidised over a period of some 2 billion years apparently in bursts, resulting in enormous loss of iron from the sea as $Fe(OH)_3$ (see Figure 1.9) though much remains in the crust. Of the other elements, cobalt, nickel and then zinc were released from their sulfides in turn before copper and cadmium. Meanwhile, the elements, vanadium, molybdenum and perhaps to some extent tungsten were oxidised to vanadate, molybdate and tungstate making the first two relatively more available, but the availability of tungsten may not have changed much, as it does not precipitate as WS_2 (see Chapters 5 and 7). Observe that release of metal ions from sulfides of Earth is very far from complete.

Very briefly we must now refer to the origin of the input of oxygen to the environment. As stated, it was initially due to the photolysis of water using the energy of the Sun and note that light causes disproportionation:

$$2H_2O \quad \rightarrow \quad 2H_2 + O_2$$

While the oxygen entered the atmosphere, the hydrogen has accumulated in reduced mainly carbon, compounds:

$$2H_2 + CO_2 \quad \rightarrow \quad C/H_2 + 2H_2O$$

These reduced compounds are found in living organisms, natural gas, oil and coal. Now clearly this situation is not one of equilibrium and forces us to look again at our general assumption that as oxygen partial pressure changed the elements in the ecosystem, have come to equilibrium in the sea and air. Two processes have in fact made the approach to equilibrium incomplete in the cases of several elements. The first, mentioned above, is life which traps carbon in reduced states while releasing O_2 to the atmosphere. At best this could come to a cyclic condition, which we describe under the description of a steady state in Chapter 3 where two processes are balanced (see also water circulation in Section 1.9)

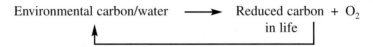

This scheme does not take into account the rate of loss of reduced carbon as buried coal, oil and gas. We do not know either if life could take a form that can exist at higher oxygen levels perhaps at lower temperatures, implying again that the present state of our atmosphere is not the final one. We shall analyse the nature of such steady states in Chapter 3 and return to the problem in Chapter 11. Note that CO_2 levels in the sea are fixed by the solubility of certain carbonates, and hence by solubility products, the pH of the sea, the CO_2 in the atmosphere and the temperature.

When we come to other non-metal elements, we note immediately the state of nitrogen that is deeply involved with all life cycles and is also being lost very slowly from Earth's atmosphere as well as undergoing oxidation to nitrate. These states of nitrogen are not in equilibrium with the oxygen of the atmosphere and here the process of change towards equilibrium is inhibited by barriers to rate of change in the gases. On the other hand, the nature of sulfur and selenium aqueous chemistry would appear to come to equilibrium quickly with dispersed metal precipitates in the sea.

The second set of processes, which has not yet come to equilibrium, is the oxidation of mineral sulfides. We can assume that the concentration of free iron was at equilibrium with the O_2 pressure and the pH of the sea, as sulfide was removed. Slowly, most iron of exposed sulfides became ferric and precipitated at equilibrium as hydroxide. Using equilibrium considerations and solubility products of other metal ion sulfides, we are able to estimate what the free metal ion concentrations in the sea would be at equilibrium as affected by the changing presence of hydrogen sulfide and of oxygen. The metal ions of interest are mainly those from Mn^{2+} to Zn^{2+} in the Periodic Table other than iron. We can estimate that their free-ion concentrations will change as the oxygen partial pressure rises and with the oxidation/reduction ratios of SO_4^{2-} to H_2S. As the sulfates are all highly soluble, the concentration of all these metal ions increased and this increase is not prevented by hydrolysis, in to contrast Fe^{3+}, since their hydroxides are quite soluble at pH 7.5. The ultimate limitation of their concentration to around 10^{-3} M arises from the almost equal value of the solubility products of all their carbonates. When we contrast this limit with the known concentrations in the sea (Table 1.7), we see that the composition of the sea has not come to be saturated with these M^{2+} transition metal carbonates. This is not due to a lack of resources of these elements since there is an, admittedly slow, constant supply from the deep ocean trenches. Note however that after release into the sea the balance of oxidation to reduction of all these elements is ultimately fixed by the O_2 concentration, variable with depth in the sea. The oxidation in sequence of these elements had a profound influence on the evolution of life, but it is far from complete. Most importantly, the progression of element states and availability had to follow the path towards oxidation in the environment as in Figure 1.14. This will allow us to use oxidation relative to reduction together with solubility products at equilibrium as major guides to the behaviour of the elements in the environment (and even inside organisms in part) throughout this book. Finally, accidentally-increased, novel input to the environment of reduced material from the core could swing the steady state back to much less oxic times and this could also result in lowered light absorption. The long-term future over millions of years is, therefore, very uncertain. To have a full impression of the way in which the whole system has developed we have to consider the emergence of life and the development of the absorption of sunlight by living systems, for it was this that generated and still generates the slow production of oxygen. The fluctuation of oxygen partial pressure by a factor of 2, ~15 – 30% of the air, from 450 to 200 million years ago (see Berner in Further Reading), before reaching the present atmosphere level, ca. 21% in volume (see Figures 1.9 and 1.12), is of little evolutionary consequence compared with the enormous changes on going from an anoxic to a highly oxygenated atmosphere and the resulting impacts on the availability of many elements in the upper part of the sea, including oxygen itself. In subsequent chapters, we will see how these surface changes interacted with organisms and to what effect.

1.13. Conclusion

The purpose of this chapter is to describe the formation of Earth and its place in our planetary system and then to provide a description of its geological development over a period of almost 5 billion years. The initial stages are thought to have occurred at close to equilibrium giving a reduced core and a solid mantle of more oxidised chemicals. The mantle separated the core and much of itself from any of the chemicals that accumulated later on or above the surface. These chemicals were again mostly reduced, though carbon was mainly CO and CO_2. It is here that the original sea and atmosphere formed. Their chemical interactions *with the solid surface* then developed relatively quickly often at, or near, equilibrium, but isolated from the deeper mantle and core, and are mostly describable by solubility products of salts and solubility of atmospheric specific gaseous chemicals in the sea at given pressures. At this time, and subsequently, both the Sun and the deep Earth gave energy to Earth's surface, while the Earth suffered some loss of volatile materials to its surroundings. As a consequence, these immediate surface zones were held at an approximately fixed temperature by heat conduction from the interior and radiation from the Sun. This and various other features of Earth (see Table 1.5), make it quite unusual and we may not assume too readily that there must be similar planets in the universe.

Land formed very early due to eruptions from the hot mantle, which here and there upset equilibria and temperature locally for relatively short periods. Owing to the motion of the Earth, the Sun's radiation fell, and still falls (unevenly) on Earth causing air circulation which carries rain water and which, on precipitation, erodes the land so that fine deposits developed close to the sea. At the same time, there is the slower movements of the underlying mantle giving rise to both chemical and physical changes.

Into this picture of a slowly evolving Earth we then considered the even slower steady injection of oxygen over more than 3 billions of years due to the action of light on organisms. The effect *on the inorganic surface zones* was to cause a gradual transformation from a rather reducing chemical system to an oxidising one. The increased free oxygen underwent a more rapid rise in the last one and a half billion years when reaction with much of the buffering reducing materials of the Earth's sea left them exhausted there and then O_2 accumulated with oxidised chemicals, especially in the sea. All these changes took place over such a long period of time that we can assume that many, not all, development of chemicals in all three zones, surface minerals, sea and atmosphere, followed equilibrium redox considerations over time measured in billions of years. These changes were progressive with the production of reduced organic materials (in living organisms). The obvious implication is that reduced material has been introduced to *the three surface zones* at a rate that is closely related to the oxygen and oxidised chemical increase in the environment. We shall consider the origin of these two inputs in Chapters 3 and 4, but we must observe that there is a limit. It so happens that they are both mainly due to chemical reactions associated with living organisms. To understand the chemistry involved in both it is necessary for us to give an account in Chapter 2 of the very nature of the chemicals involved, especially those in the surface zones. However, the most important feature of this chemistry for our description of the evolving ecology of the planet was the effect of injected energy, from whatever source, upon these surfaces. This will require us to describe energy and its impact upon chemicals, not mainly in stationary but largely in conditions of flow, in Chapter 3 before we put together chemical and energy flow in a biological context in Chapter 4. There we shall begin to see how the evolution of Earth is a complicated interaction between changing mainly equilibrated and a few out-of-equilibrium chemicals in the Earth's surface zones and the very much non-equilibrated chemistry of life. The result was an inevitable development of Earth's surface geochemistry and biological chemistry in just one synergistically evolving ecosystem (see cover picture). We must never forget, however, that this is an account of the surface zones. Deeper in the crust and below it a very different chemical world far from the influence of light exists, and it has other rate considerations to be taken into account. For much of the book we leave this dark zone out of account, only returning to it in Chapter 11, for it is in the zones exposed to the Sun where evolution has been so great. Note that throughout this book we are not concerned greatly with events before the beginning of life except in so far as previous to life's appearance on Earth there was the development of Earth itself which provided the initial environment on which life was based. We have nothing to say about the origin of matter and energy which are treated as available on Earth. Equally we do not attempt to describe the origin of life except for pointing out its probable components and energy requirements. In essence the book concerns developments after the point of the evolution of the ecological cone, shown on the cover of this book, and which represents the beginning of life and of our ecological system.

Further Reading

BOOKS

Cox, P.A. (1989). *The Elements – Their Origin, Abundance and Distribution*. Oxford University Press, Oxford
Freedman, R.A. and Kaufmann, W.J., III (2002). *Universe* (6th ed.). W.H. Freeman and Company, New York
Mason, B. and Moore, C.B. (1982). *Principles of Geochemistry* (4th ed.). Wiley, New York
Press, F.S. and Siever, R. (1986). *Earth* (4th ed.). W.H. Freeman and Company, New York
Stanley, S.M. (2002). *Earth System History*. W.H. Freeman and Company, New York
Wayne, R.P. (2002). *Chemistry of Atmospheres* (3rd ed.). Oxford University Press, Oxford
Williams, R.J.P. and Fraústo da Silva, J.J.R. (1996). *The Natural Selection of the Chemical Elements – The Environment and Life's Chemistry*. Clarendon Press, Oxford
Williams, R.J.P. and Fraústo da Silva, J.J.R. (1999). *Bringing Chemistry to Life – From Matter to Man*. Oxford University Press, Oxford

PAPERS

Arnold, G.L., Anbar, A.D., Barling, J. and Lyons, T.W. (2004). Molybdenum isotope evidence for widespread anoxia in mid-Proterozoic oceans. *Science, 304*, 87–90
Berner, R.A. (2003). The long-term carbon cycle, fossil fuels and atmospheric composition. *Nature, 426*, 323–327
Canfield, D.E., Habicht, K.S. and Thamdrup, B. (2000). The Archean sulphur cycle and the early history of atmospheric oxygen. *Science, 288*, 658–661
Catling, D.C., Zahnle, K.J. and McKay, C.P. (2001). Biogenic methane, hydrogen escape and the irreversible oxidation of early Earth. *Science, 293*, 839–843
Chyba, C.F. (2005). Rethinking Earth's early atmosphere. *Science, 308*, 962–963
Dickson, J.A.D. (2002). Fossil echinoderms as monitor of the Mg/Ca ratio of phanerozoic oceans. *Science, 298*, 122–124
Dismukes, G.C., Klimov, V.V., Baranov, S.V., Kozlov, Yu N. Das Grupta, J. and Tyryshkin, A. (2001). The origin of atmospheric oxygen on Earth: the innovation of oxygenic photosynthesis. *Proc. Nat. Acad. Sci. NY, 98*, 2170–2175
Elderfield, H. (2002). Carbonate mysteries. *Science, 296*, 1618–1620
Garcia-Ruiz, J.M., Hyde Carnerup, A.M., Christy, A.G., Vara Kranendonk, M.J. and Welham, N.J. (2003). Self-assembled silica-carbonate structures and detection of ancient microfossils. *Science, 302*, 1194–1197 and see M. Brazier (2001). *Earth System Processes*. Edinburgh Conference

Habicht, K.S., Gade, M., Thamdrup, Bo, Berg, P. and Canfield, D.E. (2002). Calibration of sulfate levels in the Archean ocean. *Science*, *298*, 2372–2374

Harder, B. (2002). Water for the rock; did Earth's oceans come from the heavens? *Science News*, *161*, 184

Hoffmann, P.F. and Schrag, D.P. (2002). The snowball Earth hypothesis: testing the limits of global change. *Terra Nova*, *14*, 129–155 (see also Hoffmann *et al.* (1998). *Science*, *281*, 1342–1346)

Kasting, J.F. (2001). The rise of atmospheric oxygen. *Science*, *293*, 819–820

Kasting, J.F. (2004). When methane made climate. *Sci. Am.*, *291*, 52–59

Kasting, J.F. and Siefert, J.L. (2002). Life and the evolution of Earth's atmosphere. *Science*, *296*, 1066–1068

Kerr, R.A. (2002). Inconstant ancient seas and life's path. *Science*, *298*, 1165–1166

Kerr, R.A. (2005). The story of O_2. *Science*, *308*, 1730–1732.

Norman, E.B. (1994). Stellar alchemy – the origin of the chemical elements. *J. Chem. Edu.*, *71*, 813–819

Robert, F. (2001). The origin of water of Earth. *Science*, *293*, 1056–1058

Viola, V.E. (1990). Formation of the chemical elements and the evolution of the universe. *J. Chem. Edu.*, *67*, 723–730 and see also *J. Chem. Edu.*, *71*, 840–844 (1994)

Wiechert, U.H. (2002). Earth's early atmosphere. *Science*, *298*, 2341–2343

Williams, R.J.P. (1959). Deposition of trace elements in basic magma. *Nature*, *184*, 44–46

Yin, Q. (2004). Predicting the Sun's oxygen isotope composition. *Science*, *305*, 1729–1730

Zhu, X.K., Guo, Y., Williams, R.J.P., O'Nions, R.K., Matthews, A., Canters, G.W., Belshaw, N.S., de Waal, E.C., Weser, U., Burgess, B.K. and Salrato, B. (2002). Mass fractionation processes of transition metal isotopes. *Earth Planet. Sci. Lett.*, *200*, 47–62

Zinner, E. (2003). An isotopic view of the early solar system. *Science, 300*, 265–267

See also: Earth System Processes. Proceedings of the Meeting at Edinburgh, Scotland, 24–28 June, 2001 and Sudgen, A., Stone, R. and Ash, C. (eds.) (2004). Ecology in the underworld. *Science*, *304*, 1613–1637

Chapter 2

Basic Chemistry of the Ecosystem

2.1. Introduction[*]

Before we describe the biological part of the chemistry of the Earth's ecosystem to put beside the geological chemistry introduced in Chapter 1, we must give an outline account of our knowledge of the broader principles and practices of chemistry since it is not possible to understand biological systems without an understanding of a very wide range of chemistry, that is of at least 20 elements. We need to provide, in particular, an introduction to organic chemistry, especially that of relevance to biological chemistry as it is based on such chemistry, and, unlike much of geochemistry, it is not equilibrated. While much inorganic chemistry has been described in Chapter 1, we need to introduce some additional features of it including the reactions of elements traditionally classified as inorganic, e.g. metals, with organic molecules, many of which are found in organisms.

Here we define organic chemistry traditionally as that of the combinations of certain non-metals, based mainly on carbon, in compounds of moderate or large size related to many chemicals found in organisms, while inorganic chemistry covers all aspects of element chemistry outside this area. It is clear that there must be some overlap and we shall return to the importance of it later in the chapter. Once we have an outline of both chemistries, we shall be in a better position to show how the geological character of the Earth and biological organisms are interactive in one global ecosystem. Before we can do so, we need to understand one other feature of all chemistry – the distribution of energy in chemicals. This chapter will make a beginning to this topic, while Chapter 3 will be devoted to it.

While describing the geochemical system in Chapter 1, we have assumed that much of the observed surface composed of inorganic chemicals was at or moving towards chemical equilibrium even while the state of elements changed when and as oxygen increased in the atmosphere. The prevailing chemistry was energised only in particular compounds. Although this is not a fully satisfactory description, it has enabled us to give a fair impression of the Earth's surface development. We had to set aside the deeper reserves of non-equilibrated reducing materials in and below the surface. By contrast, as stated, organic and biological chemicals are out-of-equilibrium energised compounds. In order to appreciate these complications of chemistry, it pays us to take a step back and to examine basic features of all chemistry. This is best done starting from the nature of all the elements and then asking which are the possible chemistries of the compounds of the various elements including in the description the principles of both equilibrium and out-of-equilibrium examples.

---

[*]This chapter is intended to give the reader, who is not familiar with chemistry, an introduction to chemistry, which is useful in the understanding of the other chapters of this book. The important points are summarised in the last section so that those more familiar with this science can see the major features, which we shall use and they can then refer to appropriate parts of the chapter to see if there are points that require explanation. We must stress that we consider evolution of organisms cannot be understood without a basic knowledge of the inorganic chemistry of our environment and its changes with time as well as its essential involvement in organisms. We have repeated some remarks made in Chapter 1 where they are useful in developing the theme of chemistry.

We can then proceed to examine non-equilibria energetics in more detail in Chapter 3, which, together with Chapter 4 on the principles of biological chemistry, will prepare us for the discussion of selective groups of chemicals in organisms in the second part of the book (Chapters 5–10). Their energisation and their remarkable evolution follows, as described in Chapters 5–9. In fair part we shall show that organism evolution was forced by the abiotic evolution of the Earth's surface described in Chapter 1. The very novel chemistry recently introduced by mankind is examined in Chapter 10 before we attempt an integration of the whole present-day ecosystem and its future in Chapter 11.

2.2. Atoms and The Periodic Table

In order to look at the basis of chemistry, we have to appreciate the limitations on the number and abundance of different atoms, why these atoms react to form compounds and what can prevent their further reactions even when the would-be products are thermodynamically more stable than the reactant compounds. The first steps are then to appreciate the formation of compounds from atoms. As mentioned in Chapter 1, as gaseous atoms cool from high temperatures, >3000 K, and combine to form compounds, the ones that form are the most thermodynamically stable since at high temperatures there are no barriers to reaction in atomic gases. Which are the compounds formed?, in what amounts are they present? and why are they so stable? We have answered the first two questions in Chapter 1 but to answer the third question we need to look at the structures of atoms in addition to their abundances given in Chapter 1, their compounds that form at moderately high temperatures and then at the barriers to further reaction at lower temperatures when some of the compounds formed at high temperature are not the most stable. We shall have to distinguish clearly, therefore, between *thermodynamic stability* of a compound, meaning that it cannot change unless exposed to changed conditions, and *kinetic stability*, meaning that it should change spontaneously but is prevented from doing so by a barrier. The very nature of the different atoms decides both their thermodynamic and kinetic properties in compounds.

The nuclei of atoms formed, as we have said in Chapter 1, at very high temperatures, >10^7 K, in giant stars. They are made of particles of positive charge, protons, and uncharged, neutrons. Neutrons are required if more than one proton are to be combined. The two particles together make stable nuclei up to atomic mass of just over 200. Some of the most stable light nuclei are based on multiples of mass 4, two protons and two neutrons of almost equal mass, e.g. carbon, atomic mass 12 (6 of each) and oxygen, atomic mass 16 (8 of each) (Table 2.1), while heavier elements require a larger ratio of neutrons to protons, e.g. Fe, 26 protons and 30 neutrons. Above the atomic number of protons of about 90, and atomic mass of about 200, the nucleus is unstable, no matter how many neutrons are added. *Thus, in the universe the number of elements is limited—a striking observation in itself—and the abundance of them is fixed (Figure 1.1).* All the nuclei are very small, $<10^{-3}$ Å radius,

TABLE 2.1

SOME STABLE LIGHT NUCLEI OF ELEMENTS

Element	Number of protons in the nucleus	Number of neutrons in the nucleus	Number of electrons
Hydrogen, ^{1}H	1	—	1
Helium, ^{2}He	2	—	2
Carbon, ^{12}C	6	6	6
Nitrogen, ^{14}N	7	7	7
Oxygen, ^{16}O	8	8	8
Magnesium, ^{24}Mg	12	12	12
Phosphorus, ^{31}P	15	16	15
Sulfur, ^{32}S	16	16	16
Calcium, ^{40}Ca	20	20	20
Iron, ^{56}Fe	26	30	26

Note: The nucleus of each element may have more than one neutron/proton ratio (different isotopes); in the table are presented the most abundant stable isotopes of some elements and the number before their symbols represents very approximately the mass of that isotope (*mass number, A*).

and positively charged. At temperatures below 5000 K they become surrounded by electrons of negative charge to give neutral atoms (Figure 2.1). The electrons occupy much more space in a radius of 1–2 Å around nuclei and are found to be arranged in shells of units 2, 8, 8, 18, 18, 32 with a final incomplete shell of 6 as the nuclear charge rises steadily from 1 to 90, which would be followed by a completed second shell of 32 and so on, but for the increasing instability of the nucleus. A most notable and seemingly strange chemical feature is that the electron shells are based on the odd numbers multiplied by 2:

$$2 \times 1(2), 2 \times 1 + 2 \times 3(8), 2 \times 1 + 2 \times 3 + 2 \times 5(18), 2 \times 1 + 2 \times 3 + 2 \times 5 + 2 \times 7(32)$$

This remarkable set of simple numbers dominates all chemistry and in fact the whole of the Earth's geological and biological ecosystem which we wish to describe. Today these numbers can be explained by quantum theory in which electrons are represented by wave formulations making each electron a composite function of a wave and a particle. (This is beyond imagination but not to worry, as it is beyond all imagining. Our brains are not equipped to deal with the properties of either very, very small or very, very large objects. Fortunately, mathematical formulation that accurately reflects properties has extended beyond physical images and allows much coherence to the understanding of atomic structure.) We must also note that as electrons of a given element are put into an atom shell, they fill up the shells one at a time to the half-filled condition and then by pairing they make each shell or sub-shell complete. For example, in the first row of eight elements, Li to Ne, the filling is one, Li, then a second electron to make a pair (1×2), Be.

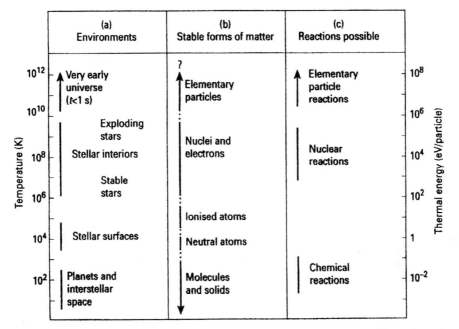

(a) Environments	(b) Stable forms of matter	(c) Reactions possible

FIG. 2.1. Energy and temperature scales for chemical and nuclear processes. The scale on the left shows temperature, and that on the right indicates the average thermal energy for the particles present. Column (a) shows typical environments with different temperatures; (b) shows the stable forms of matter present; (c) indicates the types of reaction possible (1.0 eV = 23.06 kcals mol^{-1}) (reproduced with permission from Cox, P.A. (1989)).

The following three electrons build up in singles (1 × 3), B, C, N, before pairing begins again to reach (2 × 3), O, F, Ne.

Now the arrangement in the Periods and Groups was not discovered by analysing the electronic structure of the atoms but by the study of the chemical properties since 1750–1850 AD. The study, only fully completed by around 1900 AD, showed that the Periods and Groups were related to the close repetition of physical and chemical properties of elements along each Period except for the first Period of only two elements H, He, which is unique. The second member of it, helium, is an inert (noble) gas not able to form stable compounds. The next two Periods have eight (2, 6) members from Li and Na to Ne and Ar, respectively; Ne and Ar are noble gases. The following two Periods of 18 emerge as a group of 8 (2, 6) split by an additional 10 so-called transition elements, all of similar properties. The 10 are inserted between the 2nd and 3rd member of the set of 8. The sixth Period of 32 elements has two insertions after the 2nd member of the set of 8: there is a set of 14 after this second element and then a further set of 10 transition elements before the last six of the set of 8. The set of 14 "inner transition" elements, the so-called rare earth or lanthanide series, are even more chemically similar to one another than the 10 in the transition metal series and are usually grouped together (Figure 2.2).

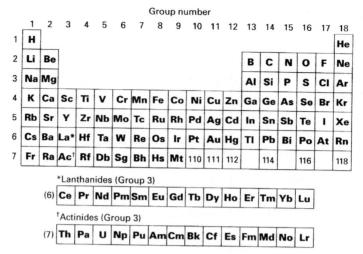

FIG. 2.2. The periodic table of chemical elements, displayed in the modern "long" form. Each element is denoted by its symbol. From U (element atomic number 92) the elements have been synthesised by man.

Notice that each Period closes with an inert or "noble" gas and that there is a systematic change of chemical and physical properties across each Period. There is also a systematic change in physical and chemical properties down each Group. This scheme, which allows the Periods to be aligned in 18 Groups in the Periodic Table (Figure 2.2) has a very strong theoretical backing today, but was based at first, as stated, on the physical and chemical properties of the elements and their compounds. By the atomic number 36, at which abundance is falling rapidly, virtually the complete range of physical and chemical properties have been realised. We shall now describe briefly these properties since only by understanding them can geological and life's chemical evolution be understood. While there are 92 elements on the Earth (actually 90 since Tc and Pm have disintegrated) only about 15–20 of the mentioned first 36 elements are essential for all life and almost every Group has a representative, often from the lightest three Periods and hence the most abundant (see Figures 1.1 and 2.2). As we shall see in Chapters 5–8, this fact alone shows that life's chemistry has taken full advantage of the wide variety, given the availability, but not the full complement of elements in the Periodic Table. (We can also show the rational of this selection and why 16 of the first 36 are not used and why and how mankind has come to use almost all; see Chapters 10 and 11.)

The major chemical distinction between elements is that between metals, about 70, and about 20 non-metals. The simplest distinction between the two is that metals in the condensed state conduct electricity while non-metals do not. The non-metals, reduced to about 15 if we remove the inert gases, are crowded towards the ends of the sequences of each Period, especially the first three. Typical and dominant in abundance (and in organisms) are C, N, O, (F) and Si, P, S, Cl, as well as H (see Figures 1.1 and 1.4), but note that Se is found in most organisms. They have at least half the number of electrons in their ultimate partly filled shell of eight, all held tightly, and have a tendency to capture electrons to become negatively charged species, anions, X^{n-}, with a filled final shell, for example oxide, O^{2-}, and chloride, Cl^-. All the metals in the first two periods of eight have any number up to three outermost (most loosely held) electrons, that is less than half of those required to complete the eight of their particular shell which are not held very tightly. They, therefore, tend to lose electrons and become positively charged species, cations M^{n+}, for example all elements Li, Be, Na, Mg, Al and also K and Ca. This ease of loss of electrons is particularly strongly so for the metals in Groups 1, 2, 3, 12, 13, 14 (not C or Si) and those of the rare earths. In this way, these cations finish with the full electron shell of the previous Period or of the sub-Period of 10. The remaining groups of 10 metal elements in Groups 4 to 11 in the transition metal series tend to lose only some electrons outside the inert gas core of the previous Period, being less willing to do so as we progress further along a Period. In fact, in any one ionised state there is a slow shift of properties towards non-metal properties through Groups 1 to 13 or 14 before there is a rapid change from metals to the non-metals of Groups 14 to 17. Of all the metal elements, Na, K, Mg, Ca, Mn, Fe, and in less quantity V, Co, Ni, Cu, Zn with the curious inclusion of either Mo or W, are largely the most abundant and/or available (see below) that are essential for life.

Since about 1940, mankind has realised that energy could be released if the very light nuclei of hydrogen could be made to react to produce deuterium or helium, where nuclear fusion would provide energy. The alternative is nuclear fission of the very heavy elements to give two nuclei of lower atomic number. Already there exist many nuclear power stations using fission but none using fusion. We return to the discussion of the potential value of nuclear power in Chapters 10 and 11.

Before concluding the section, we comment that knowledge and understanding of abundances and of The Periodic Table have given mankind an ability to appreciate and manipulate chemistry in all parts of the ecosystem, environmental (geological) and biological, in which we live (see Chapters 10 and 11). We discuss next the basic features of inorganic and organic chemistry stressing the difference between equilibrated and non-equilibrated chemistry.

2A. Inorganic Chemistry

2.3. Nature of Inorganic Chemical Compounds: Groups 1 to 3 and 12 to 17

As stated, inorganic chemistry deals dominantly with the 70 metal elements in combination with non-metals, with metal/metal chemistry and with combinations of non-metals among themselves, e.g. CO, CO_2 and CS_2 but excluding largely and especially larger compounds of carbon. Some of the chemistry of combinations of elements now becomes obvious, but it is convenient to describe first those in Groups 1 to 3 and 12 to 17 since their inner electrons are less involved in chemistry. Metals, which donate electrons easily, form thermodynamically stable combinations with non-metals, which accept electrons readily, resulting in stoichiometric salts, M^{n+} X^{n-}. The charged atoms are called ions, e.g. common salt is Na^+Cl^- and magnesium oxide is $Mg^{2+}O^{2-}$. They do not form molecules in the condensed phase but continuous ionic assemblies (Figure 2.3). The obvious examples in geochemistry are metal oxides, sulfides, chlorides and somewhat more complicated carbonates, $M^{2+}(CO_3^{2-})$, and we shall meet some parallel compounds in living cells. As solids they do not conduct electricity. When charges are small, $n \leqslant 2$, many but not all of these salts are soluble in water to give conducting ionic solutions of separate ions, M^{n+} and X^{n-}. Thus, major interest for us is in the solubilities of these salts as they control the availability of the metal elements to organisms. We discuss the relative preferences of some metals for sulfide rather than oxide below; see also their geochemistry described in Section 1.5. The stoichiometry of these salts is easily understood from their Group provided that we consider transition metals separately. Less obvious are the properties of metals themselves and metals with other metals, alloys, as well as those of non-metals themselves and of non-metal/non-metal compounds. Metals and alloys form continuous assemblies in which some of their loosely held electrons become free in the condensed state and hence they are electronic conductors. They could be represented like salts, $M^{n+} \cdot ne$, where e is an electron. Alloys, such as Mg/Al (and see brass, Cu/Zn) are non-stoichiometric, and

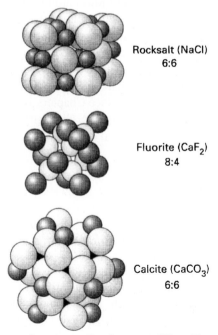

Rocksalt (NaCl)
6:6

Fluorite (CaF$_2$)
8:4

Calcite (CaCO$_3$)
6:6

FIG. 2.3. Some simple crystal structures showing space-filling. The metal ion is shaded.

they, like metals, are not thermodynamically stable in the presence of oxygen, water, hydrogen sulfide or in acids, and metals generally are not usually found on the Earth except for gold and silver (see Ni/Fe in the Earth's core) and not in living systems. We will not discuss them in more detail until Chapters 10 and 11.

The non-metals and the non-metal/non-metal compounds concerned here are formed stoichiometrically through covalent bonds between atoms and are very different from salts. Such bonds are a means of satisfying the electron demand by nuclei to reach a noble gas complement through sharing electrons in pairs between atoms so that they form molecular structures with shapes (Figure 2.4). (This is in contrast with ionic bonds where one partner donates and the other accepts electrons and with metals where, in both cases, shapes of bulk combinations in lattices are largely a property of packing ion or atom sizes.) Examples are the hydrogen molecules, H_2 or H–H, where the line between atoms represents two electrons. In this way each H has a helium-like pair of electrons where helium is the nearby noble gas. For carbon combining with carbon

$$-\overset{\diagdown}{\underset{\diagup}{C}} - \overset{\diagup}{\underset{\diagdown}{C}}-,$$

we have four electron pair bonds for each carbon, or combined with four hydrogens

$$-\overset{\diagdown}{\underset{\diagup}{C}} - H,$$

four pair bonds per carbon and one pair bond per hydrogen, thus making a solid lattice of diamond of pure carbon in the first case, or a gas, CH_4, in the second. Each arrangement gives each carbon the sense of having the eight electrons of the next noble gas, neon, while each H has a shared pair. These combinations are relatively inert, like Ne, so that they show *kinetic, not thermodynamic, stability* even in the presence of oxygen and water with which they should react. (It is the resistance to reaction that has allowed the chemistry of organisms to appear.) There are other covalent combinations of the so-called double bonds to satisfy C and O together, each with eight electrons, as in O=C=O, carbon dioxide, and even those with triple bonds, N≡N in N_2. By this means, stoichiometric substances of non-metals are made (Figure 2.4). Many of these non-metal compounds do not dissolve significantly in water, they do not conduct via electronic or by ionic means, and they are often quite kinetically stable volatile gases, such as H_2 and CH_4. Some are common on the Earth, e.g. CH_4, CO_2, N_2. However, as the end of a Period is approached some of the non-metals both in combination as hydrides and oxides, those of N, S and Cl for example, give rise to conducting aqueous ionic solutions (Table 2.2). Larger molecules are formed mainly by combining many H, C, N, O and to some degree P, S atoms, and to lesser extent the elements of Group 17, all with covalent bonds. They are dominant in biological chemistry. As stated, the systematic study of these molecules is called organic chemistry because much of the chemistry of organisms relates to them (see Section 2.13). We look at the interactions of these, some eight, non-metal compounds with many of the first 20 metals of the Periodic Table in Part 2C. They are of great biological importance.

Description of shape	Shape	Examples
Linear	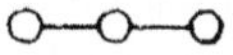	HCN, CO_2
Angular	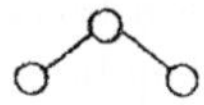	H_2O, O_3, NO_2^-
Trigonal planar	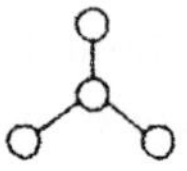	BF_3, SO_3, NO_3^-, CO_3^{2-}
Trigonal pyramidal	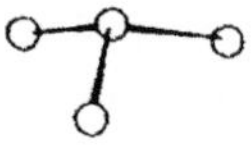	NH_3, SO_3^{2-}
Tetrahedral	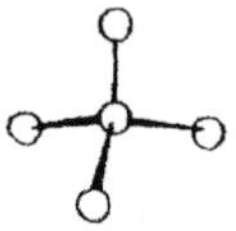	CH_4, SO_4^{2-}, NSF_3
Square planar tetragonal	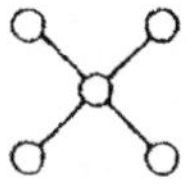	XeF_4
Square pyramidal	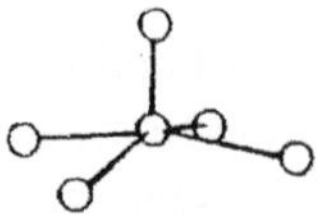	$Sb(Ph)_5$
Trigonal bipyramidal	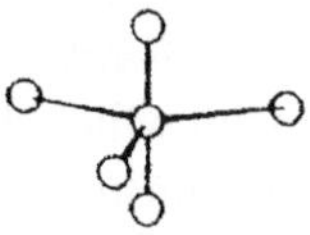	$PCl_5(g)$, SOF_4
Octahedral	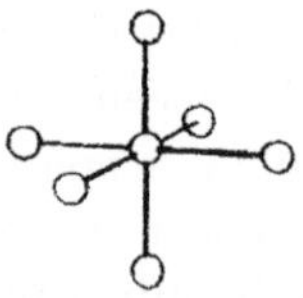	SF_6, PCl_6^-, $IO(OH)_5$

FIG. 2.4.   The description of some molecular shapes.

2.4.   The Nature of Transition Metal Compounds: Groups 4 to 11

We turn now to the transition metal atoms in three series of 10 elements and the building towards a set of 18 electrons in them which is preceded by two Groups, 1 and 2 (Figure 2.2). Many of them and the early elements of Groups 1 and 2 are of immense importance in both geological and biological metal ion chemistry. As the

10 electrons are introduced, transition metals behave first much as Groups 1, 2 and 3 metals and lose even four (Ti) electrons to form ionic lattices, e.g. TiO_2. Thereafter, many of the added electrons tend to become part of an unused "sub-shell" of a partly filled shell so that the metals lose fewer electrons when forming compounds. In fact, in the row starting with Sc and Ti we find that from Mn to Zn a dominant state is M^{2+} as in oxides MO and sulfides MS. By Zn the sub-shell of 10 electrons is stable, remains so till the end of the Period and is not used in compound formation, where the only state of chemical interest of zinc is the ion Zn^{2+} in which all 10 sub-shell electrons are paired. Groups 13 to 18 elements then complete the period of 18. This means that a great deal of general geological and biological chemistry from Mn to Zn is concerned with M^{2+} ions. However, we note that Mn can form higher oxides, MnO_2 and even Mn_2O_7, Fe readily gives Fe_2O_3, while Cu forms a lower ionic state of Cu^+, Cu_2O (see Figure 2.7). Before we describe different states of these 10 elements, we stress the features of these M^{2+} states as they are extremely important in biological chemistry.

First, there is a gradual change of M^{2+} along the series Mn^{2+} through to Cu^{2+} in which the size of the ion decreases and the affinity of the metal ion for electrons increases, which is reversed at Zn^{2+}. Consequently, relative affinity for non-metal atoms or anions, which can donate electrons to or share electrons with it, increases, but also increases more rapidly in the series of increasing donation (see Table 1.4 and Figure 2.5), for example the order

$$Ca^{2+} < Mn^{2+} < Fe^{2+} < Co^{2+} (Ni^{2+}) < Cu^{2+} > Zn^{2+}$$

This Irving–Williams series therefore applies to the binding of sulfide more markedly than to oxide (Figure 2.5 and see Section 1.5). Later we shall see that this relative affinity series applies to the ions combined with organic compounds in water where binding is largely to O, N or S atoms in molecules. It is critical to a full understanding of biological systems and their evolution. There is also a binding contribution from the polarisation of the inner partly filled electron "sub-shell", but we shall not describe this aspect (included in ligand-field theory, see Williams and Fraústo da Silva in Further Reading). Among monovalent ions, we note that, of those of consequence, Cu^+ binds strongly in water in a similar manner to Cu^{2+} while Na^+ and K^+ hardly bind

TABLE 2.2

THE SIMPLE FORMS IN WHICH MAIN GROUP BIOLOGICAL ELEMENTS OCCUR IN AERATED SOIL, WATER, RIVERS, LAKES AND SEA (OR IN BLOOD PLASMA)

Cations	Anions	Neutral species
NH_4^+, H_3O^+	HCO_3^-, CO_3^{2-}, NO_3^-	H_2O, $B(OH)_3$
Na^+, K^+	$H_2PO_4^-$, HPO_4^{2-}	CO_2, $SiO_2(nH_2O)$
Mg^{2+}, Ca^{2+}	OH^-, F^-, Cl^-, Br^-, I^-, SO_4^{2-}	N_2, NH_3, O_2

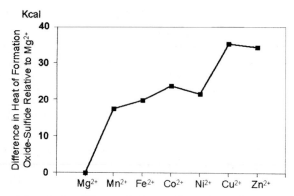

FIG. 2.5.   A plot of the relative difference in heat of formation, oxides–sulfides for a series of divalent ions.

at all. Three consequences follow *at equilibrium*, which we have observed in geo-chemistry:

(1) The binding energy of the metal *ions* with non-metal anions increases in the series to a maximum at Cu^{2+} (or Cu^+).

(2) The binding energy increases more rapidly for the sulfide anion than for the oxide or hydroxide anion (Figure 2.5). Sulfide donates or shares electrons more readily than does oxide or hydroxide.

(3) The properties of M^+ and states higher than M^{2+} are quite distinct.

Together this gives a general explanation of the finding that Mn^{2+} is found in oxide ores. While increasingly, the elements at the end of the series are found as sulfides, but note the exceptions of vanadium, in part seen as VS_4 and molybdenum which occurs as MoS_2. This distinction between oxides and sulfides is also true for ions of elements before Group 7 (Ti to Cr), which are found as oxides or as oxyanions, but those after Group 12, i.e. in Groups 13 to 16 (e.g. Sn, Pb, Bi atoms) of higher elec-tron affinity but similar stoichiometry, are again found as sulfides. (Oxyanions are negatively charged ions of small covalent combinations of non-metals with oxygen, e.g. carbonate, CO_3^{2-}, silicate, SiO_4^{4-}, sulfate SO_4^{2-} and nitrate NO_3^{-}.) It is a general rule that the metal ions, which have greater electron affinity, i.e. moving from left to right along any Period, tend to form sulfides rather than oxides. Goldschmidt, a famous geochemist, distinguished the two groups under the headings lithophiles (O-loving) and chalcophyles (S-loving) (see Table 1.3). Nowadays, the different groups of ions are called 'a' and 'b' or hard and soft (Table 2.3). The differences, not these names, are important since in essence the electronic explanations given for the dif-ferences are very similar, but care must be taken about the effects of ion sizes. These equilibrium considerations, which will be discussed again in Section 2.19, will help in determining the binding of metal ions in organisms.

TABLE 2.3

HARD AND SOFT ACIDS AND BASES

Hard (class a)

Acids H^+, Li^+, Na^+, K^+, Be^{2+}, Mg^{2+}, Ca^{2+}, BF_3, BCl_3, $B(OR)_3$, Al^{3+}, $AlCl_3$, $Al(CH_3)_3$, Sc^{3+}, Ti^{4+}, VO^{2+}, Cr^{3+}, Fe^{3+}, Co^{3+}

Bases NH_3, RNH_2, N_2H_4, H_2O, OH^-, O^{2-}, ROH, RO^-, CO_3^{2-}, SO_4^{2-}, ClO_4^-, F^-

Borderline

Acids Fe^{2+}, Co^{2+}, Ni^{2+}, Cu^{2+}, Zn^{2+}, Rh^{3+}, $B(CH_3)_3$, R_3C^+, Pb^{2+}, Sn^{2+}

Bases $C_6H_5NH_2$, N_3^-, N_2, Br^-, Cl^-

Soft (class b)

Acids Cu^+, Ag^+, Au^+, Cd^{2+}, Hg^{2+}, Pt^{2+}, Pt^{4+}, MoO_2^{2+}, Pd^{2+}

Bases H^-, R^-, C_2H_4, C_6H_6, CN^-, CO, SCN^-, R_3P, R_2S, RSH, RS^-, I^-

---

Note: Hardness is defined (see textbooks in "Further reading") as $\eta_M = \frac{1}{2}(I - A_e)$, where I is the ionisation potential of the state of M concerned and A_e is its electron affinity. I dominates for positive ions, but in aqueous solution must be considered relative to z/r (data from Ahrland, S., Chatt, J. and Davies, N.R. (1958). *Quart. Rev.*, *12*, 265–276; Pearson, R.G. (1963). *J. Am. Chem. Soc.*, *85*, 3533–3539).

2.5. Variable Combining Ratios and Spin States

The transition elements and the elements of Groups 14 to 17, both metals and non-metals, present us with problems of understanding due to the balance between willingness to bind and the desire to retain electrons in partly filled shells. While the combining ratios for Groups 1 to 4, 12 and 13 are almost always simple, those of the other Groups are not. Instead, they show a variety of combining ratios in stoichiometric compounds. For the non-metals, the variations in combination arise from the kinetic stability of many compounds. The obvious examples are the two oxides of carbon, CO and CO_2. In CO, carbon has retained some electrons to itself while sharing with others. The oxide of nitrogen NO, equivalent combining ratio 2, contrasts with the hydride NH_3 combining ratio 3; the hydride of sulfur H_2S is very different in combining ratio from that in the two oxides SO_2 and SO_3. We refer in these different kinetically or thermodynamically stable combinations to the *oxidised states of elements* when oxygen (or halides) is involved, or to the *reduced states of elements* when hydrogen is involved, and where H_2O is treated as neutral (see our previous books in Further Reading for a more general description). Importantly, a huge variety of reduced and oxidised carbon compounds arises from the various ways of bringing C, H and O together in single and double bonds, and this variety is increased by involving N and S. Such compounds are stoichiometric, kinetically stable, and are the major basis of organic and biological chemistries.

The possible oxidation states of the metals from Mn to Zn are shown in Figure 2.6. (Note that oxidation states are defined by ion charge, n in M^{n+}.) The way the maximum values increase and then decrease in the series is a consequence of

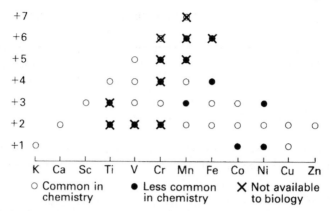

FIG. 2.6. Oxidation state patterns for the elements from potassium to zinc, compiled from a survey of aqueous monoatomic ions and aqueous oxyions, stoichiometric oxides, halides and other simple salts.

the increasing atomic number and number of outer electrons together with the consequent increasing difficulty of removing electrons from the sub-shells, especially of the later atoms. Often many compounds of one of these elements in different oxidation states are almost equally *thermodynamically stable*, while a few others are kinetically unstable and do not persist for long periods in the presence of reducing materials. Where stability is almost equal, more than one electron state of an element can exist in balance, e.g. the states Fe^{2+} and Fe^{3+} in water. This variety of reduced and oxidised states of non-metals and metals and the reasons for their persistence or lability are of great consequence for the evolution of both the geo- and biospheres, since over a period of 4.6 billion years the conditions on the surface of the Earth changed from reducing to oxidising, see Section 1.12. Note, as stated in Section 1.12, that the aqueous part of the geosphere changed at close to equilibrium with the oxygen partial pressure. Now the oxidising power of all the elements can be compared quantitatively from the (free) energies of competition for oxygen. (Free energy is defined in an appendix to Chapter 3, where we see that the (free) energy of reaction is temperature-dependent.) In the cases of carbon and iron the reaction is

$$C + FeO \rightleftarrows Fe + CO$$

A more generally useful measure for the purpose of this book is the (free) energy difference between reactions at electrodes in water at 298 K such as

$$Fe^{2+} + 2e \rightarrow Fe$$

and

$$2H^+ + 2e \rightarrow H_2$$

The free energy difference can now be measured in an electrochemical cell in volts and by putting the H_2/H^+ electrical potential at 0 V in normal (1.0 M) acid, we relate the oxidising strengths of all elements to this couple. It was in this way that Figure 1.14 was devised but based upon electrochemical potential energies at pH = 7.0. We shall use it as a pointer to the order in which elements were oxidised in the ocean (pH ~ 7.0) on the Earth as O_2 increased (see the later chapters of this book), which we shall show greatly influenced, perhaps dominated, the evolution of organisms.

There is a further difficulty with the transition metal elements. As we have stated, the shells of gaseous *atoms* build up one electron at a time forming half-filled sub-shells before completing the sub-shells with electron pairs. Now in the transition metal series, we have said that in *compounds* in, for example, the series Mn to Cu, many electrons belonging to the set of 10 drop out from bonding easily and behave as electrons in a distinct ($3d$) sub-shell of the atom where they are often unpaired as much as possible. Such atoms or ions, with a maximum numbers of unpaired sub-shell electrons are said to be in *"high-spin"* states (Table 2.4). (Spin refers to a special property of electrons, which can have either of the two signs; those with opposed spins combine in pairs.) However, *low-spin* states in compounds of these metals with certain organic molecules in which as many electrons as possible are paired are also known. Consider iron; because the electrons in forming part of the set of 10 remain as unpaired as far as possible the iron atom itself has the electronic structure (2, 8, 8) 2, 6 with four of the last six electrons of the sub-shell of 10 remaining unpaired. Now as explained above, just because the sub-shell of six of the 10-electron set tends to remain unused, the cation of iron, Fe^{2+}, can behave rather like Ca^{2+} or Mg^{2+} and form the high-spin ion Fe^{2+} (e.g. in its oxide FeO or sulfide, FeS) with four unpaired electrons of the inner six. However, iron bound immediately to certain powerful electron donor binding atoms (C, N and S_2^{2-}) has these six electrons in three pairs so as to give better binding in a *"low-spin"* arrangement. An example is pyrite, Fe^{2+} (S_2^{2-}), but there are many others in organometallic chemistry. When such pairing occurs in a reaction step it is called a spin-state change (Table 2.5). Although this may seem complicated chemistry, it also has an essential part to play in the evolution of the Earth/Environment/Sun/Life system, for we shall see that metal ion organic chemistry in organisms provides many examples of the evolution of low-spin combinations in metal porphyrins. (The theory behind this part of

TABLE 2.4

USUAL SPIN STATES FOR SOME METAL IONS

Metal	M^{2+}	M^{3+}
Mn	High-spin (d^5)	High-spin (d^4)
Fe	Low-spin, high-spin (d^6)	High-spin, low-spin (d^5)
Co	High-spin (d^7)	Low-spin (d^6)
Ni	High-spin (d^8)	Low-spin (d^7)

TABLE 2.5

SPIN-STATE SWITCHES FOR SOME METAL COMPLEXES

Cation	d-Electrons	High spin	Low spin
Mn^{2+}	5	The vast majority	CN^-
Fe^{2+}	6	NH_3, H_2O, Cl	CN^-, NO_2^-, phen, cytochromes
Co^{2+}	7	The vast majority	CN^-, vitamin B_{12}
Ni^{2+}	8	The vast majority	CN^-, F-430
Cr^{3+}	3	All	None
Mn^{3+}	4	The vast majority	CN^-, phen
Fe^{3+}	5	H_2O, F^-, Cl^-	CN^-, phen, cytochromes
Co^{3+}	6	F^-	All the others
Ni^{3+}	7	O-Donors	CN^-, many

Note: phen is 1, 10 orthophenanthroline. Note the change to low-spin, paired d electron, states gains σ-acceptor and π-donor strength offsetting spin-pairing energy.

chemistry is called "ligand field theory"; see Williams and Fraústo da Silva in Further Reading.) We must be aware also that in low-oxidation and low-spin states, metal ions may well donate electrons easily during sharing in compounds and are called "electron-rich". As we shall see, exchange rates of partners from these low-spin complexes is often also slow so that they do not equilibrate rapidly as high-spin ions do. Clearly, the chemistry of transition metals with different combining ratios and in different spin states is complicated. We shall see that all these features allowed evolution of organisms when the possible partners of the metals, both organic inside cells and inorganic outside cells, were changed with the progressive oxidation of the environment. Many of these elements were originally or became essential catalysts for organic reactions (see Parts 2B and 2C). Note that in the second and third rows of transition metals the compounds are more frequently low-spin, and these elements have found considerable use in laboratory and industrial chemistry as catalysts. They are too unavailable to be used generally by organisms, but molybdenum and even tungsten are available.

In cases where elements in compounds retain unused exposed electrons, especially those in pairs, they occupy space and are stereochemically active. They are often sites of reaction.

2.6.   Important Heavy Elements

As mentioned above, some chemistry of a few heavier elements is also of concern in the development of the geosphere and of living systems as we shall see later. A striking case is the chemistry of molybdenum (Mo) and tungsten (W), which we take here with vanadium (V). The first two elements are in the *second* and *third* series of transition metals and all three are found in higher combining ratios and with a greater preference for S rather than O, W less so than Mo (the preference is for O for Cr in the first series) and molybdenum occurs as MoS_2 but tungsten as $CaWO_4$ (cf. chromium as Cr_2O_3). Both Mo and W combined with S are essential in all life, one or the other in particular organisms, and in some circumstances V can replace Mo. Interestingly, the energies of states of both Mo and W from the 3+ to the 6+ condition are similar unlike the states of any first-row transition metals, but V states come closest from 3+ to 5+. We believe that this explains their functional value in organisms (see Chapter 5). Two other cases of biologically important heavy elements are those of the non-metals selenium, Se, found in all organisms, and iodine, I, found in very few advanced forms of life. The chemistry of selenium and iodine differs from that of the lighter elements, S and F, Cl, Br of their respective Groups, not so much in combining ratio but in stability of oxidation states and reactivity, as we shall describe later. It may well be that the use of both types of heavy elements, e.g. Mo and Se, relies on their ability to act as atom donors/acceptors in reactions such as

$$1 - Mo + O \rightleftarrows - MoO$$

$$2 - Se + H_2 \rightleftarrows 2 - SeH$$

while iodine can form a somewhat labile bond to carbon. We shall have few occasions to refer to other heavy elements in the evolution of the Earth until we come to the use of many such elements by mankind in Chapter 10.

2.7. Availability

From the above chemistry, the availability of elements in the sea both in primitive (anoxic) and modern times is explicable (see Section 1.8). As we pointed out, there are many soluble salts of sodium and potassium, e.g. NaCl and KCl. In fact, virtually all M^{2+} chlorides, nitrates and sulfates are soluble. Then there are somewhat insoluble M^{2+} (or more so M^{3+}) carbonates, phosphates and silicates, for example of Ca^{2+} (and Al^{3+}), and they give rise to minerals both geological and biological. The solid oxides or hydrated oxides, hydroxides ($O^{2-} + H_2O \rightarrow 2OH^-$), of M^{3+} or M^{4+} ions occur often since their other salts hydrolyse, increasingly so as charge increases, and this insolubility greatly limits the availability of M^{3+} and M^{4+} ions. All these properties govern the availability of metals of Groups 3, 4, 13 and 14 and some transition metals in the sea. (Notice that the pH of the sea varies between 7.0 and 8.0 so OH^- is not very available at 10^{-7} to 10^{-6} M concentration, leaving many M^{2+} ions free from hydrolysis.) The equally influential salts in the evolution of the Earth are the sulfides since hydrogen sulfide was present in the primitive (perhaps acid) sea, at 10^{-3} M level. It is replaced by sulfate today and most metal sulfates are soluble. As we have seen sulfides of M^{2+} ions vary in solubility in the series given above (see Figure 1.8), so that while in the first transition series Mn^{2+} and Fe^{2+} were always available, $>10^{-8}$ M, increasingly Co^{2+}, Ni^{2+}, Zn^{2+} and

$Cu^{2+}(Cu^+)$ were not to be found in quantity in the primitive sulfide sea (Table 1.7) though the greater availability of Ni^{2+} particularly, and Co^{2+} and to some extent Zn^{2+} in more acid conditions may be important (see Chapter 5). Molybdenum, precipitated as MoS_2, and vanadium, probably as VS_4, were also not available in the sulfide-rich sea, but tungsten was. These availabilities from sulfides and later in evolution from oxides have been profoundly important for the evolution of organisms. Note that all this discussion refers to the solubility of *simple ionic salt solutions in the solvent water treated as in balance, equilibrium, with their respective ions.* In Chapter 1, we have indicated that care is necessary with this approach. There were also certain soluble clusters, such as Fe_mS_n and Ni_mS_n, thrown up especially from volcanoes, into the early sea, but some of them are not stable in the presence of CO and H_2S. Before we turn to non-equilibrium situations, we refer the reader back to Section 1.12 where we describe the slow way in which availability changed with time due to the oxidation/reduction equilibria in the sea. This will be seen to be of profound importance for the times of geological and biological changes on the Earth.

2.8.   Non-Equilibrated Inorganic Systems: Barriers to Change

While in the above we have stressed equilibrium and fast rates of reaction as being dominant in inorganic chemistry, this is far from being generally true under all circumstances. Restrictions on reactivity are obvious where there are physical barriers to reaction as in geological layers, containment in industrial or chemical apparatus or resulting from the membranes of organisms. We have already pointed out that chemical barriers, certain bonds, can also cause resistance to change and while this is rare for insoluble ionic metal/non-metal compounds of oxygen or sulfur on aqueous surfaces, it is most usually the case between more covalent non-metal/non-metal compounds, e.g. $N_2 + O_2$ and $O_2 + CH_4$, and some non-metal/metal compounds that include organic and organometallic chemicals (Section 2.16). Here we must note that the atmosphere is not in equilibrium between the oxides of nitrogen, oxygen and nitrogen, between oxygen and ozone nor between carbon oxides, oxygen and methane, but the last of these cases really belongs to organic chemistry (see Section 2.12). It is of great interest that the rates of change of small non-metal and organic molecules themselves can be increased by the presence of the metallic elements or heavy non-metals in associated compounds (see Section 2.12). In essence, some metal/non-metal and non-metal/non-metal bonds, especially of sulfur and phosphorus, have to be described in mixed ionic/covalent language and often they have intermediate rates of reaction, for example especially in low-spin metal complexes (see Section 2.16). We return to these major non-equilibrium inorganic systems in Chapter 5.

Before we impose complex barriers to reaction, we observe that more limited barriers are found even in quite soluble salt solutions. Small ions, such as Mg^{2+} and also Ni^{2+}, do not exchange very easily even the water from around themselves with bulk water, in contrast with the behaviour of Na^+, K^+, Ca^{2+} and Zn^{2+} (Figure 2.7).

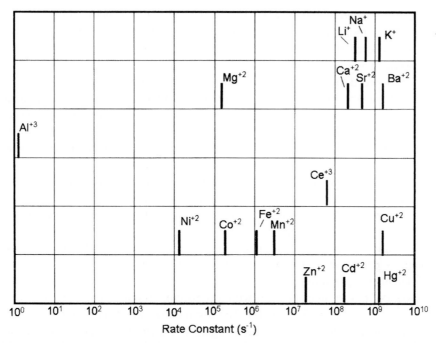

FIG. 2.7. Characteristic rate constants (s^{-1}) for substitution of inner-sphere H_2O of various aqua ions. *Note*: The substitution rates of water in complexes $ML(H_2O)_n$ will also depend on the symmetry of the complex (adapted from Frey, C.M. and Stuehr, J. (1974). Kinetics of metal ion interactions with nucleotides and base free phosphates in H. Sigel (ed.), *Metal ions in biological systems* (Vol. 1). Marcel Dekker, New York, p. 69).

These exchange rates are of extreme importance in life's evolution. Generally heavier elements exchange partner atoms more readily than lighter atoms and this is a dominant feature in the value of the heavier elements, e.g. the exchange rates are $Ca^{2+} > Mg^{2+}$, Mo(W) > Cr, P > N, Se > S and I > Cl(Br).

2.9. Non-Equilibrium Inorganic Systems: Energy Storage

In the above discussion on inorganic chemicals, we see that in the gaseous state and in solution they usually go readily to the most thermodynamically stable condition with the loss of heat. However, in the description of solids in geological chemistry (inorganic) we saw that, quite differently, on condensation, barriers were created to reaction between a high-temperature melt, the core and the mantle, and the aqueous sea and the atmosphere. Hence, the inorganic system was open to very rapid change on the surface, even between the solution and the immediate crust, where gases, liquids and life are found, but was only open to very slow change, which is still very far from complete, between solutions, the deeper crust, the mantle and the core.

The energy trapped is made visible most dramatically in volcanoes, earthquakes and also in black smokers in the deep ocean. As stated in Section 1.11, the collisions of continents caused huge mountain ranges to arise, and their drifting apart left fissures and exposed previously molten mantle/crust rock to the sea. Again the folding of the mantle has forced rocks of the surface down and the mantle rock up. The result is that some geological formations are out of equilibrium. In the flows down and up metals re-appear in transformed states (see Figures 1.6 and 1.13). In discussing evolution of the inorganic chemicals and geochemistry, we must always remember that in the (buried) non-equilibrated state there is a huge reserve of energy in the Earth, which could be transferred to surface (biological) chemistry at any time with possible disastrous consequences. When energised compounds (out of equilibrium) come to the surface they can react and use energy to drive other reactions. This process may also have assisted the start of life by reactions of such unstable combinations as $FeS + H_2S$ and $Ni_4S_3 + H_2S$.

The out-of-equilibrium condition of the minerals of planet Earth also arises through its energy relationship to the Sun. The Earth rotates on an axis and has a considerable temperature gradient changing daily and with the seasons. This causes movement of the air, which carries with it moisture, bringing rain, rivers and there-fore erosion as stated in Section 1.9. The resultant *sediments* have introduced new inorganic mineral formations, which together with organic-matter debris have formed the fertile soils. Soils are sources of minerals but also of easily accessible organic chemicals. As well as the seas and other aqueous environments, the soil pro-vides opportunity for living organisms to thrive. In fact all such material is "ener-gised" to some degree. Other sediments are formed from the mineral skeletons of animals and plants which lived in the sea. A further important example of the for-mation of an energised kinetically stable compound is that of ozone today in a pro-tective layer some kilometres above the condensed surface of the Earth. We consider that the effect of light on other inorganic compounds, such as the disproportionation of water to hydrogen and oxygen, was slight before biological intervention. Finally, as mentioned above, the oxygen in the air is out of equilibrium with small non-metal compounds such as CH_4 and N_2, and this instability becomes dramatically impor-tant when we consider organic chemicals in organisms and an oxygen atmosphere.

To conclude, despite the overwhelming importance of inorganic chemical equi-libria in the early formation of the Earth, some of the Earth's chemicals arose from trapping of energised material (largely hidden from the surface) and later from energisation of exposed materials. Some of the changes to equilibrium and some away from equilibrium are still on-going, and some substances such as water are in energised circulatory activities. Even so much of inorganic chemistry in water and on mineral surfaces comes to equilibrium quite rapidly. We shall often be con-cerned in this book with this rather rapidly equilibrated surface inorganic chem-istry and only now and then with kinetically stable energised inorganic chemicals.

Another way in which energy can be stored in aqueous solution is not only in organic compounds, as we shall see, but also in kinetically stable combinations of inorganic ions. An example is the molecule pyrophosphate, which stores energy at pH = 7 relative to two phosphates. Great use is made of this particular energy storage in organisms. Another example is that of gradients of ions across barriers.

2.10.   Reactions and Catalysis by Inorganic Environmental Compounds, Especially Sulfides

It may well be that the sulfides, inorganic and organic, were among the most important of all chemicals in the evolution of the Earth and indeed of life. We shall look at them in different ways – their original formation, their energy states, their solubilities in water and their occurrence in organic chemistry, especially in living organisms. The last has been stressed by de Duve in his books, see Further Reading. Here we are concerned with their inorganic geochemistry. The reducing nature of the planet Earth with its vast excess of metals over oxidising non-metals, oxygen and halogens, insured that some sulfur occurred initially as metal sulfides and some as hydrogen sulfide, H_2S, in contact with water. The sulfides were open in places to high temperatures and slight acidity. We have already discussed their solubilities, but now we must turn to some peculiar features of a few transition metal sulfides as potential sources of energy and of catalysts.

The sulfides formed at high temperature in the presence of excess of some metals are metal rich, especially those of cobalt, Co_9S_8, and nickel, Ni_4S_3. The ability to form these compounds is not shared with iron, which forms much FeS but also forms the more stable compound pyrite, FeS_2, in which sulfur occurs as the anion $[S-S]^{2-}$; Fe^{2+} is here low-spin. There are then the possible following reactions, which can give energy and could trigger, catalyse, organic syntheses:

$$FeS + H_2S \rightarrow FeS_2 + \text{hydrogen (bound)} + \text{energy}$$

and

$$Ni_4S_3 + H_2S \rightarrow 4NiS + \text{hydrogen (bound)} + \text{energy}$$

or

$$Co_9S_8 + H_2S \rightarrow 9CoS + \text{hydrogen (bound)} + \text{energy}$$

(bound) here refers to the fact that the hydrogen may be incorporated into carbon compounds. Before we turn to the fate of the hydrogen there is a feature of these metal sulfides in the Periodic Table, which extends to copper in Cu_2S and to MoS_2. Note that S^{2-} readily gives up electrons, much more so than O^{2-}. On going from FeS to Cu_2S the compounds are increasingly electron-rich (see Section 2.5), especially compounds of their lower valences of M^+ and even M^0, which can exist in reducing conditions. These states of ions, and especially because they are often in low-spin states in compounds, provide very powerful electron donors and acceptors, so that as well as being able to form such compounds as carbonyls they can also form somewhat unstable methyl and hydrogen bound compounds on their surfaces. The catalytic property of them is in addition to the possibility that they may well have been the source of energy for the beginnings of the organic chemistry of life. Quite curiously, if Wächtershauser (see Further Reading) is correct, *the high-spin/low-spin switch of FeS to FeS₂ could be the energy source for the origin of life.*

2.11. Summary of Inorganic Compounds Related to the Global Ecosystem

It is very important to realise that inorganic elements are not just a large part of our environment but are now known to be extremely valuable in organic chemistry and essential for life, quite apart from the use of them made by mankind (see Chapter 11). Their changing availability with the oxidising power of the environment as observed in geochemistry is in fact, we consider, the major driving factor in life's evolution. We shall see its significance both in equilibrium and kinetic properties. Outstanding among the properties of the available metals are:

(1) Their ability to form both continuous lattices alone or with non-metals in salts.
(2) The solubilities of their salts in water.
(3) Their selective binding interactions with different O-, N- and S-donors both from inorganic or organic sources affecting solubility and complex formation (see Section 2.19).
(4) Their variable oxidation states and spin states.
(5) Their very selective rates of reaction.
(6) The ability of some of them to retain energy in compounds and distribute it.
(7) Their role in catalysis (see Section 2.20).

While the properties of the available non-metals are

(1) Their ability to form molecular structures alone or with one another as well as their above ability to form salts with metals (see also (2) and (3) above).
(2) Their ability to form compounds and anions in different oxidation states.
(3) Their selective rates of reaction particularly notice the difference $Se > S > O$, $P > N$, $I > Br > Cl > F$ and $Si \gg C$.
(4) The general ability to retain energy in certain compounds due to kinetic stability.
(5) Their role in catalysis especially of S, P and Se in certain compounds.

At the extreme, inorganic compounds are usually very different from organic chemicals made from non-metals, to be described next, but they may exhibit intermediate and even similar behaviour, as we shall see later in this and subsequent chapters.

Most importantly when the two are put in compounds a wealth of new properties are found, which is described in brief at the end of this chapter.

2B. Organic Chemistry

2.12. Introduction to Organic Compounds of Ecological Relevance

Organic compounds, classically defined, are those compounds that have carbon as their central element in association with a variety of other non-metals but mainly H, N, O, S and halogens. We can also include the non-metals Se, P and Si, but other members of the Periodic Table are metals or "metalloid" and form rather different compounds with carbon, so that until relatively recently they were not included among organic combinations (see Section 2.16). The traditional reasons for having two separate parts of chemistry are therefore disappearing, but it is still perhaps useful for didactic reasons to keep much of the chemistry of complicated molecules based on carbon separate from the seemingly simpler inorganic compounds. Of course, organic as a word implies a relationship to living organisms, and the original organic chemicals of C, H, N, O, S and P were thought to be the only class of chemical found in such organisms, but as we shall see this is known to be not so today. In this book, we can only touch on the diversity of organic chemicals, stressing their relevance to the ecosystem. Turning to their properties, in general, organic compounds are molecular and do not form continuously bonded lattices in contrast with a large number of such inorganic compounds. They are relatively volatile and many are not soluble in water, for example, N_2, O_2, CO, CH_4. Although stoichiometrically many are complicated compounds, are relatively kinetically stable (see below), only open to slow change at temperatures of our environment and have to be structurally defined for reasons given below. They readily form long-chain, electrical insulating, polymers, sometimes cross-linked, of very high molecular weight. Hence they were very difficult to characterise at first and very little was really known about them until after 1850 AD, while inorganic compounds had already been studied, albeit often in a somewhat rough and ready way, for more than 3000 years. Another feature of organic chemistry, as it had been examined until recently, is that its syntheses of compounds were carried out in organic solvents, such as hydrocarbons, alcohols, chloroform and ether since many organic compounds are not soluble in water and some are not even stable in the presence of water. By contrast, inorganic "ions" of salts were examined almost exclusively in water. Today it is known that biological organic chemistry, which will interest us, takes place overwhelmingly in aqueous media too, with and utilising many inorganic elements, which provide another reason for examining organic and inorganic chemistries together (see Section 2.16). The implication is that biological organic compounds are to a large degree of high polarity.

Organic chemical compounds are a minor part of the chemistry of the universe and of the Earth, but they have played a major part in the Earth's ecosystem evolution.

While undoubtedly many organic chemicals are to be found in outer space including hydrocarbons, fullerenes, ketones, amino acids and nucleic acid bases (see Bernstein *et al.* and Ehrenfreund and Charnley in Further Reading), most of such compounds are on Earth and are newcomers to the universe due to biological "organic" activity or more recently to human manufacturing. They have been synthesised by application of energy, which makes them distinctive from many inorganic compounds, but not, of course, from the organometallic compounds. As we now stress, it is this energisation with kinetic stability which also makes a second convenient reason to discuss them at first separately from inorganic compounds.

2.13.   Stability and Reactivity of Organic Chemicals

We start the description of molecular organic chemicals and their reactions with the statement that none of these chemicals are thermodynamically stable to reaction going to CH_4, methane, and carbon dioxide, CO_2, if they contain oxygen, or to CH_4 and carbon if they do not. Again none of these organic chemicals are stable in the presence of oxygen of the air since they all burn to CO_2 and H_2O. However, these reactions do not occur spontaneously, so that the whole of organic chemistry is based on *kinetic stability of molecules*, that is molecules which are very slow to change to truly thermodynamically stable states despite the fact that they are all energised in the atmosphere. We are familiar with many organic compounds because they are metastable. Organic chemists take advantage of the kinetic stability to make a vast range of thermodynamically unstable but relatively unreactive chemicals (see Further Reading). Within living organisms there are also very many kinds of such non-metal combinations but, as stated, the majority of them are now soluble in water (see Table 2.6).

The skill in laboratory or industrial *organic chemistry* lies in (1) choosing thermodynamically unstable starting materials, so that reactions can go downhill giving out heat, e.g. along a path such as

$$CH_4 + 2O_2 \rightarrow CO_2 + 2H_2O + \text{heat (energy)}$$

TABLE 2.6

SOME ORGANIC COMPOUNDS IN CELLS

Organic compound	Typical formula
Fats (lipids)	$CH_3 \cdot (CH_2)_n \cdot CH_3$
Amino acids	$NH_3^+ \cdot CHR \cdot CO_2^-$
Bases (adenine)	$C_5H_5N_5$
Sugars (glucose)	$C_6H_{12}O_6$
Acids (propionic acid)	$CH_3(CH_2)CO_2H$
Sugar phosphates (glucose phosphate)	$C_6H_{13}O_5PO_4^-$

but (2), by using a selected catalyst and conditions, to see that the initial part of this reaction is guided on a particular path to a desired half-way product, under circumstances in which the complete reaction is prevented and in which these half-way products can be isolated. For example,

$$2CH_4 + O_2 \rightarrow 2CH_3OH + heat$$

The intermediate product is still a thermodynamically unstable organic chemical, which is able to react further with oxygen. It is then obvious that living organisms, based as they are largely upon synthesis and degradation of specific organic compounds, require selective catalysts. A catalyst active site often needs to have reactive atoms such as heavier non-metals, e.g. sulfur or metal ions. Moreover, the active sites of these catalysts when in organisms often have to be centred on the use of *available* inorganic elements although this is not the case for industrial or laboratory organic chemistry for which a wide choice of metal is open (Chapter 10).

A further energy consideration in the execution of synthesis is the use of compartments, both in inorganic and organic chemistries, but here we stress the second. Obviously, an organic chemical reaction in a laboratory or in an industry is carried out in a protected container which maintains concentrations. It limits disorder as does a balloon controlling a gas volume. To understand the further reason for special containers, we have to be aware of the required environment for reaction. A container can control chemical conditions, for example a chemist can choose the chemical and physical nature of the atmosphere in which reaction occurs, e.g. in nitrogen gas, the temperature and the pressure. Again certain reactions will not occur in a desirable way in the presence of oxygen so it must be removed, an observation of great importance in our understanding of all cell chemistry apparently exposed to air. Finally, separate vessels are used in organic chemistry for the preparation of intermediates of complicated reactions before mixing them. This use of compartments avoids confusion of reactions forcing partial syntheses along a route by preparing separate intermediates, to be mixed later in a further compartment. Such steps require extra energy for the transfer of material. As reaction paths become more complicated and the number of chemicals in them increases, it is logically necessary to use more compartments and methods in order to organise them (see Section 3.10 on organisation). We shall see that advanced organisms function of necessity in this way using different "atmospheres", catalysts, vesicles for preparations in particular cell compartments before passing the intermediate compounds through membranes to selected sites. Note that industrial plants and cells both use flowing systems not batch systems, in their syntheses, and flow will be dominant later in this book since it is central to cellular chemistry. A difference between organic chemical reactions in the laboratory and in industry from biological reactions in cells is that in the former the initial system usually contains only two reactants in rigid containers whereas in a biological cell there may be at least 100 held in compartments by mobile membranes. Thus, a biological catalyst, an enzyme, must be very highly selective and intermediates must remain bound before the desired product is released (see Thomas and Williams in Further Reading).

Now all these manipulations require energy, differently in different compartments, and also a plan. The chemist is the planner in organic chemistry and his sources of energy are obvious in his equipment, but notice again that laboratory syntheses start from chemicals of high-energy content, so that the process is downhill, implying that careful selection of starting material is necessary and energy is applied only to increase the reaction rate and to manipulate the containment of reactants. An organism must also have a plan, but it is now self generated, including its membranes, and in addition it has to capture energy since many biological syntheses are energetically uphill. Organisms synthesise unstable organic chemicals from stable simple inorganic chemicals such as CO_2, N_2 and H_2O. While we shall show in Chapter 3 that left without any energy source all chemicals systems go finally to the state of greatest probability and stability, equilibrium, we ask in that chapter: is there a principle for the direction and final condition to which systems which capture energy will go?

It is convenient to divide organic chemical reactions between acid–base and oxidation–reduction reactions as in inorganic chemistry. In acid–base reactions the oxidation states of carbon do not change, e.g. in hydrolysis, where reaction is, for example,

$$CH_3CO.OR + H_2O \rightarrow CH_3COOH + ROH$$

In oxidation–reduction reactions the oxidation states of carbon change, for example,

$$CH_3COOH + H_2 \rightarrow CH_3CHO + H_2O$$

The two types of reaction require quite different catalysts; in the first an acid or base centre is required, e.g. Mg^{2+}, while in the second a heavy atom capable of redox state change is required, e.g. Fe^{2+}. Here the laboratory, industrial and cell chemistries are alike.

Let us now look at some further peculiarities common to organic chemistry (and cellular chemistry) but not so dominant in inorganic chemistry.

2.14.   Stereochemistry

The unstable organic molecules have fixed structures because of the nature of their bonds. The structure of methane is a tetrahedron of covalent single $-\overset{|}{\underset{|}{C}} - H$ bonds. Where the structure of even a simple compound is based on double bonds, such as in an ethylene, there are two forms labelled *trans* and *cis* geometric isomers:

$$
\begin{array}{ccc}
\underset{H}{\overset{X}{\diagdown}} C = C \underset{X}{\overset{H}{\diagup}} & \text{or} & \underset{H}{\overset{X}{\diagdown}} C = C \underset{H}{\overset{X}{\diagup}} \\
\textit{trans} & & \textit{cis}
\end{array}
$$

A further complication is that for a carbon, which has four different groups attached to it in a tetrahedral arrangement, there are two optical isomers called laevo (L-) and dextro (D-) from their different abilities to rotate the plane of polarised light. They are structural mirror images which cannot be superimposed:

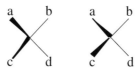

Optically active compounds, enantiomorphs, can be made in certain other ways, e.g. by forming helices. Whereas L and D forms are equal in energy, *cis* and *trans* forms are not. Biological compounds are usually optically active and virtually all amino acids have L structures, while saccharides more frequently have D structures. (One remarkable exception is D-serine, found in some bacteria and in the human brain.) The reason for these peculiarities is not known, but D/L-selection is very important in medical compounds because one form may be active and the other is not. While both types of isomers are difficult to change from one form to the other this can be achieved by applying radiation to the *cis/trans* pair and this is sometimes used in biological energy capture (see Chapter 5). Together with the size of many organic molecules and the fact that up to six different types of atoms can be used in almost any order in three-dimensional space, these isomeric forms allow a huge variety of organic chemicals of the same analytical composition, but of the same or different stability, to be constructed. It is for this reason that an organic compound cannot be described by element analysis alone. Of course, carbon is not the only central atom open to such a variety of binding and even complexes of heavy metals (in low-spin states) have optical isomers. We cannot go further into the diversity of components and their synthesis in organic chemistry, but we do need to see the special relevance to the synthesis of large biopolymers.

Proteins, made from optically active amino acids, have molecular weights of tens of thousands, as do polysaccharides made from sugars, while DNA made from bases has a molecular weight of millions. The order of units in any of the polymers can be varied so that, undoubtedly, the multiplicity within composition is one reason why so many living species exist. Naturally a major discipline, molecular biology, has been built up to find ways of analysing and of making these and other, kinetically stable, compounds of highly selected sequences and then examining their properties in terms of their structures. Nowhere outside the cells is this synthesis skill more evident than in the cells, and some cell chemicals have also internal mobility, which allows the development of many novel properties (see Chapter 4).

The next interesting feature of certain large organic molecules is that apart from local linear polymeric structures based on small units, monomers, they can fold to give shaped molecules, for example proteins. Many such molecules have very selective surfaces with little mobility and hence they can bind smaller organic molecules based on their general structure or on *cis/trans* and L--D-preferences. This high degree of selective binding occurs especially with protein catalysts, enzymes, found in all organisms. An essential feature of their catalytic capabilities is that the active groups in them are often in *constrained states aiding the catalytic act.* Proteins bind to proteins too to build idiosyncratic structures, leading eventually to the peculiar variety of shapes of organisms. The associated protein/substrate and protein/protein units are not held together by covalent bonds but by electrostatic or weaker van der Waals interactions. The binding strengths have, therefore, an enormous variety and correspondingly there is a great variety in the rates of binding, association, k_{on}, and the rates of leaving, dissociation, k_{off}. The two are sometimes in close to fast balance, equilibrium:

$$A + B \overset{k_{on}}{\underset{k_{off}}{\rightleftarrows}} AB$$

where $k_{on}[A][B] = k_{off}[AB]$. However, equally frequently, the off-rates are much slower, especially where binding is strong and equilibrium is not approached, e.g. in syntheses in a series of steps. A fascinating feature of organic chemical procedures in cells is that they often concern just the (energised) forward reaction or just the back reaction of the change A/B, that is

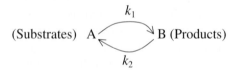

(Substrates) A ⇄ B (Products)

where k_1 and k_2 refer to quite different routes allowing control over relative A and B concentrations not possible at equilibrium. Hence much consideration goes into the selection of catalysts and physical circumstances for the forward A ⟶ B or the reverse transformation. When we come to describe the overall organic chemistry of ecosystems, including organisms that recycle materials, we shall find that the reverse reaction rate, k_2, is as important as k_1 but the reaction goes by quite a separate path using a different catalyst and hence a cyclic condition is attained. We shall also see in several chapters that proteins, even perhaps RNA, can form molecular machines for mechanical as well as chemical transformation, e.g. in pumping, and description of them returns us to the examination of their mobility.

2.15. The Importance of Temperature and Light: Rates of Organic Reactions

In Chapter 1, we have stressed the importance of a temperature of 300 K at the Earth's surface for the chemistry we observe here. A temperature of around 300 K is essential in two respects: water is kept as a liquid and the rate of change of organic chemicals is very slow in the absence of catalysts. The second point implies that, as we remarked above, biological change is almost invariably under the control of the catalysts.

While discussing temperature, it is important to remember that the value of 300 K is maintained by heat from the Earth's core, which contributes some 250 K, and light from the Sun. Hence, the Earth acts as a thermostat. Heat flow from the Earth's core is just the general transfer of heat from a hot to a cold body in an effort to bring the whole system to equilibrium – an even temperature. This heat energy is not in a form that can be retained by incorporation into unstable chemicals directly. The Sun's radiation, by way of contrast, excites electrons locally and, in so acting on a particular molecule, energy can be trapped and then used to energise synthesis of other molecules, hence driving living processes. Once energy is trapped in a chemical or in its concentration, for example in the Earth's crust or in an organic molecule, it can be used to drive other upper-hill processes so long as the overall change degrades some of the energy to heat (see Chapter 3).

The effect of light on organic chemicals can cause separation of charge, electrons from a residual positively charged molecules, and further reaction can lead to separation of products, one oxidised, one reduced. This is the same photoreaction as we have described for the disproportionation of water to hydrogen and oxygen. It will be seen that this photoreaction has become the basic energy capture reaction of organisms leading to reduced organic molecules and molecular oxygen, an enormous energy storage. Back-reaction is relatively slow except in the presence of catalysts, and the controlled use of the back-reaction, faster than it occurs in abiotic or laboratory chemistry, is the basis of much catalysed organic and biological chemistry. It is the basis of our petrol engines.

2C.   Bringing Inorganic and Organic Chemistry Together

2.16.   Introduction

The last 50 years have seen a remarkable change in attitudes to these two chemistries. The one most recognised is the so-called *organometallic chemistry*. Here, instead of the focus resting on compounds of carbon with other light non-metals, organic chemistry, or mainly on simple metal compounds with oxygen and sulfur, inorganic chemistry, it is the combination of metals and carbon compounds that is the centre of interest. A huge variety of such compounds have been made using the procedures of organic chemistry. The compounds are kinetically quite stable, so that they do not go rapidly to equilibrium states. Recently, it has been realised that metals bound to carbon and other non-metal frameworks, particularly phosphine derivatives, have sufficient kinetic stability to allow them to be used in catalysis. Note that much of this chemistry is carried out in *non-aqueous solvents* and may require the absence of oxygen. In essence it is, by method, really a new branch of organic chemistry, which is now used extensively in industry. There is no limitation here in the choice of the 70 metal elements and many have been pushed into use by mankind. One point of interest is that certain metals in them seem to be better able to catalyse particular reactions, see Section 2.19, and these differences apparently apply to cell catalysts. This type of organometallic compound may not have much functional significance in biological systems, but is seen especially in some cobalt and nickel biological compounds.

Somewhat less well recognised is the older branch of metals in combination with organic chemicals – a major part of the so-called *coordination chemistry*, which is dominant in organisms. Here, initially, laboratory studies of coordination compounds (complexes) concentrated upon metal-ion binding of non-metal centres such as N, O and S in *simple organic frameworks in water* and were included in inorganic chemistry. The first examples used the very simple molecules such as NH_3, H_2O and simple thiols. Some more complicated examples of different types of organic ligands are given in Table 2.7. In the majority of cases in these complexes, e.g. $Cu(NH_3)_4^{2+}$ or Cu(amino acid)$_2$, *exchange* of the two units, say Cu^{2+} and NH_3, is *rapid* so that they easily attain equilibrium. The detailed knowledge of these equilibria only came to the fore around and after 1950. It has been found that although many ligands, groups bound to metals in complexes, do exchange, i.e. come to equilibrium quickly, some others do not (Table 2.8). It was realised soon after this time that many biological catalysts were based upon protein-bound metal ions, some of were relatively rapidly exchanging metal ions, others in relatively slowly exchanging complexes although both used N, O and S electron-donor centres of the proteins. Often it is the low-spin transition metal complexes that exchange ligands most

TABLE 2.7

SELECTIVE LIGANDS BASED ON DIFFERENT DONOR ATOMS

O-Donor	^-O_2C–CH_2–CO–CH_2–CO_2^-
N,O-Donor	^-O_2C–CH_2–NH–CH_2–CO_2^-
N-Donor	H_2N–CH_2–CH_2–NH–CH_2–CH_2–NH_2

O-Donor
  (phenolate)

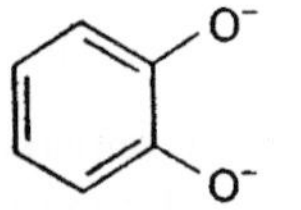

N-Donor
  (aromatic)

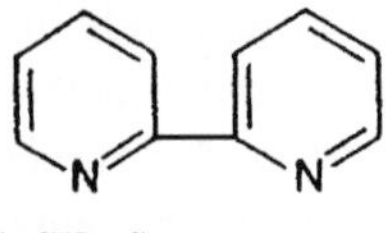

O,S-Donor	^-O_2C–CH_2–S^-
N,S-Donor	H_2N–CH_2–CH_2–S^-
S,S-Donor	^-S–CH_2–CH_2–S^-

O-Donor
  (hydroxamate)

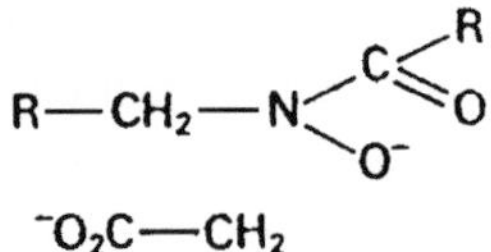

EDTA

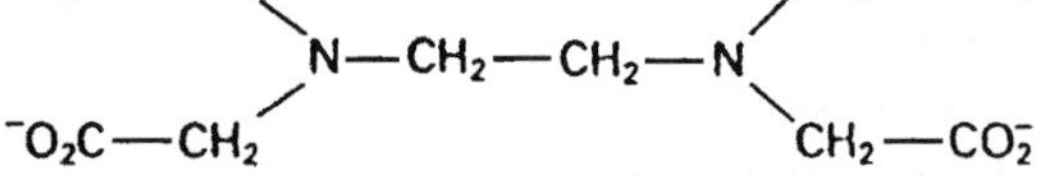

TABLE 2.8

SOME FAST AND SLOW CHEMICAL EXCHANGES

Reaction	Reaction
Ions in water	Mostly fast, e.g. M^{2+} with anions or organic molecules
Ions in cavities	Haem, B_{12}, chlorophyll; slow exchange
Organometallic compounds	Slow exchange; see covalent organic compounds
Covalent organic compounds	Slow, considerable activation energy required; S and P compounds faster
Association of organic compounds	Often fast

slowly (see Section 2.5). In essence, there is no difference in chemical principles between the very slow exchanging and organometallic centres or the properties of organic covalent compounds. However, the case of fast exchange of complexes is of very different value and is also extremely important in cells, as we see later, and the implication is that certain such equilibria are of great consequence in metabolic controls. We shall find it necessary to consider especially coordination chemistry in more detail, but only a little organometallic chemistry when we describe biological chemistry (see Williams and Fraústo da Silva in Further Reading). (Note it was the difficulty of analysing for traces of essential elements until after 1950 that led to the prevailing early view that cell chemistry does not require metal ions.)

Now we have seen that both in inorganic and organic chemistries there are both acid–base and oxidation–reduction reactions. Of the inorganic elements, some of their ions, such as Mg^{2+}, Ca^{2+} and Zn^{2+}, in complexes cannot undergo oxidation–reduction changes when acting as catalysts, they can only be effective in organic acid–base reactions such as hydrolysis. Quite differently, transition-metal ions can undergo valence change, so that all these ions in complexes can catalyse the oxidation–reduction reactions of organic chemistry such as the introduction of H_2 or O_2 into organic molecules. Typically, Fe and Cu are good oxidation catalysts. Both types of catalysts must be used in cells in separated reactions often in separate compartments, since organisms must carry out both types of reactions, hence they must also have an ability to bind separately to reactants (Sections 2.17 and 2.18). Note that in oxidation–reduction there is often a need to match closely the potentials of reaction in the organic and inorganic units (Sections 2.19 and 2.20).

2.17. Complex Formation: Selectivity

The finding above that many coordination complexes between organic molecules and metal ions are essential to life raises interest in both their quantitative binding strengths and their exchange rates. While the monovalent ions Na^+ and K^+ bind very weakly in water, they are present in the environment and in cells in higher concentration than other ions. Moreover, as we shall see they have to be pumped across cell membranes to maintain concentrations. It has been found that provided the conditions of pH are such that H^+ does not compete (cell pH is around 7.0), then a few very weak binding sites comprised of several neutral oxygen-atom donors can bind both Na^+ and K^+. Selective binding depends on creating a hole or a restricted size of an organic chelate, using O-donors of ethers, carbonyls and alcohols and now and then a single carboxylate anion. Given appropriate construction there is no competition from Mg^{2+} or Ca^{2+} or any other divalent ion or the proton, which is a special separate case as it is so small. Much of the selective binding of very simple anions is also based on similar size discrimination as it also is weak, e.g. binding of halides, hydroxide and HS^-. There are larger anions and here, while size remains a factor, shape and multiple H-bonding is involved. As size of molecules increase, we enter into the topic of the fitting of substrates to strong binding protein sites when every type of interaction except covalent bonding is important.

As we have stated, many organic binding centres have a modest affinity for divalent metal ions and come to equilibrium quite quickly – much faster than reactions involving covalent bonding changes of organic or metallo-organic compounds. The first feature of great importance in cell chemistry is then the general order of affinity of the divalent metal ions (Figure 2.8) for almost all coordination compounds at equal concentration of free ions at equilibrium, namely

$$Mg^{2+} < Mn^{2+} < Fe^{2+} < Co^{2+} < Ni^{2+} < Cu^{2+} > Zn^{2+}$$

often called the Irving–Williams series, which we have seen before in geochemistry (Section 1.5), and in precipitations in analysis (Section 2.4). Although the above monovalent ions hardly bind at all and certainly equilibrate rapidly, Cu^+ ions bind strongly and hence exchange slowly. The trivalent ions are of much lower availability and concern, and few are used in enzymes, organic chemistry in laboratories or industry, but note $AlCl_3$, and especially Fe^{3+} and Mn^{3+} in cells. Cells either reject or ignore ions from the Periodic Table Groups 3, 4, 13 and 14, and especially very heavy metal ions such as Cd^{2+}, Pb^{2+} and Hg^{2+}.

It is clear that selection of an organic partner for a divalent metal is difficult given the dominant position of Cu^{2+} (or Cu^+) in the above series (see Figure 2.8). The major way around the problem of selective binding used in chemistry is to remove each metal ion in turn employing its strength of binding, starting from copper. This is the qualitative analytical approach, which was taught to all students even before 1950. The simplest procedure for removal, used in analysis, is not by formation of soluble complexes but by precipitation in a sequence of inorganic sulfide compounds (see their parallel geochemical differences, Section 1.5). For example, in qualitative analysis, hydrogen sulfide, with the aid of acidity to control the S^{2-} concentration, forms precipitates in the above sequence (see Figure 1.8). After precipitation at a succession of pH values, each precipitate is separated from its solution by filtration, an example of compartmental selection. This selection of elements may also be achieved with complexing or precipitating organic agents of

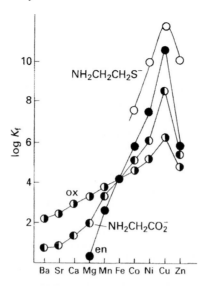

FIG. 2.8. The variation of stability constants, K, for the complexes of M^{2+} ions of the Irving–Williams series, ox, oxalate; en, ethylenediamine (note that the constants plotted are absolute, not "effective" constants at pH = 7).

particular organic chemical design, which give very high selective affinity for the metal ions in the same order and with the same pH control. The complexes so formed may also be extracted from water into an immiscible solvent in this sequence. The best way to proceed is to limit the amount of reagent available to a mixture of metal ions to the concentration of the strongest binding metal ion in the solution while choosing a reagent which binds strongly. After separation, e.g. by sulfide, analysis utilises weaker binding with reagents in the donor atom order S>N>O. It is then possible to separate all the metal ions quantitatively. This selective removal by organic chemical binding centres is best used in the order based on sulfur donors selectively for Cu and Zn first, then on nitrogen/oxygen mixed donors for Ni, Co, Fe, Mn, with use of very less N donation along this series, and finally on oxygen donors only for calcium and magnesium (Table 2.9). Selectivity is enhanced by the selected structure (hole size) and symmetry of the donor groups (Table 2.10); see Williams and Fraústo da Silva in Further Reading, and this may be particularly important for separation of high valence states of V, Mo and W. When we turn to biological systems, we shall see that closely parallel and equally effective controls over reaction donors and reagent, protein, concentrations are present and that the equilibrium basis of selectivity remains much the same as in analysis – an invariant feature of organisms. *There is universality in cell chemistry related to the equilibrium stability constants in Figure 2.8.*

Now interesting as is this binding to organic molecules since it applies to the bindings in proteins, it is not applicable to the binding of small molecules such as

TABLE 2.9

LIGANDS PREFERRED BY DIFFERENT METAL IONS IN SIMPLE COORDINATION
COMPOUNDS

Metal ion	Ligands
Na^+, K^+	Oxygen-donor ligands, neutral or of low charge (-1)
Mg^{2+}	Carboxylate, phosphate and polyphosphate (total charge $\geqslant -2$). N-donation (special)
Ca^{2+}	Carboxylate, phosphate (less than Mg^{2+}), some neutral O-donors
Mn^{2+}, Fe^{2+}	Carboxylate, phosphate and nitrogen donors combined, (thiolate)
Fe^{2+} (special)	Unsaturated amines (particularly porphyrins)
Cr^{3+}, Mn^{3+}	Phenolate (e.g. tyrosine), hydroxamate, hydroxide
Fe^{3+}, Co^{3+}	Carboxylate, N-donors, polypyrroles, (thiolate)
Ni^{2+}	Thiolate (e.g. cysteine), unsaturated amines, polypyrroles
Cu^{2+}, Cu^+	Amines, ionized peptide $>N^-$, thiolate
Zn^{2+}	Amines, thiolate, carboxylate

Note: The case of Fe^{2+} is especially complicated by spin-state changes and polymeri-sation of mixed oxidation states.

TABLE 2.10

PREFERRED GEOMETRIES IN SIMPLE COORDINATION COMPOUNDS

Metal ion	Preferred geometries
Cu^{2+}, Mn^{3+}	Tetragonal > 5-coordination > tetrahedral
Cu^+	Linear, trigonal, or tetrahedral
Ni^{2+}	Octahedral > others
Co^{2+}	Octahedral > tetrahedral > others
Zn^{2+}	Tetrahedral > octahedral > 5-coordination
Mn^{2+}	Octahedral > others
Fe^{3+}, Co^{3+}, Cr^{3+}	Octahedral > others

Note: In this table all metal ions are in high-spin states and liganding atoms are small
O, N donors. S-donors favour lower co-ordination numbers. Ligand-field theory, that
is polarisation of and binding by the core electrons and orbitals of the metal ion compounds, can explain the above observations; see inorganic chemistry textbooks in
"Further Reading".

O_2 and H_2, which is critical for some cell chemistry. For such reactions particular
transition metals, which can undergo valence state change, are essential since in
effect the combination with O_2 raises the oxidation state and that with H_2 lowers it
much as we have noted in organic chemistry. Quite different metal ions are therefore required in many oxidation–reduction reactions than those for acid–base reactions, and we shall see this is true in both laboratory and cell chemistries. In
oxidation/reduction reactions the stability of the complexes have to be considered

together with their oxidation/reduction potentials. For example, only in a reducing atmosphere can Fe^{2+} be kept in solution at pH = 7.0, since in oxidising conditions and even in the presence of complexing agents, Fe^{2+} usually forms complexes of only moderate stability and is oxidised to Fe^{3+}, but note that there are some ways of maintaining bound Fe^{2+} using ligands for which the ion has stronger relative affinity at pH = 7, e.g. orthophenanthrolines with two N donors or in certain proteins. Similar considerations apply to manganese chemistry in different states. Copper too is readily reduced to Cu^+. In this case the affinity of both oxidation states for several donors is comparable. Once again most of these oxidation/reduction reactions of metal complexes come to equilibrium quickly, but this is not the case for redox reactions in most organic molecules.

2.18. Matching Redox Potentials of Inorganic and Organic Chemicals

The redox potentials, given in Table 2.11, give a numerical value for the reducing power, labelled negative, where oxidised and reduced molecules are present in equal concentration for some organic reactions. It shows that the reduction of various organic molecules requires very low potentials. We have considered that when the Earth formed, the oxidation–reduction potential of the sea was not much below the average value of -0.2 V. There is then a requirement for an energised process of reduction if CO or CO_2 are to form $[CH_2OH]_n$ and then organic compounds generally. In Chapter 3, we shall consider from where this energy comes and in Chapter 4 we shall indicate the likely earliest cycles of redox potentials in organisms driven

TABLE 2.11

Oxidation/Reduction Potentials of Some Reactions

Reaction	Potential (V) at pH = 7
Hydrogen → H^+	-0.42
Acetate → acetaldehyde	-0.60
Acetaldehyde → ethanol	-0.20
Pyruvate → lactate	-0.19
Glutathione (ox) → glutathione (red)	-0.23
Fumarate → succinate	$+0.03$
Quinone (ox) → quinone (red)	$+0.10$
Nicotinamide (ox)→ nicotinamide (red)	-0.32
(NAD^+)	NADH
Oxygen → H_2O	$+0.82$
Fe_n/S_n (ox) → Fe_n/S_n (red)	-0.2 to -0.5
$Mo(VI) \xrightarrow{(a)} Mo(V) \xrightarrow{(b)} Mo(VI)$	(a) -0.2; (b) -0.2
$Fe(OH)_3$ → Fe^{2+}	$+0.1$
Cu^{2+} → Cu^+	$+0.2$
FeO → Fe^{3+}	$+0.8$

by energy. To be most useful in catalysis, usually by metal ions, the catalyst should have oxidation potentials close to those of the organic reaction to be catalysed (see Table 2.11). When the environment became oxidising, a quite different set of redox potentials of compounds arose and quite different metal catalysts were necessary (see Table 6.1). The levels of redox potentials of metal ions are managed by the organic ligands which bind them. Organic and inorganic chemistry in life had to develop together.

2.19. Electron and Proton Transfer

There is an interesting feature of oxidation–reduction, which is not applicable to hydrolysis reactions. Metals themselves are excellent conductors of electrons so that it is possible to use a battery connected by wiring to bring about chemical oxidation and reduction in separate vessels. Long-range electron transfer is also possible in semiconductors containing metal ions and there are many man-made devices using such materials. In living organisms, it is also advantageous to keep a reaction separate, say of oxidation of hydrogen, separate from another, say reduction of oxygen as in a fuel cell, and here the medium for electron transfer between reaction sites has to be proteins with bound oxidation/reduction centres such as metal ions as periodic links. The two following reactions in cells are in essence separated by a membrane containing "conducting" proteins:

$$2H_2 \rightarrow 4H^+ + 4e$$

$$4H^+ + 4e + O_2 \rightarrow 2H_2O$$

The advantage gained is that the processes of electron and proton flow, downhill in energy from the first to the second (protons can also pass through membrane proteins), can be coupled to driving an up-hill reaction using the energy difference between the above two reactions of 1.24 V (see Table 2.11). This is the principle of energy transduction for the coupling of removal of hydrogen from carbon compounds in one place and its oxidation by oxygen in another, as in the above equations, in the overall reaction:

$$CH_4 + 2O_2 \rightarrow CO_2 + 2H_2O + energy$$

where CH_4 acts on one side of the membrane and O_2 reacts on the other. In fact, pairs of oxidation–reduction reactions from Table 2.11 can be coupled so that their difference in energy, here in volts, can be used to drive any equal up-hill voltage change of another reaction, see Williams in Further Reading. We shall see that these couplings through proteins and the machinery including "wiring" are essential in all cells. The wire for the $O_2 + H_2$ reactions is made up of mainly Fe^{2+}/Fe^{3+} ion couples for activating $2H_2 \rightarrow 4H^+ + 4e$ and $O_2 + 4H^+ + 4e \rightarrow 2H_2O$ at each end of a chain, while other Fe^{2+}/Fe^{3+} centres in proteins act as the link in the wire for electron and proton diffusion. The conduction occurs despite the fact that proteins are bad conductors as they do allow adequate electron tunnelling as in impurity semiconductors (see Edwards *et al.* in Further Reading) and proton hopping. We shall see in more detail in later chapters how the electron/proton flow and energy drop are used to drive reactions.

2.20. The Importance of Rates of Exchange

It is apparent that in this "coordination" chemistry we must separate equilibria from slowly changeable energised states. Equilibrium implies no spontaneous change, and no evolution is possible in such conditions. Understanding equilibrium is, however, necessary in the analysis of organisms, since it gives knowledge of the stability of some states in a time-independent balance, also independent of catalysis, and hence of any other set of reactions. As stated before, it is the very different non-equilibrium states that allows evolution, and the time scale, rate of change, i.e. kinetics of reactions, is dominant. Understanding kinetics requires us to know the excess energy of the reactants over the products, if any, and the reason why the reaction is not spontaneous, i.e. what kinds of barriers there are to the reaction. Now, it is the case that there are limits to the on- and off-rates of complex ion formation between metal ions and binding compounds. The on-rates are controlled ultimately by diffusion and no ion or compound in water, except H^+, can diffuse more rapidly than the diffusion limit for on-rate binding to an organic molecule of 10^{10} s^{-1} at molar concentrations, see Fig. 2.7. The relationship of the equilibrium constant, K, to the two rate constants, $K = k_{on}/k_{off}$ in the reaction $M + X \rightleftarrows MX$, means that for strong binding, say, $K = 10^{10}$ M^{-1}, k_{off} has a fastest possible value for dissociation of 1 s for a maximum $k_{on} = 10^{10}$ s^{-1}. Clearly many processes such as those of metabolism in cells are faster than this and do not allow equilibrium exchange if K is so large, e.g. for copper ion reactions. On the other hand, if $K < 10^7$ M^{-1} then exchange, e.g. of Ca^{2+} and Fe^{2+} ions, can be in the millisecond range, which is fast enough for equilibrium to be established in many reactions, while intermediate exchange rates for Zn^{2+} ions only allow equilibrium to be attained over minutes or hours, e.g. in slow functions. We shall see that functionally these observations make possible different uses in controls of both tightly (even non-exchanging) and weakly bound metal ions and molecules (see Table 2.12 and Figure 2.7). Recall that such considerations affect acid–base binding of substrates and proteins to metal ions and also oxidation–reduction binding of substrates such as O_2 and H_2. The transfer rates of oxygen atoms from MO groups in V, Mn, Fe, Mo, W and Se are also particularly important. We return to these rates in Chapter 4. We need to discover how to control them in both bindings and chemical reactions by adjusting the height of the on/off barriers of binding and of activation by especially the catalysts, and the effects of temperature and physical diffusion barriers. Note that change of rate can be brought about also by the direct local application of energy.

TABLE 2.12

IDEALIZED BINDING AND RATE CONSTANTS

Ion	k_{on} (s^{-1})	k_{off} (s^{-1})	K	Function
Na$^+$, K$^+$, Cl$^-$	$>10^9$	$>10^7$	$<10^2$	Electrolytic message
Ca^{2+}	10^9	10^3	10^6	Mechanical trigger
Mg^{2+}	10^5	10^2	10^3	Phosphate transfer
Zn^{2+}	10^8	10^{-2}	10^{10}	"Hormone" communication
Cu^{2+}	$>10^8$	$<10^{-7}$	$>10^{15}$	No exchange
C, H, N, O	Covalent, enzymic control			Structure building
HPO$_4^{2-}$ (RPO$_4^{2-}$)	10^9	10^6(?)	10^3	Trigger
Protein PO$_4$ (phosphorylation)	Slow	Slow	Weak	Kinetic control faster than C, N, O bond breaking
H$^+$	$>10^{10}$	Slow to 10^{10}	Huge range	Catalysis, energy store
OH$^-$	$>10^{10}$	Slow to 10^{10}	Huge range	Catalysis, energy store

From what has been said, our first priority is, therefore, to understand equilibrium, which, as will be seen in Chapter 3, can be discussed in terms of the balance between order and disorder, and then to enquire about (organised) change when we consider kinetics and barriers to reaction. These problems are all related to the energy content of materials and the impact of external energy upon them. In the synthesis and activities of organisms an over-riding concern is with the overall energy uptake. Is it restricted to a limit?

2.21. Selective Action of Metal Ion Complexes in Catalysis

Throughout this book a major stress is on catalysis in organisms. Catalysis is confined to non-metals and metal ions of attacking power, either as Lewis acids or in oxidation/reduction and this excludes the simplest ions such as Na$^+$, K$^+$ and Ca^{2+} (and Cl$^-$ among anions). The transition metal ions and zinc are the most available powerful catalysts. The metals in a transition series are known to have selective binding properties, exchange rates and oxidation/reduction states, which can be put to use in catalysis in quite different ways (Table 2.13). It is noticeable that especially the complexes of metal elements

Fe Co Ni

Rh Pd

Ir   Pt

together with Se and S are used to catalyse transfer of H or carbon fragments in organic chemistry and we shall note this use of complexes of Fe, Co, Ni, Se and S

TABLE 2.13.

COMMON ELEMENTS OF SPECIAL CATALYTIC VALUE*

Catalytic Value	Element
Hydrolysis	All metal ions $M^{3+} > M^{2+} > M^+$
	"b"-class > "a"-class, e.g. $Zn^{2+} > Mg^{2+}$
Oxidation/reduction	
(a) H transfer	Fe, Co, Ni (pyridinium ions)
(b) CH_3 transfer	RS^-, Zn^{2+}/RS^-, Co, Cu, Fe
(c) O transfer	Mo, W, S (low potential), Se, Fe, Mn
	(higher potential)
(d) CH_3CO transfer	RS^-, Ni/Fe
(e) NH_3 transfer	Aldehydes
(f) e transfer	Fe_nS_n, Ni, Co, Cu

*We do not refer to elements heavier than Se except for Mo and W since the others are not used in biological catalysts. Note the different use of early transition metals in O transfer (and N transfer), while later transition metals are used in H- and C-fragment transfer (see Section 2.20).

in organisms. The metals in this group are able to form M–H or M–H or $M-\overset{\diagup}{\underset{\diagdown}{C}}-$ bonds due to their ability to hold on to electrons going formally to lower oxidation states. On the other hand, major catalytic complexes of central atoms for O or N transfer are to be found in the series

$$Ti \quad V \quad Cr \quad Mn \quad Fe \ ... \ (Se)$$

$$Mo$$

$$W$$

(Note that Fe and Se are versatile.) Immediately we see that this second group of elements can form MO, oxenes, moving between higher oxidation states due to the readiness with which they lose electrons in reactions. Many of them have quite a high redox potential but Mo, W and Se do not and so the two groups have different values in reactions. We shall be concerned with the parallel uses of elements in reactions in cells. Acid catalysis rates by divalent metal ion complexes, by way of contrast, follow the Irving–Williams binding order unless stereochemical effects prevail, and we shall observe particularly the different values of zinc and magnesium in cells, due to their different acid strengths. These two are particularly important in acid/base catalysis since they have no redox properties.

A striking feature of the employment of the catalytic elements in cells is that they are not only taken in, or rejected, and bound selectively but they are tuned for use within organic chelating centres of proteins. The tuning is seen in the binding strengths and exchange rates, in the redox potentials and in the relaxation rates so that essential kinetic parameters approach the best-of-all states of the elements. One of the nicest examples is the selection of molybdenum for the transfer of both O- and N-atoms at low redox potential (see Williams and Fraústo da Silva in Further Reading and see Table 2.13).

2.22.   The Special Nature of Hydrogen

Hydrogen as an element can form stable covalent association with especially carbon in organic compounds or it can form the ion, H^+, in equilibrium with O-, N- and S-binding centres. This versatility is increased by the fact that it can be transferred as H^-. Hydrogen is, of course, central to all organic chemistry acting as a non-metal as we have seen in early sections of this chapter, but its functions as H^+ in both equilibrium exchange (acid/base exchange), in catalysis and in bioenergetics, are equally important. There is also the possibility of the development of man-made engines utilising hydrogen as fuel instead of oil. In some respects, then, the properties of the hydrogen atom allow it to act as a typical non-metal, like carbon, whereas in others it acts in very similar ways to a metal ion forming a cation and also as a halogen in forming an anion. For these reasons it is put in the middle of the Periodic Table, at the top, and in a class of its own. Its functions in the environment/organisms systems is absolutely crucial since it binds quite well at a pH $= 7.8$ that is the pH of the sea and in much of cellular biology buffered at pH $= 7$. Its homeostatic condition has a controlling balance on many biological processes. In addition, it is central to energy transduction in cells that is in the synthesis pathway turning light energy into a chemically useful form, see Williams in Further Reading.

2.23.   Summary of the Basic Chemistry Relevant to Our Global Ecosystem

It is useful to have a summary set of conclusions of this chapter on chemistry so that quick reference can be made to the chemical factors that have dominated our ecosystem. They are:

(1) All chemistry is limited by the abundance of the elements, restricted to 90 in total (two of the natural 92 have disintegrated completely). The properties of the elements are correlated in fixed Groups and Periods of the Periodic Table.

(2) The Earth formed as a set of elements and compounds, ultimately distributed in zones, with a large gradient of temperature and consequently providing an enormous store of energy.

(3) The zones are the metal core, the non-conducting mainly oxidic mantle and crust, the aqueous and the atmosphere zones.

(4) The interior zones – the core and the mantle – at high temperature interact very little with the surface, although here and there and now and then there is input to the surface from volcanoes and sea-floor "black smokers". The surface also receives dust and debris from extra-terrestrial bodies and more rarely from more sizeable meteorites.

(5) The surface has three interactive parts: the solid crust, the aqueous components – oceans, seas, rivers, lakes – and the atmosphere. As stated, the thickness of these zones put together is only little more than 1–2% of the Earth's radius. The elements are in compounds in these layers, the environment,

which equilibrate to a large degree and are limited in availability by solubility and oxidation/reduction conditions.

(6) The surface is a very minor part of the entire Earth, but being heated from the interior and energised by the Sun's radiation can not only maintain an approximately constant temperature but it can allow activated chemical reactions that create especially energised organic compounds not at equilibrium. These processes gave rise to life, apparently starting in the aqueous phase.

(7) All biological chemistry is therefore limited by the abundance of elements, but especially by their *availability* on the Earth in its aqueous part, namely the oceans, which has changed with time.

(8) All organic chemicals are energised non-exchanging molecules in isomeric forms due to selected geometry. Those formed by the *light non-metals H, C, N, O* are especially kinetically stable. The organic chemicals of organisms are a special set of such small molecules and polymers of various kinds soluble in water. In essence they are formed by the action of light.

(9) The inorganic elements in aqueous solution reactions, both acid–base complex formation, precipitation and oxidation/reduction, frequently come rapidly to equilibrium when no more reactions are possible. The implication is that in the environment and in organisms many of their properties cannot change unless circumstances change, for example the introduction of new components.

(10) The reactions of the organic chemicals (see (7)) in chemical laboratories and organisms are generally controlled by bound inorganic metal elements, catalysts. Heavy non-metals such as sulfur and phosphorus are also advantageous as carriers and catalyst centres of small organic groups for synthetic purposes due to their intermediate reactivity between light non-metals and metal ions. Catalysis is at the heart of biological activity.

(11) Overall these conclusions concerning chemistry mean that an ecological system of the following restricted kind can be generated and can evolve.

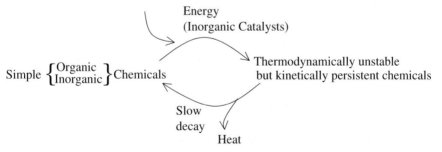

In the fully cyclic condition the system is element neutral.

The next chapter will discuss the nature of energy and the ways in which it can be incorporated into chemicals using the basic principles of chemistry and geochemistry set out in Chapters 1 and 2 so as to create what we know as a system called life locked into the environment, the total ecosystem. (*Note*: Heat is given out in small amounts even in the forward step but we shall ignore it here and elsewhere.) Importantly notice that equilibria limit the diversity of particularly inorganic compounds and complexes but are not usually relevant to the discussion of the properties of organic compounds although they are relevant to certain organic chemical complexes. The equilibrium restrictions make it easier to follow change in the ecological system using inorganic element chemical markers in both the environment and cells rather than to attempt analysis of the huge multitude of organic chemicals in cells. This is especially true as we are using thermodynamics as the basis of our study.

Further Reading

BOOKS

Atkins, P.W. (1996). *The Periodic Kingdom*. Phoenix, London
Barrett, J. (2003). *Inorganic Chemistry in Aqueous Solution*. Royal Society of Chemistry, London
Cox, P.A. (1989). *The Elements: Their Origin, Abundance and Distribution*. Oxford University Press, Oxford
de Duve, C. (1991). *Blueprint for a Cell – The Nature and Origin of Life*. Neil Patterson Publishers, Burlington, NC, USA.
de Duve, C. (1995). *Vital Dust: Life as a Cosmic Imperative*. Basic Books, Harper Collins Publishers, Inc., New York
Emsley, J. (2001). *Nature's Building Blocks*. Oxford University Press, Oxford
Gray, H.B. (1995). *Chemical Bonds – An Introduction to the Atomic and Molecular Structure*. University Science Books, New York
Holum, J.R. (1995). *Elements of General Organic and Biological Chemistry* (4th ed.). Wiley, New York
Hornby, M. and Peach, J. (1993). *Foundations of Organic Chemistry*. Oxford University Press, Oxford
Martell, A.E. and Hancock, R.D. (1996). *Metal Complexes in Aqueous Solution*. Plenum Publishing, New York
Mingos, D.M.P. (1998). *Essential Trends in Inorganic Chemistry*. Oxford University Press, Oxford
Schriver, D.F., Atkins, P.W. and Langford, G.H. (1999). *Inorganic Chemistry* (3rd ed.). Oxford University Press, Oxford, Chapters 1–4, 9 and 18
Williams, R.J.P. and Fraústo da Silva, J.J.R. (1999). *Bringing Chemistry to Life – From Matter to Man*. Oxford University Press, Oxford
Williams, R.J.P. and Fraústo da Silva, J.J.R. (1996). *The Natural Selection of the Chemical Elements – The Environment and Life's Chemistry*. Clarendon Press, Oxford

PAPERS

Bernstein, M.P., Sandford, S.A. and Allamandola, L.J. (1999). Life's far flung raw materials. *Sci. Am.* *281*, 42–49
Edwards, P., Gray, H.B. and Williams, R.J.P. (2006). To be published
Ehrenfreund, P. and Charnley, S.B. (2000). Organic molecules in the interstellar medium, comets and meteorites. *Annu. Rev. Astron. Astrophy.*, *38*, 427–483
Huber, C. and Wachtershauser, G. (1997). Activated acetic acid by carbon fixation on (Fe, Ni)S under primordial conditions. *Science*, *276*, 245–247
Thomas, J.M. and Williams, R.J.P. (2005). Catalysis – progress and prospects. *Phil. Trans. Roy. Soc. (London)*, 363, 765–792. (Discussion "Catalysis in Chemistry and Biochemistry", *Phil. Trans. Royal Soc. London, 2005*)
Wachtershauser, G. (1988). Origin of life and iron sulfides. *Microbiol. Rev.*, *52*, 482–486
Williams, R.J.P. (1961). The properties of chains of catalysts. *J. Theoret. Biology*, *1*, 1–13

Chapter 3

Energy, Order and Disorder, and Organised Systems

3.1. Introduction

In Chapters 1 and 2, we have described in very basic outline the geological chemistry of the surface of the Earth and the essential chemistry, for our purpose, of the chemical elements and their compounds. We must turn now to the energy supply to the surface chemicals which drives the whole global ecosystem. At first we are concerned with limitations upon the sources of energy but an understanding of changing energy flows into chemicals is also essential for an appreciation of any energised system and its evolution since it is only kinetically stable. Once principles are described we shall ask in general terms how could any particular organised system of chemicals such as the observed global ecosystem arise and become sustained by energy flows. The central theme of this chapter is then the way energy can be retained in chemical systems in organisms and the environment together for long periods of time. The discussion of the chemical components of the ecosystem have been divided into two parts.

Those of the environment were already described in Chapter 1 and those of the biological chemistry of organisms will be outlined generally in Chapter 4. Subsequent chapters bring all the components together with the energy flow in real organisms.

3.2. Energy

At a simple qualitative level we all know that energy is a capacity held within a system which can be applied to move objects or to change the state of substances, hence doing work. We shall see that there are many kinds of energy (Table 3.1) and since the Earth's ecosystem is always changing, hence work is done, and energy is being applied continuously. From where does the energy come and what forms can it take? We know that objects in the Earth's gravitational field are energised and will fall until they can fall no further when they are prevented by a barrier, resting on a surface. We also know about electrical energy, associated with the storage of charge, e.g. electrons or ions, which creates an electric field, and that charge will flow from a source to a sink, given a suitable medium. We further recognise the energy of light, which we sense as it heats objects. This is radiant energy, not due to position in a gravitational or an electrical field. Chemical energy is recognised too as the energy that can be released, for example in a fuel cell in which natural gas, methane, is burnt to give out heat and light:

$$CH_4 + 2O_2 \rightarrow CO_2 + 2H_2O + \text{energy (heat and light)}$$

As explained earlier (Section 2.13) this chemical energy is due to the instability of the arrangements of the elements in $CH_4 + 2O_2$ compared with those in $CO_2 + 2H_2O$. (Chemical energy is, in fact, an electrostatic storage of energy in

TABLE 3.1

TYPES OF ENERGY

Gravitational potential energy (associated with the attraction between separated masses)
Electrostatic potential energy (associated with the interaction between separated charges)
Kinetic energy (associated with directed motion of masses)
Chemical energy (associated with transformations of chemicals, changes of internal potential, electrostatic energy)
Radiant energy (associated with electromagnetic radiation)
Electrical/magnetic energy (associated with electrical currents)
Nuclear energy (associated with the nuclear binding forces)
Surface energy (associated with the surface of liquids and solids)
Mechanical energy (associated with stress in solids) and including pV energy
Thermal energy (associated with temperature, random kinetic energy of masses, i.e. random motion)

Note: There is really a fundamental underlying unity in energy, which is always conserved although it can be interconverted amongst its different forms. All types of energy conversion, except nuclear, are associated with biological as well as human machinery.

TABLE 3.2

SOME MODES OF ENERGY STORAGE IN COMPARTMENTS

Compartments	Storage	Barrier
Reservoir vs. ocean	Pressure under gravity	Dam
CH_4/O_2 vs. CO_2/H_2O	Chemical energy	Chemical bond
Centre of Earth core Fe/air, O_2	Chemical energy (and heat, temperature)	Rocks, Mantle
Biological liquids/external chemicals	Pressure, chemical energy	Lipid membranes
Metal plates in electrical cell	Electrical potential	Air
		Plastics
Aqueous solutions	Electrolytic gradients	Plastics
		Lipids

relatively unstable bonds.) There is also energy storage in a gradient, say a pressure, or a charge or concentration difference, between two divisions, compartments, in space separated by a barrier (Table 3.2). For example, the popping of a balloon throws fragments of the balloon around as the internal energy of its gas flows outward. Finally, there is energy in a temperature gradient, which we recognise by the flow of air mixing a colder and a hotter region.

Now, if we release a material object held energised in any field – gravitational, electrical, or a thermal gradient, for example – there will be spontaneous motion, or flow, of the object as it attempts to reach a stable condition. Flow has a direction controlled by the fields. We say that the stored energy, described above, is converted into *directional kinetic energy of motion or flow*. This is not the same as the energy connected with temperature, thermal energy, which is due to the mean kinetic energy of molecules in *random motion*. Controlled flows, currents of material, are a major activity in nature, in organisms and in industrial plants, for example water-powered electrical generators. Energy must have been put into them from one of the above sources but note that energy also flows through materials. Before we examine these flows we need to analyse in detail the concept of the fixed energy content of unchanging collections of molecules, since this examination will assist us in the study of ecosystems. It requires us to look at ordered, structured and randomly disordered systems, their motions, their temperature and their pressure or concentration.

3.3. Order and Disorder: Equilibrium

We all recognise order very readily. It is defined by a static arrangement of objects in space. For example, atoms are ordered in crystals and the ultimate order of all material has shape as in a crystal. This static order has a simple description – it is seen to have a particular (regular) arrangement in space independent of time. Disorder is somewhat harder to define since objects can be disordered in space but they can also be disordered in time in all their random motions, with no flow in any direction. There is a vast number of possible random arrangements in space and time of atoms in translatory motion, while those atoms attached to one another in molecules can also have rotational and vibrational disorder of numerous different motions (Figure 3.1). Clearly molecules in a gas are disordered, but in a liquid they are less so. At a variety of increasing temperatures one substance such as H_2O may be an ordered crystal, ice, then in succession a liquid, water (a drop or a pool) and a gas, water vapour (Figure 3.2). There is more disorder, more random distribution in space and time per fixed quantity, the higher the temperature, more random kinetic energy, and the lower the pressure, more randomness in space. Temperature and pressure (concentration) are in fact a measure of the *equilibrated random disordered motion* of all kinds. At any given temperature and external pressure and with all other conditions fixed, a substance is said to be in a particular *physically stable* thermodynamic state. If conditions are not changed, no energy is applied or removed, no flow exists and the state is clearly unchanging (see Appendix 3A for the thermodynamic treatment of such equilibrium states).

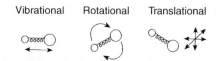

FIG. 3.1.   The three kinds of motion of a molecule.

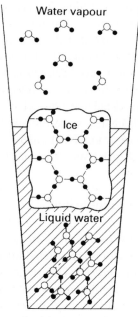

FIG. 3.2.   Increasing disorder: ice < liquid water < vapour.

Atoms can also be ordered by forming compounds which restrict their motion, so that a compound of many atoms such as a polymer of A_n units is more ordered than n.A separate A units. Even in a reaction such as the formation of ammonia

$$N_2 + 3H_2 \rightleftarrows 2NH_3$$

there is less order in $N_2 + 3H_2$ than in $2NH_3$ since there are more units to the left to be disordered. The compound NH_3 has more bound atoms per molecular unit; however, it has also a higher binding, *chemical stability*, related to the electrostatic interactions of electrons and nuclei. At low temperature, NH_3 is the more stable state. In fact, there can be nearly 100% NH_3 if this molecule is at a low enough temperature. At higher and higher temperatures, as disorder of all molecules increases, the weighted disorder of the state $N_2 + 3H_2$ is more favoured. At these higher temperatures the thermodynamically *stable chemical state* of the system of all three molecules is a balanced mixture of N_2, $3H_2$ and $2NH_3$, i.e. chemical and physical disorder and order coexist just as a physical balance exists between solid, liquid and vapour states of water, different at different temperatures. This balance is called a *physical/chemical equilibrium* condition between order and disorder, where order is related to binding energy and disorder is again quantitatively related to *entropy*, a thermodynamic function of state defined at equilibrium. (The energy associated with all such disorders, needed to create it from a fully ordered system, is the product of the absolute temperature and this equilibrium thermodynamic state function.) The higher the temperature all equilibria move to a more disordered distribution of units. Conversely, at all temperatures it is the electrostatic forces of various kinds between molecules and atoms that hold them together. It is these forces which give time-independent structure that requires no maintenance. An equilibrium system cannot flow, much though atoms or molecules exchange positions randomly and isotropically.

Consider two biochemical examples of the folding of a polymer, a protein, RNA or DNA, and that of formation of a virus from a combination of proteins and RNA or DNA. In each case we need to note that the stable state is dependent upon the aqueous solution conditions, the temperature, and the pressure. On fixing these conditions a balance or equilibrium is observed. In the case of the protein, RNA or DNA it is between folded and unfolded forms, so there are cases of largely folded and largely unfolded forms of particular biopolymers. The balance in the solution is dependent upon the difference between the two conditions in binding energy of all the components (including those of the medium), which bring about order, and in the energy of the disorder within them. In the case of a virus all these factors come into play for its different biopolymers in different states, but there is the additional consideration of the binding together of different biopolymers and their disorder within this bound state. It is a common error to attempt the understanding of biochemical systems from structures alone, though this is of functional interest. Before proceeding note that all equilibrium restrictions mean that the systems have no flow and are not living. In addition some of the bindings of molecular components within an organism do equilibrate.

Experience shows that there are also other observable conditions of materials besides this variety of stable equilibrium states (Table 3.3). One is a frozen *metastable* state such as we observe in organic chemicals, many in biological cells, in air, or when we isolate 100% NH_3 at any temperature as mentioned above. They should change but they do not. All organic chemicals should react with oxygen and all NH_3 should decompose (be oxidised) to some extent. Metastable states are in *stationary energised* states, not in equilibrium, trapped by kinetic barriers to change. Such states like equilibria are not living. Another state is a condition of material in constant flow, which we shall call a *steady* state energised away from equilibrium (Figure 3.3). The steady state must have either a constant flow of the starting material in a given direction, which can be downhill (Figure 3.3(a)) or uphill in energy (Figure 3.3(b)), or a cyclic flow of materials (Figures 3.3(c) and (d)). Both uphill and cyclic flows require constant energy input to maintain them while downhill flow requires energy to have been put into material initially before flow starts. All the three conditions may have "maintained structure" in the sense that the flow is constrained. Laboratory and industrial organic chemistry are (overall) in downhill flow, but biological chemistry has uphill and downhill parts and is, in part, cyclical. In principle, equilibrium systems cannot do work; metastable systems can do work for some time when released from barriers while steady states can do a fixed amount of work continuously as long as energy flow is supplied in the uphill case and constraints maintained. It is very important to appreciate that a *truly cyclic steady state is constrained and only needs energy*, not material, to be supplied, while it does work (Figure 3.3(d)). These considerations are also important in many parts of the ecological system on the Earth and we shall be very concerned with those cases in which energy is supplied and degraded to heat continuously and material may or may not cycle, as in machines.

TABLE 3.3

DEFINITIONS OF SYSTEMS CONDITIONS

Stable	A condition of a system which cannot change spontaneously. It does not receive or donate material or energy. (The system can be dynamic in cycles only in non-classical mechanics, e.g. electrons in orbits.)
Stationary	A condition of a system which could change spontaneously but does not since barriers prevent it. It does not receive or donate material or energy. The system has excess energy over a stable state.
Steady	A fixed condition of a system through which supplied energy and/or material flow continuously in a direction so that it is in flow at constant rate. The system has excess energy over a stable state.
Developing	A system which changes steady state due to the loss or accumulation of energy and/or material in it. The system has changing excess energy over a stable state.
Cyclic steady state	A system which does not change as reactants go forward to products that cycle back to reactants. It requires energy but does not develop. The system has optimal retained energy over a stable state.

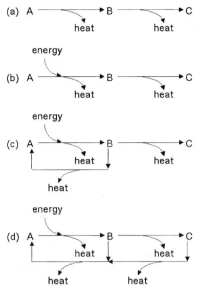

FIG. 3.3.   Different modes of flow in systems. (a) and (b) are steady non-cyclic flows, (c) has a cyclic portion and a portion lost to the cycle while (d) is a completely cyclic system. All flows which produce non-cycling products, C, are effectively polluting.

3.4.   Some Steady States and Organisation

We shall consider material in overall *controlled* continuous energised flow as an organised system. Organisation has structure due to boundaries within which flow in a direction is maintained. The flowing material does not have any *internal* structure in its parts. The sense that the flow is controlled by boundaries, physical apparatus or by fields of force (Table 3.4) immediately means that its content cannot be a simple solid; it must be, at least in part, a powder, a liquid or a gas or, in another context, electrons or radiation. The physical boundaries can be ordered solids or liquid crystal phases with internal order, while the fields can be electrical, gravitational, thermal or any interactive gradient. The structures can be stable or metastable and energised. The general sense of physical organisation is clear from common experience of the distribution systems for the supply and use of gas, water or electricity in everyday equipment in a house. The sense of flow control by fields is seen in planetary systems and in clouds (see below). Living systems, organisms, are based on energised controlled systems of flow in confinement due to both such devices, that is limited by physical boundaries and by fields. Thus, living systems have overall "*structure*" due to boundaries, or *constraints* (morphology), which is a characteristic of the ordered flow which may help to *form* them, but they also have *internal organisation* of flow. Much of this flow in organisms is internally controlled due to transfer of chemical potential or concentration-dependent messages, which we shall connect later to "information" or instructions. A particular

TABLE 3.4

STRUCTURING FIELDS IN THE ENVIRONMENT

Structure	Created by
Radiation fields (temperature gradients)	The Sun The hot core of Earth
Gravitational fields	Earth (general), contours of land
Magnetic fields	Earth's magnetic poles (convection currents in the Earth's core)
Material barriers	Contours of land Immiscible substances (lipids) in membranes Earth's mantle (oxides)
Chemical gradients	Across Earth's layers of core solids Sea depth Atmosphere height Concentrations in liquids behind barriers

case of internal control rests in the genome, a metastable "structure", in cells, but see our comments on the meaning of information in Sections 3.12 and 3.13. To understand life we shall need to see more clearly how "information" is related to the controlled internal flow in organisms. The words *order* and *organisation* in cells or organisms must now be clearly seen to cover two different "structural" features of our observable environment. As we have described in Chapter 2, individual molecules have ordered structure but no organisation. Study of ordered, even energised, structure alone, as in much of molecular biology, cannot describe living organisms since that study is mostly of static molecular structure, order, in isolated molecules (not of their states) and not of the essential controlled flow within boundaries, organisation. Organisation has dynamic form but may well be informed from outside. Since order is easier to study than organisation the present stage of investigation of organisms is largely one of the necessary study of ordered molecules, but it must be followed by a study of the states of molecular ensembles and then of the dynamics of energised organised systems partly linked to some static structures.

As stated, the immediate difference between a well-ordered and a well-organised system is that an organised system requires energy to maintain its flow as well as boundaries to contain it (Figure 3.4). Ice or water in a pool are in a stationary state (a stable or a metastable state), a stream of water is in a steady state since energy is constantly used in a controlled way to maintain the supply of water in it (Figure 3.5) but note that it can also be structured by barriers (see Table 3.4), which can be said to form the flow system. Organisation fails once there is a failure of supply of energy or if structural containment is lost – for example of a river or a living organism. In addition as a continuous steady state of flow as in a stream

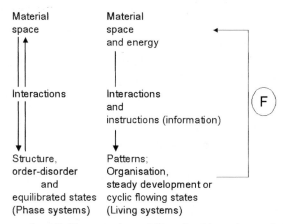

FIG. 3.4.   An attempted parallel between the variables of stable stationary (left) and flow (right) systems and their properties. Ⓕ is feedback.

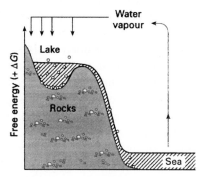

FIG. 3.5.   The obvious physical barriers to water flow are the banks of lakes and rivers and dams constructed by man.

or an electric current we shall consider also a *cyclic* steady state as in the following scheme:

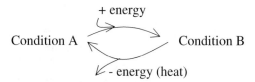

where the arrows represent flow, but now we stress that the way back is different from the way forward. In a flow process more "effective" energy $(+)$ is constantly degraded to less "effective" energy $(-)$ and so long as A is constantly replenished by B, A need not be supplied from outside. (The term "effective" here is used in the sense of the capacity of the energy to produce especially chemical work; it, therefore, relates to the "quality" of the available energy which is suitable for a given task.) We shall show in Section 3.5 that this degradation of energy leads to an increase in thermal disorder, entropy, and hence the *overall irreversible processes* in an organisation are thermodynamically viable as a consequence of the tendency to produce this *thermal* disorder. As an example, the flow of water on the Earth is in a cycle – sea, air, river, sea – while high-frequency light is degraded irreversibly to lower frequency heat in this organised cyclic activity driven by solar energy. The path forward differs from the path back. We shall show that the Earth's ecosystem, including the geological and biological chemistry (see Figure 3.7), which we wish to describe, is struggling to be *fully cyclical* in material but will always be irreversible as energy is constantly degraded. Note how we write a cyclic steady state with separated flows by curved arrows in opposite directions, while in a equilibrium stable state we connect the two sides by two horizontal opposed arrows $\rightleftarrows$. The distinction we are making is very important since the rates of going from one side of the equation to the other in an equilibrium condition are *equally altered by catalysts* – they cross the same barrier (Figure 3.6). In effect, catalysts are substances which increase both forward and back reaction rates but are not changed overall in themselves so that the equilibrium composition does not change. In a cyclic steady state away from equilibrium added catalysts can change one rate, say forward, relative to the other, say backward, when the composition of the whole cyclic steady state changes. We shall be mainly concerned in this book with the evolution towards a catalysed cyclic steady state, that of the total Earth ecosystem (Figure 3.7) and therefore with rates of both forward and back reactions separately in which different catalysts can be used. It is "element neutral" and non-polluting if fully cyclic. We shall also consider variation of rate of energy input. We state again that equilibrium states are all "dead" at an entropy maximum, while steady states are not at this maximum. They are of the essence of living systems though we do not call all

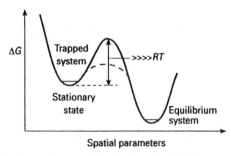

Spatial parameters

Fig. 3.6.   One kind of stationary state, which is not an equilibrium state, is defined by a system that is not in equilibrium with its surrounding material and is not gaining or losing material or energy since the barrier is too large for change at a given temperature, full line. Dashed line is for the same process catalysed to lower the barrier allowing change, that is flow.

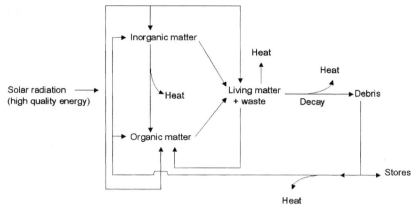

Fig. 3.7.   An outline of the ecological system of flow of material and energy (see Jorgensen *et al.* in Further Reading).

of them living except in metaphors – "where streams of living waters flow" – and they produce thermal entropy continuously.

The ultimate change to a state of stability, equilibrium, is that to which all systems out of equilibrium will go when the input of energy to a steady state fails *no matter how quickly or slowly*, that is to a condition of balanced compromise between ordered and disordered units. Given all the possible combinations between two or more chemical elements there is, therefore, a unique final state which results from the greatest release of low quality, less "effective" energy, heat, in the change, and this heat raises the temperature of the combination and of the environment resulting in the maximum possible thermal disorder under the prevailing conditions, when no further change can occur. Here no material or energy is put into the system from outside – it is an *isolated* system. The statement that isolated systems tend to go to a state of maximum thermal disorder (entropy) is a law, the second law of thermodynamics, and this state is the ultimate fate of our ecosystem, when the Sun dies, and possibly of the universe. It is the general principle which drives the system of Sun, Earth and organisms, but it must be seen to be consistent with the continuous creation and loss of *limited time-dependent "structure"*.

We wish, therefore, to explore the different situation where energy is put into a flowing chemical system continuously until, if possible, it reaches a fixed complete *cyclic steady state*, when again no further change can occur. *An important first feature of it is that it is self-sustaining and self-productive, and not necessarily reproductive.* Is there a rule concerning entropy that governs this state? We shall use four examples to illustrate the problem and then offer a general conclusion, which we shall use throughout the book. The first example illustrates a very clear physical system, the planets; the second concerns physical transformation as above, the change from liquid to vapour and the reverse, but now with continuous energy input – the formation of flowing drops and molecules in clouds. The third describes chemical transformation with continuous energy input – the formation of an ozone layer from oxygen and its reversal to oxygen around the Earth. These three are of consequence for the whole of the present day ecosystem as we shall show later. The fourth describes heat applied unequally to a liquid creating particularised flow patterns. In the first three of the four systems there are gradients due to external conditions which structure, constrain, the flows (see Table 3.4) but no physical barriers, while in the fourth the system is contained physically. (It is most important to observe that there is not just kinetic energy in these flows but the energy is moving through easily observed metastable structures which as we see later includes all the organic molecules and gradients of cells.) These are examples of persistent cyclic *steady* states where the continuously irreversible energy flow is of high- to low-quality energy, e.g. light to heat, resulting in an overall continuous thermal entropy increase. In these and in all the cases of cyclic steady states which we shall discuss, the energy supply is assumed to be unlimited and constant. Obviously, this is not true and it must fail eventually when the systems decline towards an equilibrium state, but the assumption is a very useful approximation in the description of Earth over several billion years (see Chapter 1). Great interest in steady states concerns the availability and flow of material (see Chapters 1 and 2), and energy which can flow in it, the way flow is structured and the rates of forward and back reactions for they decide the geological/biological ecosystem we observe as it moves of necessity towards a cyclic condition. (Note that the process of change cannot conflict with the laws of equilibrium thermodynamics.) Only when we describe biological systems do we need to consider reproduction.

3.4.1. THE PLANETARY SYSTEM

This is a very simple example of *apparent* "constant flow" of planets round the Sun. It can be treated as close to perpetual motion with no material changes. We describe the system first before we explain why it is a simple but not a quite correct case. The planets flow in orbits under two forces or fields – gravity and centrifugal "force". The velocity is assumed to be fixed. Hence the systems have *form*, "structure", which is constantly repeated, but no information passes between the objects as everything in the system is part of the form and without any of the pieces, planet masses, the Sun and gravitational forces, there would be no flow. The total of all the parts is needed to maintain the flow characteristics of the "formed" whole system. It is clear that the system is not to be described as random and it has an evolutionary cause even though generated by accident, and has allowed life.

The case is unusual and not quite correct, as no energy has been put into it since its initial creation, so that it is not in a true cyclic steady state. Friction, no matter how small, will cause the flow to stop; it must produce some thermal entropy. Unlike a true cyclic system it is not truly time-independent and would require energy input to be so. We treat next a physical system where energy input is clear.

3.4.2. CLOUDS

Inspection of clouds shows them to be a simple consequence of organised water flow with physical internal change where thermal gradients, wind and gravitational forces create fixed "field" boundaries of organisation locally. Clouds are composed of flowing metastable water droplets dispersed in air (Figure 3.8). In a temperature gradient from cold to hot the droplets fall under gravity to levels of warmer air. There they evaporate and within the hotter air the molecular vapour rises until it reaches again the level of the lower condensation temperature where it reforms droplets which fall and so on repeatedly in a cycle. Notice that the two gradients help to form the cloud and are an essential part of its organisation. We shall describe the cloud generated by these field gradients and energy input from them as having "maintained form" in a cyclic steady state. Note how different it is from a static structure of a liquid such as water in equilibrium with water vapour even though there is exchange.

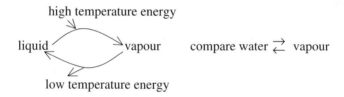

A cloud of droplets moves in the wind, a horizontal temperature gradient, which helps to shape it and it alters slowly and may be disperses or falls as droplets of rain later. However, for some period the shape of a cloud is fixed and describable, for example as cotton-wool cumulus, streaky cirrus, etc. Each type of these clouds is self-sustaining and the water/energy system in them is self-producing. We see these particular moving shapes from time to time again. *In effect they are different types of cloud classes formed with inevitable shape in different fixed physical gradients.* Their extreme differences are accounted for at a low level by turbulent wind, due to the presence of land, giving rise to cumulus, and at another high level by streaming wind, giving rise to cirrus. Wind arises from horizontal temperature gradients and is a third

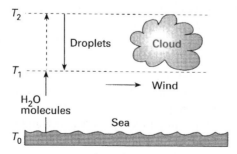

FIG. 3.8. The pattern of a cloud is decided by the temperature and its gradient, $T_0 \geq T_1 \geq T_2$, the wind speed, and gravity.

force creating the maintained form. The clouds represent shape ("structure") not derived from internal binding, order, but shape derived from physically maintained energised flow (initially from the Sun) in external gradient confinement, the whole being a natural, spontaneous organisation in a cyclic steady state. We can give a good account of the origins of these major types of cloud but not of the multitude of their varieties or "species" which appear by chance circumstances (see organisms in Chapter 4). Notice that added chemicals can help, catalyse, the formation or disruption of clouds, e.g. added sodium sulfate. (Such substances catalyse drop formation to larger drops, rain.) Catalysts assist rates of development of form but act differently on back and forward reactions. *These* (and other) *cyclic steady states must evolve to new forms if more energy or new chemicals are allowed to enter them.* The cloud is clearly dependent on rates of physical changes of state, not equilibria between states. The energy input here is of heat at a higher temperature at lower heights causing vaporisation, as opposed to lower temperature heat lost at a greater height on condensation. Heat flow is always unidirectional from a high-temperature source to a low-temperature sink (see textbooks of thermodynamics). We explain why this flow of energy is accompanied by an increase in rate of thermal disorder (thermal entropy generation) as well as dynamic structure (loss of entropy) in Section 3.5. (To appreciate the variety and speciation of clouds see Trefil in Further Reading.)

Now a cloud disperses as it enters a higher temperature region and the droplets fall as rain in a lower temperature region. If we include clouds, rain, structured land, rivers and the sea in this pattern of water movement, the cyclic systems of cloud species are components of a larger overall ecosystem cycle involving many parts. The whole system including weather patterns has a causal explanation and it also has a part to play in life's evolution. Observe here the additional feature of physical barriers, e.g. banks of rivers, which give form to the self-sustaining flow. (*Note*: Looking for explanations of the exact nature of the weather on a particular day can be looked upon as similar to looking for explanations of individual species of clouds or of living things – impossible – however, general changes of cyclones and anticyclones can be understood.)

3.4.3. THE OZONE LAYER

Organisation can also arise spontaneously within chemical as well as physical changes. An example is the ozone layer, which is formed on the upper edge of the oxygen-containing atmosphere by the cyclic steady-state reaction as shown in Figure 3.9. The heavier metastable ozone falls under gravity, but as it does so it comes to a level where the oxygen around it is screened off from UV-C radiation by the upper O_2, and no O_3 is produced. The ozone decays at this lower height, absorbing UV-B radiation while giving out heat, and returning to oxygen.[*] Hence there is always a *formed* ozone layer in an oxygen-containing atmosphere, but it is not in

---

[*]Actually the process is more complex (see, for example, Baird in Further Reading).

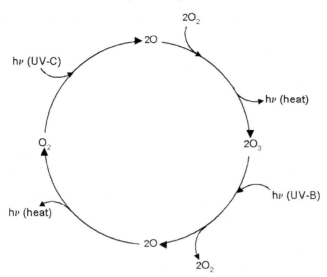

FIG. 3.9. Photochemical formation and non-catalytic destruction of ozone. UV-C radiation (200–280 nm wavelength); UV-B radiation (280–320 nm wavelength). Note how high-quality energy (UV radiation) is converted into lower quality energy (heat). Catalysts such as freons or nitrogen oxides can destroy ozone (e.g. $Cl + O_3 \rightarrow ClO + O_2$).

chemical equilibrium and is not at its equilibrium height, which should be low down. The layer is self-sustaining and structured by the directional flow of the Sun's radiation and by gravity, both of which provide gradients (see Table 3.4). The process of decomposition can be altered by injecting catalysts increasing the back reaction rate, and in effect we know that Earth's ozone layer has been seriously depleted by man through the release of refrigerant gases, freons, which reach the upper atmosphere. While here the rate of degradation of light to heat is increased, we stress that the addition of extra chemicals can either increase or decrease the energy content of the final steady state. A most interesting case is one in which the product catalyses its own production or its decomposition – an autocatalytic system – when energy throughput is increased. Note that the overall ozone layer has no ordered internal structure but it is part of an organisation of fixed cyclic chemical flow in a zone (or compartment) with radiation and gravity gradients. There is again the irreversible flow of light to heat with overall increased thermal entropy generation (see Section 3.12) as the Sun's energy constantly irradiates Earth. Again we note that this flow in the environment has an important influence on the late evolution of life.

All the above processes concern the rate of degradation of higher quality energy radiation to lower quality energy radiation, either a drop of temperature or a degradation of light to heat. We can say that the systems have constantly self-maintained form due to *the rate of energy flow* (see Sections 3.12 and 3.13). Before continuing, if we look back to the evolution of Earth described in Chapter 1 we see that flow was only possible if high-quality energy (from the Sun) was supplied to generate the change of state of the chemicals involved on the surface. This is a basic feature of our organic ecosystem too. Interestingly, when we consider the ecosystem as a whole all the above cycles are involved since as we shall see the flow, of water especially on land, and of the ozone layer allow life to evolve in particular ways (see Chapter 4) and rotation of the Earth round the Sun gives us climate variation.

3.4.4. THE BÉNARD CELL

One further example of a physical system is worth mentioning – the Bénard cell. The cell is a flow within a flat circular dish, which contains a thin layer of water. Heating the water under some controlled circumstances gives a more rapid degradation of heat through a pattern of hexagonal convection cells, Bénard cells, before turbulent boiling. Here a controlled input of energy through creation of an organised pattern gives a more rapid degradation of energy, and hence of thermal entropy production. The flow is structured by the temperature gradient and the physical presence of the circular dish. Note that here, as in each case we have given, it is the rate of thermal entropy production which we are describing as there is no change in the total energy within the flowing steady state. In each case there is also an associated shape. It may well be that intense storms such as those in the Caribbean Sea arise from atmospheric cells attempting to degrade energy faster by convection than by random diffusion as in the Bénard cell.

In the context of an organism we can compare all these flows with those of the organic compounds in a cell looking at them from the point of view of cycles such as

Provided that light energy is constantly introduced and its degradation is continuous the cycle too can have form if it is contained. The introduction of new chemicals or energy sources will also change this cycle. We are all familiar with the cycles of other elements than carbon between cells and the environment such as those of N, P, and S, and we shall have to consider at least 20 such elements (see Figure 3.10). When all are put together we have the possible cyclic ecosystem of Fig. 3.7, connected to the constraints of cells. We shall argue that this system has a causal origin, absorption giving an increase of degradation of energy, as does its evolution, much as the four above flows can be traced back to the effects of energy degradation in materials. We must not look for explanations of very small differences between such cycles, species, but between those with markedly different

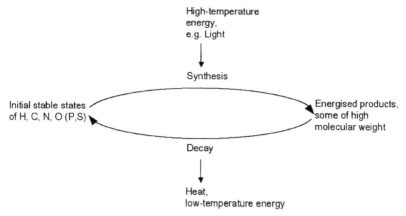

FIG. 3.10. A simplified cyclic state of non-metal elements related to biological systems.

components and constraints, which later we shall call chemotypes. Notice these systems are not yet necessarily coded, are not reproductive, but they are persistent.

A cycle can be simple, as above, but it is also possible to conceive of series of reactions starting from basic material, X, with an environmental input of energy which produce 2X in each reaction sequence. The process will be open to error unless it is coded, but it is self-expanding (not reproductive) and not totally cyclic.

$$X \xrightarrow[\text{Material}]{\text{Energy}} 2X$$

This reaction scheme applies to the cases of a nuclear explosion, non-cyclic, and of increase in cell population and disease, which may appear to be expanding reaction patterns, but are also self-limiting (see Eigen in Further Reading). We also draw attention to autopoietic systems described by Bitbol and Luisi in Further Reading.

3.5. Radiation Energy: Calculating its Disorder and Amount of Flow

Much as the structure of atoms beyond a qualitative account of electrons and nuclei has entered a stage of scientific sophistication (described by wave mechanics) that has lost any image known to our senses, so it is difficult to provide a description of energy. There is no use offering confusing talk about waves and particles in separate images which are familiar to all of us when the mathematical expressions which describe both "matter" and "energy" have the two images simultaneously embedded in them. Here we limit ourselves to making bare statements. We have said that we can count atoms and electrons as if they were particles and hence discuss their order and disorder in space and time much though they behave in certain ways as waves. Today all scientists accept this fact of nature. We have to accept too that we can count packets, quanta of radiation, as if radiation comes in particles with mass, much though it behaves most frequently as waves. The distinction between atomic matter and radiation then loses descriptive meaning. High-energy radiation comes in large quanta (large "mass particles"), e.g. UV radiation, visible light or high-temperature heat; low-energy radiation comes in small quanta (small "mass particles"), low-temperature heat. This approach is used since it satisfies experimental tests and so we accept the lack of an image. We can then calculate not only the energy of a certain amount of radiation reaching the surface of Earth per second from the Sun (or the deep Earth) in the same units as that of heat energy, but we can also state that the number of outgoing small quanta "particles" of heat generated is much greater, some 20 times, than the number of incoming large quanta "particles" of sunlight radiation from the Sun of equal total energy. We can then write

One quanta of high-frequency radiation → many quanta of low-frequency radiation

   or

One UV or light quanta (more effective energy) → many heat quanta
                                                                    (less effective energy)

Here effective implies ability to be absorbed (and permit chemical change) and this process generates a rate of entropy production dependent on the quantity absorbed. Thus, the statistical probability (statistical entropy or disorder) of the overall system including the environment is increased in this energy flow process from Sun to Earth at a given rate, being dependent on the flux of the Sun's energy absorbed and degraded to heat. This type of high energy entering a system as particles and leaving as dispersed low-energy heat particles is the major factor creating organisation, especially in life, that is essential in much of the flows we observe. Now during the course of the transitions (degradations) the energy is retained in processes which result in a change of material flow patterns and syntheses (clouds from vapour, O_3 from O_2 that is in metastable structures). The absorption and the consequent increased conversion rate of the high to the low-energy quanta are inevitable although the absorbing material holds the energy for a period in a high-energy, more organised, even more ordered in part, state of that material. Put in another way, there is a more effective degradation of light to heat in a given time through energy absorption. Both the increase in combination, O_2 to O_3 or of vapour to droplets and clouds, bring about some increase in material order, but the overall cycles including the radiation flow, dissipation of heat to the environment, always provide an overall increase in the production of (thermal) disorder. This energy absorption, therefore, raises the steady-state energy content of materials and so causes the general possibility of creating uphill physical/chemical change, e.g. of organisms, as we shall see.[†] It will often be the case that the material which absorbs energy will cycle, e.g. $3O_2 \rightleftarrows 2O_3$ when in the overall cyclic steady state there is

---

[†] These changes are in accord with thermodynamic laws in that while the created energised units have some more order and some energy is retained in them the overall processing of the system is toward thermal disorder in each cycle. The whole system moves towards equilibrium through the degradation of energy and does so more effectively than in the absence of absorption. Much though the timing of the creation of order is unforeseeable such processes continue, so that even stars and planets are still being formed as energy is degraded.

no material change. This is a kinetically stable condition, and cannot evolve. Note again, however, that any steady state is very sensitive to additions of material to it, which will be seen later as a major cause of evolution but can cause regression to a state of lower energy content even though the rate of energy degradation is increased; see the addition of freons to the ozone layer.

As such material steady-state processes originate ultimately in a faster (a more efficient) rate of change to more units (small energy quanta) and a more rapid approach to a state of higher statistical probability than would occur in the absence of light absorption, they occur spontaneously. The rate of approach to an overall equilibrium (including energy degradation) is actually increased. (Energy can also be obtained from unstable chemicals of Earth where it is held in high-frequency electronic states of chemical bonds.) These energised processes are then inevitable and are found everywhere on Earth, both abiotic and biotic, and they add somewhat to the temperature of the environment. The ecosystems of geological and their biological chemistry which we shall discuss are driven by such considerations (see Figure 3.7).

3.6. Optimal Rates of Energy Conversion and Optimal Retention of Energy in Cyclic Steady States: Content of a System

Consider again the ozone layer or a cloud. Why were they limited in content, as observed, since their development ceases with fixed form? Obviously, a major consideration is the amount of effective energy they can capture and this must be equal to the amount of heat released in a given time in their cyclic steady states. In any whole cyclic system the material in the final state (say $O_3 + O_2$) is the same as that in the initial state and all the energy taken in must be constantly given out but in a more degraded form. Of course, this is true only after the full cyclic steady state is realised. In the case of ozone it is the energy *gradients* which in part decide the thickness of the layer in the absence of catalysts. When we write

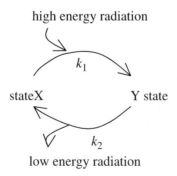

high energy radiation

k_1

stateX   Y state

k_2

low energy radiation

the optimum balance is struck when $k_1[X] = k_2[Y]$, where [X] and [Y] are concentrations and k_1 and k_2 are rate constants, while heat goes to the environment at constant rate. (Note that this is not an equilibrium condition.) The limit of the system in the steady state corresponds clearly to *the optimum overall retention of energy*[‡] *in the state Y*, that is at a certain O_3 content for the $O_2 \rightleftharpoons O_3$ system. This condition we shall call generally *the optimum state of an organised system*. It is also the system that releases *the most heat in a given time (a rate related to $k_2[Y]$) and thereby generates an optimum rate of creation of overall thermal disorder* in given conditions. Another way of putting this is to say that any energised state Y formed from X, and which can degrade to X, will build up to *a cyclic steady state which gives the optimal conversion rate of higher quality energy to lower quality energy*, that is an optimal rate of production of thermal entropy. This system will have a fixed shape of its organisation with or without conventional order. We take this as a general principle and look later at the ways in which the optimal retention of energy could be attained in a more complicated chemical system, the ecosystem of our concern. What we cannot say is how long it will take to reach such a cyclic steady state starting from some condition X in the presence of radiant energy (see Corning and also Kay in Further Reading). Note again that it is not possible to describe the reason for the exact shape of any one cumulus cloud but it is possible to describe the cause of the formation of cumulus clouds generally. Finally, we are describing here the limiting condition of a system to which no new sources of energy or chemicals are introduced, and this system cannot evolve. Observe also that from the system it will often be possible to isolate metastable structures and to examine them as if they were permanent, e.g. proteins from biological cells, but over-emphasis on such study misses the central feature of ecosystems which is energised flow.

3.7. Shape of Organised Systems and Energy: Maintained Form

Both a cloud (see Figure 3.8) and the ozone layer have shape but no containment defined by a physical barrier to diffusion exists. Their shape arises from restrictions to flow imposed by gradients. As stated above, in the case of a cloud the vertical gradients are of gravity and temperature and the horizontal gradients are also of temperature, producing motions of the air and winds, and affected by Earth's motion. It is more obviously true that many flowing systems of chemicals in a cycle have a shape *by being confined in a physically restricted space* by seeable barriers. A river is an example. There are also well-known cyclic flows of heated liquids in a container, giving rise to patterns which occur repeatedly (see Section 3.4.4). These patterns or shapes are reproducible (but note that there is no code), but not self-reproducing.

[‡] In any process such as the cycle of material the conversion of energy is to work, useful constructs is limited by thermodynamic reasoning to a maximum amount (not 100%). This maximum thermodynamic efficiency cannot be achieved by any machine working at a real speed and which operates under constraints. The resultant work output, we shall refer to as optimal insofar that waste is avoided. As the constraints in the ecosystem are often ill-defined the reader will observe a certain looseness in the use of the words efficiency and effectiveness (fitness) throughout this book (see Section 4.7 and Appendix 4C).

From all the above observations we wish to draw a general conclusion. A cyclic steady state is unchanging in different aspects and is self-sustaining. It has a fixed energy input and equal output and a fixed material content. The energy input has to be of a higher quality (frequency) than the output. This essential change of energy quality occurs at a fixed rate commensurate with the rate of the cycling of materials. The flow of material can be controlled by fixed physical structures and/or by fixed fields, gradients of charge or material, and has therefore a "maintained form", no matter that material in it is in constant motion. The constructs observed have not the same origin or appearance as equilibrium thermodynamic structures, which are also found around us. As stated, much of Earth's fixed inorganic surface may well owe its origin and presence to certain different equilibria between chemicals at a fixed temperature and pressure (Chapter 1), and this surface then gives fixed structural and gradient constraints to flowing systems. Even when new materials, such as oxygen, are added to these environmental inorganic chemicals on the surface the whole changes relatively quickly to new equilibrium conditions (see Chapter 2A). Radiation has little effect upon them unless it changes their physical conditions, e.g. creating air, water and sediment flows. This then results in types of shape, contours and see section on clouds (Section 3.4.2), which have physical/chemical explanations (see also Thompson in Further Reading). In contrast, change of radiation and/or addition of new chemicals will change considerably a steady state of chemical flow as long as the chemicals in it are constrained by barriers to change, and it remains away from equilibrium. These barriers, such as membranes, need not be made of chemicals at equilibrium but can be metastable constructs open to cycling.

Now on Earth there are some chemicals, notably those made largely from small abiotic C/H/O/N molecules called "organic chemicals", which retain absorbed energy for considerable periods (see Sections 2.12 and 2.13). The chemicals when exposed to energy unavoidably entered into flow systems of slow synthesis and decomposition. They will re-occur time and time again under identical conditions. Such flows may well have been the precursors of life. Now, as was the case for both the production of a cloud and of the ozone layer, the organic chemical flows will evolve, even if slowly, towards some cyclic steady state (Figure 3.10) as long as decomposition yields the starting materials *and no new materials are introduced*. There will be an optimum condition of formation and decomposition. However, as we shall see, the development towards this cycle is complicated by the very nature of the products first produced and dispersed into the environment if they are interactive with one another and/or are only slowly interactive with the original reactants. Such interactions are progressive. They could lead to a completely non-cyclic steady state unless the secondary as well as the primary product chemicals themselves can either readily decompose back to the initial chemicals or gradually change the initial conditions so that everything is ultimately in a new cyclic condition. The evolution of this system could then build up a range of inorganic and organic chemicals, in effect a new chemical system containing as much or as little of the contents of the original one as circumstances allow. It is important that we do not just look at the materials in these flow systems since equally critical is the uptake of and the use of energy in them which can also develop (see Figure 3.10). Of course, it may take a very long time for the old and the new chemicals to even approach a final cyclic steady state which can be of higher or lower energy content, but the content will be optimal for this new system eventually. There is no guarantee that a cyclic state can be attained, as we shall see.

We trust the reader is convinced that patterns of flow can arise with shapes, structures, but it must also be made clear that these shapes are fickle, open to a multitude but limited variation, within what we have loosely described under "structure" types. For example, almost every cumulus cloud is slightly different and can be classified down from major types to "species" and individuals. The appearance of a given cumulus cloud could be called an accident within a general impression that certain conditions produce clouds of this kind. We cannot expect to detail the causes of the appearance of a particular cloud though we can explain its type in general terms. We are distinguishing, therefore, random collections of happenings, all of which have equal probabilities, with happenings which have different probabilities, but where the differences are so small as to be unlinkable to causes, from types with large differences in happenings, which can be given causes. While all within each type are somewhat similar in kind. In discussing biological/environmental evolution we cannot expect to give an explanation to individual species though they may well not arise from completely random variation. We can look for an explanation of the causes of large groups of rather dissimilar species, types, and why they have changed systematically with time. The origin of such groups of types, to be called chemotypes later in this book, is a much simpler logical task than that of the origin of species.

3.8.   Evolution of a System going away from Equilibrium

Now, whatever else occurs, the creation of a flow system of some permanence (Table 3.5) is about the use of energy to create new energised steady states of material. The general chemical effect of energy absorption is to create charge separation, i.e. separation of oxidised and reduced materials. To all intents and purposes, therefore, the expanded ecological evolution which has occurred on the Earth's surface can only be due to its increasing absorption of energy causing redox separation as is seen between organisms (reduced) and the environment (oxidised). Three such natural sources of energy are high-energy heat and chemical energy from the interior of the Earth and high-energy light from the Sun, all of which would be dispersed constantly as low-energy heat in the atmosphere in the absence of chemical absorption and reaction but with lower efficiency. Gravity acting on all material is a fourth, but it, like low-energy heat, cannot generate chemical synthesis directly much although it too affects physical flow. Among them, light is the dominant source for the above chemical change which, followed by degradation, increases energy conversion to heat (see Figure 3.11). Everything we observe of chemical change is, following the lines of the previous discussion, seeking a cyclic

TABLE 3.5

THE EVOLUTION OF FLOW SYSTEMS

Year ago	Flow systems[a]
15×10^9 ↓ onwards ↓	Big Bang: expansion of universe – creation of "space" and "time" Formation of "mass" "energy" Turbulent flow Galaxies, stars Basic atoms in stars Heavier atoms in stars (nuclear reactions; see Fig. 8.2)
5×10^9	Solar system Earth Cycling of planets (stationary states) Geochemical reactions
4×10^9	"Living" steady states Evolving developmental states on Earth
200	Man's industry

[a] Flow involves bulk transport or local rearrangement of particles to form atoms and then larger associations (see text). Ecological systems started with "living" steady states.

steady flowing state in this basic "structured" or constrained irradiated environment (fields due to gravity and radiation), overwhelmingly due to the input of light. Generally, we may consider that the passage of light to heat is catalysed by transfer in matter; see again the sections on Ozone layer (Section 3.4.3). We include here a system in which the product such as ozone itself is catalysed in either production or decomposition and also autocatalytic systems. Extra constraints were necessarily added by the creation of more or less permanent physical structural barriers such as to a stream by hills or to metabolism by a cell membrane, but the last is self-synthesised. We must then ask how can energy absorbed be increased and new constraints evolve with a fixed energy source (see Chapter 1). (In the last chapters of this book we shall show that new energy sources can also be introduced.)

If there had been no life on the Earth there would have been a certain heat production from the absorption of energy coming to the surface from the Sun. Much energy, perhaps 30% or more, would have been immediately reflected (or scattered) by the Earth's surface, the atmosphere and clouds. (This is often called the "albedo".) The atmosphere and clouds account for 19% of energy captured, and another 51% is captured by the surface, melting of ice, evaporation of water, driving abiotic photosynthesis and increasing the temperature of the Earth's surface. It is re-radiated as longer wavelength radiation in part by the so-called greenhouse gases (CO_2, H_2O, CH_4, N_2O, etc.) and finally dispersed as heat, which warms the atmosphere; see global warming (Figure 3.11). We have already discussed the resultant flows of air, clouds, rivers and chemicals, which must have been present

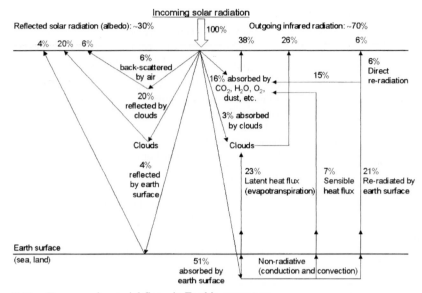

FIG. 3.11.   Energy and material flows in Earth's ecosystem.

immediately on the Earth in the absence of life. Undoubtedly, with time, different physical and chemical flows, would have been initiated one or more of which led to life. We must expect the process to appear and develop, towards that shown in Figure 3.10, and we know today that about 2% of the retained energy on the Earth is used by life and it has increased continuously from the abiotic start, see cover picture. The question then is how could such a chemical energised system, not yet to be called a living system, have evolved? There was no competition for light at first and we may presume that all chemicals that were required to form the ener- gised system were in plentiful supply in the environment. We must recognise, as stated above, that the only way we can account for evolution is if the chemicals produced became involved in a flow system that did not produce an all inclusive cycle, which is a terminal condition like that in the cyclic ozone layer.

Consider a structured flow system of chemicals X, which can absorb light and give products Y + Z, but in which not all the products are involved in the cycle in contrast to Fig. 3.10.

$$X \underset{k_2}{\overset{k_1}{\rightleftarrows}} Y + Z \text{ (waste)}$$

light

heat

We know that the product components Y will build up until the steady-state system, which is not fully cyclic, attains the conditions $k_1[X]=k_2[Y]$ and light energy in = heat energy out. Under these conditions, [Y] reaches an optimal value but [Z] increases continuously as it is not part of the cycle. Let us assume that X is a large reservoir so that it is not appreciably changed by loss of Z. The chemical composition of Y is not the same as that of X so the system evolves with a new environment including inactive (energised) Z but the X/Y system cannot evolve from the cyclic condition. Now the energy throughput may be slow if the product Y is very stable but the X–Y system can evolve in rate if Z is a catalyst for formation or decomposition of Y, that is an autocatalytic system (compare an increasing addition of a freon to the ozone layer). The system is still not fully cyclic but note that an autocatalytic system restrains the catalytic activity to the very materials that are needed in synthesis of the catalyst, Z. This is in fact a dominant feature of cellular activity.

Let us assume next that Z can absorb energy, say light, more or less as efficiently as X, and further that this new energy enters the system $X \rightarrow Y$ by Z giving energy to it. The system evolves using increasing energy and hence Y increases but the system is not fully cyclic as Z will still accumulate. We shall see that improving energy capture is a very important step in evolution but even more important is the material, chemical, influence of Z on X, the original environment.

Now, if we assume that Z decays back to X or a part of X, at some rate not necessarily the same as that of Y, then eventually Z will also cycle and Z will effectively increase the adjusted process $X \rightarrow Y$ whether or not Z absorbs light or catalyses X/Y changes. If all the energised chemicals Y plus Z decay to X, X need not be a large reservoir and the process $X \rightarrow Y + Z$ continues until it reaches a completely cyclic steady state of novel material content. There will also be an increased rate of light $\rightarrow$ heat conversion. X and Y are still not changed chemically and evolution is only in the quantity of X and Y now plus Z in a combination of X, Y and Z which cycles. It is clear that only if Z unmodified or somewhat modified can, in some way, enter chemically, as well as physically perhaps (see signalling) and increasingly into the organised $X \rightleftharpoons X + Z$ cycle, that the system can evolve materially continuously towards a new energy retention state which will have a new eventual optimal value. This corresponds to *chemical evolution* of the system from $X + Y$ to $X + Y + Z$ in a cycle. As stated, evolution after introduction of novel chemicals, here Z, proceeds to an increasing energy and chemical flow towards a final limit, a fully cyclic condition. To this system we need to add the extra complication that Z may interact with the original environment, X, or with Y to produce many other new chemicals in the environment. These too can drive evolution so that the rate of evolution and what evolves is dependent on these reactions as well as the production of Z and it can take a very long time to attain cyclic balance. However, if as mentioned, the products are catalytic of their own formation, that is they are autocatalytic, then the X, Y, Z system is again increased in rate of energy throughput, though it may still be slow to move towards a final cycle state. The catalysts now limit the direction of change to special pathways. Additionally, as noted, Z or its products may absorb more light energy than did X alone. Just such a series of changes on the Earth involving incorporation of waste, Z, in many different ways, will be shown to be systematic, not random, and that they have not yet reached a final state. In fact, this will be seen to be the very nature of our ecosystem, the evolution of which is, we consider, an unavoidable chemical sequence. (We must be constantly aware however that Z the novel chemical could act to reduce as well as increase the content of the cycle if it interfered with energy absorption, or is a pollutant or only increased rate in the X/Y system; see Chapter 11.)

Since the organised steady states present in life (and also in an industrialised society) as well as in clouds or ozone layers are cooperative or at least interactive with the physical and chemical environment, on which they all depend, the whole can be represented schematically as heading towards an optimal cycle and Z will be particularly effective if it acts in all three of the above ways, as catalyst, an absorber of energy and as a chemical part of the system as in Fig. 3.10. Due to physical space confinement or due to field gradients (see Table 3.6) all such steady-state systems, e.g. even organisms, can have spatial definition, and long before genes characterised "living species" they were characterised by their limiting shapes; compare clouds (Section 3.4.2). They are all systems of maintained form and an enclosing membrane is particularly valuable in this respect as it prevents dispersion. In fact, the way we all describe many identifiable objects including organisms in every day language is by their shape, morphology, which is not only due to order but to maintained form. The analysis of shape of organisms was also the starting point for the study of biology and evolution but the form will change as the environment changes. Later we wish to show that the organisms formed dependent "chemotypes"[§] of species using different particular sets of chemicals

TABLE 3.6

TYPES OF GRADIENT CAUSING FLOW

Field gradient	System open to flow
Gravitational	Layering of Earth, sea and atmosphere; separations in chemistry; biological polarity of masses in cells
Chemical concentration	Storage across boundaries, batteries
Electrical	Storage of charge (many electrochemical cells and all biological membranes)
Mechanical tension or pressure	Deformation of structures
Radiation causing a temperature gradient	Excited absorber; convection currents in all materials

[§]By analogy with phenotypes and genotypes.

and many living shapes arose in succession, e.g. the plants and animals of today. At all times the system creates an ecological energy store, Y + Z, which develops as the environment changes. Before going on it is essential that we introduce more complicated multiple flow systems and internal information and then add reproduction as a dominant way of increasing the survival of living organisms and *in part* controlling form.

3.9.   Form and Information: Multiple Component Systems

We have seen that single substances, X, when exposed to a source of energy will degrade the energy while cycling through energised states, Y + Z. We also observed that two systems can exist independently such as those of water and oxygen/ozone. Now the flow of organic material on energisation in cells is different in that the flows are of many starting components, are cooperative, and yield various large molecules of complex composition. The question we then ask is starting from several very small H, C, O, N, P, S compounds in the environment why should any system grow in multiple flow complexity and kinetic stability so that it persists as we observe in organisms? Let us consider a system of several small environmental molecules which are able to pick up energy and create many combinations of them. We take it that these combinations are all more complex molecules such as amino acids, sugars, etc. How can this whole system of many components become cooperative? They will do so if they are themselves able to absorb energy, in fact more easily than their precursors, and this energy can then be used both to increase further the size of each of the molecules and they could drive, catalyse, the initial system to increase the numbers of molecules of their own kinds. The system is then increased in energy uptake and is autocatalytic. Such selective catalysts are usually large molecular units, as are pumps, and are based on proteins (see Mossel and Steel in Further Reading). Now for it to be reasonably stable, remember it is just kinetically stable, it will only persist if some of the molecules, lipids, make a membrane to contain the system, which is clearly then a cell, efficient in local energy absorption. The persistence of the system is increased if it also eliminates any trapped material in the cell that reduces cell stability. Equally it must maintain osmotic and electrical stability, that is it must use energy to the advantage of survival, persistence, by pumping out unwanted and pumping in wanted components. All its contents will decay at some rate, but the cell membrane prevents much loss (though it also cycles slowly) and hence much material will cycle internally, which helps survival though some is lost. However, the losses can also be replaced by the inward diffusion of basic small molecules from the environment. Now the more energy the system can capture and therefore the more it forms the molecules which capture energy, the more it expands. This expansion is helped later if the molecules it produces itself and which are rejected into the environment assist the intake of more energy and/or material from the environment or catalyse their synthesis. The system can also grow if some of the catalysts actively destroy the molecules which are not wanted—a step which can supply both energy and material to the synthesis of required molecules. Hence both degradative and synthesising catalysts are required to make energy-capture devices. There is one step further which the system can take to help growth and survival. The composition of the energy giving molecules and the most useful molecules for maintaining the cell membrane, e.g. lipids, and for catalysis, e.g. proteins, should have particular concentrations, so that it is advantageous to growth and survival if the synthesis of the required initial small molecules and the larger ones is in proportion to need—that is to satisfy the particular overall composition. It is this step which introduces cooperativity. There is now a requirement to give information to the catalysts to control their activities. The controls can be by the binding of the small molecules, feedback controls (see Section 3.11). We see that such a complex system will develop sooner or later given a somewhat random beginning since the complex system is more effective in energy capture, in growth and in survival, so that it will dominate. It is important to see that only a *selected set of component molecules* will meet the requirements best and presumably these are the ones we observe in cells (see Chapter 4). Of course, an individual cell grows but sooner or later it decays and so a cyclic system is approached which moves towards optimal energy flux, i.e. energy degradation. The whole system has a fitness since it is strongly self-sustaining and will occur again and again, but it is not yet reproductive. Many such systems could coexist as long as energy and the basic environment are not depleted. If we now add that some molecules in the synthesis programme could become coded and self-reproducible and others can act as expression machinery of the code we have a cell system which is even more efficient and self-sustaining. The code ensures also that only the selected set of molecules are produced. In Chapter 4 we shall show that the whole system just described is, as it had to be, that of organisms as we observe them, and there we shall detail the components resulting from energising the environment.

Before we examine further the effect of introducing reproduction we note the essentials of a fit system are (a) an energy source, (b) environmental elements (20), (c) selective catalysts, (d) containment, to which we add an expressible reproducible code for guiding synthesis and that the system is then totally autocatalytic with all the pathways including those guiding expression and is linked by feedback controls. Such a system must change with the environment and must do so in selected ways according to environmental change. It has great survival value. It is the presence of the code that is the basis of reproduction, but note that reproduction is not the essence of the flow system, much though it increases survival of it. We should also be aware that any coded system limits the variance of the system since it informs it in a predetermined specific way. (Remember also that initially a code can only confine or limit a system which existed before it.) A code cannot be the only source of information, since it is a conservative limitation on evolutionary rate and diversity. Extra information will be seen to come from the environment which is not conservative. If this is so, how did the code change so as to be compatible with the observed drive to optimal energy retention of the ecosystem when the environment changed? The code must have been forced to do so, however reluctantly, since, as we shall see, there was and is only one *material* route for organisms to go following the changes in the material environment, that is by simultaneous inevitable and parallel change of its type of chemistry with changing interactive energy and chemical organisation, see the cover picture. This implies that the way to change is to improve *the overall efficiency of* the whole chemical organisation, using the limitations and advantages of environment change, which will lead towards optimal energy retention and degradation. We are in effect describing survival of system fitness (not of the fittest species), where survival can be seen to be dependent on either the ability to maintain or increase activity in single cells or organisms, may be of a large size, or in the ability to produce single small cells rapidly. However, these two can coexist and could be cooperative as we shall see. We can think of all this biomass as forced to move towards an optimum use of energy absorption and of energy degradation in a final cyclic steady state of a total ecosystem. The force is survival strength and will increase the rate of energy intake which will happen just because it can occur no matter how the increase is brought about, and then, since the synthesised biomass is energised, it will decay leading towards a material cycle, but we must be aware that this fully cyclic state may never be reached much though the flow is inevitable. A very clear conclusion is that the code must be flexible but how is this managed? There is also no possibility of predicting the time of evolution much though we can now date it and state its direction.

Apart from having a code another way to increase survival of a system of increasing number of chemical activities is to increase the use of space by increasing number of interactive compartments, as we explain below. A major consideration here is that we must examine not just the materials in different activities but the materials and energy which go between them in compartments in mutual *signalling* in addition to obtaining information from the code and the environment. The compartments are part of a unified whole and hence must be linked by control molecules. The molecules are the *messengers* of the organisation and their evolution is as important as the evolution of new activities, catalysts, the code, and compartments themselves since they are the essential components of *information transfer* for control.

3.10. Organisation and Compartments

It is now important for us to explain the nature of systems of many compartments and chemicals. Why should systems evolve not only new chemistry but do it in many compartments rather than in a simple single compartment? The question applies equally to the manner in which industrial plants or organisms develop. Any compartment is, of course, based upon a division of space, either by physical boundaries or by fields (Table 3.7; see also Tables 3.2 and 3.4). We saw that abiotic cycles of water (clouds) and oxygen (ozone layer) formed in compartments containing droplets or ozone, respectively. Here each system has one component, controlling fields, with no physical barriers or information transfer.

TABLE 3.7

MATERIALS USED TO MAKE COMPARTMENTAL BARRIERS

Type of compartment	Materials
Geological compartments	Minerals[a]
Man-made compartments	Metals[a]
	Minerals[a] (clays, glasses)
	Plastics (wood, paper, etc.)
	Composites
Biological compartments	Organic polymers (proteins)
	Composites
	Fats (membranes)

[a]High-temperature preparation.

Consider the example of the manufacture of a complicated physical product, e.g. a car of many components. At first all the parts of a very basic car were made in small numbers in one place. Today we are used to large car-manufacturing industries, producing very many sophisticated cars. As more and more units were introduced into the manufacture the difficulty of all work being done in one place increased. We observe, however, that the industry today does not attempt to make all the components of the fully assembled products, cars, in the same place. Instead *specialist* separate factories make particular components which are *transported* to an assembly plant. It is also necessary to supply each factory with elementary materials from different sources by transport means and with energy from distant sources, e.g. a power station using coal, oil, gas, or electric power transmitted, e.g. through cables. Now each source (compartment) generates parts in a maintained flow, but to make sure that all the component factories work cooperatively there has to be communication from a centralised command office. This office receives messages from all the factories (compartments) and then sends messages to them making demands. As stated above, messages are either in the form of material or radiation but they are the basis of *information transfer*. In factories computers manage a great deal of this control today, and computers are more effective than people. (Later in Chapter 11 we compare brains and genes.) Only in this way can the assembly of the whole product work efficiently. The compartments cooperate in an overall organisation.

Major considerations in the development of separate organised factories for each component rather than one factory for all components are that very different materials and different skills are used in each compartment or factory. Specialisation locally and supply of different bits to different places is the received wisdom because it avoids confusion, increases the capacity to construct the whole and is more effective in the use of resources, but it demands much-enhanced organisation using transport, improved messages and feedback structures. Notice that in the whole procedure it is desirable to approach optimal energy and material use (efficiency) so as to have constant overall *balanced flow* of production of parts in manufacturing. We have to realise that organisation of several compartments is a more efficient (economic) way of using resources, energy and materials, than in a central single compartment where the risk of destructive interference, confusion, is higher, and this is especially true as more components are added for the creation of products. In fact, increase in use of compartments is unavoidably associated with effectiveness and is a natural part of evolution, i.e. survival. After we have looked at biological components in Chapter 4 we shall discuss efficiency generally in more detail (see Appendix 4C). (When we consider life we shall have to remember the additional factors that it has to reproduce and adapt. Here simplicity, a single compartment, has advantages and we should not be surprised to see that such simpler forms persist even when compartmentalised organisms evolved.)

Consider now a second example of an industrial synthesis of a chemical of some complexity, for example a polymer formed from two or more monomers. Each monomer is prepared in a separate apparatus or plant and the final polymerisation is carried out in yet another manufacturing plant. The synthesis of each monomer is carried out in a physically and chemically protected different environment and selected catalysts are used separately in each plant to avoid unwanted products and to give carefully controlled rates of reaction. The bringing together of the monomers in balanced amounts requires a transfer or transport step. The chosen catalysts also control the final synthesis. All steps are in continuous-flow processes rather than by batch, and flow rates are monitored by sensors connected to feedback commands to the flows. Once again there is specialisation and particularisation of activity locally. The whole organisation has a central command, again a computer with *information*, carried by messenger systems, from operators, which monitors all the production lines and adjusts them. See, for example, an oil cracker plant. Energy input is necessary and is regulated similarly.

Examination of both the operations of the physical car manufacture above and this chemical industry leads one to recognise their similarity to a complicated biological cell or organism which has also many separate reactions to coordinate. Organisation in spatial compartments is then of the very essence of both the chemistry of modern industry and of advanced organisms for the reason that it generates optimal use of resources, energy or materials, which, in the same compartment, could damage the whole. This is especially true if a system produces waste which can be damaging to the original workings but which can be made useful when treated separately (see Sections 3.8 and 3.9). The more the inputs the more difficult single locations are to manage and hence to sustain. Even so, all organisation starts in one compartment but must evolve into many if effective advantage towards the desired end of a fully cyclic system is to be achieved. This discussion will form a basic part of our view of the inevitability of the evolution of life – increasing chemical complexity met by increasing organised compartmental activity, always keeping in mind the need for reproduction and adaptation to changed circumstances. We can also consider simple cells and complex cells as separate compartments in a whole system when both together give the greatest survival strength.

Examining both industrial and biological complicated systems we see that organisation has evolved into a flow activity in compartments in informed systems which need (1) materials, (2) energy, (3) spatial constraints, (4) transport (pumps), (5) catalysts (rate enhancers), (6) central command (a code), (7) internal communication in each compartment and (8) communication between compartments. Items (4)–(8) are linked cooperatively by energy and material controls. Finally, compartments can also be used to store materials and energy for subsequent use. As stated, the advantage of such a system is increased efficiency and more economic use of resources in the making of energised materials, but we shall have to look carefully at the disadvantages as well as the advantages of such compartment evolution, i.e. of increased complexity.

In the above we have changed the nature of the flow systems. At first we discussed only *maintained form* in a constantly energised flow system of a single material. Subsequently, we have introduced cooperation via controls over several flows by messages and a central control in one compartment. Lastly we have included coordinated flow in compartments where management of energy and material throughput increased production, in the case of organism chemicals, in part by preventing interaction between steps of reaction to avoid mutual interference. In particular selected catalysts could be kept apart. The overall system is now one of *informed maintained flow* where several activities inter-communicate. We know how this is done in our industrial society, but how is it done in cells? We know the direct steps of chemical reactions as discussed in Chapter 2, but what are the messenger systems?

3.11. Organisation Messengers Feedback and Codes

There is a way in which a chemical flow system can be controlled in a final state but which is not obviously optimal in energy storage. Consider that the product can interact with its own mode of production in a different manner from the ways described above in Section 3.8, in one compartment.

For example, let us consider that Y contains (or can create) a product Q which binds, signals, to informs and regulates a catalyst, E_1, of the overall $X \rightarrow Y$ process (Figure 3.12) but is not engaged in the process itself. The $X \rightarrow Y$ reaction can then be enhanced or inhibited by this reaction. While Q could be an enhancer or inhibitor of

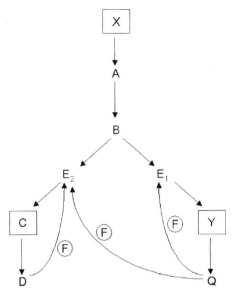

Fig. 3.12. An outline of a feedback system in which X to C and Y are regulated by downstream products so as to control the balanced formation of C and Y from X. Ⓕ represents feedback.

rate it is in effect a messenger carrying a message about its concentration and that of Y to a catalyst operator. In organisms Q acts on a catalyst, an enzyme. How could this be desirable in an approach to optimal energy use (linked in biology to optimal reproduction) if the rate $X \rightarrow Y$ was inhibited by such feedback? We only need to think that X may be equally required to produce other chemicals, such as C, from the same materials and energy sources using another catalyst, E_2 (see the scheme in Figure 3.12), and in controlled synchrony with Y production. For example, we can think that C is required to make a structure to prevent Y from diffusing out of the system and losing energy. The synthesis of Y then has to commensurate with that of C if the whole system is to survive and degrade energy. It is then essential that reactions to or from C are also controlled in rate by Q or another product of Y or C, e.g. D, at other catalysts. In fact, there is in most organisations a need to control many different products in certain ratios and this demands feedback management of supply and energy to each process. *The feedback control can apply to pumping, effectively catalysts for transfer between compartments, or to valves in any chemical process.* It may be said that the reaction $X \rightarrow Y$ is controlled by instructions (information such as the concentration of Y, here to its production through Q, but Q could also interact with other reactions and compartments in a network) creating an informed system of maintained flow. Interestingly, components themselves in a flow creating form can also act to inform the flow and to inform other flows by feedback, as we shall see. We also note that the more different flows of material there are, the more different message systems between them are required. This is a fundamental feature of evolution and is in keeping with the idea of survival of fitness now in material use and energy degradation.

The question of coordination of activities in production also arises in the nature of the central command. Here information from all compartments must reach the control centre and the centre must send instructions (information contained in messages) to control each activity. We all know how this is done by responses in industrial plants through electronic messages to and electrical driven feedback controls from computers. We shall have to see how this is managed by chemical (or electrical) messengers in cells to and from their central coded control centre (DNA plus genetic apparatus) from and to reaction pathways and even from the environment. Central coordination within organisation will help the drive to integrated optimum retention and use of energy and material and must change with evolution. Note that feedback by chemicals implies binding to a synthesis apparatus, an enzyme (catalyst) or another receptor in the case of cells, which is dependent on its ordered structure (see Figure 3.13). Information, which is of necessity initially present in a plan and its equipment, is transferred to give *informed activity*, flow of material and energy, but this can only be so if the receiver understands the message and energy is employed to generate a consequence. ("Understanding" in biological systems lies in binding ability, see Section 3.13.) Later we show that a cell is yet more complicated since use of codes in the genetic machinery is not simple (see Figure 3.13). Now, the discussion above has dealt largely with material flow and we need to appreciate energy flow as well. Once again it is essential to observe that as more and more components and compartments are added communication becomes more complicated, so that *the mode of energy signalling also has to advance with the complexity of organisation.*

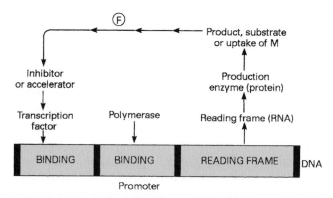

FIG. 3.13. A simplified version of a genetic construct which requires not only a reading frame to generate the m-RNA and then the protein but requires a binding region for the polymerase machinery and an instruction region for the binding of a transcription factor, both of which must be bound in appropriate form before reading can occur. The form of the transcription factor which may interact by feedback Ⓕ with an element, M, can be adjusted (by [M]) to stop reading. The whole is necessary for cell operations not just for that of DNA reactions.

3.12.   Energy Sources and Controlled Distribution of Energy

In Section 3.2 we described the sources of energy which can be utilised in chemical change. The way they have changed with time is shown in Table 3.8. There are certain ways in which chemicals can absorb energy for use, which we shall describe in detail in Chapters 5 and 6, but this energy must be distributed amongst many different synthesis paths and maintained gradients and even in different compartments in cells. We show how this is done in detail also in Chapters 4 and 5. Here we have to see that like the distribution of common materials the distribution of energy has to be controlled if syntheses based on several paths are to function equally well. This is manageable if one or more of the chemicals produced subsequent to the absorption of light, the major source of energy in our ecosystem, acts as a general distributor. For example, light can lead to the production of pyrophosphate, P_2(ATP), which can drive all condensation (water-removal) reactions, which includes many major paths in cellular synthesis, by picking up water giving two phosphate ions $2P_1$ (see Chapter 2).

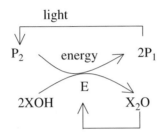

TABLE 3.8

SOURCES OF ENERGY

Period (years ago)	Energy Sources
Initial (4.5×10^9)	(a) The Sun (b) Basic unstable chemicals in the crust (c) Chemicals stored at high temperature in the core (d) Gradients of materials
After say 1 billion years (3.5×10^9)	(a) As above (b) Some oxidised materials, some SO_4^{2-}, Fe^{3+}; very little O_2, H_2O_2 (c) Biological debris
After say 2 billion years (2.5×10^9)	(a) As above (b) Further oxidised materials; modest O_2, H_2O_2
After say 3 billion years (1.5×10^9)	(a) As above (b) Further oxidised materials, almost 1% of final O_2 pressure
Today	(a) As above but final O_2 pressure (b) Man's fuels (c) Atomic energy (d) Wind, tide, hydro

Now, if the product X_2O inhibits the process at, say, the enzyme E, when it reaches a certain concentration, the use of energy is controlled in this reaction. P_2 can then be used in a different reaction elsewhere as we described in Section 3.11, making an overall distribution of energy to many simultaneous processes.

We have now considered how energy and the material changes are knitted together into a cell's cooperative activity through controls. It is very important to observe that each chemical can have an energy and information content related to its bonds, its concentration and the fields to which it is subjected. We are describing the ecosystem in thermodynamic terms. We have done this together with the introduction of information and feedback and we now wish to clarify the meaning of information in cells and organisms since it is often misleadingly related to computer-like activities.

3.13. Information Defined

Information is a difficult concept and is defined generally as "that which is conveyed in any type of message, represented by a particular arrangement or sequence of things such as numbers, letters or symbols". It is often taken together with the word "signalling". This is easily understood in a written communication between two people. When we analyse further we see that there has to be more than this sequence since both a source and a receiver, which must have a language in common, are required, so that the message can be sent and understood, and that energy is expended to construct and send that which is conveyed. Moreover, a receiver is only informed in the sense that it responds, so that information leads to action, immediate or delayed. Energy is therefore needed by the receiver even to retain information as a memory. (Note, however, that any "language" in a message has unique all or none *symbolic units* and may have a concentration or strength (emphasis) placed on the units, an *intensity* component as well; see below.) In a biological system we need to understand how this analysis of information is related to molecular interactions. For information to be used in an organism there needs to be a source in cells of releasable chemical or physical messengers and responsive receivers, receptors, *which recognise by binding*, as well as energy supplies in order to store and respond. Information is a quality embedded in these relationships. We can relate this idea to that of an *"informed system"* by using the definition of *"an energised system which has the capability to send a message to an energised receiver which can recognise the message (by binding), and using energy, causing it to act"*. We must not confuse this manner of creating activity with that of internal contribution of material or energy to *form* a simple system, e.g. water in a cloud, that is the simplest creation of form, where no information is exchanged.

We need, therefore, to recognise that maintained form can arise from the action of fields, incorporation of material particles, molecules, or radiation, and that information alters the quality of forms by interaction between two or more such factors (see

Figure 3.12). It is not related in any necessary way to simple statistics, as is sometimes said, much although some kinds of simple information can be defined in statistical terms. Let us consider the case where we might apply a statistical approach,
for example in an electronic computer A, which recognises two symbols, 0 and 1,
generated by another computer B, and then carries out a process. Both computers
have the same language. The two energised computers and the messages constitute
an informed system. Any description of information in the computer B includes the
fact that there is an emitted energised source of 0 and 1 in any sequence and that a
given sequence, say 0.1.0.0.1.1, has an informative quality, a meaning at the receptor computer A, giving rise to a particular energised output. It is clear that we can
relate the probability of that sequence occurring relative to all randomised ways of
distributing 0 and 1 in a sequence of six digits. In one sense this type of message carries information related to a statistical probability which was analysed by Shannon
and is related to the so-called "entropy of information" in some books on communication theory. Although the mathematical expression that defines such probability is
isomorphous with that derived in thermodynamics it is in fact not related to any
thermodynamic entropy, a state function, since it is not related to general functions
of the state of the system, e.g. the temperature. It is a pure number. We shall, therefore, avoid any suggestion that such statistical information is related to thermodynamic quantities although it is describable by an expression analogous to that given
for a molecular statistical probability. Now, in a more complex chemical system the
quality of what is conveyed by a particular arrangement of things, chemical units, is
not a "yes"/"no" sequence analysis of 0 and 1 or of any number of symbols such as
the letters A, C, T, G of fixed intensity. (These letters are those for the bases of DNA,
the known coded molecule of all cells but this is of no consequence here.) Consider
the sequence to be of chemical units ACCTGA. The sequence from the source creates a chemical field (the information) through which the receiver units (other A, C,
T, G units in transcription, amino acids in translation) are attracted by chemical
affinities and joined together, linked, driven by energised enzymes. But note that the
genetic machinery may also be considered a receiver since it acts in response to
changes in the conditions of the cells. The reception of this chemical message cannot be compared with that of any other message in just an all-or-nothing way as the
interaction between messenger and receiver has an intensity dependent upon the
reading machinery involving thermodynamic binding constants under
physical/chemical conditions, i.e. the state of the system involving concentrations,
temperature, other interactions, etc. This has a binding probability associated with it,
but totally different in kind from the specific chance of a randomised number in a
computer. Moreover, the reception of a conveyed chemical sequence, a message such
as this, is open to different interpretations in that different outcomes could result at
somewhat similar receptors due to binding constant differences and conditions of the
system. (Humans can also *interpret* messages differently (see Chapter 11).)
Information in the type of chemical machinery we deal with in organisms is a property of the receiver as well as the source and of the energised machinery which causes
and receives the flow. In biological systems, then, we must use the word information

(and the word gene) only with great care, but we can still describe the overall effects of sending and receiving messages as giving rise to an informed system which has also a formative quality, e.g. a modified structure and dynamic function. Now we need to re-examine feedforward and feedback in this context since receivers in a cell, the effectors of the metabolites and ions, are in communication with the coded source in the feedback sense. The whole generates an integrated informed and formed system, where the making of A from B and the reverse are not included:

A informs the making of B

Feedback from the concentration of B informs A

(Note again that messengers and components of form can be, but do not need to be, one and the same as those transferring information – see coenzymes in Chapter 5.) Using this approach, any system exposed to any field, chemical gradient, etc. which arises from a stored source may cause the system to be informed through the energy of back interactions, some of which must be used to cause flow. Thus the point of binding can be to a membrane or a catalyst. A body such as a cell is then informed by chemical gradients initiated at, say, the coded (DNA) machinery, which leads forward to control of metabolism. The chemical gradients can be of energy carrying units such as pyrophosphate (ATP). Metabolism, material and energy content, back react so as to inform expression by the coded machinery, but they are also open to being informed by the environmental conditions and chemicals. This arises since the cell membrane and its activity is informed about the environment on one side and by the metabolites and ions which bind to it on the other. All bindings linked to action give information. Life has arisen and evolved since the very nature of its chemicals is open to interactive information from two major sources, the genetic code and the environment. (The environment also gives the components to *form* organisms.) If the sources change with time then the system evolves and is driven specifically by the change to generate new informed systems of maintained flow, material and energy. The DNA expression must then change to match the environment so that the changed code represents a viable organism in that environment. We shall see that one large part of evolution is the increased use of *external information* as opposed to "simple" use of *internal (DNA) information.* It is the environment that is linked to a causative, not a random process of change and to adapt (evolve) DNA must respond in a way helping to generate a parallel internal cell chemistry. *It is the whole interactive ecosystem of the environment and cells which evolves systematically* (see Figures 3.14, 11.7 and the cover). It is now worth looking at chemical binding in a little more detail since it relates back to the basic chemistry of Chapter 2, and to the fact that use of thermodynamics is necessary to understand organisms.

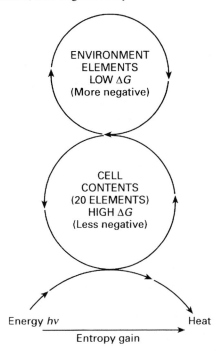

FIG. 3.14.   The total thermal entropy drive which underlies the energy connection between the cycling of elements in life and environment while energy is degraded.

3.14.   Cell Organisation, Equilibrium and Kinetic Constraints

There are many molecular species in a cell and as we have just discussed it is necessary for them to bind to one other to create not only activity but to generate control. In Chapter 2, we have discussed binding in terms of equilibrium constants and have given the expression for the binding constant K at equilibrium:

$$K = \frac{[AB]}{[A][B]}$$

Equilibrium is satisfied if both the rates of formation of AB and that of its dissociation to A + B are faster than other processes affecting their concentrations. Many molecular associations and dissociations and those of metal ions with organic molecules are fast, and usually this implies that K is relatively small, say $<<10^{10}$ M. Hence in a cell we need to observe these binding constants, for example in the Irving/Williams series, since they are not adjustable during evolution at constant temperature (see Section 2.17). These binding constants (also called stability constants when AB is a metal complex) are not involved in energy uptake, of course, but are extremely important since they impose limitations on complicated metabolic activities (see Appendix 4A). We may say that they too, like the genetic code, decrease the variance of the system, and the same is true, of course, for solubility products. Equilibrium constants are not affected by controls of the kind discussed under information.

Equilibrium considerations other than those of binding are those of oxidation/reduction potentials to which we drew attention in Section 1.14 considering the elements in the sea. Inside cells certain oxidation/reductions also equilibrate rapidly, especially those of transition metal ions with thiols and –S–S– bonds, while most non-metal oxidation/reduction changes between C/H/N/O compounds are slow and kinetically controlled (see Chapter 2). In the case of fast redox reactions oxidation/reduction potentials are fixed constants.

It is very important to realise that if many processes share a component in reaction pathways or controls then this is readily achieved if the binding of that component is at equilibrium with closely the same binding constant at sites within the different paths and controls. This is often true of components such as metal ions, coenzymes and energy-distributing molecules such as ATP.

Kinetic constraints over flow are, however, the major consideration in biological reaction systems. This is especially so since the cells are both energy requiring (synthesising) and energy giving at the same time. They must take material, H/C/N/O atoms, in particular, and transfer them from one molecular condition to another in one set of catalysed reactions (synthesis) while decomposing molecules in another, all within a cell. At the same time coupling allows chemical energy to be transferred to synthesis from degradative reactions, i.e. decomposition, as well as initial input from light. We have then the balance of flow

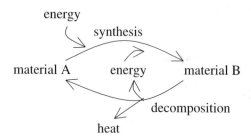

where equilibrium is avoided by using kinetic reaction barriers, and reactions must be controlled to give a sufficiency of both material and energy to synthesis. The energy in the centre of this diagram is due to reactions such as glycolysis, the breakdown of sugars to pyruvate inside cells, which generates some heat but also the synthesis of some pyrophosphate derivatives (such as ATP) (see also mitochondria in Chapter 7). The pyrophosphate derivatives are then used to further organic synthesis largely by removing water arising from condensation. Although the energy in the centre is from a different source than the initial supply, light, it is indirectly derived from it. It is important to see that this breakdown is in an extra context, since while light produces sugars the breakdown of sugars gives rise to aldehydes and ketones to which nitrogen, ammonia, is readily bound leading to the major biopolymers, proteins and nucleotides, driven by pyrophosphate. We shall be examining this scheme in more detail later but make two points here: (1) the two reactions, synthesis and degradation, are best separated by control over them by two different catalysts, enzymes, but (2) because both synthesis and decomposition separately must be correlated to give a variety of products via numerous pathways, the control of synthesis and decomposition requires exchange between all pathways employing *carriers* of material, energy and information. We shall see how cells manage this by using not just different catalysts, enzymes, but also *different carriers of material, energy and information in synthesis from those in decomposition*. The two can then also become differently controlled and energised though they remain linked. (In fact, as stated the binding to a multitude of control sites to manage the whole is readily achieved if these bindings for each control component are all similar and at mutual equilibrium; see free metal ions, mobile coenzymes and ATP (energy) binding in cells described in Chapters 4 and 5.) Remember that the catalysts, enzymes, are products of the reactions and are also controlled by them.

3.15. Informed Cellular Systems

From the above analysis we can derive the following relatively simple diagram:

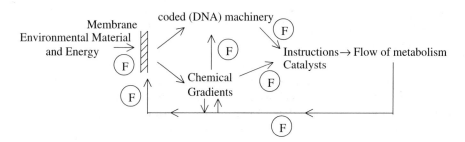

The whole is a single cross-informed system, and ⒻF is feedback or feedforward information to increase or reduce flow. Thus we see the six ways in which biosynthesising systems can evolve using informed action:

(1) by increasing absorption of energy;
(2) by utilising new materials of the environment, especially those produced as waste by the initial system and, maybe of necessity, in new compartments;
(3) by creating new catalysts especially in an autocatalytic system;
(4) by devising increasingly effective feedback controls and especially through modified coded controls, in one case changing reproduction codes, and through DNA machinery changes;
(5) by utilising space more effectively; and
(6) by devising new suitable message systems.

We shall return to the ways the environment and DNA interact in Chapter 4 and frequently in later chapters. Information connected to DNA is far from simple and not the only kind and we must always remember that equilibria as well as information, internal or external, are different limitations on the way a flow system can behave.

3.16. Ways of Looking at Ecological Chemical Systems: Summary

The way a biological chemical system can be inspected is open to separation into parts, but on re-assembly the parts are able to create unsuspected (emergent) properties. Taking the parts first, conventional textbooks describe biomolecules and their assemblies, their composition, structure and properties (see Chapters 2 and 4). These features of chemicals are time-independent, including here molecular sequences used in describing evolution. The energy content of the individual molecules is then described by sums of potential energies of bonds at the absolute zero of temperature. The molecules are metastable in the aqueous system. The description switches in kind when temperature and pressure are applied to the chemicals since here it is bulk properties of large assemblies of molecules that are examined. Temperature and pressure (concentrations) are, of course, involved in thermodynamic equilibrium and rate concepts. Equilibrium considerations allow balanced conditions and through transition state theory, downhill rates of change of energy-rich materials in many pathways to be treated. Here statistical (entropy) procedures can be used in theoretical discussion of equilibrium thermodynamics as well as to the kinetics of change. All of these studies are then based on *random distribution* of energy in space and time in an ensemble of molecules and in the absence of fields. Here we are not concerned with the overall limited structural form of the system. These considerations are important in inorganic aspects of cell biology, and especially within much of the environment, as we shall see later. However, equilibrated or downhill systems running to equilibrium cannot evolve in complexity, in contrast with cells, organisms and ecosystems, which are open, out-of-equilibrium, energetically uphill systems, not largely subject to such constraints, and which evolve. This forces us to adopt a quite different approach to biological systems in an environment.

We have made the overall statement that all systems that change increase *thermal* disorder or rates of production of it, e.g. by absorbing higher frequency energy and passing it through material before releasing it as heat. This is not contradicted by the statement that part or parts of the global system may become more organised, even ordered, in their flow while greater overall thermal disorder is created continuously. These flow systems may have "self"-created form which accounts for a way of looking at evolution through morphological features of organisms. This is substantially different from the view expressed by Schrödinger in his seminal book *"What is Life"* (see Further Reading).

Turning from just considerations of material components then we look next in a little more detail at energy. Missing usually from courses in chemistry or biochemistry is the description of energy applied to systems in bulk fields. The study in these courses of such fields, such as gravitational and electric (perhaps we should also add magnetic) fields, is usually confined to static, not to dynamic situations in which objects flow. On the whole, Newtonian dynamics, e.g. the description of planetary motion, and simple electric current phenomena, e.g. Ohm's law, are usually taught in physics, but they are also central, although not always revealed as such, to teaching in chemistry, e.g. in the examination of electrons in fixed orbits. While chemistry courses go deeply into the energy associated with such electron motion and the related electronic vectorial momentum (apparently a form of perpetual motion as here the movement of electrons around a nucleus does not lose energy), they usually leave on one side bulk momentum, simple flow, and its application in bulk chemical dynamic properties. When an approach is made to biological systems, for example, by scientists trained in the above manner, consideration is given overwhelmingly to the static (equilibrium) energies, molecular structures and downhill rates of change in isotropic systems. This makes it impossible to understand the appearance of many formed features around us such as external water flows quite apart from living organisms. The biological systems, like clouds, arise from organised non-isotropic flow of material and energy, while both are transformed, and have to be treated quite differently by the dynamics of energised flow systems. That is why we started our description of systems from clouds, which is illustrative of simple abiotic guided flow. Within energised flow there can arise also new chemical forms, such as the ozone layer. While equilibria are relatively easily described by classical thermodynamics, the non-equilibrium steady-state flow systems and their shapes are not, but this cannot be an excuse for putting to one side the whole of their functional dynamics for they are the basis of life and much of industry. In this book our concern is with the flow connected to and within biological systems, that is complex flows of many components, which are coordinated but it is clear that the analysis applies also to mankind's industrial activity as well as to some abiotic flows on Earth. Therefore, if we are to understand and manage life on Earth, it is necessary to study flow. Very importantly, in all cases it will evolve and will attempt to become completely cyclical, to avoid self-damage via waste. The cyclic condition is final when, as in equilibrium conditions, it can evolve no further. Such a cyclic condition would be an end point of evolution, a fixed maintained optimal system with fixed shape or form. We have described this condition as one of optimal retention of energy and rate of thermal entropy production within a flow system. A system can only change if energy or material input to it changes.

Now we have also shown that in all complicated cases of many flows they are, by the requisite nature of efficiency (for survival), constrained, autocatalysed, and informed in the optimal state. We have connected the constraints to the overall description of the direct or indirect interactions between flowing components, which relates to passage of energy or material in space and time. For the form of many different flows to become fixed and where flows share energy and chemical components it is essential that information from the status of each path passes between them. This means that the sender of information requires storage of messengers and energy while the receiver also uses energy in interpretation. We have stressed that the complexity of the flow systems does not allow us to treat each and every observed case, species, individually, but we can describe in general terms the classes of species, "chemotypes", and their evolution which, as we shall explain, are systematic, causative, and not random in their relationships.

In the next chapter we shall show that while the general chemical principles analysed in Chapter 2 and in this chapter apply to all the elements and their compounds, the biological systems we know about are restricted by particular physical/chemical conditions, e.g. energy supply and a limited number of elements available to them from the environment. The elements then form a very special set of autocatalysed controlled compounds, i.e. components, in very restricted pathways. After Chapter 4, which will outline this essential limited biochemistry common to all organisms, we shall show that the very primitive earliest biological cellular system *was forced to develop towards today's conditions of organisms by self-generated (organism-produced) changes of the availability of some elements in the environment in a systematic sequence* (Chapters 5–9), but that the basic components may well be an optimal set (see Chapter 4). This sequence is based on chemical necessity since biological activity is reductive and hence oxidation of the environment was inevitable. The environment then back-interacts with the organisms which adapt. This is the general cause of evolution. At the same time we must always be aware of the sources and ways of increased capture of energy and its degradation and the way they too changed, and this will also be a topic in Chapters 5–9. The geochemical or environmental availability limitations upon the ecosystem do not apply directly to mankind's laboratory and industrial chemistry, nor does the limitation of a code, and we shall return to a wider exploration of energised chemical flow systems in Chapters 10 and 11, much although this is an evolved consequence. Evolution in the broadest (our) sense is not limited by activities such as those seen in cells, nor by a biological code and neither by cognitive as opposed to metabolic activity. It is however knitted together by thermodynamic necessities. At each stage of *biological evolution* the energised cells explore the way towards an end point – a fully cyclic system with optimal absorption of energy utilising whatever materials and energy they and the changing environment provide. We shall see that the driven nature of this evolution required more and more use of elements in *more and more compartments* and more and more application of novel information transfer, imposed message systems, acting especially on catalysts and the code. The compartments will be shown to increase, not only internally in cells but also eventually externally of necessity. All of this development includes much of the activities of mankind. Now biological systems have the peculiarity that they developed one set of coded molecules which greatly assist survival and reproducibility in all forms of life, but the coded molecules (and the associated machinery) then had to evolve though they resisted evolution. As stated, by definition a code is essentially conservative and we must enquire how it was forced to change. In a chemical sense biological evolution, and genetic control, had to follow slowly the faster directed, though uncontrolled, inorganic changes of the environment as it changed towards equilibrium. The whole system of organisms therefore evolved in one ecosystem (Figure 3.14 and see cover) organisms following behind that of the Earth's geochemical surface changes (see Sections 1.5 to 1.12).

In order to appreciate the interlocking of cell chemistry with the environment we need to look particularly at the nature of the components of biological systems separately much as we look at the inorganic chemistry of the environment in Chapter 1. We shall do so by examining at first their non-metal and then their metal compounds as in Chapter 2 in order to simplify description before putting the two together. These component units will then be examined as they changed in evolution (Chapters 5–10), but there we must maintain inclusion of the energetics as in this chapter. All of the evolution has therefore to be taken in the context of thermodynamics within chemistry and we shall adopt a thermodynamic name for the groups of particular organisms ("Chemotypes"), which contain special elements either in a particular limited number of selected oxidation states, and/or in particular compartments and/or in selected concentration conditions, and/or in manners of organisation. All through later chapters the chemical principles of Chapter 2 as well as the thermodynamics in this chapter must be kept in mind.

Appendix 3A. A Note on Equilibrium Thermodynamics and Equilibrium Constants

Throughout this book the term energy is used in a rather flexible way, often in generic terms but in several, more specific, circumstances referring to the available, useful or usable energy, which is a quantitative description of that part of the total energy stored in a given system that can be used to produce work. There are two parts of the energy which we separate: that which is independent of temperature (U or H) and that which is related to disorder and is temperature-dependent, S (see Section 3.3).

The available energy (A) is defined as the difference between the total energy (U or $H = U + pV$) and the product of the absolute temperature (T) by the entropy (S), which measures the amount of the thermal energy that cannot be used at that temperature to produce work:

$$A = H - TS$$

This expression reduces to $G = H - TS$ for systems at constant pressure, where the state function G is Gibbs' *free* energy, a particular case of the function *availability*, A. A similar expression can be derived for systems at constant volume, when the Gibbs' free energy (more commonly used in "controlled mass" chemical processes) is replaced by Helmholz's free energy (less used in common chemical processes but of necessity in "controlled volume" processes, e.g. flow systems through a reactor in engineering).

From changes in free energy in standard reference conditions it is possible to calculate equilibrium constants for reactions involving several reactants and products. Consider, for example, the chemical reaction $aA + bB = cC + dD$ at equilibrium in solution. For this reaction we can define a "stoichiometric" equilibrium constant in terms of the concentrations of the reactants and products as

$$K_c = [C]^c[D]^d / [A]^a[B]^b$$

The change in Gibbs's free energy (defined, as stated, for a system at constant pressure and temperature) when a moles of A and b moles of B are converted into c moles of C and d moles of D is $\Delta G = c\mu_C + d\mu_D - a\mu_A - b\mu_B$, where μ represent "chemical potentials", i.e. "free energy per mole" of the substances, defined in terms of concentration of a component X as

$$\mu_X = \mu_X^o + RT \ln[X]$$

where μ^o is the chemical potential in a standard reference system ($T = 298\,K$, $p = 1$ atm, and the concentration is 1 M).

At equilibrium there is no free energy change, $\Delta G = 0$, hence $c\mu_C^o + cRT \ln[C] + d\mu_D^o + d\,RT \ln[D] - a\mu_A^o - a\,RT \ln[A] - b\mu_B^o - b\,RT \ln[B] = 0$
or

$$RT \ln ([C]^c[D]^d / [A]^a[B]^b) = a\mu_A^o + b\mu_B^o - c\mu_C^o - d\mu_D^o = -\Delta G^o$$

Hence

$$\Delta G^o = -RT \ln ([C]^c[D]^d / [A]^a[B]^b) \text{ or } \Delta G^o = -RT \ln K_c$$

Since ΔG^o can be calculated from the values of the chemical potentials of A, B, C, D, in the standard reference state (given in tables), the stoichiometric equilibrium constant K_c can be calculated. (More accurately we ought to use "activities" instead of "concentrations" to take into account the ionic strength of the solution; this can be done introducing the corresponding correction factors, but in dilute solutions this correction is normally not necessary – the activities are practically equal to the concentrations and K_c is then a true thermodynamic constant).

Now the change in free energy for the reaction above is related to the changes in enthalpy (H) and entropy (S) in the same reaction by the expression $\Delta G = AH - T\Delta S$ ($\Delta G^o = \Delta H^o - T\Delta S^o$ for the standard reference state).

Remember that G, H and S are all thermodynamic *functions of state*, i.e. they depend only on the initial and final states of the system, not on the ways the last is reached. As we have seen, for $\Delta G = 0$ the reaction has reached equilibrium (and in *isolated* systems ΔS has reached a maximum). If $\Delta G < 0$ the reaction was spontaneous, but if $\Delta G > 0$ the reaction could not have taken place unless energy was provided from other coupled source. If the source is external then the system is not isolated; it is *closed* if there is no exchange of material or *open* if there is such exchange. In both cases the environmental changes must be taken into account.

If this is done it can be demonstrated that

$$-\Delta G = T_o\Delta S_{total}$$

where T_o is the temperature of the (large) environment and the subscript *total* refers to the sum of the entropy change in the system considered and the environment.

This equation, Keenan's equation, can also be used for Helmholtz's free energy. It clearly shows that we can discuss any process in terms of the available, usable, useful or free energy, effectively used, rather than in terms of entropy changes times the temperature, which correspond to the left overs of such process.

A final observation is in order: the quantitative application of the equilibrium thermodynamical formalism to living systems and especially to ecosystems is generally inadequate since they are complex in their organisation, involving many interactions and feedback loops, several hierarchical levels may have to be considered, and the sources and types of energy involved can be multiple. Furthermore, they are *out-of-equilibrium open flow systems* and need to be maintained in such condition since equilibrium is death. Leaving aside very simple cases, in the present state of the art we are, therefore, limited to general semiquantitative statements or descriptions (e.g. ecosystem "narratives").

Further Reading

BOOKS

Baird, C. (1995). *Environmental Chemistry*. W.H. Freeman and Company, New York
Craig, N.C. (1992). *Entropy Analysis*. VCH Publishers, New York
Thompson, D. (1961). *On Shape and Form*. Cambridge University Press, Cambridge, (1st abridged edition, 1966)
Denbigh, K. (1966). *The Principles of Chemical Equilibrium* (2nd ed.). Cambridge University Press, Cambridge
Eigen, M. (1992). *Steps Towards Life – A Perspective on Evolution*. Oxford University Press, Oxford
Hamblyn, R. (2001). *The Invention of Clouds*. Picador Press, London
Kline, S.J. (1999). *The Low Down on Entropy and Interpretative Thermodynamics*. DCW Industries, Inc., Palm Drive, CA
Küppers, B (1990). *Information and the Origin of Life*. MIT Press, Cambridge, MA
Price, G. (1998). *Thermodynamics of Chemical Processes*. Oxford University Primers no. 56. Oxford University Press, Oxford
Prigogine, I. and Stengers, I. (1984). *Order Out of Chaos*. Bantam Press, New York
Schneider, E.D. and Sagan, D. (2005). *Into the Cool-Energy Flow, Thermodynamics and Life*. The University of Chicago Press, Chicago
Schrödinger, E. (1967). *What is Life*. Cambridge University Press, Cambridge (1st edn., 1944)
Trefil, J. (1987). *Meditations at Sunset: A Scientist looks at the Sky*. Charles Scribner's Sons, New York
Ulanowicz, R.E. (1997). *Ecology, the Ascendent Perspective*. Columbia University Press, New York
Weber, B.H., Depew, D.J. and Smith, J.D. (eds.) (1988). *Entropy, Information and Evolution*. MIT Press, Cambridge, MA
Wicken, J.S. (1987). *Evolution, Thermodynamics and Information*. Oxford University Press, New York
Williams, R.J.P. and Fraústo da Silva, J.J.R. (1996). *The Natural Selection of the Chemical Elements – The Environment and Life's Chemistry*. Clarendon Press, Oxford
Williams, R.J.P. and Fraústo da Silva, J.J.R. (1999). *Bringing Chemistry to Life – From Matter to Man*. Oxford University Press, Oxford

PAPERS

Bitbol, M. and Luisi, P.L. (2004). Autopoesis with or without cognition: defining life at its edge. *J.R. Soc. (London) Interface*, *I*, 99–107

Corning, P.A. (2001). Control information: the missing element in Norbert Wiener's cybernetic paradigm. *Kynetics*, *30*, 1272–1288

Corning, P.A. and Kline, S.J. (1998). Thermodynamics and life revisited. Parts I and II. *Syst. Res. Behav. Sci.*, *15*, 273–295 and 453–482 (Lucid discussion of many questions related to energy, entropy, information and evolution, a critical analysis of the different points of view and vast bibliography)

Corning, P.A. (2002). Thermoeconomics: beyond the second law. *J. Bioecon.*, *4* (1), 57–99 (A follow-up of the previous reference.)

Jorgensen, S.E., Patten, B.C. and Straskaba, M. Ecosystems emerging: towards an ecology of complex systems in a complex future. *Ecol. Model.*, Introduction *62*, 1–27 (1992), (1) Conservation *96*, 221–284 (1997), (2) Distribution *117*, 3–39, (3) Openness *117*, 111–164, (4) Growth *126*, 249–284

Kay, J.J. (2000). Ecosystems as self-organising holarchic open systems: narratives and the second law of thermodynamics. In S.E. Jorgensen and F. Mukker (eds.), *Handbook of Ecosystems Theories and Management*. CRC Press, Boca Raton, pp. 135–160

Keenan, J.H. (1951). Availability and irreversibility in thermodynamics. *Br. J. Appl. Phys.*, 2, 183–192

Mossel, E. and Steel, M. (2005). Random biochemical networks: the probability of self-sustaining auto-catalysis. *J. Theor. Biol.*, *233*, 327–336

Schneider, E.D. and Kay, J.J. (1994). Life as a manifestation of the second law of thermodynamics. *Math. Comput. Model.*, *19* (6–8), 25–48

Schneider, E.D. and Kay, J.J. (1994). Complexity and thermodynamics: toward a new ecology. *Futures*, *24* (6), 626–647

Schneider, E.D. and Kay, J.J. (1995). Order from disorder: the thermodynamics of complexity in biology. In M.P. Murphy and L.A.J. O'Neill (eds.), *What is Life – The Next Fifty Years*. Cambridge University Press, Cambridge, pp. 161–172

Skär, J. and Coveney, P.V. (eds.) (2003). Self-organisation: the quest for the origin in evolution of structure. *Philos. Trans. R. Soc.*, *361*, 1045–1317

Ulanowicz, R.E. and Hannon, B.M. (1987). Life and the production of entropy. *Proc. R. Soc. Lond.* B 232, pp. 181–192

Weber, B.H. (1998). Emergence of life and biological selection from the perspective of complex systems dynamics. In Van de Vigver *et al.* (eds.). *Evolutionary Systems*. Kluwer Academic Publishers, pp. 59–66

Weber, B.H. and Deacon (2000). Thermodynamic cycles, development systems and emergence. *Cybern. Hum. Know.*, *7* (1), 21–43

Weber, B.H., Depew, D.J., Dyke, C. Salthe, S.N., Schneider, E.D. and Wicken, J.J. (1989). Evolution in thermodynamic perspectives: an ecological approach. *Biol. Philos.* *4*, 373–405

Williams, R.J.P. and Fraústo da Silva, J.J.R. (2002). The systems approach to evolution. *Biochem. Biophys. Res. Commun.*, *297*, 689–699

Williams, R.J.P. and Fraústo da Silva, J.J.R. (2004). The trinity of life – the genome, the proteome and the mineral elements. *J. Chem. Edu.*, *81*, 738–749

CHAPTER 4

Outline of Biological Chemical Principles: Components, Pathways and Controls

4.1. Introduction

In Chapter 1 we have given an outline of the geochemistry of the Earth and its evolution in chemical terms. This is purely a limited inorganic chemistry at or moving towards equilibrium in the vast majority of its surface activities. The Earth is irradiated by the Sun's energy, but this has had little *direct* effect on *geochemical inorganic activity* except through the cycle of water and its consequences and later the creation of the ozone layer. It has, however, contributed to an approximately fixed temperature on Earth's surface even while some gases were being lost continuously (Table 1.1). There are then very few minerals in energised states on Earth's surface due to the direct effect of the Sun's radiation. Some energised minerals have been thrown up mainly by volcanic activity and they were one possible immediate source of the required energy for the first forms of life. However, the Sun's radiation has *indirectly* affected slowly and considerably Earth's surface chemistry through the chemical changes produced in and by living organisms, which is mainly "cellular" *organic* compounds with the slow release of dioxygen giving products of its reactions in the environment. There has also been the deposition, on death of organisms, of debris such as organic compounds in soils, and as gas, coal and oil. We described the environmental chemical changes from the primitive to the present day times in Section 1.14. Before we explained the origin of dioxygen and organic compounds in Chapters 2 and 3 we needed to discuss in brief the important chemical principles of non-equilibrium states of energised chemicals, pointing out that organic chemicals together with oxygen fall into this category. The organic chemicals were incorporated into cells together with many inorganic environmental elements generating the combination which produced life. To explain the differences between equilibrium and non-equilibrium states we also described in Chapter 3 the way energy interacted with matter to produce both stationary and *flowing energised chemicals*, that is, energised with respect to the original environment and even more so with respect to any changed environment produced by the flow. The major flow system for producing any such chemicals remains that of living organisms. In fact, they were the only such organic systems known until mankind developed industry. We stressed that we shall examine the surface geological and biological chemistry open to light and the atmosphere in this and in Chapters 5–10, and only in Chapter 11 do we refer again to other zones of the Earth.

Now in both Chapters 1 (on the environment) and 2 (on the major chemical reactions), the flows examined were the one-way (downhill energetically) transformations of energised chemicals in the direction of equilibrium though at intermediate steps material could be trapped.

$$A + B + C \text{ (etc.)} \rightarrow X + Y + Z \text{ (etc.)} + \text{heat}$$

These flows of chemicals can be in a steady-state condition, provided the reactants $A + B + C$ (etc.) are part of a very large resource and the reaction is relatively slow. Energy, as applied here, that is even in laboratory organic chemistry, generally acted so as to increase the rate of reaction by increase in temperature. In Chapter 3, we also introduced *uphill steady states of flowing chemicals taking up energy* and these are a major part of the combination of interactive geochemistry and the biological chemistry of organisms. It is the biological chemistry of particular components of cells and their interactions, produced by this energisation, that are to be described as an outline in this chapter. *We shall be dealing here with components of biological chemistry as a whole* and we shall not refer to any particular groups of organisms or to the divisions of space in them, i.e. compartments, which are so much a part of evolution, leaving those matters to Chapters 5–9. Much of the chapter is therefore about the functions of particular units of cells and not of organised activity and is close in content to traditional molecular biology. We shall ask repeatedly about the suitability, i.e. fitness, of individual types of unit, nucleotides, proteins, metal ions and so on for the tasks they perform. We leave on one side largely geological chemistry much though this is a partner in life's evolution. From the outset we must stress that in contrast to the general chemistry described in Chapters 1–3, we have to see that biological chemical systems can only use geologically available elements and energy from two sources, the Earth and the Sun, and not the full ranges of either chemicals or energy resources open today for use by mankind (see Chapter 10).

To start our approach we divide this chapter into parts. In the first part we outline the general chemical element content of all organisms (Sections 4.2–4.4); next, we look at the uses of non-metal elements and their in small molecule combinations (Sections 4.5–4.8); while in the third part we extend this description to their major biopolymers (Sections 4.9–4.13). Section 4.14 is a summary of these sections. In Sections 4.15 and 4.16 we examine the metal ion content of cells and combinations of these ions with organic molecules. The final sections integrate these descriptions with those of the principles of bioenergetics outlined in Chapter 3.

4.2. Organisms: Their Classification as Thermodynamic Chemotypes

We observe immediately that all organisms are based on *controlled energised chemistry essentially in physically confined and organised flow systems*, that is, in cells. Elements are concentrated as free ions or in energised compounds in the cells. To this end, we shall describe cellular evolution as being at all times within an energised advancing environment (see Chapter 1). Now the organisms evolved (a) in chemical content and use of chemicals, (b) in the ways they obtained and used energy, (c) in the space they occupied, and (d) in their organisation. In Chapters 5–10 we shall examine evolution with a broad outlook following in order these factors, stressing the differences in the early and later prokaryotes, single-cell eukaryotes, multi-cell eukaryotes, animals with nerves, and mankind with respect to these four chemical and thermodynamic characteristics (see Table 4.2). Looking at all such organisms, either in parts or as a whole, one primary need for understanding by chemists is then their composition. This might appear to be an unreasonable approach in that every species is somewhat chemically different and especially since the conventional way of looking at organisms is by functional specialisation – for example, modern plants are stationary and photosynthetic while animals forage and digest. However, we shall find that even in this example, the plants use closely similar elements and compounds but in different proportions from those used by animals and by bacteria. Most importantly, we wish to examine if the whole of life and the development of its chemical properties, i.e. of all the organisms together, had one overall chemical compositional direction from the origin of life. By introducing this simplifying approach to chemical composition at first, we proceed to give at the end of this chapter a very general impression of the chemical thermodynamics of organisms in given environments.

Immediately we see that we cannot use the conventional ways of describing organisms, for example, by morphology (see Tables 4.1(a) and (b)), or by molecular

TABLE 4.1(a)

EXAMPLE OF CLASSICAL DIVISION OF ANIMALS

Division	Total Numbers	Examples[a]
Domain	3–5	Eukarya
Kingdom	10^2	Animalia
Phylum	10^2	Chordata
Sub-phylum	10^3	Vertebrata
Class	$>10^3$	Mammalia
Order	10^4	Primates
Family	$>10^4$	Hominidae
Genus	$>10^5$	Homo
Species	2×10^6	Homo sapiens

[a] The examples are the series starting from the bottom to which an organism belongs.

TABLE 4.1(b)

PLANTS

Division	Example	Characteristics
Kingdom	Plantae	Organisms that usually have rigid cell walls and usually possess chlorophyll
Subkingdom	Embryophyta	Plants forming embryos
Phylum	Tracheophyta	Vascular plants
Subphylum	Pterophytina	Generally large, conspicuous leaves, complex vascular system
Class	Angiospermae	Flowering plants, seed enclosed in ovary
Subclass	Dicotyledoneae	Embryo with two seed leaves
Order	Sapindales	Soapberry order consisting of a number of trees and shrubs
Family	Aceraceae	Maple family
Genus	*Acer*	Maples and box elder
Species	*Acer rubrum*	Red maple

sequences of DNA (genes), RNA or proteins. In our classification, mankind, for example, though a single species, is as different from all other organisms including animals as plants are from animals. (This difference is obvious in all four given characteristics but not by morphological or genetic analysis.) We therefore need a new label for our classification of organisms, which we choose to be that of *chemotypes* (see Table 4.2), where a chemotype differs markedly in any one of the above four characteristics, that is in the use of chemical elements, of energy in chemicals, of space distribution and of organisation of chemicals. As with some other classifications we shall only make distinctions between the major chemotypes and some considerable minor variations (see Figure 4.1). For example, in Table 4.2, we see that as we descend the columns of our classification of organisms, the use of organic chemicals in oxidised states increases and so does the use of certain elements such as zinc, copper, calcium and halogens. Additionally, later we shall see that distribution in compartments, i.e. use of space and organisation, increases systematically down the table, while all groups advance with time. Because the number of slightly different types of organisms also increases down the columns and differences inside the major groupings diminish, as they do in any conventional scheme (see Tables 4.1), it is harder to distinguish clearly between them as chemotypes and we shall not attempt to do so. A great advantage of our scheme is that we follow the chemotypes with the

TABLE 4.2

THE MAIN CHEMOTYPE GROUPS IN THIS BOOK

Major Chemotypic Groups	Novel Chemical Characterisation and Sub-groups
Prokaryotes	
Autotrophs	Energy from energised minerals
Chemoautotrophs	Energy from oxidised sources, Fe^{3+}, SO_4^{2-}, NO_3^-
Photoautotrophs	Energy from light, Mg, Mn
Aerobes	Tolerate and use O_2 from moderate to high atmospheric levels
	Use of Mg^{2+}, Fe^{2+}, Mo(W)
Eukaryotes (single cell)	Cholesterol in cell's membrane; compartments, nucleus, filaments, Ca^{2+} signals
Plant-like	Photosynthetic, sedentary
Fungal-like	Dependent on plants, sedentary
Animal-like	Dependent on digestion, mobile
	Use of Cu, Zn
Eukaryotes (multicellular)	
Plants	Sedentary photosynthetic organisms
Fungi	Filamentous, sedentary
Animals	Mobile foraging organisms with senses and later a nervous system, use of Na, K, Cl
Mankind	Multiple new uses of chemicals and energy sources. Use of external space
	Fully developed self-conscious brains

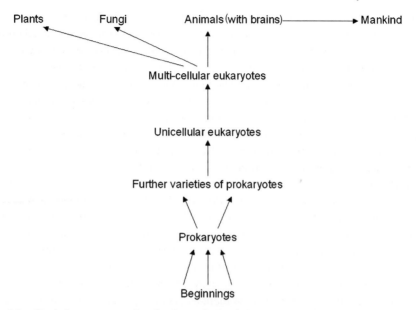

Fig. 4.1.   Evolutionary connection for the analysis of chemotypes which we shall use.

parallel but previous changes in the environment so that we can see evolution as a complete ecosystem, and not just in terms of organisms*.

Note that our classification can still be used alongside broad general considerations of changes in shape (morphology), of genes (genomes) or of proteins (proteomes) but we shall be mostly concerned with metal ion and small molecule non-metal content, the metallome and the metabolome, respectively, as they relate to chemical content changes. (The concept of *phenotypes* will only be developed in Chapter 11, when we see how individuals can arise, that is, within human beings, although there are clearly minor variations of phenotype in all the chemotypes.)

We can express the use of all the different units in evolution in the language of thermodynamics. While the genome is defined by a DNA sequence so that each base has a singular *intensive* property as in a computer code of symbols, by way of contrast, the protein content of a cell is an *extensive* property being concentration dependent and therefore varies under circumstances such as temperature and pressure although

---

*There are probably more than five million species on the Earth many of which are very similar. Moreover, as many as now exist have probably been lost. In our opinion it is not to be expected that an explanation can be given for particular species although they are connected by morphology or DNA/RNA or protein sequences. We shall turn to the problem as to why there have been and are so many species in Chapter 11. We wish to look at evolution from the point of view of very general chemical and physical principles which is the same thermodynamic approach that we used in the analysis of formation of clouds or the ozone layer in Section 3.4. We observe immediately that the major groups in Table 4.2 have all advanced and have not displaced one another.

each protein also has a fixed sequence. In this respect the RNA content is also to be likened to the protein content or proteome, to the substrate content, the metabolome, and the metal ion content, the metallome. All these concentration-dependent terms are *extensive* thermodynamic properties of a cell system via the expression of the DNA but not of the DNA itself. Expression itself is a thermodynamic concept. By using the word *chemotype*, as opposed to *genotype*, we are therefore using an all-embracing thermodynamic concept based in part upon *the concentrations of elements in the energised genome, the proteome, the metabolome and the metallome.* As stated we are then forced to describe also *the spaces* (volumes) which are under consideration, *the energy* which is put into both compounds and concentrations since many of the elements are not in equilibrium with their surroundings, as well as *the internal organisation*, and *any relationship to the environment.* It is energy use which also typifies the dynamics of the system, the organised controlled flows due to pumps, synthesis and degradation and information transfer in confined spaces. The concept of a chemotype is then very different from that of a genotype since it is environmental, material- and energy-dependent. The DNA genotype allows the definition of a species but does not fit easily with the activity of an organism, particularly, the chemical activity within an organism. Note again the comparison between mankind and gorillas which are hardly distinguishable in genotype or general physical shape. This release from the dependence on sequence information in the genotype or on morphology in the discussion of evolution is extremely important as it allows us to see that it is not chance which guides evolution but a dominant thermodynamic drive towards equilibrium via energy degradation in sophisticated chemical pathways – organisms.

4.3. Organisms: Their Generalised Element Content

The first chemical observation we wish to make is which are the available elements commonly required by all organisms, that is usually treated under their inorganic chemistry in later sections. We can then consider whether or not organisms could have arisen in other ways by the use of these or other elements. Some of the chemistry of organisms as described will be simplified and very familiar to biochemists but it is necessary to give this description for the more chemically oriented reader. In this book treatment has to stress also the inorganic chemistry of cells in order to maintain the link with the environment in our approach to the evolution of Earth's ecosystem. Inorganic chemistry is almost absent from most biochemical or biological treatises.

We start with the analysis of organisms by observing that the *cytoplasms* of cells of all organisms (all chemotypes) are reducing and energised and have closely comparable non-metal element content, that is of H, O, C, N, P and S (Table 4.3). This is not surprising since they are all 80–90% water and they all make the same polymers: proteins, nucleic acids, fats, lipids and saccharides, water being 98% of their mass. All of these components are made overwhelmingly from these six non-metal elements. Moreover, there is but one underlying basic synthetic *organic* chemistry

TABLE 4.3

CONCENTRATION (MOL L^{-1}) OF FREE IONS AND MOLECULES INSIDE THE
CYTOPLASM OF ALL CELLS AND IN THE EARLY SEA

Ion	Inside Concentration	Outside Concentration
Na$^+$	10^{-2}	$>10^{-1}$
K$^+$	10^{-1}	10^{-2}
Cl$^-$	10^{-3}	10^{-1}
Mg^{2+}	10^{-3}	10^{-2}
Ca^{2+}	$<10^{-5}$	$<10^{-2}$
Mn^{2+}	10^{-7}	10^{-6}
Fe^{2+a}	10^{-6}	10^{-6}
Ni^{2+a}	10^{-14}	$<10^{-9}$
Zn^{2+a}	(10^{-15})	$<10^{-12}$
ATP^{4-}	$\sim10^{-3}$	Zero
HPO$_4^{2-}$	$\sim10^{-3}$	$\sim10^{-5}$
HCO$_3^-$	$\sim10^{-3}$	$\sim10^{-3}$

^{a} Limited by the presence of sulfide.
Note: The inside here is the cytoplasm which has not changed to a large extent to this day and the number for heavy metal ions are estimates.

in the cytoplasm, *reduction of environmental C/N/O/S compounds followed by condensation polymerisation*, in which these elements, obtained from very simple chemicals in the environment, are largely converted into substrates and then incorporated into the above polymers containing the reduced chemical groups

$$-OH, \quad \diagdown CH_2, -CHO, -CH_2OH, -COOH, -NH_2, -SH$$

while P remains as phosphate (see Table 4.3). The requirement for H, C, N, O and P is absolute in that only they were sufficiently available and functionally adequate to make *kinetically stable reproducible polymers, soluble in water*, of the above kinds (see Chapter 2). Sulfur in such compounds was mainly needed as (1) a catalyst, (2) a carrier, and (3) a metal-binding centre (see below). Phosphate, often used in condensation reactions, is also additionally able to form mobile hinges in macromolecules and gives charge and solubility to them. Its properties help to allow controlled (informed) chemistry in molecules such as DNA, RNA and via protein phosphorylation. As pyrophosphate (e.g. in ATP), phosphorus is also the major element involved in *energy distribution*. We have explained the special chemical kinetic value of S and P relative to C/H/N/O in Section 2.18. For reasons of availability and chemistry it is difficult to see any alternative elements which could have been used in the above separate functions. We can conclude safely that all life had to be, in large quantitative part, composed of *one set of basic organic chemicals* (free or combined) in water solution which has a rational basis and that this is particularly true of the essential cytoplasm of all cells. Again containment of the cytoplasm and later separation into compartments is made by a smaller set of non-aqueous, mainly C/H compounds, lipids in membranes, in all cases. Clearly, the synthesis of all these organic compounds, started from $CO(CO_2)$, $NH_3(HCN)$, H_2S and HPO_4^{2-} and, after reduction, required energy in one-way reactions mostly introduced via ATP. *Is this the only way polymers of life of precise sequence with the necessary functions can be made in water?*

Before we explore these organic chemicals further we note that no life form (chemotype) has been found without several basic essential *inorganic* elements: Na, K, Ca, Mg, Cl, Mn, Fe, W(Mo) and Se (see Table 4.3 and Figure 4.2). In fact, they are all present in virtually all if not all life in rather similar *free* amounts in the cytoplasm (see Figure 4.3), though in different amounts in compartments within later organisms (see Chapters 6–10) and in different combinations generally. Note that some elements, notably Na, Cl and Ca (perhaps Mn), were initially rejected to a large degree (Table 4.3). There may be one or two other elements we should add to the initially required group, e.g. some V, Zn, Ni and Co to a total of at least 15 and at the most 20 elements. Some of these additional elements, e.g. Ni and Co, have been much used in early life but are now less employed; others are more used now, e.g. V and Zn, or even introduced later, e.g. Cu and I. There are some other additional elements such as Si, B, Sr, and Ba which only have special value in particular later chemotypes. We can safely conclude that the cytoplasm of all organisms needs one minimum set of inorganic elements in almost fixed free concentrations (Section 4.15), as well as one set of organic chemicals (see Sections 4.8–4.14). The reason for *the fixing of the energised concentration* of many of these chemicals in particular compartments must also be examined since we believe that

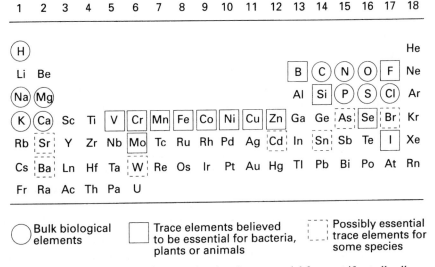

FIG. 4.2.   Periodic Table of the elements showing those essential for most if not all cells.

the nature of life and its later evolution can only be understood if we appreciate the *quantitative* roles of these 15–20 elements in a system, an approach quite different from that based upon structural properties of isolated molecules such as DNA, RNA or proteins. Again we must always remember that they are all in an energised system of *organised flow*. Note that the above description is in keeping with our simplifying assumption in this chapter that all life has one underlying chemical thermodynamic principle.

In the above list there are some elements which either have been or are now commonly available on Earth and which were largely missing from the cytoplasm of very early cells, notably Ca, Co, Mn, Cu, Mo and Zn. We can explain this in the following way. When Earth formed, Cu and Zn did not condense as metals with Fe, Ni and some Co, and could not form oxides as all oxygen was bound to other elements (see Chapter 1). At these very early times any Cu, Zn and Mo ions on the surface were in fact removed as water-insoluble sulfides. They were not readily available then and were not thrown up in available forms or amounts by early volcanic action. It may have been the case too that the more available tungsten made use of the chemically similar molybdenum, in an H_2S solution, of low priority at first (see the Periodic Table, Figure 4.2 and Figure 2.2). Of the other largely missing metals the least abundant in the Fe, Ni, Co central core is Co being only 1% of the iron of Earth. (Note that cobalt has an odd atomic number.) Cobalt is as useful as iron or nickel in mankind's catalysts, especially in organo-metallic chemistry (Section 2.10), but its scarcity (see Figure 1.1), may have meant that it was very little used in organisms until it appeared later incorporated into coenzyme B_{12} in low concentration (see Chapter 5). The fact that Ni was used in very early organisms despite its insoluble sulfide may indicate that life started in somewhat *acidic* reductive conditions which would have made its sulfide more soluble. The case of Mn is interesting in a different way since although it is moderately available as Mn^{2+}, this ion does not bind strongly to organic side chains containing either N- or S-donors or to sulfide as do Fe and Ni; in other words, there may have been no sufficiently strong donors for it in the cytoplasm. There is, in fact, little evidence for its involvement in the *cytoplasm* of the most primitive or even of recent cells. It is a large ion, not too unlike calcium, and may be it was rejected with Ca^{2+}, by cells since large cations precipitate organic anions too readily and are damaging to DNA. In addition to the above elements, Al and Si were always available, but they have very limited possibilities to react with organic material mainly because Al is only available at 10^{-13} M in water at pH 7, and silicon, although available at 10^{-3} M as $Si(OH)_4$, is not able to form kinetically stable compounds in water with carbon compounds. Silicon, as amorphous silica, is used later in many external minerals of diatoms, plants and animals. Again aluminium and indeed all *metal* elements in Groups 3, 4, 13 and 14 are virtually absent from all cells to this day since, as well as forming insoluble compounds, their ions tend to block organic catalysed reactions. Finally, note that Na and Cl though present in the cytoplasm are in much reduced concentration relative to the sea. A summarised general idea of cell element content is given in Figure 4.3 and Table 4.3. Note how we have linked biochemical and geochemical features together in this paragraph and we do the same in the next.

In order to reach the concentrations shown in the figure, through uptake and rejection of the ions, energy, again mostly from ATP, was needed by cells mainly by application of selective cellular pumps. Applying our knowledge of inorganic and geological chemistry to the possible beginnings of life it would appear that one possibility is that the highly energised state of the reduced mantle and crust of iron and nickel and their sulfides provided the major initial source of energy for pumping and for reductive catalysts (see Wächtershauser in Further Reading). Some of these Ni/Fe sulfides were not at equilibrium with the environment. We indicated the way their energy could be applied and utilised in Section 3.12. The subsequent entrapment of these two metals, Fe and Ni, with sulfur derivatives bound to carbon compounds in an enclosed compartment, could have led to repeated (not initially reproductive) catalysed and energised reductive chemistry which could well be somewhat different from one compartment, vesicle, to another on Earth's surface. The vesicles could then have arisen from the general reduction of carbon to give the oils and lipids of membranes. We do not rule out the possibility that the original mineral Ni/Fe compounds themselves were part of the essential physical compartment membrane (see Martin and Russell in Further Reading), and we do not exclude light from consideration though we believe it came to be used later. There are other possible inorganic sources of energy from gases such as H_2, H_2S and CH_4 and they cannot be disregarded (see Pace in Further Reading).

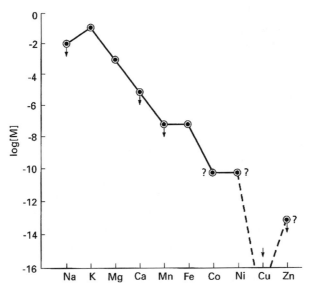

FIG. 4.3.   An approximate profile of the free element content of the cytoplasm of all organisms – the free metallome. Downward arrows show outward pumps. Note how closely the sequence follows the inverse of the Irving–Williams binding constant sequence (see Appendix 4A).

The number of coincidental physical and geochemical thermodynamic constraints on the elements on the Earth which allowed life to start, and to continue as chemical flows, is therefore considerable. We summarise them as follows:

(1) The sizes and nature of the Earth and the Sun, Earth's element contents in zones over Earth's history (Chapter 1), and their physical developments (Table 1.1).

(2) The distance of the Earth from the Sun.

(3) The annual and daily rotation patterns of the Earth.

Furthermore,

(4) As a result of (1) (2) and (3), the closely fixed temperature of the Earth for 4.5×10^9 years, assisted by an atmosphere, allowed water to form and remain as a liquid on the surface as the sea. Both the atmosphere and the sea were restricted in element composition (see Chapter 1).

(5) The somewhat reduced state of the original chemical components of the Earth led initially to the availability of only certain elements in the sea, the atmosphere, and from the crust. Equilibrium was approached between the contents of the sea, the atmosphere, and the surface it contacted.

(6) Selected energised elements from the crust and the mantle in compounds were ejected in volcanoes or ocean's "black smokers", including reduced metal sulfides. They could have led to the initial catalysed and energised, carbon-based, reductive chemical synthesis, using a fixed pattern of other elements in selected amounts. Likewise, certain reducing gases could have assisted the start of life; note particularly the presence of (CH_4), NH_3, HCN, H_2O, H_2S and H_2Se as well as the two oxides of carbon, CO and CO_2. The assumption is that perhaps light was not the initial source of energy for life since light needs transduction to become usable chemical energy. The formation of spatially confined volumes was essential and arose due to production of hydrocarbon chains in lipids.

(7) The formation of a compartment and the trapping of energy in gradients of elements, in chemicals, in their spatial confinement followed. This, like the items under (4)–(6), was inevitable and created life. The confinement limited diffusion and otherwise unavoidable dispersion while controlling flow. Surfaces may have been used as part of the compartment traps, but the main feature was the production of oily lipid membranes (see (6) above and Segré *et al.* in Further Reading).

(8) The composition of living cells is related to the availability and usefulness of 15–20 elements (see Figure 4.3) and, as we shall see, changes in availability profoundly influenced evolution of chemotypes. However, it is probable that the collection of elements in life at any one time is a unique set.

(9) The absorption of energy into the synthesis of *reduced* chemicals in cells inevitably produced an ever-more *oxidised environment* which is not only the source of material evolution in life but also with respect to the chemicals in cells becomes a huge energy store.

(10) There is one general way of introducing chemical energy into synthesis and use of it in pumps which is to transduce any environmental source to pyrophosphate. Once again is this a uniquely suitable route?

TABLE 4.4

COMPONENTS IN A TYPICAL SIMPLE (BACTERIAL) CELL

Component	Numbers	Variety
Proteins (proteome)	2.5×10^6	3,000
DNA (genome)	1(2)	1
m-RNA	4,000	2,000
Ribosomes	$>10,000$	2
Peptidoglycan	1	1 (wall)
Metabolites (metabolome)	3×10^8	10^3
Lipid	2×10^6	2?
Water	5×10^{10}	1
Metal ions (metallome)	10^7	12

Note: See Harold in Further Reading for data on E. coli.

The probability of life emerging is, of course, unknown and perhaps unknowable. It is extremely unlikely in a very local part of the universe but may be high overall. Any judgement here is exceedingly speculative. The safe starting point is that the life we know on the Earth is based on the above particular mix of non-metal, in organic chemicals, and metal elements which arose by the very nature of the Earth, but this mix would be inadequate unless energised, limited in space and with its reactions organised (see Section 3.7). Given the changing conditions of Earth's chemistry and that it was energised later by the Sun (Chapter 1), and our general hypothesis of how energy-capturing systems change (Section 3.8), we have to consider how life would progress towards optimal energy capture. First, we must look at the value of the original available elements which were taken into or rejected by organisms. Afterwards, we shall look at the way the originally available elements could form the essential compounds we observe in all life (Table 4.4); then we must ask about organisation. (see Section 4.20).

4.4. The Functional Value of the Elements in Organisms: Introduction to Biological Compounds

We turn now to the essential ways the elements are used in biological chemistry. Rather than taking all the elements together, we shall look at two sets of them separately – the non-metals and the metals – indicating the common types of compound in which they are found in all cells. This separation makes a clear connection to the description of the chemistry of the elements in Chapter 2. We shall see that their conditions are energised in different ways. We shall look at these compounds and their use in cells while asking whether or not there is likely to be equilibrium between their binding sites within any aqueous compartment, also across membranes and even with their small ion or molecular states as found in the environment. This is a considerable concern in thermodynamics (see Section 3.14). The parallels with the different chemistry of non-metals and metals (in organic and inorganic chemistry) as introduced in Chapter 2 will be clear, but as stated there we shall have to bring these chemistries together before the end of the chapter.

The compounds of a cell are often divided, as mentioned before, into the genome (DNA), the RNAs, the proteome, the metabolome and the metallome, the metallome being in part free ions and in part in combination with organic molecules (see Williams and Frausto da Silva, 2004, in the references). The first four are composed of non-equilibrated energised non-metal organic compounds. Table 4.4 gives an outline of a number of major types of synthesised molecules present in a simple bacterial cell but not their all-important concentrations which are also energised. There is much in common with all cell cytoplasms. Elements in them do not exchange readily, so the elements do not equilibrate between their sources and their compounds. Taking these non-metal compounds, we start by examining the basic synthesis and functions of them in flowing living systems, and then *ask if any other possible selection of the non-metal elements and their compounds could have been made to meet the required overall function which is to form a system of effective energy capture on the route to optimal energy retention*, including its degradation (see Section 3.9). This point concerns all of biotic chemistry. All the time we shall ask about the possibility that at first the basic reactions could have arisen spontaneously without a code no matter what advantages a code provided subsequently, since *form* can arise as described in Sections 3.4–3.9. (Note that, by definition, a code cannot arise before there is something to code, though it could arise simultaneously with that something.) However, when we describe *informed cellular* systems we must include a code (see Sections 3.9–3.12). We follow this study of non-metal chemistry with that of the individual uses of the energised analytical content of *metal elements in all cells especially through their concentration*, and of elements rejected by cells, and subsequently we describe their combined use. *Again we ask if these were the only possible elements for their selected uses.* After we have put metals and non-metals together in one system we turn, later in the chapter, to the discussion of which equilibria exist in cells. (Note again that equilibrium constants are not under any biological control.) The next step is to look at control via signalling over pathways, i.e. flows of all non-equilibrated chemicals individually and cooperatively as we build up a picture of the partially cyclic life/environment organisation which could have led to the simplest cell types – prokaryotes (see Chapter 5) before both the environment and organisms together developed new chemical systems (Chapters 6–10).

4.5. Non-Metal Chemistry and Its Basic Biological Pathways: Coding

We must give first an outline of the non-metal pathways which we observe in all cells. We start here because we know nothing about their abiotic chemistry but assume that cellular life arose from it. We shall assume that the basic requirement of all metabolism is the energised and catalysed synthesis of polysaccharides, lipids, proteins and nucleic acids. These are polymers (see Table 4.5), formed from monomers, all of which could have always arisen when energy was applied to the

TABLE 4.5

OVERALL INCORPORATION OF NON-METALS C, H, N, O, P, S

$$CO_2 \text{ (in } H_2O) \text{ or } CH_4 \xrightarrow{\text{energy}} HCHO \xrightarrow{\text{polymerisation}} [HCOH]_n$$
The polymers are polysaccharides, note reduction

$2HCHO \rightarrow CH_3CO_2H \qquad N_2 \rightarrow NH_3$
$CH_3COOH + CH_3COOH \rightarrow CH_3 \bullet CO \bullet CH_2COOH$
Acids and keto-acids $+ NH_3 \rightarrow$ amides $+$ amino acids
Condensation of amino acids $\rightarrow$ proteins

$HCN + HCHO + NH_3 \rightarrow$ purines and pyrimidines (bases)
Condensation of bases, sugars and phosphate $\rightarrow$ nucleotides
Nucleotides $\rightarrow$ RNA $+$ DNA
$HCHO + H_2 \rightarrow$ lipids
$H_2S + HCHO \rightarrow$ thiolates and thioethers

available C/H/N/O/S/P simple compounds in the environment. As indicated above, the available non-metal compounds, starting materials from the environment, were H_2 (H_2S and CH_4), CH_4, CO (CO_2), NH_3 and HCN; H_2S and HPO_4^{2-} existing in the atmosphere or in the sea. The steps of *common synthesis* of mostly energised, small molecules, the metabolome, that are easily envisaged and then related to that in the most primitive, and in all other, cells are the following (see textbooks on Biochemistry):

(1) The formation of formaldehyde HCHO by reduction, e.g. CO $+ H_2$. The condensation of formaldehyde leads to sugars such as ribose and glucose. Glycolysis from glucose then follows one degradative path in cells, with NADH as a coenzyme, leading to energy generation (ATP) (see below), while the alternative pentose shunt using NADPH is the pathway of H-transfer for reduction and synthesis.

(2) The formation of acetate $CH_3CO_2^- + H^+$ from CO_2 and CH_4. The acetyl group CH_3CO- is the original building block of other carboxylic acids, by the reverse citrate cycle (Figure 4.4), and of lipids in cells.

(3) The addition of ammonia to the variety of acids derivable from either the breakdown of glucose, glycolysis, or of the pentose shunt reaction products, ribose and NADPH, and from the citrate cycle, gives the amino acids (see Table 4.7 and Figure 4.4) Polymerisation of amino acids in cells gives proteins. In some of the amino acids sulfur and selenium can be incorporated easily. We assume NH_3 was present. (Note that Se is in a coded amino acid not in Table 4.7.) Some selective metal-binding properties can be seen in Table 4.7, but amino acid carboxylates can bind all.

(4) The polymerisation of HCN plus reactions with NH_3 and H_2O gives the nucleotide bases in cells (Figure 4.4). These reactions are followed by condensation with sugars and phosphate. We assume HCN was present.

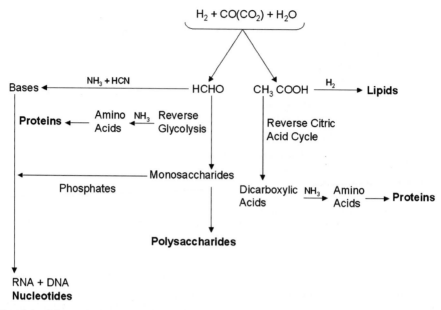

FIG. 4.4.   Non-metal material input to cells.

(5) The uptake of phosphate by condensation with its own and other –OH groups is a special step which requires energy. In fact, the large demand for phosphate by cells has limited its availability in all waters. Energy, i.e. adenosine triphosphate (ATP), is needed in parts of all other steps. A major activity of cells is the removal of the unit of water (H_2O), while driving combinations of organic chemicals by condensation. The drive is due to energy storage in polyphosphates such as ATP. Later, we shall see that the same energy source drives the uptake or rejection of metal ions, pumps, and many other mechanical devices.

Note that many products are from basic pathways or their combinations and that their syntheses need energy. We do not know how the basic units came about but we can see that the resultant major units are of considerable kinetic stability. Were there any other comparable molecules? Now the reactions also need catalysts largely derived from the combinations of organic molecules, proteins, plus metal ions (see Section 4.15), and again we cannot say how their frameworks came about but observe how several simple pathways from the same basic starting materials cooperate in forming the complicated molecules catalyses through these (see Figure 4.5 and Section 3.9). We stress that such an autocatalytic system limits the pathways. We know that in a simple prokaryote today there are only some 10^3 types of small molecule in the metabolome, and many if not all are related to these starting materials. Could a much smaller number have generated organisms? Note that some of their reactions are to

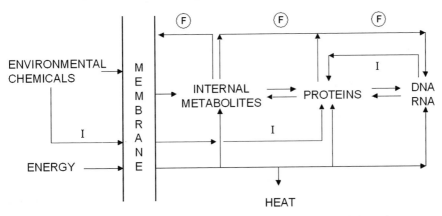

FIG. 4.5.   Basic connections in the ecosystem: F, feedback of material; I, information.

TABLE 4.6

MAINTAINED PATHWAYS THROUGHOUT EVOLUTION

Pathway	Example
Syntheses and degradation of saccharides	Carbon incorporation (Mg) Glycolysis (Mg)
Dicarboxylic acid reaction sequence	CO_2 incorporation in incomplete cycle
	Completed later to give energy capture,
	i.e. Krebs cycle (Fe, Mg)
The pentose shunt	Reducing equivalents from sugars go to
	synthesis via NADPH, see also CO_2 uptake
Nitrogen incorporation	Formation of NH_3 (Mg, Fe, V, Mo)
	in symbiotic bacteria
Hydrogen reactions	H_2 as a reductant (Fe, Ni) in anaerobic archaea
Energy: electron/proton flow	Energy capture related to ATP formation
in membranes	(Fe, NADH, flavin, quinones)
Amino-acid synthesis	Products of glycolysis and Krebs cycle + NH_3 (Fe)
Protein synthesis	Formyl initiation (Fe) and methionine initiation
	(Fe, Co) on completion
DNA, RNA, syntheses	Nucleic acid pathways (Mg, Zn, Fe/S, B_{12} or Fe_2O)
Synthesis of fats	β-carbon oxidation/reduction (flavin, Fe)
Exchange of ions in membranes	Osmotic and electrolyte control
	(Na, K, Mg, Ca, H)
Pumps (ATPases)	Most elements as cations or anions

be classified as acid/base, e.g. condensation, and some as oxidation/reduction. Figures 4.4 and 4.5 and Table 4.6 summarise the sequence of primordial series of biological reactions which are closely common to all cells but *the exact steps* of the reaction pathways in Table 4.6 are not precisely those of today. It is very important to stress that the central units of all kinds, $H_2C(OH)_2$ (hydrated formaldehyde),

$H_2N \cdot CH_2CO_2H$ (glycine), HCN and $PO_2(OH)_2$, are the simplest possible units which can give the required kinetically stable linear polymers in water by the removal of the elements of water from them. We shall see that all natural polymers are made from such simple C/H/N/O/P building blocks. We can ask again: could any other monomers have arisen and could they replace these and build an alternative set of useful polymers? Note the considerable kinetic stability of the peptide bond, the phosphate diester and the ether links of polysaccharides in water at pH 7. Later, we shall observe the reasons for the remarkable and essential catalytic properties of the proteins. The very limited number of non-metals capable of giving this required kinetic stability in combination makes it very improbable that there are in fact other possible element combinations for, or methods of synthesis of, biologically suitable polymers (see Sections 4.9–4.14). For example, it is not possible to base stable linear polymers on sulfur or silicon monomers, and notice the much greater abundance of H, C, N, and O over S and P.

We consider then that the effects of energy upon the non-metals once in a cell were inevitable eventual events leading to those products which were and are the most kinetically stable and therefore provided the best starting capability for retaining energy for a considerable period. *All the further development of cell chemistry is then to be seen as further steps toward the optimal manner in which to generate energy flux through survival strength in a system of inevitable instability and decay.*

We shall now digress for some paragraphs while we ask if we can see a possible explanation for the usefulness within complexity of the non-metal cell contents in Table 4.4 before we return to our main theme in Section 4.6. We start from the position that the above four pathways (1) + (5), (2), (3) and (4) are found without exception in all forms of life as we know them, but it may well be that the four developed separately before they coalesced in a single organism. Whether we call such cellular or vesicular reaction paths living or prebiotic does not concern us. Important points are (1) the need to reduce and to keep materials reduced while removing oxidised products, (2) the necessity for condensation reactions in many steps, (3) the requirement of energy, not so much for some simple reaction products but for their polymerisation by condensation, (4) the need for catalysts made from the products of basic pathways and metal ions, and (5) the system is then autocatalytic and controlled to ensure that the pathways are a limited interactive set. (We describe the incorporation of essential inorganic units later.) The catalysts are essential for a high-energy content, synthesis, and a high-energy flux, degradation. We have to consider that these monomers generated the polymers in random sequences in the beginning, and then slowly there developed controlled connections in pathways and limitations of sequences of units in products, including a code. They dominated because of their kinetic strengths, that is their ability to retain energy. We are assuming that the maintained *form* arose before *informed* systems (see Chapter 3). We must always be aware too that all organic compounds, being energised, decompose and therefore have to be replaced; moreover many are deliberately hydrolysed using catalysts. The cell system of chemicals is one of cyclic flow and the decomposition of organic molecules too can be useful in that the energy released can be put into the system but it never is sufficient so that extra energy supply, more so than novel incoming material supply and new confinement, is always essential. Life is locked into the environment.

If we ask which pathways came first then it must be the case that carbon (CO, CO_2) assimilation led the way to lipids, sugars and carboxylic acids. Lipids met the first need – membranes for confinement. Sugars led to glycolysis and perhaps separately to the pentose shunt,* while perhaps quite separately again CO_2 uptake by the reverse citrate cycle gave carboxylic acids. These paths could give three separate C/H/O synthetic routes which could coalesce later to give new possibilities (see Table 4.4). It is then possible that different bases and different amino acids were produced, requiring nitrogen, sulfur and phosphate uptake, and became associated in different vesicles. In effect we know that the three pathways mentioned above, plus a combination of two of them lead to different amino acids to this day, that is glycolysis (giving Ser, Gly, Cys, Val, Ala, Leu), the reverse or forward citrate cycle (giving Asp, Asn, Lys, Thr, Ile, Met, Arg, Glu, Gln, Pro), the pentose shunt (giving His), and the combination of the pentose shunt plus glycolysis (giving Phe, Tyr, Trp). In the reverse reaction of metabolism of amino acids, i.e. catabolism, much of this distinction is also observed today. See Table 4.7 for their formulae and for their characterisation. (We note that most of the simplest possible side chains with a variety of functions (see Table 4.7) are observed.)

A further feature of steps in the synthesis of proteins from amino acids is that in all known cells they became associated with particular series of triplets of bases of RNA/DNA (Figure 4.6 and Table 4.8). These triplets when joined in sequence became the coded information directing the linear sequence of the amino acids in the proteins. Such specific *information transfer* from a series of bases to a series of amino acids, i.e. translation into proteins, requires an explanation. Given its obvious complexity we have to consider how such an informed system developed from the simpler maintained flows hypothesised above. The simplest hypothesis is *that parts of the triplet coding of synthesis arose together with particular different metabolic paths*, which led to association of groups of amino acids and bases in particular vesicles. The loose association may well have been of one or two bases in nucleotides and a few amino acids before any code emerged and the triplet code evolved. First we note the correspondence between bases and amino acids but we do so largely to bring to the reader's notice the difficulty in envisaging the origin of reproductive life. For this reason the origin of life is not a major concern of this book but only its evolution.

One suggestive piece of support that leads to the idea that certain different RNA informed pathways of protein synthesis from amino acids arose in different vesicles is the correlation between the different first and, to some degree, the second

---

*While glycolysis degrades glucose to pyruvate and NADH, the pentose shunt leads from glucose to pentose and NADPH.

TABLE 4.7

THE NATURAL PROTEIN AMINO ACIDS

Name	3-Letter Symbol	Symbol	Side Chain	Character	Metal Binding
Aspartic acid	Asp	D	$HOOC-CH_2-CH \overset{COO^-}{\underset{NH_3^+}{}}$	Acid (polar)	Mg, Ca
Glutamic acid	Glu	E	$HOOC-CH_2-CH_2-CH \overset{COO^-}{\underset{NH_3^+}{}}$	Acid (polar)	Mg, Ca
Tyrosine	Tyr	Y	$H-O-\langle \bigcirc \rangle-CH_2-CH \overset{COO^-}{\underset{NH_3^+}{}}$	Neutral (non-polar)	Fe
Alanine	Ala	A	$CH_3-CH \overset{COO^-}{\underset{NH_3^+}{}}$	Neutral (non-polar)	
Asparagine	Asn	N	$H_2N-CO-CH_2-CH \overset{COO^-}{\underset{NH_3^+}{}}$	Neutral (polar)	
Cysteine	Cys	C	$HS-CH_2-CH \overset{COO^-}{\underset{NH_3^+}{}}$	Neutral (non-polar)	Cu, Zn, Ni, Fe
Glutamine	Gln	Q	$H_2N-CO-CH_2-CH_2-CH \overset{COO^-}{\underset{NH_3^+}{}}$	Neutral (polar)	

Name			Structure	Property	Metal
Serine	Ser	S	$HO-CH_2-CH{\overset{COO^-}{\underset{NH_3^+}{}}}$	Neutral (polar)	
Threonine	Thr	T	$CH_3-CH{\underset{OH}{}}-CH{\overset{COO^-}{\underset{NH_3^+}{}}}$	Neutral (polar)	Mg
Histidine	His	H	(imidazole side chain) $-C-CH_2-CH{\overset{COO^-}{\underset{NH_3^+}{}}}$	Basic (polar)	Cu, Zn, Mn, Fe, Ni
Arginine	Arg	R	$H_2N-C{\underset{NH_2}{\overset{+}{=}}}-NH-CH_2-CH_2-CH_2-CH{\overset{COO^-}{\underset{NH_3^+}{}}}$	Basic (polar)	
Lysine	Lys	K	$H_3\overset{+}{N}-CH_2-CH_2-CH_2-CH_2-CH{\overset{COO^-}{\underset{NH_3^+}{}}}$	Basic (polar)	
Glycine	Gly	G	$H-CH{\overset{COO^-}{\underset{NH_3^+}{}}}$	Non-polar hydrophobic	
Isoleucine	Ile	I	$CH_3-CH_2-CH{\underset{CH_3}{}}-CH{\overset{COO^-}{\underset{NH_3^+}{}}}$	Non-polar hydrophobic	

(Continued on next page)

TABLE 4.7 (Continued)

Name	3-Letter Symbol	Symbol	Side Chain	Character	Metal Binding
Leucine	Leu	L	CH_3–$CH(CH_3)$–CH_2–$CH(NH_3^+)COO^-$	Non-polar hydrophobic	
Methionine	Met	M	CH_3–S–CH_2–CH_2–$CH(NH_3^+)COO^-$	Non-polar hydrophobic	Cu, Fe
Phenylalanine	Phe	F	C_6H_5–CH_2–$CH(NH_3^+)COO^-$	Non-polar hydrophobic	
Proline	Pro	P	(pyrrolidine ring) CH_2 COO^- / C–NH_2^+ H	Non-polar hydrophobic	
Tryptophan	Trp	W	(indole)–CH_2–$CH(NH_3^+)COO^-$	Non-polar hydrophobic	
Valine	Val	V	CH_3–$CH(CH_3)$–$CH(NH_3^+)COO^-$	Non-polar hydrophobic	

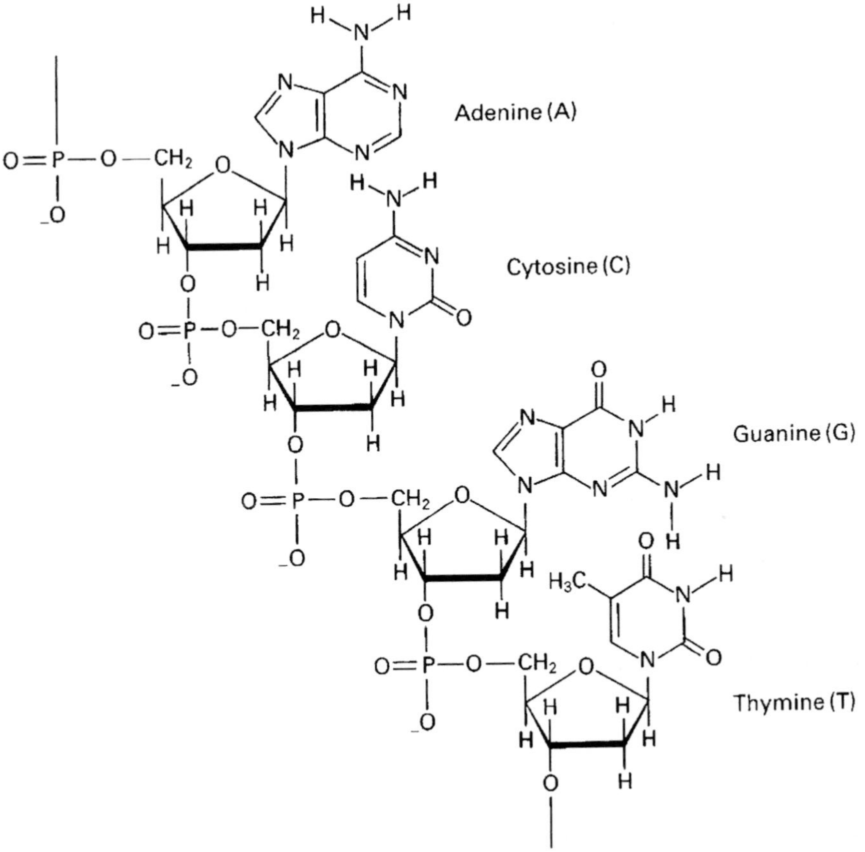

FIG. 4.6. A short string of single-strand DNA giving the formulae of four bases. In RNA, thymine (T) is replaced by uracil (U) and deoxyribose is replaced by ribose.

nucleotide bases of RNA coding for the particular amino acids produced by the four major amino acid synthesis routes. The four bases are adenine (A), guanine (G), cytosine (C) and uracil (U), where A and G are purines and C and U are pyrimidines. See Table 4.8 and note that uracil of RNA codes opposite thymine of DNA (see Figure 4.6). We observe that all three aromatic amino acids Phe, Tyr and Trp, synthesised from the pentose shunt plus glycolysis, are coded by the initial base U; of the amino acids made from glycolysis intermediates, serine and cysteine also have the dominant first coded base U while glycine is GG, and of the remaining three, also from glycolysis, valine and alanine are coded by the first base G while leucine is coded by CU or UU. Observe that none of these amino acids has as its first code the letter A, though one has a first base A(Ser) in an alternative code and histidine from the pentose shunt is coded by CA. There is a strong bias toward U/G coding. Amongst the citrate cycle-derived amino acids from oxalo-acetate, the first six indicated above are derived from aspartate and are coded by the first base A,

TABLE 4.8

THE GENETIC CODE

	U		C		A		G	
U	UUU	Phe	UCU	Ser	UAU	Tyr	UGU	Cys
	UUC	Phe	UCC	Ser	UAC	Tyr	UGC	Cys
	UUA	Leu	UCA	Ser	UAA	Term	UGA	Term
	UUG	Leu	UCG	Ser	UAG	Term	UGG	Trp
C	CUU	Leu	CCU	Pro	CAC	His	CGU	Arg
	CUC	Leu	CCC	Pro	CAC	His	CGG	Arg
	CUA	Leu	CCA	Pro	CAA	Gln	CGA	Arg
	CUG	Leu	CCG	Pro	CAG	Gln	CGG	Arg
A	AUU	Ile	ACU	Thr	AAU	Asn	AGU	Ser
	AUC	Ile	ACC	Thr	AAC	Asn	AGC	Ser
	AUA	Ile	ACA	Thr	AAA	Lys	AGA	Arg
	AUG	Met	ACG	Thr	AAG	Lys	AGG	Arg
G	GUU	Val	GCU	Ala	GAU	Asp	GGU	Gly
	GUC	Val	GCC	Ala	GAC	Asp	GGC	Gly
	GUA	Val	GCA	Ala	GAA	Glu	GGA	Gly
	GUG	Val	GCG	Ala	GAG	Glu	GGG	Gly

The 64 triplet codons are listed in the 5′→3′ direction in which they are read. The three termination (term) codons are given.

except for Asp itself which is coded by the first two bases GA, while of the last four, derived from α-ketoglutarate and related to glutamic acid, three are coded by the first base C but for glutamic acid the code is GA, as for aspartate. Note that there is bias towards the bases A/C and that U is not the first letter and also not the second letter of the code for the series of amino acids derived from the citric acid cycle, except Ile and Met. The third code letter is generally varied, suggesting that it came later. Table 4.8 summarises these observations.

Note that the strength of the correlations is increased by the fact that the citric acid pathway is today isolated in mitochondria derived from a distinct early life form and linked to both aspartate and glutamate, in which A and C are dominant amino-acid carriers, while glycolysis and the pentose shunt are cytoplasmic, where U and G are more dominant amino-acid carriers.

Now, this tentative description of the development of a correlation, later to become information from bases to the synthesis of proteins, by no means solves the problem of the origin of this code nor does it bring into focus the fact that the very proteins which were produced are responsible for the synthesis of the basic metabolic units, formaldehyde and acetic acid and then the amino acids and bases and finally the polymers by catalysts which are the polymers themselves. We do state, however, that the set of reactions quite probably give the most kinetically stable products. Now, the amounts of the different amino acids, lipids, saccharides and bases had to be controlled since all their syntheses depend on the same environmental materials and the same sources of energy and they are synthesised proportionately, which requires control or information transfer. Note immediately that once reproduction arose, not assumed above, the base pairing in double DNA(RNA) helices is GC and AT(AU) to give the coding for all the amino acids in one cell. It is also the case that all nucleotide bases are made from glycine and aspartic acid today. The final necessarily informed coded system of a cell is very complex and it is worthwhile for us to look again at the nature of informed systems of organic molecules in the light of what cells can produce from the environmental chemicals. (Much of this section is wide open to experimental enquiry but itself rests solely on hypotheses.) We return now to our theme that the content and space of a cell had to be organised to be efficient.

4.6.   Informed Systems of Organic Molecules

In Chapter 3, we have stressed the distinction between maintained *form* of a flowing system and *informed* flows within a system. A single flow creating *form* of one or more components is driven by the continuous uptake and degradation of energy whether it is cyclic or not, but it is uncorrelated with any other flow. In an informed system of many flows the flow of any component is also governed by transfer of messages to and from other flows, acting on catalysts and coordinating them. The messages have, of necessity, energised sources and receivers (see Section 3.11). We now draw attention to the known complex set of messages, signals, in all cells between central coded molecules (DNA/RNA), proteins (also coded, see below), metabolism including substrates and energy supply in nucleotide triphosphates (NTPs), ions involved with the proteins in synthesis and degradation, and environmental resources which supply the cell with essential elements through the outer membrane. This set of chemicals was eventually made into an interactive whole by feed-forward and feedback links. We reproduce the overall scheme in Figure 4.5. It shows that many products in one reaction step of a path act forward and/or backward either to inform its pathway and/or to inform other pathways. There are also the following steps, all of which involve binding with conformational changes which generate information transmission:

(1) The environmental inorganic and organic chemicals bind to the insides and outsides of selective and regulated membrane pumps (proteins) and act as on/off switches of the pumps.
(2) The DNA/RNA machinery operates to produce the proteins including the above pumps and catalysts under feedback control from simple metabolites and ions through their binding to transcription factors.
(3) The DNA/RNA machinery (including polymerases, energy and transcription factors) produces proteins, especially enzymes (catalysts) which produce substrates which back-react, bind (inform) the membrane, the active proteins,

and as stated in (2) the DNA/RNA coding machinery, while other proteins (catalysts) act degradatively.

(4) The inputs of transduced energy, perhaps initially of pyrophosphate, later NTP, and from gradients of elements as ions are distributed to and control many pathways.

Although we do not know how such a system came about we can make the following statement. *The major active components of cells are proteins and as coordinated catalysts of energised flows of synthesis, of degradation, and of pumping of the environmental chemicals they ensure that energy uptake and degradation overall is faster than it would be in the environment without these controlled catalysed cycles of chemicals in cells.* This is the essence of autocatalytic life which became coded, compelling the limitation to pathways already established. We can now summarise the combined network of communication between the active units of flow, stressing *concentration dependence* as well as sequences in information. Information about the *concentrations* of internally collected environmental ions and molecules, some with energy content, is transmitted via transcription factors to the DNA/RNA code which subsequently informs the *amounts* of synthesis of coded proteins related to the introduction and reactions of the collected units. The proteins carry in their structures the translated code for these initial and subsequent substrate syntheses. The substrates of these syntheses also inform by feedback, via their *concentrations*, the proteins which carry out these very syntheses, as well as the DNA/RNA. They also inform the membrane so that uptake starts or stops according to the *needed concentration*. Meanwhile, the membrane is informed externally about *concentrations* in the environment by selective interactive binding of its contents to the uptake machinery. The proteins in the membrane are governed in synthesis by these same environmental chemicals, through some indirect *concentration-dependent* interaction with the DNA/RNA transcription factors. Remember that the controlled *levels* of NTP are needed for DNA and RNA synthesis by protein catalysts, while proteins are synthesised by RNA catalysts. The whole synthetic activity is driven by the available energy, NTP in its internal *concentration*, which can also act as a source of information by binding, without energy transfer of reaction, to DNA/RNA or to the protein machinery, enzymes and pumps, to start or stop activity. A major information system is clearly the transfer of bound phosphate by enzymes, i.e. kinases, to both substrates and proteins and to water to provide energy. (Note. Remember the link to the environment for energy and material which we do not stress in this chapter yet as we are deliberately leaving on one side the roles of metal ions until later.) *The whole is a thermodynamic irreversible system of controlled flows* and it is homeostatically fixed in a given steady state.

We see that the functional information in this network concerns concentrations and acts through selective binding constants, which being thermodynamically fixed factors, are only partly linked to a code (see Appendix 4A). The sources and the receivers of the information have the common language of binding specificity given by the nature of atoms in the messengers, which can be an environmental ion, a substrate or a large molecule, and the binding is temperature dependent. The pathways are secured by the extreme selectivity of coded catalysts, i.e. enzymes, in them. Now all the compounds in the pathways degrade slowly and some pathways are deliberately degradative and produce energy. Energy is also obtained from outside and it too has to be cooperatively shared. There is a consequent multitude of energy flows. All this transfer of information between flows means that *in any steady state the cell has a precise dynamic homeostatic description,* a thermodynamic quality, which we label for convenience under the *analytical headings* of the genome, the proteome and the metabolome that make up the non-metal components of the chemotype. The cell content cannot be obtained from analysis of the DNA since concentrations also are needed in each compartment. The remarkable feature is the singular underlying cytoplasmic nature of the whole chemistry of all cells despite biological diversity. Is it a unique system?

The coming together of the major pathways is shown schematically in a simplified way in Figure 4.7. This diagram represents the major organic synthetic pathways of all cells to this day. They form a basic requirement of all organisms and are shown in much more complicated diagrams in wall charts of pathways to be seen in all biochemistry departments. Reflecting again on this non-metal chemistry, we observe that its general purpose is the building from monomers of large molecules – lipids, polysaccharides, proteins and nucleic acids – to form the basic set of shaped building blocks of membranes, walls associated with them, protein assemblies, catalysts, ribosomes, genetic machinery and so on. The separate organic molecules are *not* in equilibrium with their monomers internally or between themselves or their basic chemicals obtained from the environment. The structured units make up kinetic divisions, centres of binding and activation of small molecules, proteins (enzymes) and control units (RNA/DNA). They are sometimes in very large assemblies, but they remain mainly in water except the lipids, which form membrane phases and contain some other molecules including special proteins.

Since none of these organic molecules is thermodynamically stable they can only be maintained by constant synthetic activity and energy input for repair or resynthesis. The essential small molecules of the environment, CO_2, N_2, H_2O and H_2S, for such syntheses have to diffuse into the cells, but HPO_4^{2-} for several purposes has to be pumped inwards costing energy. To a first approximation the first four equilibrate across membranes, so that in early organisms they have approximately the same concentration as in the primitive sea. Phosphate became and is today richer in cells than in the environment. The capture of larger but still small molecules such as monomers from the environment also depends on pumps, protein machines, driven by ATP or Na^+ and H^+ gradients. The very noticeable feature of these pumps and the subsequent mechanisms for trapping the units in larger molecules is that every step requires energy and has to be selectively specific. Most steps are also informed in more than one way. The internal steps also require metal elements for one purpose or another (see below). Clearly, since the organic molecules are all unstable, cells had to reproduce if they were to maintain integrity,

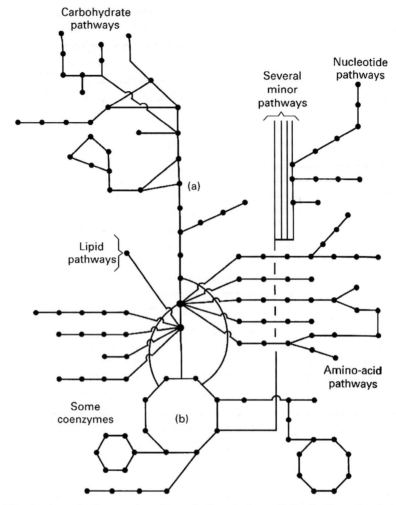

Carbohydrate
pathways

Several
minor
pathways

Nucleotide
pathways

(a)

Lipid
pathways

Amino-acid
pathways

Some
coenzymes

(b)

FIG. 4.7.   A scheme for the organic pathways in the cytoplasm of all cells illustrating their interconnected nature (after Kauffman; see Further Reading). (a) Glycolytic; (b) Krebs cycle pathways.

sustainability, in the environment. Importantly, we know of no other ways of using
chemical elements than this selection to make the kinds of polymers required for
the reproductive synthetic chemistry of life, i.e. proteins, lipids, DNA and RNA.
We repeat the question "Is there in fact any other possibility given the conditions
of Earth?" It is important to recognise that the pathways which gave the above
materials using energy are also efficient in generating survival between synthesis
and decay (see below and Appendix 4C), as well as being contained. After that we
need to show that the four different types of polymers have the very essential features necessary for function in living systems, which will help to establish that they
are an irreplaceable set. Finally, we shall consider the question of binding constants

and exchange rates of small molecules and ions to the large molecules for syntheses and informed control.

(Note that we shall not introduce the idea that these and/or other required organic compounds found on the Earth came to the Earth from outer space although some are found there. Such additions to the Earth go no way in helping us to understand how *an energised system* of reactions evolved in confined volumes, cells).

4.7. Pathways and Efficiency

Our belief is that an energised ecological system develops such that for given conditions it moves towards optimum retention of energy in chemicals so as to secure the overall optimal increase of thermal entropy production. The implications are two-fold: first, the energy taken in is made use of in creating increasing concentrations of unstable energised chemicals by synthetic activity which then degrade only relatively slowly in concentration initially but later equally in a cycle; second, the processes or pathways of reaction should be such that they produce effective chemicals in this respect. Such a system could be described loosely by saying that it is optimally efficient in the conditions. Now the word "efficiency" of a system has an important thermodynamic meaning which is often confused with this qualitative use of the word. Efficient in either case does not refer to the 100% transfer of energy in a process from one form to another which is the theoretical transfer of energy and material from one side to the other of an equilibrium. Efficiency *of a machine which does work* refers to the ability of the machine to transfer energy of all kinds from one set of materials to another not at equilibrium. Energy put into a machine can never be converted 100% from input to output and there is an ideal maximal value for the transfer, for doing work, given by thermodynamics. The practical (or kinetic) efficiency of a machine is then given as a percentage of this ideal. In Appendix 4C, we elaborate using a particular example of one machine. We shall use the word efficient in this book somewhat loosely as referring to the closeness of approach of optimal practical efficiency to ideal or maximal thermodynamic efficiency of a cyclic system. *This efficiency correlates with optimal survival fitness and with optimal rate of energy degradation, i.e. thermal entropy production.*

Starting from the example of many pathway systems in Chapter 3 we showed that efficiency was linked to the machines driving the networking of paths that used the same starting materials to produce jointly catalysts, messengers and codes. Specifically, we have shown so far that the chemical pathways involving small molecules give rise to proteins, DNA/RNA, saccharides and lipids, which as a group indeed are very efficient, perhaps the most efficient of all such polymers, to retain energy and catalyse decomposition in water at pH 7.0. In Sections 4.9–4.14, we shall examine them in more detail. Examination of glycolysis, the citrate cycle and other pathways show that they are extremely effective ways of producing useful energy while providing the intermediates for these syntheses (see Cornish-Bowden in Further Reading). In Chapter 5, we shall show that the use of light as an energy source is also very efficient. Thus, when a steady state of this chemistry is obtained it has a very effective rate of material production and energy degradation all driven by self-produced machines. Can chemists devise alternative schemes managing all these basic needs for an increased rate of moving towards a cyclic system, which involves not just material but has to involve energy absorption and degradation efficiently? If this is not an efficient (optimal) system there must be another system which can add to or displace it (see Section 3.8). We consider that adding to the initial system of organisms is the basis of ecological evolution.

Rather than looking in more detail at pathways in this chapter, we now wish to illustrate the way in which the large molecules produced by them provide the essential functional features in the machinery for efficient energy retention and degradation, that is for creating thermal entropy (heat) from light. The set of proteins and RNA(DNA) are made by proteins and RNA(DNA) with fidelity, but why are these biopolymers so functionally effective? We now go to some length to convince the reader the value of the particular polymers found in cells forcing again the question: are they a unique set for life? One part of effectiveness is, of course, containment.

4.8. Structures and Maintained Flow: Containment

As we have observed in Chapter 3, all organisation, that is coherent flows of material, requires containment and a maintained energy supply. So far, we have described the major non-metal flows of chemicals in all cells. We next consider the ways in which these flows are contained by fields (energy restrictions on processes or reactions in space – see clouds and ozone formation in Section 3.4) or by physical barriers. Treating the *physical* barriers of cells first we observe that in cells there are molecular aggregates in the form of three products of the metabolic paths: *lipids*, the major content of membranes (Figure 4.8), underlying *proteins* such as spectrin and other proteins which form internal parts of membranes and cytoplasmic filaments which guide flow; and *polysaccharides* which provide extra open external surface structures i.e. walls (often associated with metal ions) and which allow flow of small molecules and ions into all cells. There could have been no development of organisms without this in and out and confinement of energised flow, much as rivers would not exist without land barriers and energy. We do not need to comment on the precise nature of these chemicals here, but their evolution, especially changes of membrane lipids, is extremely important, as we shall see in Chapters 5–9. The best set of lipids for all kinds of cells was not discovered immediately (see Chapter 7). The in and out flow is due to pumps – machines energised by ATP or gradients of ions.

The internal physical flow along filamentous structures, made from actin, myosin, tubulin and similar proteins in later cells, can be of small or large molecules, or even of vesicles, so that movements on internal surfaces become more

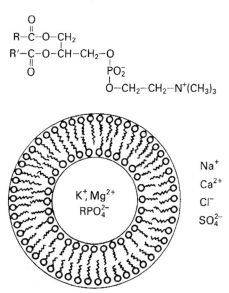

FIG. 4.8.   (Below) A diagram of the bilipid layer membrane of a vesicle or a cell with (above)
a typical lipid, phosphatidylcholine. Large molecules and ions cannot penetrate the membrane as
illustrated by the ions surrounding and inside a cell, but the distribution is reversed in vesicles
(see Chapter 7). The ions create chemical and electrical field gradients across the membrane.

important the larger and more "structured" the cell interior is. Many of these filaments can be broken down and remade stressing the dynamic character of cells.
While all the internal filaments are made from proteins, the best extracellular set
of structural biopolymers are combinations of proteins and polysaccharides.

Now, in any cellular system, the very selective catalysts in the cell, the enzymes,
act like local field gradients in that they direct internal reaction paths along selected
routes. These routes may have controlled inputs of energy and controlled rates due
to feedback. All these properties are products of binding interactions, effectively
local field constraints. It is these enzymes together with filaments and membranes
which "structure" the internal flow of synthesis and degradation. Remember that
the reactants in the pathways also control their enzyme concentration by feedback
to the genetic machinery. Virtually, all effective machines including energised catalysts are proteins (see below and ribozymes).

It is much more difficult to describe the relationship of the bulk field gradients,
easily recognised in the flow of water in clouds and of oxygen in the ozone layer
described in Section 3.4, to that of the gradients controlling the chemical flow in
cell liquids. The effects of electric fields due to charge distribution in various parts
of the cell is an obvious possibility.

We turn to the nature of the biopolymers asking: what properties in their structures make them such an essential part of organisms? In what follows, we try to
show why the four types of unit, i.e. nucleotides, amino acids, saccharides, and fatty

acids, were selected for their intrinsic properties as expressed in polymers (see Mulkidjanian *et al.* in Further Reading), apart from their ease of synthesis. The first question is why is there only one genetic coding scheme with but minor exceptions?

4.9.   The Selection of Coded Molecules: DNA(RNA)

We see that the organic chemistry of organisms is based largely on *a limited set of controlled special syntheses* of coded polymers with given linear sequences of amino acids, nucleic acid bases and some saccharides in water, but the whole of the coded instructions for these syntheses appear to rest in a single molecule of DNA in each cell (but see later chapters). *Reproduction* of the coded long-chain sequence of DNA has to be based on pre-synthesis of itself as a soluble primary *linear* molecular structure. (A *continuous molecular* code cannot be easily read, certainly not reproduced, if it is two- or three-dimensional; but see the brain in Chapter 10 and electronic circuitry generally.) DNA provides a linear series of bases for guiding synthesis, with, necessarily a primary structure open to exposure at some stage. There are, in contrast, hundreds of types of RNA and protein molecules, often non-linear and folded and multiple copies of each one. (Note that the monomers of DNA and RNA differ and are both based on three units, the aromatic moieties, sugars and phosphate.)

DNA must not fold into three-dimensional form unless it is readily unfoldable, at least in part at a given time, so as to be read for transcription to RNA by an energised machinery, and it must be reproduced. Hence, the units in DNA have to be such that their surfaces do not allow self-interactions except in helical duplexes of limited strength. The open structure is then, in large part, due to the particular deoxyribose-phosphates (*note they are negatively charged*), which hold particular bases giving DNA as a water-soluble linear single-strand form. (Transcription becomes much more complicated than this in more folded eukaryote DNA, as we explain in Chapters 7–9.) DNA is neither capable of giving powerful catalysts nor the dynamics associated with the driving of flow since it is unable to generate the variety of states required of the components of such active machines, in part because it does not give variable folds (see proteins below). DNA is the basic linear coded polymer for all *reproduction* due to the fact that the component bases are of relatively low water solubility, are separated and flat, and their edges can hydrogen-bond parallel to their flat surfaces giving unique pairings, double-stranded helical DNA. Only four bases (two would be inadequate and six too many) are required to give sufficient variation for a useful code. DNA does not bind redox-active metal ions but binds K^+ and Mg^{2+} and organic and protein cations to help neutralise the anionic charge and develop mobile secondary structure. Is it possible to devise an alternative polymer with these coding functions fixed in a cell? Here the work of Eschenmoser is of great interest (see Schöning in Further Reading). In Chapters 5–8 we look in detail at the structure and character of DNA and its associated apparatus since it has evolved considerably. How did it change in keeping with environmental change; a fundamental question for the understanding of evolution? (We must worry later about the possibility of gene transfer between cells, which if common would spoil the whole idea of species, but does not affect the idea of chemotypes, at a given time. Such transfer was common at the beginning of life and is probably so now when the idea of prokaryote species in the wild may not be tenable. Chemotypes, related to genotypes in a limited way, may be a more useful classification.)

We must also be very careful when we consider the relationship of information to DNA. It is not just related to expression but to the control of expression. The first fact is that only 2% of *human* DNA expresses proteins through RNA. Some of the rest expresses other kinds of RNA, but a large proportion apparently expresses nothing (see Freeland and Linder in Further Reading). Some percentage of human DNA may consist of accidentally trapped inactive "viral" genes or parts of them (see Villarreal in Further Reading). The information content is further complicated in that much of this "modern" DNA is heavily methylated which clearly requires a way of modifying DNA which is not a feature of directly reproduced base sequences. Methylation is an added-on feature due to enzymes and coenzymes (vitamin B_{12} and folic acid). The methylation is environmentally sensitive, e.g. upon oxygen for its removal, implying that *DNA is dependent on the environment for expression.* The methylation of DNA is just one feature of the complexity of DNA information and its readability. Thus, the proteins, histones, of the chromatin nucleosomes in eukaryotes (the folded form of DNA) are often methylated as well as acetylated or phosphorylated in selective ways, which also alters the ability of the protein machinery to express the coded information. The whole relationship of gene expression is no longer a matter of simple DNA sequences; it is necessary to include DNA and protein modifications to the interpretation and expression of DNA sequences. To these complications, we must add the requirement for the transcription factors and the polymerases which need to bind and to be energised. Information connected to the DNA is therefore linked to the cell machinery chemical and mechanical, and then to the environment via the cell contents (even K^+ and Mg^{2+}) – this topic is often included under the name of epigenetics (see Jablonka and Lamb, Turner and Caporale in the references, and Chapter 11). Note that we are using thermodynamic language in the description of active DNA and not just that of internal sequence. Finally, a very important feature of DNA is that it is open to mutation and transposition without which there can be little evolution of coded organisms. DNA is the unique *centre* of reproduction today, but is far from a sufficient unit for that purpose. In recent times it has become apparent that what was thought of as junk DNA when transcribed as RNA becomes part of the feedback system that informs DNA activity. Information in cells is complicated.

4.10. RNA and the Possible RNA World

As stated in the previous section, there is one type of DNA in a cell but there are several types of RNA in multiple copies. They are: (1) ribosomal RNA which is involved in the catalytic machinery of protein production (Figure 4.9); (2) messenger

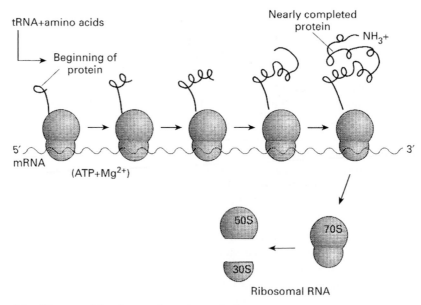

Figure label:

tRNA+amino acids

Beginning of protein

Nearly completed protein

NH$_3$+

5′
mRNA

(ATP+Mg^{2+})

3′

50S

30S

70S

Ribosomal RNA

FIG. 4.9.   Diagram of the ribosomal protein-synthesis apparatus.

RNA which holds the coded sequence of triplet bases for protein production; (3) many transfer RNAs (Figure 4.10), which bring individual amino acids to the ribosome for protein production; (4) some catalytic RNAs involved in self-cleavage, e.g. in splicing, and (5) many small s-RNA and 'interfering' RNA$_i$ (see Matzke and Matzke in Further Reading. Today RNA is produced by transcription from DNA, but all messenger RNA is open to modification before it is translated into proteins. All these RNAs require bound Mg^{2+} and/or K$^+$ for activity and many are not of linear structure. The variety and weak catalytic activities of RNAs in folded forms and their properties have led to the idea that maybe RNA came before DNA and even before proteins as the first cell catalysts (Figure 4.11), see Orgel in Further Reading. One problem with this "RNA world" is that a particular messenger RNA represents only a very small fraction of the protein complement of the cell while DNA represents all of it and must be self-replicating. We have to imagine that messenger RNAs were made selectively in different vesicles (see above), but together with given sets of proteins. This could be in agreement with the proposal in Section 4.5 that certain RNAs and amino acids were formed in separate vesicles, but all four bases are required for reproduction. Note also that DNA has stop and start signals for messenger RNA production.

    There are further theoretical objections to this RNA-world hypothesis, especially since we know of no organisms based on RNA without DNA and proteins. Moreover, RNA is not able to act as a catalyst for many essential simple molecule reactions, e.g. of CO, H$_2$, N$_2$, nor for most oxidation/reduction changes, and RNA

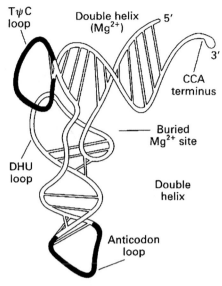

FIG. 4.10. Schematic diagram of the three-dimensional structure of a tRNA.

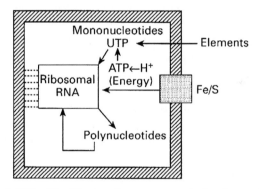

FIG. 4.11. Proposed RNA life. The membrane was made by the production of oils. The monomers were made from basic available forms of elements, H_2O, CO_2, NH_3 and HPO_4^{2-}. The source of energy was, for example, an energised Fe/S particle trapped on the membrane. The system reproduces when the duplex polymer, say $[A\cdot U\cdot A\cdot U]_2$, is released as two single strands only for each to be trapped in a new synthesising vesicle. The improbability of such a scheme emerging is very high, yet some such event happened. Note that it is not necessary to have a simple $[A\cdot U]_n$ unit and a variety of units of any ordering A and U is reproducible with fidelity on second copying.

itself needs a complex set of protein-dependent synthetic steps to make its ribose and its bases. All these reactions require energy, and we do not know if energy could be captured by an RNA-based machine. Again RNA by itself would not be useful in producing a membrane, which had to be made to enclose the system, or pumps. For these reasons, we prefer to treat RNA (with DNA), proteins and membranes as arising simultaneously among a multiplicity of energised molecules, which came together by chance via chemical interaction, perhaps in groups. Once they did so, they formed a unique energy capture and retention system, possibly coincidental with the emergence of reproductive life. As stated, the need for a code is for reproduction, repair and control, allowing much greater energy capture in chemicals than in a simple maintained flow. We return to the possible roles of RNA in Chapters 7, 8 and 11 since recently it has become clear that the back interaction of small RNAs on DNA can cause changes in expression and even in inheritance (see epigenetics). In a curious sense, the thermodynamic drive of life we have described needed a code to evolve but the code is open to manipulation in interpretation by the cell's machinery. (Note that all RNAs are associated with concentration dependence, a thermodynamic feature; DNA itself is not, so that back combination with DNA is a complicated form of informing thermodynamically as are the forward effects of RNA.)

No matter whether polymers arrived in turn or simultaneously, the more interesting point is whether or not there are any possible alternatives to the bases of DNA/RNA code. Pursuing this line of thought from bases to the required amino acids, the whole is a very small limited set (see Tables 4.7 and 4.8). They have hardly evolved in 3.8×10^9 years of life's existence. Looking at them, it is apparent that the variations in the four bases in sequences provide adequate structural and activity possibilities in the translated products. The resultant amino acids shown in Table 4.7 have among them various shapes and sizes, hydrophilic/hydrophobic properties and simple acid/base properties for proton and metal binding but *per se*, except for cysteine, they carry little catalytic potential for redox reactions. (We shall see that it is the function of the other part of the environment, i.e. the metal ions, to provide this function.) They are costly to make in energy terms and difficult to bring together, as indeed are the bases of RNA and DNA. It is only when we remember the synthesis system and the kinetic stability of all the polymers and when we look at folded proteins (see Figure 4.12(a)), that we realise how the translated amino acids inside protein folds have such powerful functional value relative to RNA. Proteins, in fact, serve many functions in a cell including, as already mentioned, structures (of membranes) and protection (and control) of DNA. Other major functions concern the machinery of catalysis, transcription, pumping and electrical and electrolytic devices. Before we describe particular functions of proteins we must note that their synthesis (dependent on RNA) and folding is not a simple process in a cell.

4.11. Proteins: Folding, Catalysts and Transcription Factors

The extremely important feature of proteins is their idiosyncratic folds. When describing proteins the problem of all their folds is not solved just by the thermodynamic stability of the interactions between the amino acids much though this may be approximately true if we take the majority of proteins (see Lesk in Further Reading).

The fold energy can depend in degrees on the interaction with other chemicals including the solvent, other proteins, RNA or DNA, the ionic milieu, often in aggregates or on membranes. In eukaryotes, proteins are often exported from cells after they are glycosylated, a process involved in folding. (This is but one use of polysaccharides which by themselves are found in the open meshes of eukaryote multi-cellular organisms. We do not stress their functions here as they do not seem to have universal value). There is the possibility too of protein misfolding, which can lead to serious diseases. In the synthesis of proteins the folding is guided by the mode of production from ribosomes and especially so if the ribosome is on a membrane (see Figure 4.9). Later, as they leave the membrane, their folds may be guided by "chaperones", proteins which assist the folding of other proteins. Hence it is not a general truth to say that the fold of a protein is decided by its sequence. Proteins can also be active when unfolded, for example, those with many charged amino acids in their sequence form a protein which will not fold (see Williams (1977) in Further Reading), while all must be unfolded to be destroyed by cell systems, which is extremely important for certain proteins in the cell cycle. In a cell there are means of unfolding proteins, for example, by the ubiquitin system. The fold certainly depends on the complement of different amino acid types (see Table 4.7). Finally, a protein can be modified chemically, either reversibly or irreversibly. The dynamics of the structures are also fundamental to function; the fold is not entirely fixed.

In a prokaryote cell there may be as many as 3,000 proteins in the proteome, which can range in concentration from a very few per cell to millimolar. Here we divide them into a few groups of very different functional importance, but we are not able as yet to describe adequately concentrations in particular steady states of the proteome. We begin with water-soluble catalytic proteins, the enzymes. A major feature of biological chemistry is the *required controlled catalysis of metabolism* by enzymes, a part of organised flow. Recall that enzyme catalysis is an essential feature of energy flow using as substrates components made of the same basic elements as in metabolic paths (see Section 3.11). Enzymes have a special folded structure suitable for binding highly selected substrates, activating them in special ways and releasing products (Figure 4.12(a)). The selection is of the stereochemistry of the substrates down to the level of optical isomers (Section 2.14). Hence, they have very specific surfaces and surface dynamics. Moreover, these surfaces and their dynamics are open to adjustment by agents related to other pathways besides the one in which a particular enzyme is involved. Often the adjustment alters the catalytic rate by feedback from product concentration as well as from energy carrying molecules (ATP) and mobile coenzymes and ions. The result of this 'allosteric' control over many enzymes by molecules and ions common to more than one pathway is a process of coordination of pathway activities, hence of informed organisation. As stated above, this coordination is furthered by a balanced production of the enzymes themselves by feedback from substrate or product concentrations to DNA transcription factors, themselves proteins bound to DNA but not catalysts (see Figure 3.13). Such factors inform the DNA of the need to start or stop synthesis of the enzymes appropriately and are required features of

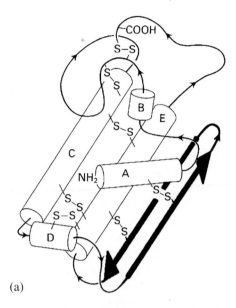

(a)

FIG. 4.12(a).   An outline structure of a protein (here the enzyme phospholipase A_2), showing
α-helical runs of amino acids as cylinders (A–E) and anti-parallel β-sheet runs as heavy black
arrows. Disulfide cross-links are shown (the enzyme is extracellular), and runs of no α/β secondary structure appear as thin lines. The structure is relatively immobile, and binds calcium in
a constrained loop. (Reproduced with permission from Professor J. Drenth.)

an effective informed system (Section 3.15). Naturally, the pathways as well as
being informed internally are informed externally by the environment, and both
increase in evolution although external information gradually becomes more and
more important (see Chapters 5–9). It is the network of cooperating flows which is
so powerful in energy absorption, retention and degradation.

There is another feature of enzymes in that their catalytic action is dependent on
selected chemical atoms either of their side chains or of metal ions held in special
electronic or geometric relationships by the fold. Additional catalytic centres are
introduced by special coenzymes, small reactive organic molecules often related to
vitamins. Inspection of the catalytic side chains, coenzymes and attached metal ions
reveals that they are therefore often of heightened, special reactivity relative to these
groups or atoms when unbound or in simple compounds and exposed to a solvent
such as water (see Tables 4.9 and 4.10). Enzymes then have very effective attacking
atoms. This is not the place to discuss this "constrained (entatic) activity" in detail,
but we note the value of "unusual" –OH groups of serine hydrolytic enzymes, controlled properties of coenzymes and of many unusual stereochemistries and electronic conditions of metal ions, copper and haem iron, for example, at active sites
(see Section 4.15). Some substrates and some coenzymes dissociate easily or
exchange readily and hence their concentration controls activity. Their homeostatic
concentrations are managed by feedback to synthesis of enzymes or to pumps and

TABLE 4.9

SOME SPECIFIC METAL ION CATALYSES

Small-molecule Reactant	Metal Ion	Examples
Glycols, ribose	Co in B_{12}, Fe	Rearrangements, reduction
CO_2, H_2O	Zn	Carbonic anhydrase
Phosphate esters	Zn	Alkaline phosphatase
	Fe, Mn	Acid phosphatase
CO	Ni, Fe	Formate/acetate formation
N_2	Mo(Fe)(V)	Nitrogenase
NO_3^-	Mo	Nitrate reductase
SO_4^{2-}	Mo	Sulfate reductase
CH_4, H_2	Ni(Fe)	Methanogenesis, Hydrogenase
$O_2 \rightarrow H_2O$	Fe	Cytochrome oxidase
	Cu	Laccase, oxidases
O-insertion from O_2 (high-redox potential)	Fe	Cytochrome P-450
SO_3^{2-}, NO_2^-	Fe	Reductases
$H_2O \rightarrow O_2$	Mn	Oxygen-generating system of photosynthesis
H_2O_2/Cl^-, Br^-, I^-	Fe(Se)(V)(Mn)	Catalase, peroxidases
H_2O/urea	Ni	Urease
Small peptides and esters	Zn	Hormone control (peptidases)
Ethanol	Zn	Dehydrogenase

TABLE 4.10

EXAMPLES OF CONSTRAINED[a] SITES IN ENZYMES

Site	Constraint
Blue copper in azurin	Trigonal distorted toward tetrahedral
Copper (type III) in oxidases	Open-sided pair of tetrahedra
Nickel in hydrogenase	Distorted square pyramidal
Nickel in urease	5-coordinate, one long bond
Iron in heme enzymes	Often 5-coordinate Fe(II)
Iron in peptidases	Distorted tetrahedal Fe(II)
Cobalt in B_{12} enzymes	One long bond (octahedral)
Manganese in superoxide dismutase	Strained trigonal bipyramid
Zinc in carbonic anhydrase	Open-sided 5-coordinate

[a]The sites are sometimes called entatic.

Note: All data from A. Messerschmidt *et al.*, see references.

carriers which are often controlled by selective equilibrated binding chemistry at enzyme sites and transcription factors, all of similar binding strength (see Section 4.16). These characteristics are again features of the high efficiency of cells in controlled informed chemistry. A central feature of the communication network is the values of the binding constants to proteins as well as the concentrations of binding agents, all thermodynamic quantities (see Section 4.18 and Appendix 4A). However, the most important feature of proteins is that they form the molecular machines, i.e. the working parts, of cells.

4.12. Proteins: Biological Machines in Water

As stated, many proteins control the shape of a cell membrane and its contents while others are involved in internal structures such as filaments – actin–myosin, tubulins and so on in later cells. Structural proteins have been described briefly in Section 4.7, so we turn to the nature of proteins in machines – *dynamic proteins, special enzymes using energy and doing, for example, mechanical work.* To understand machines, we must first note that protein structures fold in different classes. There are three major secondary structures of proteins that come together differently in tertiary protein structures of different functional use but with similar energy content – α-helices, β-sheets (parallel and anti-parallel) and turns (Figure 4.12(b)). We have described already the proteins (enzymes) acting on substrates and coenzymes, and those binding metals will be described later. The value of proteins here is in selective binding, especially at the turns, which allow the formation of "holes" for substrate and coenzyme binding. The kinetics of access and leaving of substrates are also controlled by the protein fold type and its surface mobility. The α-helical proteins are more mobile while the β-sheets are more rigid giving greater selectivity of their surface turns. The β-sheets are, therefore, mostly used in enzymes, but controls attached to them may well be α-helices which are of great functional value as adjustable rods for transfer of information and energy. Hence, α-helices find the greatest use in the mechanical devices such as transcription factors, histones, triggers, allosteric units, pumps, motors and contraction devices, i.e. generally in the moving parts of molecular machines. Observe that insofar as a catalyst passes through a cycle of states while transforming substrates, it is a machine. The combination of platforms (β-sheets) with mobile rods (α-helices) and turns can easily be seen to allow the possibility of driving the functional mechanical devices required in many enzymes and pumps (Figure 4.12(c) and Figure 4.15), Appendix 4C; see Walker, Junge *et al.*, and Williams (1980) in Further Reading. Consider the parallel with the components of a car engine. There are moving parts, rods and valves, and basic platforms and cylinders, typical of all mechanical devices, connected to energy input and work. Can we devise polymers other than proteins, which will act on such a molecular scale in water as highly selective catalysts, as structures and/or as mechanical devices? Of course, it is essential that all these activities are coordinated as they are by feedback controls – information. Given the variety of functions they can perform, are they unique polymers? Now some of them are not in aqueous solution but in membranes where they can be placed if the fold creates a hydrophobic surface again needing special amino acids in a sequence.

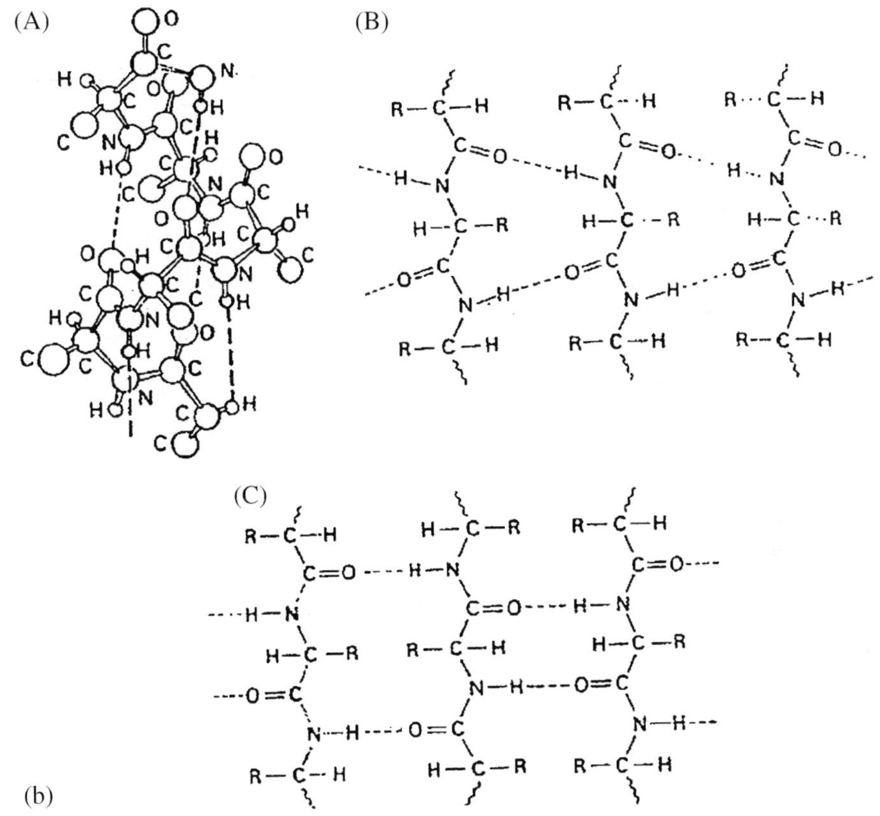

(A)

(B)

(C)

(b)

FIG. 4.12(b).   The secondary structures of proteins are largely based on helices (A) or sheets, parallel (B) or anti-parallel (C).

4.13.   Proteins in Membranes

4.13.1.   PUMPS

Transmembrane proteins face for example, both the inner solution, e.g. the cytosol, and the environment. As pumps, they have in them binding sites for environmental components, organic and inorganic on the outside, and others on the inside where they bind to essential organic substrates, NTP (energy) and essential inorganic elements as well as to other internal proteins (see Figures 4.12(c) and 4.15). Two functions follow: the first is maintained flow, as in the action of a pump; the second is response to internal or external concentrations, information, since even the very objects which are pumped can be used to switch the pumps on or off so that the internal concentrations are fixed for a given external concentration so as to maintain an activity state. The pumped ions or molecules are in steady-state flow, but the binding interactions are close to equilibria, different on each side of the membrane. It is

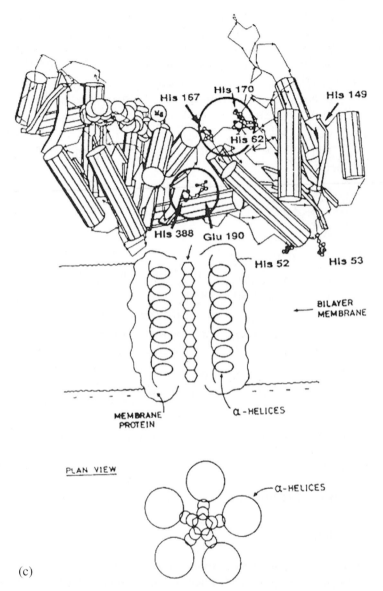

(c)

FIG. 4.12(c).   Illustration of the connection between a channel bundle of helices in a membrane and model of an energy drive of a pump, here phosphoglycerate kinase (PGK) which in principle can transfer phosphate to ADP giving ATP. It is conjectured that any ions entering the channel, full of water, from an energised gradient could first force a screw twist on the membrane helices and then that this twist would be imparted to the kinase, e.g. PGK, causing phosphate transfer. It is easily seen how protein phosphorylation could be linked instead of substrate phosphorylation. Equally, the screw twist of helices can be used to generate release of ATP. The basic ideas are those of a mechanical device (see also Figure 4.12(a)). The reverse action is a membrane ion pump.

the helical structures and their movements in the membrane which are the heart of the machinery (Figures 4.12(c) and 4.15). The pumping activity, informed by binding with internal and environmental chemicals, and the development of this activity will be seen to be a major factor in the evolution of the geochemical/biochemical ecological system. A third function of membrane pumps can be included – energy capture and transduction which we describe next. Note the importance of phosphate in all these properties which arise from its binding as an anion, its kinetic potential in exchange reactions and its ability to carry energy in pyrophosphate. The last one is exploited in a major group of enzymes, i.e. kinases, for control of metabolism as well as in pumps and in energy capture.

4.13.2. ELECTRONIC AND ELECTROLYTIC DEVICES, ENERGY TRANSDUCTION

To some extent, we have to see a membrane as a separate phase since it is not just a division between an inner cytoplasmic compartment and an outer aqueous compartment. Its enzymes bind both mobile and immobile coenzymes for transfer of electrons and protons in such phases. They are then linked to the fundamental capture and transduction of energy. The immobile coenzymes include Fe_nS_n, haem, and flavin centres, while a major mobile coenzyme in membranes is a quinone. The Fe/S centres are on the internal fringe, while haem and flavin are embedded in the membrane. Quinones form a membrane pool and are in rapid exchange with their binding sites of activity. Together the proteins and coenzymes act to create a potential of electron and H^+ activity along and across the membrane, that is in electronic and electrolytic devices from light or oxidative energy. The basis of electron flow has been described by Edwards et al. and Williams (1961) in Further Reading. The protons flow back through a protein machine (ATP synthase) for synthesis of ATP (see Figure 4.15), and hence are the major source of energy transduction to usable chemical form. The drive is along the membrane but the transduction is inside based on mobile helical structures (see Mulkidjanian et al. in Further Reading). Later we shall note that electrolytic potentials, i.e. electric fields, which are not dependent on single ions but are bulk cooperative phenomena, will be seen to be major units of information transfer along membranes and storage in nerves and the brain and not just of energy transduction.

Taking a final overview of proteins we have to observe how remarkably suitable they are as semi-soft materials. The different variety of sequences and the different ways their folds enable them to act in a variety of ways within the temperature range of water may well be unique. Remember that their value rests not just in structure but in structure associated with thermodynamically controlled features, i.e. concentration, mobility, and temperature. These structures are dynamic and are an essential feature of physical flow, e.g. of electrons and protons and metabolic activity and as such their connectivity is of the essence of energy uptake and degradation.

4.14.   Summary of Non-Metal Functions in Cells

In the above, we have seen the ways in which, in principle, non-metal organic chemicals can be used to make the majority of the energised chemical components of a cell. We consider that the resultant *thermodynamic* flows of material and energy lead to the one major way in which energy can be optimally retained in cells in relation to the environment, while energy is degraded there or in the environment later. (The other major parts are in the concentration gradients of simple ions in cells, in particular metal ions, and in the oxidised chemicals of the environment.) For the energy retention to be optimal the cell requires selected molecules made for a variety of purposes from the available $CO_2(CO)$, CH_4, H_2, NH_3, HCN, N_2, O_2, H_2O, HPO_4^{2-} and HS^- (SO_4^{2-}) in water solution. (Note again the word "optimal" is to be seen in the context of the rate of thermal entropy production as well as that of survival fitness.) The molecules had to form a kinetically stable machinery and an autocatalytic system for relatively long periods of time, and a reproducible self-synthesising unity inside a restricted volume as explained in Section 3.9. We must see that only large molecules and polymers with a membrane could generate such a cell. Moreover, we have been concerned to show that a minimum of four polymer types of molecules were needed: DNA(RNA) to provide a code for reproduction and repair (Note that DNA and RNA are combinations of a saccharide, bases and phosphate; Figure 4.5); folded proteins to generate local structure, including catalytic surfaces, transcription responses, conducting interiors, and machinery (see Figure 4.12(c) and Appendix 4C) (Note that sulfur was incorporated to act as a catalyst centre and to hold certain metal ions), insoluble phospholipids to form membranes (Figure 4.8) and external polysaccharides (Figure 4.13) to stabilise walls or to give an open protective mesh outside the cells. The non-metals in particular oxidation states, dominantly reduced, in compounds, energised, concentrated and organised in space, therefore form a large part in the composition of all chemotypes and allow a great variety of organisms within very limited pathways of metabolism. The very different kinetic properties in different oxidation states allow a variety of activities of H, C, N and O quite different from those of S and P in compounds, and together they are the only available lighter non-metals of functional value, except trace Se, for the above functions (but see uses of halogens late in evolution). We ask again: are the organic chemicals, common to all cells, DNA/RNA, protein, polysaccharides, lipids and metabolome a unique set enabling life?

Now we will return briefly to Sections 3.8–3.11 and 4.6–4.8 where we considered the general problem of multiple flows, here of H, C, N, O, S and P. We observe immediately that all the products are from the same small molecule environmental sources and are required to be formed in relatively fixed amounts using the same source of energy and a series of intermediates. Controlling all the processes to bring about optimum cellular production are feedbacks between them and linked with *the code which generates proteins, and here we note particularly enzymes, i.e. catalysts*. The catalysts are made from the amino acids, the synthesis of which they themselves manage, while the amino acids control the catalysts so as to maintain a *restricted balanced set of reaction pathways in an autocatalytic assembly*. It is also the feedback controls on both the DNA (RNA) from the same units used in the

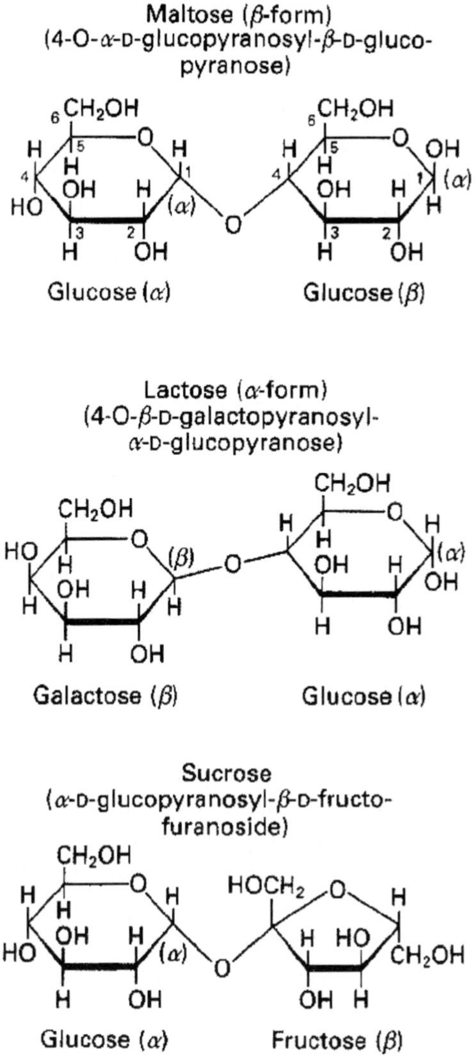

Maltose (β-form)
(4-O-α-D-glucopyranosyl-β-D-gluco-
pyranose)

Glucose (α)   Glucose (β)

Lactose (α-form)
(4-O-β-D-galactopyranosyl-
α-D-glucopyranose)

Galactose (β)   Glucose (α)

Sucrose
(α-D-glucopyranosyl-β-D-fructo-
furanoside)

Glucose (α)   Fructose (β)

FIG. 4.13. A simple saccharide, glucose, in three disaccharides, showing an ether link to itself and two other monosaccharides.

metabolic pathways which are so striking. Thus throughout this book we have to observe changes in catalysis and controls, especially, as well as of DNA, which despite their minor amounts in cells are often extremely dependent on metal ions (see below). *The essential feature of all the activity is that it is a holistic network.* We shall have to see all cells and organisms together in this light too.

We have largely left on one side some functions of one element often thought of as a nonmetal, i.e. hydrogen, since it shares functions of both metals and non-metals (see Section 2.22). We have therefore placed a paragraph on hydrogen biochemistry after the description of metal ion functions in cells (see Section 4.17).

Before going on to describe the functions of the metals we observe that among heavier *non-metals* only selenium, chlorine and other halogens need any further comment. Selenium is found in some hydrogenases in even the most primitive life forms and may be it was required initially since it is a more effective catalytic centre than sulfur although much less available. (Compare tungsten with molybdenum later.) Its amino acid selenomethionine is coded in early DNA! Later it is involved in essential oxygen-atom chemistry. At the same time chlorine and other halogens, X, could not be combined as $-\overset{\textstyle\backslash}{\underset{\textstyle/}{C}} - X$, for example, in organic molecules until cells had access to areally high oxidising potential, but cells cannot help but contain some chloride ion though much is rejected when the environment is the sea. We shall see that no halogenation was used until late in evolution after oxygen became available. The changes in the use of these elements and of C and N are all part of the systematic evolution of chemotypes.

We shall now explain why an "inorganic" metal content of organisms was also essential. Of course, it is obvious that all sources of all elements were "inorganic" initially and, as light was not easily captured, energy could have been taken initially only from the "inorganic" environment. More generally, in a biochemical content, inorganic elements refer to metals (see Table 4.2).

4.15. Why were Metal Ions Required?

In this section the essential requirement for life of different metal ions will be described according to functional use. Before giving such description it is very important to observe their very different energisation from that of non-metals. The non-metals above were trapped in kinetically stable organic compounds, while free *concentrations of metal ions* are trapped as such by pumping into or out of cells and there they frequently *equilibrate* with partners internally or externally. Their concentrations as free ions are very specially controlled in compartmental kinetic traps described in Chapter 5. A very important but obvious use is in structures, e.g. of Mg/K in DNA/RNA, of Ca, Mg and Zn in some proteins, and of Ca in polysaccharides and membrane surfaces, but their most striking value is in catalysis and controls. Without metal ion properties cellular life could not exist. The various metal elements used in cells are shown in Figure 4.2 and we have already observed that the rejection of the ions H^+, Na^+, and Ca^{2+} (and Cl^-) provides a source of energy for uptake and/or a source of information about the environment. We now discuss the different vital catalytic roles of metal ions turning to why certain ions are rejected later.

4.15.1. Reduction and Oxidation Catalysts

As noted above, the primary chemistry of life had to be reductive in order to lower the oxidation states of carbon and nitrogen from CO_2 (CO) and N_2 (HCN) since only on their reduction, i.e. combination with hydrogen, can organic polymers such as proteins, lipids, nucleotides and sugars be synthesised. It was then necessary for cells to have catalysts for the reactions of all available small molecules – H_2, CO, CH_4, (N_2) (H_2O) (O_2) – as well as ions such as SO_4^{2-}, NO_3^-, (HPO_4^{2-}) and so on. Most of these reactions cannot be catalysed by C/N/O/H/S compounds, e.g. RNA or proteins *per se*, so certain *available metal ions* in combination with proteins were essential (see Section 2.10). The only elements capable of catalysing the reduction (or the oxidation) of these small molecules are transition metal ions (see Table 4.9). Looking at the whole range of metal–enzyme catalysts in all organisms, ancient and modern, there is a striking feature in that while iron, nickel and cobalt are mainly used in reduction reactions, iron, copper and manganese are used largely in oxidative transformations, and molybdenum, tungsten, vanadium and selenium are used in a balanced way in oxygen (and sulfur and nitrogen) atom transfer at lower potential. We noted earlier that the chemical catalytic uses in H and C reactions is of later elements in transition metal series, but is of earlier elements in O and N reactions in organic synthesis (Table 4.9); see laboratory and industrial uses in Section 2.21. Of these elements, we know that Fe, Ni and W (or Mo) were used reductively in cells at the very earliest times. There are no *primitive* simple electron transfer centres for reducing reactions containing any metal other than Fe in Fe/S centres, while nickel and iron were the original main centres for reductive substrate chemistry, for example, in H_2 and CO chemistry. Note that nickel and iron in sulfides, M_mS_n, as M^{3+}, M^{2+} or M^+ ions found in these compounds are electron rich, while simple iron and manganese oxides are not (see Section 2.16). Electron-rich metal centres are very valuable in reductive catalysis since they form associations with electron-poor unsaturated organic molecules such as CO and CN^-, donating electrons to them. Initially, there were not many available electron-rich inorganic mineral compounds other than those of nickel (and cobalt) sub-sulfides, i.e. FeS_2, pyrite, and FeS. Electron richness is increased in all low-spin metal ion organic complexes, often kinetic traps, and we shall see that all three metal ions, Fe, Co and Ni, in such states evolved in complexes (porphyrins) and found uses a short while after the history of organisms began (Section 5.2). (Their low-spin complexes, except of Ni, were not available at the earliest times since they require special binding ligands.) Later in evolution oxidising agents became available and electron deficiency rather than richness became valuable. We note how the metals manganese (zinc) and then copper increased in value as they became available, while nickel and cobalt became progressively less used. The availability of iron decreased markedly, since it became precipitated as $Fe(OH)_3$ in the environment, but its uses increased somewhat requiring novel ways of capturing iron by the synthesis of sequestering agents. In Chapter 5, we shall indicate why the selection of elements in primitive life was so limited and in later chapters we describe the expansion of use to other metals as oxidising conditions were introduced in the environment (see Section 1.11). (Note it is the elements employed most in early evolution (Mg, Fe, Co, Ni, Mo), which are found in special kinetic traps, e.g. porphyrins for all but Mo.) The limitation of solubility in the two different conditions applies not just to the environment but also to the free element concentrations in any compartment in a cell (see Figure 1.8). Thus, a cell compartment has imposed upon it equilibrium conditions of free element bindings – the free metallome is fixed by complex ion binding, solubilities and redox potentials outside and inside cells and by organic synthesis of ligands and pumping considerations inside cells. We have described these equilibria in Sections 2.17 and 3.14 and see Section 4.18. *The overall impression is that the selection of metal ions for acid/base or oxidation/reduction reactions was an inevitable consequence of their bindings as well as the usefulness of their chemistry, hence they form an integral part of all chemotypes*; however, in proteins they are also specially constrained opposite function (Table 4.10 and Section 2.20, see Williams and Fraústo da Silva in Further Reading). However, relative to the sea most metal ions in a free state are reduced in concentration since they are poisonous at higher concentration. It is readily seen that these elements are just as essential to the thermodynamic system of the coupled organisms/environment as are the non-metal elements. We ask, was a different selection of metal ions possible? (see Appendix 4A).

Apart from individual sites, series of metal ion sites provide electron conduction paths, vital in energy transduction in all organisms and leading to proton transfer, and Mg^{2+} in chlorophyll is essential for light capture (see Section 4.17).

4.15.2. ENERGY AND GROUP TRANSFER

In Chapter 3 we described the possible external sources of energy required for life. Here we shall assume at first that the most primitive form was not light but the chemical energy "stored" in unstable minerals. Such minerals were the metals and metal excess sulfides and iron sulfide in their reactions with water or hydrogen sulfide to produce hydrogen (see Wächtershauser in Further Reading) or were stores in the out of balance of states of non-metals such as S_n/H_2S.

The earliest transduction of energy could have been (1) by the direct reaction of these sulfides plus acetate to give thio-acid compounds, CH_3CO-S-, and (2) from the generation of oxidation–reduction reactions of these same energised minerals with reduced carbon compounds across or along membranes to give (energised) proton gradients from hydrogen (see Section 5.11). The proton gradients were then used either directly in the uptake or rejection of elements or they were employed to make pyrophosphate derivatives, e.g. the ATP described earlier which is really Mg.ATP. Such reactions require long-range electron transfer, where Fe_nS_n centres are particularly important. Later, Fe haem and Cu complexes are linked to both photo- and oxidative energy transduction (see Williams (1961) in Further Reading).

Apart from reductive steps, the formation of many organic compounds, especially the biological polymers, requires transfer reactions of fragments using energy, pyrophosphate derivatives (Mg.ATP), all of which need a different catalyst. The transfer of groups such as acyl and alkyl as well as of energy requires somewhat reactive centres. As stated earlier, in general, all rates of acid–base transfer increase down the Groups of the Periodic Table N $<$ P $<$ As; C$<$ Si $<$ Ge; O $<$ S $<$ Se; F $<$ Cl $<$ Br $<$ I. This feature of transfer by mainly heavier non-metal inorganic chemistry is carried over into biological chemistry, mostly in coenzymes (see Section 5.4). The transfer of carbon is usually from an RS-carbon centre, of hydrogen is from an S or Se centre, of phosphate is from a phosphate centre, and of energy by S or P transfer-based reactions, but metal ions are also frequently involved (see below). Note that the original transfer of oxygen or nitrogen and possibly sulfur atoms uses W, V or Mo, while later Mn, Se, Fe, and Cu, are used, but transfer of hydrogen involves Fe (Ni) especially, while transfer of phosphate is catalysed by Mg. Again and again, we have to consider if these uses of elements in various mechanisms and pathways were the only possible ways for the system we call life to evolve, given environmental availability. The matching of function with the known chemical potentialities of the elements is extremely suggestive that there was but one effective way.

We must remember too that all the gradients of ions (or molecules) across membranes, represented sometimes by electrical potentials, are *bulk* sources of energy, not of specific chemical use but are of general value in uptake/rejection and signalling (see Chapter 9).

4.15.3.   CONDENSATION AND HYDROLYSIS REACTIONS

The limited choice in the primitive environment of catalytic metal elements for oxidative/reductive and transfer catalysis was matched by the very limited choice of hydrolytic or condensation metal catalysts available from the environment for reactions of substrates based purely on O-centred donors. Here metal ion binding to such O-donors is clearly required and since such binding is generally weak a high concentration of the metal ion was needed. We find, not surprisingly, that condensations were and are very largely restricted to catalysis based on magnesium ions, Mg^{2+}, since there was no other available cation with adequate (weak) binding ability and yet sufficient concentration to attach itself to such centres (see Figure 2.8, for example, in sugar and phosphate reactions). This restriction to Mg^{2+} also applies to the catalysed chemistry of energy use and transport by pyrophosphate (Mg) or ATP.Mg, where the phosphate groups are O-donors in their metal complexes, and hence to virtually all pumps and syntheses by condensation driven by these pyrophosphates. It is very usual for Mg^{2+} to be used in hydrolysis too but we shall note in Chapter 5 one or two exceptions, where Fe^{2+} is used, and in Chapter 7, where Mn^{2+} is employed. Later in evolution Zn^{2+}, when it became available, was much used externally with Ca^{2+} which, though always available, was rejected even by early cells as it collapses DNA and has insoluble salts. Both metal ions became catalysts of hydrolysis outside cells.

Summarising, for initial and later reductive, hydrolytic and condensation reactions and initial and later oxidative pathways it was different chemical necessities that drove the selection of the elements available in different environments and from which the catalysts of life evolved. It could well be that simple iron, nickel, tungsten (or molybdenum) and magnesium–protein complexes in different roles were the only possible and the first essential catalytic metal ions. We discuss the case of the choice of Mo or W below. The requirement for catalysis is at the heart of biological flow.

Now magnesium with potassium are also essential for the stability of the structures of DNA and RNA compensating its negative charge. In particular, later in evolution the ends of DNA, telomeres in eukaryotes, require K^+. The folded forms and catalytic activity of RNA are absolutely dependent on Mg^{2+} for their structures and no known RNA is without this cation (see Section 4.10 and Figure 4.10). Mg^{2+} and Ca^{2+} are also extremely valuable in stabilising membranes and walls. Under conditions of environmental and cellular organic chemistry no other cations can carry out these functions in which reasonable concentration, structural flexibility and very fast exchange (equilibria) are necessary.

4.15.4.   OSMOTIC AND ELECTROLYTIC BALANCE IN CELLS

At the beginning of this chapter and in describing the principles and practices of organic chemistry in the laboratory, in industry or in life, we explained that containment of reactions in vessels (cells) was necessary. Unfortunately, a cell, unlike laboratory or industrial vessels, will burst if the concentration inside it is not kept commensurate with that outside it. Similarly, charge inside the vesicle must be close to neutrality or the electrostatic repulsion will cause rupture. Now, the metabolic paths of organic chemicals in water are largely made up of anions in order to maintain solubility. The anions are mostly carboxylates and phosphates, often of side chains of small and large molecules. To avoid osmotic and electrostatic breakdown the concentration and the total charge content of primitive cells of both environmental and organic ions had to be controlled to match closely their values in the neutral sea water external to the cell (see Figure 4.8). The major osmotic component of the sea is sodium chloride and it was clearly necessary to expel this salt more than ten-fold from inside the cell to reduce cellular osmotic pressure. This leaves the electrostatic anion problem unsolved but it was resolved by admitting potassium ions. Potassium is the only sufficiently available ion in the sea for this purpose. (Note that binding is not desirable here.) At the concentrations of K^+ inside the cells the equivalent amount of Na^+ would bind considerably more strongly giving another reason for its rejection. Thus, gradients of Na^+, K^+ and Cl^- became a fundamental feature of cell systems for all time. In fact, the almost fixed ionic composition of cell cytoplasm of the vast majority of organisms, ancient and modern, is a remarkable feature of evolution. Once again evolution had no choice but to make a virtue of necessity. We note that these properties of cells are associated, like membrane potentials, with bulk fields and not local molecular thermodynamic properties. It is not possible to treat these functions in terms of the properties of single ions or molecules. The gradients are used to energise pumps and later in evolution we shall see how and why these ions were critical for fast message systems in nerves and for the functioning of the brain.

4.15.5. CONTROLS OF METABOLISM

The very fact that the metal ions equilibrate relatively rapidly makes them very useful in all forms of information transfer. We shall see in Chapters 5–10 how this function grows in evolution starting from internal communication in the cytoplasm, to linking the environmental and the cell compartments, to long-range communication between cells (nerves), and to integrating whole body activities with a memory (brains). The progression is systematic and could only use those metal ions with an exchange speed commensurate with an activity. We shall note the earliest roles of especially Mg^{2+} and Fe^{2+} in general metabolic controls with the later addition of functions of Ca^{2+}, Zn^{2+}, Na^+ and K^+. We stress that the metal ions play an equal role in controls with non-metal molecules in the integration of cell activity from the level of transcription down to that of basic metabolism and uptake.

4.15.6. PREVENTING INHIBITION: REJECTION

Now, in addition to sodium and chloride, calcium had to be removed from all cells since at a concentration $>10^{-3}$ M it forms precipitates or incorrect structures with many inorganic and organic anions, so that it would have damaged biological metabolism. (Rejection of poisons of all kinds is necessary, but energy wasteful for survival and growth of cells.) In effect, we observe very low free calcium in all prokaryote, $<10^{-5}$ M, and eukaryote cells, $<10^{-7}$ M. The outward pumping of calcium, sodium and chloride (and even manganese) together with the uptake of potassium and phosphate are inevitable requirements that had to be established in "cells" effectively before or as they were coded.

Other elements such as Ni^{2+}, Co^{2+}, Cu^{2+} (Cu^+) and Zn^{2+} would also be poisonous if their free ion concentrations exceeded about 10^{-10} M since they could then compete with Mg^{2+}, Mn^{2+} and Fe^{2+} in essential roles (see Chapter 2). Inspection of cells throughout all of the history of life shows that this free level is not exceeded (see Figure 4.3). At the same time, all of these elements are now essential in cells, so that there has to be controlled uptake and rejection to match the synthesis of the functional partners of the ions. The need is met by a set of pumps and transcription factors, which have suitable selectivity and high binding constants for the monitoring of the ion concentrations and then the synthesis of their organic molecule partners – often proteins and coenzymes. In this way, all metal ions are built into an informed total activity with non-metal chemistry. (See how the inverse of the *universal free ion concentrations* in cells is related to the Irving/Williams stability order, Section 2.17 and Figure 4.3.) The very low concentrations of free ions requires carriers to equilibrate them in the cell and other proteins to store and buffer their activity, e.g. for Ca, Fe, Cu, Zn, Ni, Co and Mo, and this equilibration often extends to their enzymes.

We must not miss in this description that the pumping of gradients in and out of cells of inorganic elements requires a considerable amount of energy (ATP). Together with the concentrated synthetic organic chemicals in a cell and the oxidised external environment this energy is then a considerable contributor to energy storage. These gradients became extremely useful later in uptake and signalling (see Chapters 6–9). We shall note again and again the progression in evolution from recognition and rejection of a poison, e.g. Na^+, Ca^{2+}, Mn^{2+}, $Cu^{2+}(Cu^+)$, and Cl^-, to its later functional value, often of its gradients. In conclusion, we stress that the control of concentrations of about 12 metal ions is an essential requirement of all organisms and is a thermodynamic feature different in different chemotypes. The concentration of attention on the DNA and genetics in modern biochemistry is hiding fundamental features of life limited yet permitted by the environment. There is a "fitness" of life in the environment.

4.16. Combining Metal and Non-Metal Chemistry: Structures and Activities

All the metal ions discussed above except Na^+, K^+ and Ca^{2+} have major uses only in combination with organic ligands, often proteins and protein/protein combinations. For example, many of the primitive enzymes contain chains of electron transfer which allow the connection of the electron donor site (say a metal sulfide) to a metal catalytic site for reduction. In bioenergy capture there are chains of metal sites in proteins for removal of energised electrons to release protons, so creating a gradient which is the source of Mg.ATP production. Chains of electron transfer catalysts, metalloproteins, have an enormous importance in all life; see Williams (1961) in Further Reading. We shall see such chains in photosynthesis (Figure 5.2), and oxidative phosphorylation as well as in the handling of many small molecule substrates such as H_2, CO, N_2, NO_3^-, SO_4^{2-}, O_2 and so on. Outstandingly, it was a great variety of Fe_n/S_n centres in proteins (Figure 5.2), which formed the very primitive chains, while later they were augmented by iron porphyrins (see Chapter 5), and later again by copper enzymes (Chapters 6–8). For long-range electron transfer primitive sites, Fe_nS_n centres or iron haem compounds, are almost always about 10–20 Å apart, held selectively in place by proteins (see Section 4.11), allowing distances of up to 100 Å for electron transfer in chains. The terminal catalytic sites of the chains are usually composed of Fe, Ni and W (Mo) for reaction with small substrates. Later there were added Mn and Co and finally Cu sites for oxidative/reductive chemical reactions apart from electron transfer. Many of these metal centres must be held in an ordered array. Again we can ask: was it possible to achieve such effective and efficient functions in any other way given the availability and limits of the elements of the Periodic Table? We shall describe in Chapter 5 how later in evolution additional specific novel complexes of inorganic elements and organic binding (chelating) agents bound in proteins and protein combinations appeared in the evolution of the later prokaryotes and have been retained in evolution by all organisms. However it is clear that very early in evolution binding of ions in proteins was already highly refined (see Section 2.20). The challenge is for chemists to show that alternative thermodynamic systems of life could have occurred if this was indeed possible. Observe again, for example, the unique value of molybdenum in some of these catalytic chains making it (or tungsten) essential for all life. Is there any other element or compound that can perform its function of O-atom transfer at low redox potential? As an element or in complexes, it has a unique range of almost equal redox potential valence states (see Table 6.1 and Fraústo da Silva and Williams (2001) in Further Reading).

In concluding the sections on metal ion involvement we must observe again that the metal ions involved in pumps, enzymes, transcription, etc. have concentrations of both free ion and complexes under feedback control. Much as non-metal cell chemistry is under general pathway selective controlled autocatalysis, these metal concentrations are under pump controls and they control the selective synthesis of binding proteins. Since they regulate the synthesis of their partners and the pumps, which are also regulated by external ions, they are under autocatalytic control in all respects. There is however an important factor, equilibria, missing largely from the non-metal chemistry as well as feedback kinetic controls. Before we introduce them we must refer to the non-metal element hydrogen also involved in equilibria but in a different form when under kinetic control (see Section 4.18).

4.17. The Biological Properties of Hydrogen

In Section 2.22, we pointed to the peculiarities of hydrogen as an element stressing that it was multi-functional. So far in this chapter we have seen it as only an element combined covalently with C, N, O, S and Se in organic compounds, and we stressed the presence in the initial atmosphere and aqueous solutions of the hydrides of all these fine elements. Now we must return to its other roles which also are general to all biological solutions since these solutions are some 80% water (H_2O) and water dissociates at equilibrium into H^+ and OH^-. H^+, the proton, is an acid capable of acting as an attacking group while OH^- is a base equally capable of being an attacking reagent in hydrolysis. Together they also buffer, at equilibrium, the aqueous environment at pH 7 for the most part interacting with acidic and basic species such as carbonate in cell compartments. It is very important to maintain buffering (homeostasis) at this pH, $[H^+] = [OH^-] = 10^{-7}$ just because of the attacking powers of these two agents. As a result of buffering, the proton is also bound to groups such as $-NH_2$ at equilibrium but not to other organic groups generally, e.g. to $-CO_2^-$ of proteins. Now there is a third way in which hydrogen can act: it can be transferred as the hydride anion, H^-, not at equilibrium with bound H or H^+. The change from H^- to H-bound to H^+ is vital in biological chemistry reaction (see Williams (1961) in Further Reading).

$$H^- \text{ or H-bound} \rightarrow H^+ + \text{electron(s)}$$

This reaction is central to the capture of light energy and indeed of the capture of the energy from oxidation of $-\overset{\diagdown}{\underset{\diagup}{C}} - H$ compounds. The oxidation itself is often by oxygen liberated from water by the reaction due to light

$$2H_2O \rightarrow 4H^+ + O_2 + 4 \text{ electrons}$$

The H^+ here forms a concentration gradient across or in membranes which as an energy source can be used to drive condensation reactions via pyrophosphate formation (Mg.ATP). During this reaction step, the proton flows through proteins due to its small size and fast exchange from organic side chains. Its many steps may be close to equilibrium although the gradient exists across the membranes. We must see H^+ and e flow as currents in parts of electrolytic machinery.

The many functions of hydrogen then allow it to be involved in many organic and inorganic acid/base and oxidation/reduction reactions. The danger is that its general role will allow its presence to be taken for granted, though the very presence of water (H_2O) is the basis of life.

4.18.   Cell Organisation and Constraints: Equilibria

In the above section, we have shown that the whole apparatus of a cell is organised by thermodynamic and kinetic constraints on concentrations of all its chemical components. We know, in fact, that individually and cooperatively the organic and inorganic molecules and ions are controlled in a cell in a given state provided that external conditions of material and energy availability are fixed. This is known as a homeostatic steady state and not an equilibrium condition. Now there are two kinds of constraints, which we mentioned in Chapter 3. The first is equilibrium, which applies when combinations of components are in balanced concentration with their free entities

$$A + B \leftrightarrows AB \quad \text{and} \quad A + M \leftrightarrows AM$$

Such internal thermodynamic equilibria where A is a protein are found for non-metal components, including free coenzymes and substrates where B is a small molecule, or where free M is an ion of either a non-metal, e.g. Cl^- or HCO_3^-, or a metal, e.g. K^+ or Mg^{2+}, or is H^+, and they are involved in, even necessary for, catalysis, pumping and cooperative controls of many metabolic paths. All such combinations reach equilibrium, as long as exchange is fast, where a fast rate can be taken as, say, 10^{-3} s for dissociation in cells. Note that equilibria with defined binding constants for AB or AM formation in any system reduce the number of variables and hence AB and AM concentrations are defined by those of free A, B and M, leaving two independent variables for each equilibrium. In some cases, the restrictions due to equilibria are more limiting, for example where solubility products act as restrictions when free M(B) or A becomes the only variable. Here there is no concern with rate constants themselves so long as rates are again fast. Note that all *equilibrium constants* are temperature dependent but they are independently fixed for given interacting pairs in *all* life forms and provide fixed unavoidable thermodynamic constraints, e.g. in the formation of Mg.ATP in all cells.

Furthermore, many such equilibrium constants must be approximately the same for all sites of either a given B, or M, binding individually in one compartment of a cell if the different processes dependent on a particular component, either B, or M, are to act together. Thus, there are many important binding constants fixed for all cells since all have similar metabolism and the types of organic binding centres for selective combination with each metal ion are fixed. They are to be considered as invariables together with separate equilibrium constants in the environment. We consider their values in Appendix 4A. There is no required equilibrium between the environment and the compartment contents as the cell contents are energised or separated in different compartments but there is often steady-state exchange

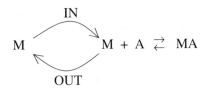

governed by rate constants even across membranes.

Finally, we note that certain oxidation/reduction processes are also at equilibrium in cells, for example, between Fe^{3+}/Fe^{2+} and the $-S-S-/H_2S$ couples, but most such reactions are kinetically managed; see Appendix 4B.

4.19. Kinetic Controls and Networks and Their Energetics

As we have stressed in Chapter 3, there is another feature of systems of reactions in which the forward and back reactions are on different pathways, which is quite different from equilibria (Figure 4.14). The forward and back reactions can have different catalysts and even carriers of reactants and information. We see this, for example, in the synthesis of saccharides, say gluconeogenesis and its reversal, glycolysis, for which pathways there are distinct enzymes and controls. Again the carriers of hydrogen-reducing equivalents are different in oxidation, i.e. NADH, and reduction, i.e. NADPH (see the pentose shunt). Such differences appear in protein synthesis and degradation too where Mg.ATP is made in degradation while Mg.GTP is mostly used in synthesis. It is this separation of catalysts and coenzyme carriers of material and energy which allows greater control over the composition of the system than is possible in a less complex organisation and of course than at equilibrium where kinetic control is lost. Moreover, as stated above, while some pathways operate close to one set of oxidation–reduction potentials, others act at another, and combinations of such different couples are used in energy capture

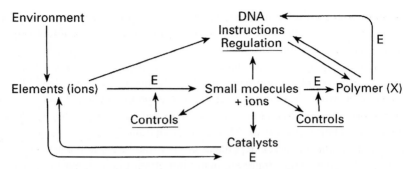

FIG. 4.14.   The polymers (X) acting on DNA are transcription factors activated frequently by small molecules or ions with a wide variety of time constants. E is an enzyme. Instructions are passed to a polymer (X) or directly to DNA. Energy requirements not shown (see also Fig. 3.13). *Note.* All these cycles are element neutral, non-polluting.

across membranes. The implication is that each carrier must only react in redox changes at selective enzyme sites but it can act also as a control in equilibrium exchange. A further remarkable development in evolution was to create separate binding groups for each of the four metal ions, Mg^{2+}, Fe^{2+}, Co^{2+} and Ni^{2+}, such that there was no exchange and so these metal (porphyrin) complexes make possible use of metal complexes in oxidation–reduction reactions in the same way as the non-metals are used in non-equilibrating, rate-controlled, traps (see Chapter 5).

Looking back at Table 4.4 the anaerobic cyclic steady state is from the first product of $CO(CO_2)$ reduction, formaldehyde (HCHO), to $[CH_2OH]_n$. This step requires a kinetic pathway in redox potential from -0.2 to -0.6 V versus the H_2/H^+ potential at pH 7, where sulfur is the oxidised waste:

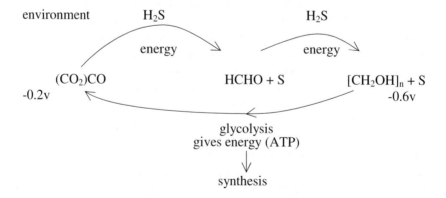

Under oxidising conditions the range of potential is increased from O_2/H_2O at a potential of $+0.8$ V externally to around -0.2 V (CO_2/CO potential) in the cytoplasm. The average of all redox potentials of the major reactions in the cytoplasm has then remained virtually unaltered over all time but this average is not an equilibrium; it is one covering a range of different non-equilibrating flows of oxidation/reduction within the cytoplasm. Other cell compartments are at different potentials perhaps even approaching the O_2/H_2O potential, e.g. the outer side of mitochondria. However, all these pathway rates are restricted by feedback controls between material and energy carriers while the individual carriers can equilibrate.

The final restriction on the cell is not information transfer within a pathway or between pathways but of its transfer from DNA(RNA). This transfer is complicated in that it requires a reading machine going in one direction along DNA(RNA) and an energy source. The machines are DNA/RNA transcriptase at DNA or translational enzymes in ribosomes, i.e. RNA/proteins (see Figure 4.9). We examine them in more detail in Chapter 5. Here we note that both are controlled by information from cell contents, ions and molecules, via transcription factors, proteins. When we looked at clouds in Chapter 3 we saw how static physical force fields control flow shape and content, while here the control is through chemical force (concentration) fields and filament tensions as well as by physical membrane boundaries. However, we must never forget that all flow demands constant input of energy and that control over energy distribution in a cell is extremely important (see Chapter 3 and Section 4.5). The whole of a cell is a massively holistic constrained machinery of chemical and energy flows. We trust that the combination of the ways organisms are homeostatically controlled in their activity allows the reader to see that any group of them can be classified as belonging to a chemotype based on element content combination, energy intake mode, use of space and organisation separately, and that this approach gives a very different thermodynamic insight from the genotype or any other sequence characterisation or from morphological features.

4.20. Summary

We have now completed the description of the basic thermodynamic chemical principles of all cells describing especially their component chemicals (Table 4.11), and their interactions but we have not given any specifics about the uses of space and hence of spatial organisation. It has been our concern to show that the overall selection and use of some 20 elements and of energy in the synthesis of molecules in all chemotypes is very "economically" effective, perhaps unique. While we have often dwelt on molecular or ion-specific features in this chapter we remind the reader that the only form of life we know is intensively cooperative in all of them, selection and use of components, use of energy, synthesis of containers, of catalysts, of coded molecules, and of degradative pathways. This is apparently the only system of flows that could be sustainable and could grow and then evolve (see Section 3.9). It is controlled, limited and autocatalytic and its evolution is based on the separation of reduced from oxidised material, the inevitable consequence of energy absorption. In this very general approach to biological systems we see that evolution of an energised chemical system is driven by the fact that it is an unstable product and ultimately "trying to be" cyclic. In such a cyclic system, energy uptake would be optimal that is consistent with optimal survival of material flow and optimal heat

TABLE 4.11

THERMODYNAMIC CHARACTERISTICS OF ALL CELLS

1.	A genetic linear coded *molecule* of four bases in DNA, the genome
2.	A controlled transcrible *concentration* of messengers, RNA
3.	A controlled translated *concentration* of active components, proteins, the proteome
4.	A basic set of metabolic pathways with controlled *concentration* of particular small molecules, the metabolome: note also Cl^-
5.	A membrane *containment* based on lipids
6.	A controlled *concentration* of free metal ions, the metallome. Note especially free $K^+ > Na^+$, $Mg^{2+} > Mn^{2+} > Fe^{2+} > Co^{2+} > Ni^{2+} > Cu^{2+}(Cu^+) < Zn^{2+}$ maintained at equilibrium with bound forms in the cytoplasm, and a required presence of Mo(W) complexes
7.	A low *concentration* of Ca^{2+} ions
8.	An energy distribution system based largely upon proton gradients and organic pyrophosphate *concentrations*
9.	The whole of 1–8 is based on feedback control which limits concentrations due to production of their catalysts: the system is autocatalytic
10.	Dominant features of the chemistry are energised reduction of environmental chemicals, energised synthesis, condensation polymerisation, reproduction and discharge of oxidised chemicals and unwanted elements to the environment
11.	Progressive back interaction of the environment chemicals and cell metabolism
12.	Higher degrees of organisation appear successively in evolution but all organisms evolve in their chemistry with time to some degree
13.	All cellular compounds and gradients inevitably decay generating heat, a process, light $\rightarrow$ heat, which drives the whole of the chemistry.

Note: The whole of the system is molecular machinery driven by outside sources of energy and is temperature dependent. We see the cell as a machine degrading energy from either inorganic (Earth) sources or the Sun (see Chapter 5).

production (optimal rate of thermal entropy creation, survival of fitness, see Appendix 4C). In Chapters 5–9 we shall describe real organisms and their evolution: first the most primitive cells known and then the evolution up to the organisms (chemotypes) of the greatest degree of internal organisation noting the appearance of more and more separation of oxidised from reduced material both between the cells and between compartments not described in this chapter. We shall also observe how later organisms depend on earlier organisms in modern form and how all organisms depend upon the environment as it changed. It is the total ecosystem, Sun/environment/organisms, which evolves in a systematic way to increase survival of the whole and the rate of energy absorption and degradation, that is thermal entropy production (see Chapter 3). It is the realisation and appreciation of the connection between all these factors by mankind that is gradually making for a quite new evolution paradigm, but we shall be cautious about its promise in the future since we are aware of certain limitations. We stress we have to think in a thermodynamic way about an ecosystem, and not in terms of the properties of isolated molecules or species; hence the efficiency of the chemistry of the environment as well as that of cells must be constantly in mind (see Figure 4.14). The implication is that we shall not concentrate upon life's organic molecules, including genes, or upon organism morphology but upon the *changing chemical element and energy flows* shared between biota, that is in chemotypes, and abiotic chemistries. A major problem for us is then that the very variety of the organic chemicals and their kinetic stability makes them difficult to analyse in the context of our objective, namely to illustrate the directional character of the evolution of our ecosystem, especially as these chemicals vary from species to species as well as within chemotypes. We shall only be able therefore to refer to features of this chemistry in general terms such as the changes in pathways, the introduction of coenzymes, and the general nature of biopolymers noting any special changes which they undergo with time. The difficulty of analysing this complexity contrasts with the relative ease with which it is possible to examine the inorganic elements and their changes in cells and the environment with time, much of it limited by equilibria. It is the case that these changes are not very species dependent, until the advent of mankind, but are seen strongly in the differences between chemotypes. Consequently we shall constantly refer to basic element analytical features in evolution. Nevertheless we do consider that there must be a general overall directional change in the organic chemistry of cells to match the changes in the environment and the inorganic element changes. (This chapter does attempt to show that features of both the organic and inorganic chemicals in cells have a fitness which is perhaps unique). It is this progression which leaves us with little doubt that there was but one way in which evolution could occur due to the very nature of the internal energised chemistry of cells and the effect of the selection of elements in them upon the environment which back interacted with the cells. Cellular evolution is the result of environmental stress, (see cover) leading to novel survival mechanisms. In this process, the cycles of all elements between the cells and the environment become neutral, eventually with no changes and no pollution. Mankind's activities, including global warming, must be seen in this light. It is also this consideration which allows a proper view of risks in the future (see Chapter 11).

We emphasise that the next chapters refer only to the surface of Earth to which light and the atmosphere have access. This is a common restriction in the discussion of evolution but we shall have to examine also the geological and biochemical zones in (and beneath) the deep sea (in Chapter 11), where it appears that evolution could be taking a somewhat different and as yet less advanced route but based on the same principles. We emphasise that each chapter adds new uses of elements, of energy, of space, and of organisation with species variation as new chemotypes evolved. The thermodynamic characteristics of all cells are given in Table 4.11.

Appendix 4A. The Magnitudes of Equilibrium Constraints in Cell Systems

In Chapter 2 we have described the nature and strengths of metal ion binding sites in the equation for complex formation in chemistry

$$A + M \leftrightarrows AM$$

We now look at the values of the free M concentration and hence to the binding strength to selected A synthesised in the cell. The constants are closely common to all cells in their common compartment, their cytoplasm. The values, suited to metabolism, can be put in series in which Na^+ and K^+ bind poorly and only to a few of the weakest donors based on neutral O-donor centres while other metal ions bind more strongly to O, N and S donors of proteins or small organic molecules in a well-recognised order, i.e. in the Irving/Williams series (see Section 2.17):

$$Mg^{2+} < Mn^{2+} < Fe^{2+} < Co^{2+} < Ni^{2+} < Cu^{2+}(Cu^+) > Zn^{2+}$$

This *equilibrium* series is common to all cell cytoplasms and we need to see how it is managed through the synthesis of A. A cell requires selective use of each metal ion and consequently it must bind each one differently to a given donor site, A, at equilibrium. We showed that the very nature of ion selection requires that each free M concentration should be controlled by pumps and each A should be produced such that for the tightly bound metal ions there is little excess A over MA (Section 2.17). This condition can be achieved by generating a controlled amount of A commensurate with the M^{2+} present, which is brought about by feedback control via transcription factors to DNA expression. The cell metabolism is then related to the activity of each MA which becomes of a fixed value. Here the selectivity order demands that the A produced for the strongly binding M has very strong selective donor sites. Reference to Section 2.17 shows that these donor sites must be related to the very chemical donor atoms and cavity sizes in molecules much as is used selectively in chemical analysis. Now the input of M to a cell is controlled by pumps which also have to be selective in exactly the same manner as all other bindings of M. The overall set of exchange interactions which settle the production of MA are then dependent on the following constraints: (a) feedback from free [M] or a unit related to it in the cell to its selective pump which stops further uptake of M; (b) feedforward to DNA(RNA) by [M] via transcription factors which control production of A, carriers of M, and the pumps while the product of the MA activity limits the activity itself and may also limit A production. The variety of controls being at equilibrium demands that each binding to the different proteins must have similar binding constants since all are to be operative simultaneously. Moreover, the binding constants for different metal ions have to be in the above series for the required selectivity to be achieved. The observed concentrations of free metal ions are related to the inverse of these binding constants (see Figure 4.3). Of necessity, for the strongest binding ion, i.e. copper, its concentration has to be less than about 10^{-15} M, while for the weakest binding, i.e. Mg^{2+}, a binding constant of about 10^{-3} M is required. All other M^{2+} ion concentrations fit into the sequence of the inverse of the Irving–Williams binding constants (see Figure 4.3). Hence, we find in all cell cytoplasm very similar and highly selected MA binding groups with fixed series of free M concentrations, but not of MA since cells produce different A for different purposes. We have called the free [M] in the cytoplasm the free metallome. Only in this way can a cell become homeostatically controlled in a cyclic steady state. *We see that there are common thermodynamic equilibrium features of all chemotypes, i.e. of all living organisms, largely independent of the environment, hence the data in Fig. 4.3 are close to invariant. This is a striking limitation on evolution, quite different from that of a code.* Similar restrictions apply to fixed solubility products and redox potentials (see Section 2.18).

Restrictions also exist on availability to the cell due to the binding of the metal ions (in the same equilibrium series) in the environment outside cells. Primitive conditions of the sea included the presence of sulfide at around 10^{-3} M HS^-, which precipitated ions such as Cu, Zn, Cd and Mo reducing their concentration of free ions to such low levels that cells could hardly incorporate them at all (Tables 1.7 and 4.3). (Note however that they could take up small units such as M_nS_n in the case of Fe and Ni.) When the advent of oxygen removed sulfide the contents of the sea changed (see in Table 1.7), and solubilities now depended on hydroxide and carbonate. We consider that these changes, among other effects of oxygen in the atmosphere, led to the enforced evolution of cellular life through changes in free ions. Thus, two different equilibria inside/outside limit the way in which cells can take up these ions. It is very interesting to see the values of free ions outside cells when organisms evolved into multi-cellular species with their own extracellular fluid (see Chapter 8). Note that similar considerations apply to anions such as Cl^-, SO_4^{2-} and MoO_4^{2-}.

Since the elements in the environment changed during evolution due to oxidation, e.g. sulfide, oxidised to sulfate, became unavailable in the sea, cells had to adjust somewhat their chemical binding agents to ensure that the increased availability of the heaviest metals from Co^{2+} to Zn^{2+} and Cd^{2+} did not damage their internal activities, especially of Mg^{2+} and Fe^{2+}. For example, protection from Cd^{2+}, not use of it yet, has evolved. We shall find that in an oxygenated atmosphere novel protein pumps and transcription factors appeared to control the concentrations of transition metal ions (see Section 6.8), but leaving the free metallome in the cytoplasm virtually unchanged. Now, outside this compartment much of biological chemical evolution took place under different restrictions, in fact in different compartments, vesicles and extracellular fluids and so helped the major chemotypes to evolve (see Chapters 6–10) and mankind became a special case. (Note that when cell differentiation occurs we have to deal with a rather different problem.)

The very great reduction of free ion concentration in cells, due to high binding constants for the ions at the end of the above series, means that some of them cannot change partners very rapidly. This equilibrium binding chemistry prevents their use in processes requiring very *fast* change of metal ion partners and such exchange is permitted only for Na^+, K^+, Cl^-, Mg^{2+}, and Ca^{2+}. The implication is that Cu^+ and Zn^{2+} exchanges are only of value in very slow processes, and carrier exchange is needed to assist equilibration. Fe^{2+} and Mn^{2+} exchange remains very valuable in processes which are relatively fast since these cations are in relatively fast exchange. We shall find the use of differential binding and rate constants central to biological control, mechanisms, homeostasis and rates of response. It is then necessary to see that only certain metal ion rates of exchange are useful in accordance with the lifetime of the organism. Bacteria may take 1 h to divide, while higher organisms may take more than 1 year even after a growing-up period of more than 10 years. Thus it is possible for organisms of long lifetime to use stronger binding and slower exchange still at equilibrium. We shall see how successively Fe^{2+} and Mg^{2+} and then $Cu^{2+}(Cu^+)$ and Zn^{2+} exchange became useful in evolution. Inorganic elements as well as non-metal organic compounds are involved in exchange controls over considerable parts of metabolism and are not just catalysts. There are thermodynamic and equally genetic factors to be considered in evolution.

As stated, these equilibrium considerations apply equally to some non-metal chemistry, especially to small substrates and mobile coenzymes, but here as well as pumps and transcription factors there is the control of small molecule synthesis itself. In fact, the distinction between rates involving weak and strong binding applies to proteins and RNA as well in their associations. There is another difference between metals and non-metals in that non-metals are stored in kinetically stable non-equilibrated compounds such as sugars, fats and proteins, while metal ions, with the exception of porphyrin complexes, are stored in buffers and precipitates while both may be held free in membrane-evolved vesicles at arbitrary levels. Finally, note that hydrogen can behave here as elsewhere like a fast exchange metal ion or as a very slow exchanging non-metal, the first in association most frequently with N-, O- or S-binding while the second is associated with C-binding.

Appendix 4B. Equilibrium Redox Potential Controls

We consider that many ion and substrate redox reactions in cells are also fast enough so as to come into equilibrium. Of particular importance *for the cytoplasm of all cells* are the two couples Fe^{3+}/Fe^{2+} and R–S–S–R/2RSH (see Table 2.11). (The particular sulfur compounds are micro thiols in bacteria and glutathione in later cells.) Given that this equilibration takes place despite the presence of relatively powerful reducing organic compounds, which are cycling in separate kinetic steady states, we can estimate that the average steady-state potential of the cell cytoplasm is somewhat below −0.2 V. This is in keeping with the observation that the iron is overwhelmingly present as Fe^{2+} and thiols are kept largely in the RSH state in cells. Since these redox exchange reactions in any cellular compartment are fast and common to cells, their equilibrium constants, standard redox potentials, are also fixed features unchangeable by evolution of new organisms (see Section 4.15.2). However, there is no equilibrium between the NAD/NADH and many similar couples with Fe^{3+}/Fe^{2+} (see Table 2.11 and see below), so that these redox reactions are kinetically "compartmentalised" without membrane restrictions.

A quite different set of oxidations/reductions, but not fast, were the *equilibria* which governed *the change of the environment*, that is *external* oxidation/reduction potentials. They involve elements such as S, Se and metals but not all C or N couples. Their slow change in value was due to the slow release of oxygen by organisms, but the realisable *inorganic* equilibrium redox potentials within the environment are quickly attained, are independent of life and are governed by fixed background constants constraining environmental evolution within the whole ecosystem. While the environment redox potential of the molecules undergoing fast redox reactions increased from say -0.2 towards $+0.8$ V, the "equilibrium" redox potential of the cytoplasm remained at or below -0.2 V so that oxygen and external products of its reactions created a great hazard for organisms (compare Tables 2.11 and 6.1). The resultant whole is a very large energy store.

In summary, certain equilibrium constants of complex formation, of solubility products and of redox potentials form a set of fixed values that must be looked at in the context of the compartment which contains the components and which controlled evolution in fair part, against a background of rising amounts of environmental oxidised elements. The other factors were the rates of synthesis as dictated by supply of energy and of reactants in the environment.

Appendix 4C.   Molecular Machines – Efficiency and Effectiveness

The flow of material and energy in organisms is frequently dependent on molecular machines producing work or converting one form of energy into another. At the prokaryotic single-cell stage there are already molecular motors, e.g. bacterial flagella, which have grown in complexity to multicellular machines and eventually to the external machines of mankind. The essence of such a molecular machine is that it does work while it goes in a cycle itself, as for example in a pump, but it can also convert energy in one form, say a gradient of charge, to an energised chemical, e.g. ATP. This is the basis of the transduction of energy, initially from reactions of inorganic chemicals and then from Sun's radiation to a proton gradient, followed by a flow of protons, through a complex molecular motor, ATP-synthetase (see Figure 4.15), part of which rotates synthesising ATP from ADP and phosphate. Of course, the machine must return to its initial situation, hence the process is cyclic, and in this as in other cases, we are often concerned with the thermodynamics of the process, its effectiveness and efficiency. We take this case, the ability to use Sun's light energy to make ATP, as an example, trying to clarify the somewhat ill-defined concepts currently employed when dealing with bioenergetics and molecular machines. Naturally, a desirable requirement for the conversion of radiative energy into chemical energy is an *effective* machine, one that captures as much light energy as possible and makes it usable in a new chemical form, ATP, with the greatest thermodynamic *efficiency* under the prevailing conditions.

If we look at evolution as a whole we will note that over 3.5 billion years, plant life has increased the ability to absorb energy from the Sun. However, only part of this energy is used to make ATP through a series of steps; another part is used to make chemicals in the plant and a third part is used directly to oxidise the environment since O_2 is produced in the process. This is the essence of oxygenic photosynthesis described in subsequent chapters and we may be interested in studying the

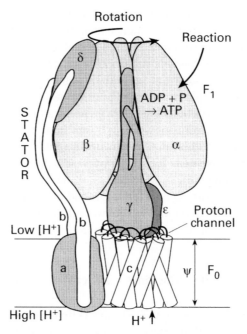

FIG. 4.15.   The ATP-synthetase (see references to Appendix 4C). The synthesis works by protons from a gradient forcing 360° cycles of the helical bundle c(γ), the motion being driven by a
weakening of binding (energy gain) and passage of ions down a gradient. Energy is transmitted
to rotation of the extra-membrane proteins, $\alpha\beta$, causing oscillations in the sites of ADP + P_i giving ATP such that ATP is formed and released in the full rotation. Note ions such as Na^+ can also
drive such a synthetase. The machine is held in the membrane by the stator, b. See also an alternative synthesis or pump machinery in Figure 4.12(c), see Walker in Further Reading.

thermodynamic *efficiency* of the overall process or of its parts, which is a measure
of the energetic *effectiveness* of the machinery used for a purpose, but other factors,
e.g. rate, need frequently to be considered since the purposes can be multiple and
requirements different, for example, maintenance, growth, development and reproduction of organisms, all of which require the ability to build biomass storing energy
in chemical bonds. The timing and the corresponding requirements of each may differ (see below rates of reaction). As far as thermodynamic efficiency is considered,
that is the ratio of the useful output (in energy terms) to the total energy input, note
that no *real* process is 100% efficient as we have already stressed in Section 4.7.
There are always "losses" of several kinds starting with the light capture system,
which evolved considerably from the first bacteria to modern plants, in which some
heat (degraded energy) is generated inevitably in the forward steps, e.g. in the production of charge separation, proton gradients, in the motion of the rotor and other
components of ATP-synthetase, and in subsequent downhill chemical reactions of
hydrogen to reduce carbon oxides and of oxygen with reduced chemicals from
plants (see Chapters 5 and 6). All these losses of the available, "high-quality" usable

energy obviously decrease the thermodynamic efficiency of the process. A similar problem concerns the subsequent use of oxygen and bound hydrogen in oxidative phosphorylation, production of ATP when a proton gradient is again an intermediate and a very similar ATP synthetase is involved (see Chapter 6).

Centring our observations on the machine itself, i.e. ATP-synthetase, whose rotor works with almost 100% thermodynamic efficiency (see Kinosita *et al.*, in References), it is important to note that the functioning of a machine is not related to an equilibrium process. Systems at equilibrium cannot do work by definition, since there is no energy change in an equilibrium system such as A + B $\leftrightarrows$ AB in which there is an energy equality in the two sides of the equation. It should also be stressed that a machine is not required for downhill reactions moving towards equilibrium although a catalyst may be required. However, if the catalyst cycles through states it does consume energy, released as heat on reverting to the initial state. To drive an uphill reaction moving away from equilibrium a source of energy is required and the same applies to a machine which cycles to do work. The *effectiveness* of a machine is therefore one aspect we need to appreciate apart from its energy conversion *efficiency*. As mentioned above, all machines doing work have a *maximal* thermodynamic efficiency of transfer of energy to do work which corresponds to a complete cycle of machine operations going infinitely slowly so that all steps are at virtual equilibrium. This efficiency is limited by the conditions of operation, e.g. the nature of the source of the energy input, its "quality", the temperature difference between the source and the sink in the case of heat engines or processes, etc. According to Carnot the maximum efficiency is given by the expression $(T_1-T_2)/T_1$, where T_1 is the absolute temperature of the source and T_2 is the absolute temperature of the sink, but this is not applicable in all cases. This efficiency can only be reached under ideal (theoretical) conditions and these can vary and affect the entire process. We may, therefore, refer to *optimal* efficiency as the efficiency that can be attained in the best possible real conditions. (Note that some authors use *optimal* and *maximal* in exactly the reverse order, but this is just a matter of semantic preference.) A different question is that of the *rate* with which a given process is achieved. Equal thermodynamic efficiency can be achieved by different ways or routes, but some can be faster than others and this may be important in terms of the "economy" of the process if the variable time is relevant, which may well be the case in linked biological processes since they all must be tightly coordinated. A too slow individual step may lead to a loss of effectiveness of the overall process, but a too fast individual step may not be compatible with the rate of other linked steps generating wastes of material and/or energy. This becomes even more critical when different cells, different organs, different organisms or the whole biota and the environment are concerned, that is when we consider ecosystems which, as we will see, develop and evolve cooperatively but in which energy and material transfer occurs between different hierarchical cellular levels (see Kline and also Corning in References). The problem is far from simple and it is hardly surprising to find that concepts established in different disciplines and for particular cases cause some confusion and do not receive universal agreement.

Now it is useful to give some real examples of molecular machines, many of which are to be found in membranes. The machines are for many purposes, following the above conversion of one form of energy, Sun's radiation, to another, ATP, suitable for general use in biological reactions. Examples of such uses are the import and export of materials, ions or molecules, which is the case of pumps, motion of organisms as is the case of flagella in some bacteria, or synthesis of biopolymers, as in the case of ribosomes. Note that every machine needs moving parts and when associated with a membrane usually requires a channel for passage of material. The channels may or may not be controlled in open or shut states on demand, and the control can be a simple switch and/or a rachet-like device driven by energy. Those with rachets cannot be as efficient as free-running machines.

The membrane part of a molecular machine is usually a bundle of protein helices crossing the membrane making a controllable cylindrical channel and the helices are mobile so as to create a selective path with or without open or shut channel controls. Connected to the helices there must be a part of the machine capable of supplying energy to make a gradient or using the gradient energy to make ATP. The part of the machinery outside can be of a different protein construction but it too must have moving parts. We have described one pump-like machine in general outline in Figure 4.12(c) and mentioned already another case of an energy production machine – ATP-synthetase (Figure 4.15, see Junge *et al.* and also Walker in Further Reading). Motion of flagella in some bacteria also depends on energy supplied by a proton gradient. Other machines work outside membranes. For example, muscles are machines as are the synthesising units in the production of proteins from amino acids in ribosomes. They are usually energy driven by ATP or other NTPs, but notice also molecular motors and pumps which can be driven by gradients of ions or ATP.

To conclude this brief note (for details see texts of Biochemistry) we stress that the thermodynamic efficiency of molecular motors can be quite high, approaching 100% but never reaching this thermodynamic limit see Everett and also Neilson and Crawford in References to Appendix. We must always be aware of the heat losses in any real process and this is true all the way from the simplest molecular machines to multi-molecular constructs to man-made machines.

Finally, the efficiency of molecular machines has to be seen with the effective selection of components and then within the whole evolution of life in organisms. The drive is always to use the available energy as much as possible, which is degraded progressively in the process increasing the rate of *thermal entropy* production in the overall ecosystem while creating interactive organisms of optimal survival strength.

References to Appendix 4C

Corning, P.A. (2002). Thermoeconomics: beyond the second law. *J. Bioeconom.*, *4*, 57–88

Everett, D.H. (1959). *An Introduction to Chemical Thermodynamics*. Longmans, London

Kinosita, K., Yasuda, R., Noji, H. and Adachi, K. (2000). A rotary molecular motor that can work at near 100% efficiency. *Philos. Trans. Act. Royal Soc. London B, 355*, 473–489. See also *Proc. Biochem. Soc.* (2005) Meeting "Mechanics of Bioenergetic Membrane Proteins: Structures and

Beyond. An extensive description of ATP-synthetase is given by Boris A. Feniouk at www.biolo-gie.uni-osnabrueck.de/biophysik/Feniouk/FAQ.html

Kline, S.J. (1999). *The Low-Down on Entropy and Interpretative Thermodynamics*. DCW Industries, Inc. Palm Drive, CA

Neilson, J.H. and Crawford, R.A. (1972). Efficiencies of thermodynamic processes. *J. Phys. D: Appl. Phys.*, *5*, 28–42

Further Reading

BOOKS

Caporale, L. (2003). *Darwin in the Genome*. McGraw-Hill, New York

Cornish-Bowden, A. (2004). *The Pursuit of Perfection: Aspects of Biochemical Evolution*. Oxford University Press, Oxford

Fraústo da Silva, J.J.R. and Williams, R.J.P. (2001). *The Biological Chemistry of the Elements* (2nd ed.). Oxford University Press, Oxford

Harold, F.M. (2001). *The Way of the Cell*. Oxford University Press, New York

Jablonka, E. and Lamb, M. (1995). *Epigenetic Inheritance and Evolution*. Oxford University Press Inc., New York

Kauffmann, S. (1995). *At Home in the Universe*. Oxford University Press, New York

Lesk, A.M. (1991). *Protein Architecture*. IRL Press, Oxford

Messerschimdt, A., Huber, R. Poulos, T. and Wieghardt, K. (eds.). (2004) *Handbook of Metalloproteins – Vols. 1–3*. Wiley, Chichester, U.K.

Stryer, L., Berg, J.M. and Tymoczko, J.L. (2002). *Biochemistry* (5th ed.). W.H. Freeman, New York

Turner, B.M. (2001). *Chromatin and Gene Regulation – Molecular Mechanisms in Epigenetics*. Blackwell Science, Oxford

Williams, R.J.P. and Fraústo da Silva, J.J.R. (1996). *The Natural Selection of the Chemical Elements – The Environment and Life's Chemistry*. Oxford University Press, Oxford

Williams, R.J.P. and Fraústo da Silva, J.J.R. (1999). *Bringing Chemistry to Life – from Matter to Man*. Oxford University Press, Oxford

PAPERS

Freeland, S.J. and Hurst, L.D. (2004). Evolution encoded. *Sci. Am.,* April 2004, 56–63

Junge, W., Lill, H. and Engelbrecht, S. (1997). ATP-synthase: an electrochemical transducer with rotatory mechanics. *Trends Biochem. Sci.*, *22*(41), 420–423

Linder, P. (2004). The life of RNA with proteins. *Science, 304*, 694–695

Martin, W. and Russell, M.J. (2003). On the origin of cells: a hypothesis for the evolutionary transition from abiotic chemistry to chemoautotrophic prokaryotes, and from prokaryotes to nucleated cells. *Philos. Trans. R. Soc. London*, B *358*, 27–85

Matzke, M. and Matzke, A.J.M. (2003). RNA extends its reach. *Science, 301*, 1060–1061

Mulkidjanian, A.Y., Cherepanar, D.A., Heberle, J. and Junge, W. (2005). Proton transfer dynamics at membrane/water interface and the mechanism of biological energy conversion. *Biochem. (Moscow), 70*, 251–256

Mulkidjanian, A.Y., Cherepanov, D.A. and Galperin, M.Y. (2003). Survival of the fittest before the beginning of life: selection of oligonucleotide-like polymers by UV-light. *BMC Evolutionary Biology, 3*, 1–7 and references therein

Orgel, L.E. (2000). Self-organizing biochemical cycles. *Proc. Natl. Acad. Sci. (USA), 97*, 12503–12507

Owen-Hughes, T. and Bruno, M. (2004). Breaking the silence. *Science, 303*, 324–325

Pace, N.R. (1997). A molecular view of microbial diversity and the biosphere. *Science, 276*, 734–740

Pennisi, E. (2003). DNA's cast of thousands. *Science, 300*, 282–285

Schöning, K-V., Scholz, P., Guntha X. Vu, Krischnamurthy, R. and Eschenmoser, A. (2000). Chemical etiology of nucleic acid structure. *Science, 290*, 1347–1351

Snyder, M. and Gerstein, M. (2003). Defining genes in the genomic era. *Science, 300*, 258–260

Sergé, D., Ben-Ali, D., Deamer, D.W. and Lancet, D. (2001). The lipid world. *Origins of Life and Evolution of the Biosphere, 31*, 119–145

Villarreal, L.P. (2004). Are virus alive? *Sci. Am., 291*(6), 77–81

Wächtershäuser, G. (1988). Origin of life and iron sulfides. *Microbiol. Rev., 52*, 482–486

Walker, J.E. (1998). ATP synthesis by rotary catalysts (Nobel lecture). *Ang. Chem. Int. Ed., 37*, 2309–2319

Wickelgren, I. (2003). Spinning junk into gold. *Science, 300*, 1646–1649

Williams, R.J.P. and Fraústo da Silva, J.J.R. (2004). The trinity of life: the genome, the proteome and the mineral chemical elements. *J. Chem. Educ., 81*, 738–749

Williams, R.J.P. (1961). Possible functions of chains of catalysts. *J. Theoret. Biol., 1*, 1–13

Williams, R.J.P. (1977). Flexible pharmakamoleküle und dynamische rezeptoren. *Ang. Chem., 11*, 805–816

Williams, R.J.P. (1980). One first looking into nature's chemistry. *Chem. Soc. (London) Rev., 9*, 281–305

Williamson, M.P. (ed.) (2005). Systems biology: will it work. *Trans. Biochem. Soc., 33*, 503–542

Chapter 5

First Steps in Evolution of Prokaryotes: Anaerobic Chemotypes Four to Three Billion Years Ago

5.1. Introduction

In this book we present an explanation of the evolution of the geological/biological ecosystem of the Earth's surface. There was little problem in describing the changing geological chemistry when looked at in isolation and assuming that to an initial atmosphere and aqueous solutions of C, N, O, S and Se hydrides and the oxides of C, oxygen was slowly added (Chapter 1). To a first approximation, we considered that the subsequent (inorganic) chemistry *on the surface*, under fixed conditions of temperature and pressure and in the presence of a gradually increasing oxygen content over billions of years, moved slowly and continuously from the relatively reduced beginning state to the oxidised condition of today, much of it close to equilibrium. Not all processes are complete even now since there is a huge reservoir of reduced material locked deep inside the Earth and very slowly it is removing surface-oxidised substances. However, by focussing concern on the faster changes on the surface, we shall be able to ignore the slow, longer term problem beneath it, although at the end of the book we must return to it.

A much more difficult task faced was the description of the evolution of the biosphere. To proceed, we have given in Chapter 3, a description of energy and the way in which it can be absorbed in systems of chemicals, mainly organic chemicals (see Chapter 2), which gave rise to life. In Chapter 4, we were able to give a general outline of the energised components of the biological chemical systems and described their main functions and pathways in cells, treated largely as one thermodynamic compartment using molecular machines to do work. Apart from absorbing energy the chemicals concerned must be contained and we observed that all life is cellular. The central chemistry of all these cells has to be *reductive* in order that the synthesis of the required chemicals, especially biopolymers, is possible. We have suggested that these basic components were bound to arise as the initial unguided search for a product that would self-synthesise the catalytic activity necessary if a system of organic chemicals was to form a cyclic feedback set of controlled flows. This was seen to give the strongest survival system, to take up, use, retain for some time some 20 elements and energy from the environment and then decay in an ever-repeated cycle (see Section 3.9). *The system was controlled, restricted and autocatalytic and the cellular chemistry in the necessary reductive mode created an oxidised environment.* The two chemistries, the reduction in cells and the oxidation of the environment, driven by an approximately constant energy input to the Earth from the Sun at an approximately fixed temperature, were interactive within the ecosystem, which became energised in a new way. The interaction, the inevitable attack on reducing cell material by the oxidised environment, must lead towards an eventual new steady state. The optimal steady state would be that in which the reactions are completely cyclic with no change in material content while energy is continuously degraded (see Chapter 3). Such a state is one of optimal retention of energy in chemicals in the whole ecosystem. Now, given the slow increase in O_2 presence due to the redox buffering of the environment and the very nature of the chemistry of reproductive organisms, especially the conservative nature of DNA, the initial reduced environment and the organism progression, the system was doomed to reach any final state extremely slowly, if at all. It is the first steps in this progression and the reasons for the way in which it has developed that we shall tackle in this chapter and then in the four following chapters we look at conditions which are increasingly aerobic before we describe the recent intervention of mankind in Chapter 10. Unlike most accounts of evolution, we stress again that we concentrate attention on the approximately 20 or so elements involved in life since, in this way, we keep in mind the connection with environmental change. This chapter starts the account therefore, not from the origin of life, but with the first types of protocells and cells, prokaryotes, we believe to have existed (see Figure 5.1). Apart from considering the particular chemistry in the cells and their energy uptake we shall be looking at their organisation. Note that we do not believe that a thorough insight into organisms or their evolution can be obtained from the analysis of sequences or structures of molecules or from the morphology of organisms, which are largely put on one side here, so we consider a thermo-dynamic system. It is energy input to stable chemicals that forced the initial separation of reduced (organisms) and oxidised (environment) components. Throughout this and the next chapter, both on prokaryotes, we shall at least at first assume that the cells were isolated, had no internal compartments and, although there is some evidence to the contrary, that diffusion within the cell is relatively rapid.

5.2. First Steps: The Evolution of Prokaryotes: General Considerations of the Origins of Anaerobes

The suggestion, it is no more than that, which we put forward in Chapter 4 is that the different anaerobic, autocatalysed, reductive, metabolic pathways seen in the earliest cells we know about developed in separate energised vesicles, protocells, where they were produced cooperatively with certain bases of the nucleic acids, RNA and DNA. By coalescence, these pathways came together to form coded RNA/DNA prokaryote cells which could occur in many forms. Both coalescence and gene transfer (see Woese (2002) in Further Reading) were major later steps of their further evolution and we shall observe these activities repeatedly in later stages of the development of prokaryotes (see especially the beginning of eukaryotes in Chapter 7). For these reasons, it is doubtful if the concept of species is useful at first, and because of our approach to the developing ecosystem we shall refer to rather larger groups of species of prokaryotes as the earliest chemotypes (see Section 4.2). As the environment and possible use of it and energy became more varied, different chemotypes evolved to reduce the need for extra complexity in one cell and increase complexity in the sum of the many interactive chemotypes. Presently we have no clear-cut fossil evidence as to what such cells were like at first, 4.0×10^9 years ago, though we and others surmise similarly about all of their minimum requirements as in Chapter 4. Instead, we use the evidence of the present-day nature of anaerobic cells as a starting point, assuming they were close to such primitive cells. Three features of the metabolism are therefore taken as starting points: first, cells had to have a source of energy possibly from unstable minerals; second, in the cells a combination of catalytic inorganic and organic chemistries had to be reducing in order to produce and protect the four classes of polymers essential to all life, nucleic acids, proteins, lipids and polysaccharides, using basic chemicals from the primitive environment; and third, cells were already very sophisticated, internally cooperative, and had an informed set of autocatalytic, cooperative pathway reactions needing 15–20 elements while a few elements were rejected. Cells therefore already had considerable organisation. Today's evidence allows us to assume that there were at very early times *at least* two major chemotypes of single-cell prokaryotes, Archaea and bacteria (Figure 5.1). Starting from the simplest features of them, the only start that we can be confident about, we shall see how some extra features were introduced by and in these early cells, giving rise to new chemotypes, before we begin to introduce in Chapter 6 environmental change which, we shall state, drove later evolution (Figure 5.1).

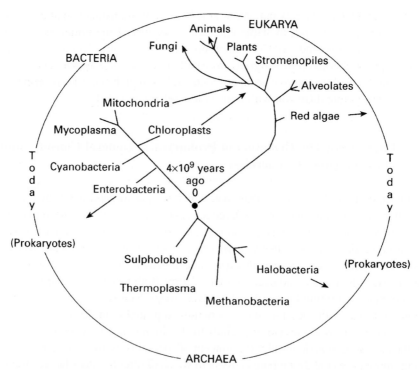

Fig. 5.1.   The three domains of life in a radial time sequence based on genetic (RNA) analysis.
The distinction between anaerobes and aerobes is not made here and the branching to eukaryotes
is left unclear. All the domains advanced with time but in very different ways (see Woese, C.
(1998) in Further Reading, and Sogin, M. (1993), *Science*, 260, 340). Note the general gene
transfer in prokaryotes and to eukaryotes (see Chapter 7).

In our account, it is important to note again the major components described in
Chapter 4.

(1)   Most of the approximately 20 elements required at the earliest times are
      those which are still employed today though there are important changes
      of use and a few new additions. The small molecule non-metal content is
      discussed under the heading of the metabolome and that of metals under
      metallome. The metal ions are overwhelmingly combined with organic
      molecules but some are more or less rejected.
(2)   There is but one major set of reductive routes from non-metals to small
      organic molecules in the metabolome including molecules for energy distri-
      bution. They gave rise to the biopolymers also by one basic set of pathways.
(3)   The major biopolymers, in the cytoplasm particularly, are the same in the
      earliest cells as they are today; DNA (RNA), proteins, polysaccharides and
      lipids. The cell protein content is discussed under the heading proteome.

(4)   The DNA and RNA undergo changes but very many basic features of their functions are unchanged. The DNA is discussed under the heading genome.

(5)   The RNA structures and protein folds mentioned in Section 4.11 are limited in number and most of the basic varieties, but not all, are maintained throughout evolution (see Sections 7.8 and 8.10).

(6)   The processes of energisation in cells are, by the nature of the chemicals synthesised, effective in retaining more reduced chemicals than the initial environment, so oxidising equivalents had to be rejected to the environment. The first process was probably the formation of elementary sulfur from H_2S.

Despite the conservative nature of the chemical system of life, it has advanced greatly in the range of environmental oxidation/reduction potentials which it has withstood by progressive and to some degree, matching changes in its own chemicals and organisation. We see this as the inevitable part of evolution of what we call chemotypes, a classification, we must remember, based on examination of chemical thermodynamic quantities such as, for example, component concentrations, energy uptake, sizes of compartments and cooperative interactions.

(7)   The driving force behind the ecosystem is the degradation of energy whether it comes from the unstable inorganic materials deposited on Earth (the material and energy from the hot, deeper levels of Earth are very important here) or (later?) from the Sun. The initial step is energy absorption separating reduced (life) from oxidised (environment) chemicals. Life is an accelerator of energy degradation.

5.3.   The Two Classes of Recognised Early Prokaryotes

Against the above background, the following main features distinguish the two known primitive chemotypes, Archaea from bacteria (see Table 5.1):

(1)   A different cell wall in Archaea; bacterial walls are made from peptidoglycans

(2)   Different membrane lipids (ether linked in Archaea, phospho-ester linked in bacteria)

(3)   Different coenzymes and use of porphyrin derivatives (see Section 5.6)

(4)   Archaea are usually extremophiles, living at extremes of salt, of temperature or of pH. (At high temperatures ($80°C$), they often obtain energy from H_2, CO, CH_4 and H_2S

(5)   DNA in Archaea is protected by histones

(6)   RNA, RNA polymerases and ribosomes are distinctly different

It is the features of bacteria that came to dominate temperate environments giving rise to photosynthetic bacteria and to the many subsequent uses of the developing

TABLE 5.1

DIFFERENCES BETWEEN ARCHAEA AND BACTERIA[a]

	Archaea	Bacteria
Membrane	Ether linkages	Ester linkages
Wall	More proteinaceous	Peptidoglycans
Gene structure (DNA)	Circular, histones bound, synthesis of DNA at many points (see eukaryotes)	Circular, synthesis at one point
Pathways (not common)	Methane formation for energy transduction Pyroglycolysis	Not known Glycolysis
Coenzymes (not common)	Coenzyme M (S) Factor F430 (Ni)	Coenzyme A (S)
Replication, transcription and translation to proteins	Similar to *eukaryotes*	Different from eukaryotes and Archaea
Ribosomal RNA	Different from eukaryotes and bacteria	Different from eukaryotes and Archaea
Folding chaperons	Prefolding as in *eukaryotes*	Different heat shock proteins (chaperones)
Physical conditions	High salt or high temperature tolerated	Live in more moderate conditions
Photopigment	Bacteriorhodopsin	Chlorophylls (later)
Use of metal ions	Greater use of Ni	Similar use of other elements

[a] Note the relationship with eukaryotes in Chapter 7.

chemical environment as discussed in Chapter 6. The Archaea appear to have developed less, for example, without chlorophylls in photosynthesis, and have largely remained in more extreme habitats such as high-salt or high-temperature environments (see Kashefi in Further Reading). However, Archaea may well have led in a major way to the development of eukaryotes, cells with many compartments (Section 7.1). Despite these differences, there is in chemical essence but one form of life in these and all subsequent cells (see Chapters 3 and 4).

Physically, all these prokaryotes are small, diameter about 1.0 μm and are of rigid, simple shape. They usually have little or no internal structure so that chemical diffusion is relatively rapid. Secondary compartments are rare but vesicles and vacuoles (even nuclei) are found in a very few large bacteria. We shall see that all the prokaryote cells have controlled, autocatalytic, internal metabolism, but are relatively little affected by external circumstances, except by shortage of nutrients.

From the very earliest times these anaerobic prokaryote cells had assembled fundamental organic chemicals as outlined in Table 4.4, but as we have stressed in Chapters 3 and 4, to be viable, each cell type had to have also certain inorganic systems in place and had developed features such as

(1)   The ability to generate and use an H^+ gradient across the membrane to supply energy for uptake and rejection of wanted and unwanted ions and

molecules. This required energy capture and assuming the cells to be autotrophs the energy came from reducing molecules such as H_2, H_2S or minerals using membrane-bound proteins to separate distantly H^+ and electrons. The movement of electrons required several metal Fe/S centres in proteins, while proton movement used organic coenzymes. The H^+ gradient led to pyrophosphate synthesis, which was transduced to magnesium-bound adenosinetriphosphate MgATP production (Figure 5.2). These transduced forms of energy (energised chemicals from gradients) were and are the basic mode of further energy distribution in biochemical pathways (see Section 5.12).

(2)   Selectivity to reject Na^+ and Cl^-, so as to manage osmotic and electrical tension, and Ca^{2+} (probably also Mn^{2+}) to prevent internal precipitation. ATP and H^+ gradients may always have assisted this rejection. At the same time, K^+ was taken up at a high concentration, and 10^{-3} M of Mg^{2+} was permitted internally. Na^+ and H^+ gradients are used to this day in the uptake of required nutritional compounds such as amino acids.

(3)   Capacity to utilise catalytic Ni^{2+}, Fe^{2+}, W(Mo) and Mg^{2+} in cell metabolism (see Table 5.2). All primitive cells had to reduce CO (CO_2) before engaging in condensation reactions. We suppose nitrogen was available as NH_3 for protein synthesis and HCN for the synthesis of nucleotide bases.

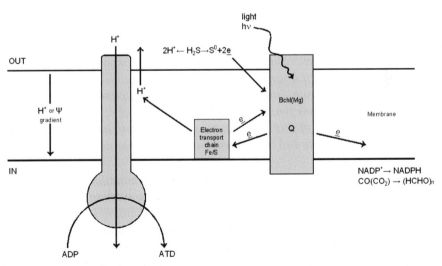

FIG. 5.2.   The photosynthetic membrane of a green sulfur bacterium. The light-activated bacteriochlorophyll molecule sends an electron through the electron-transport chain (as in respiration) creating a proton gradient and ATP synthesis. The electron eventually returns to the bacteriochlorophyll (cyclic photophosphorylation). If electrons are needed for CO_2 reduction (via reduction of $NADP^+$), an external electron donor is required (sulfide that is oxidised to elemental sulfur). Note the use of Mg and Fe.

TABLE 5.2

Some Primitive Enzymes Using Metals

Enzyme	Metals
Acetyl Co-A synthetase	2Ni, Fe/S, Co(B_{12}) later
Hydrogenase	Ni, Fe/S
Hydrogenase	Fe/S
Methyl-utilising enzymes	Ni (F-430, + CoB$_{12}$ later)
Electron-transfer ferredoxins	Fe/S
Methionine synthetase	(Co(B_{12}), later)
Dehydrogenases	Fe/S
Aldehyde reductases	W, Fe/S
Deformylase (bacteria)	Fe
Methionine peptidases (Archaea)	Fe
ATP-ases and phosphorylating enzymes	Mg
Phosphatases	Mg
2-Oxyglutarate ferredoxin	Fe/S
Cytochromes	(Fe porphyrins later)
Photosynthesis apparatus	(Mg-chlorophyll later)
External hydrolases	Ca
Aldehyde from/to carboxylate	Mo(W)

Note that some Ni/Fe hydrogenases have both CO and CN^- as metal ligands. (Methane which may well have been present is very difficult to activate.)

(4)   Capacity to utilise phosphate and sulfur (probably selenium too) compounds in energy transfer, condensation and reduction.

(5)   No copper and little nickel, cobalt and zinc were available: the last three elements were used sparingly.

(6)   As well as rejecting certain metal ions, oxidising equivalents in the form of sulfur (later oxygen) had to be rejected.

(7)   The production of proteins (enzymes) is also linked to the availability of energy, substrates and (metal) ions (see Chapter 4).

The steps were linked since (2)–(4), (6) and (7) require the energy of (1), and many steps involve units formed from the same sources. All the synthesised products and the gradients are unstable and degrade continuously.

Although there are some differences between archaea and bacteria (see Table 5.1), these basic inorganic steps are largely maintained though modified in both organisms and later throughout evolution. Much as the major metabolic products of the catalysed paths of C/H/N/O have also remained unchanged in the cytoplasm of cells, so have the above basic uses of many of the inorganic chemicals (Section 4.15). Cellular cytoplasmic chemistry changed only somewhat under pressure from environmental change. The essential, required, conservation of internal chemistry made for a series of small, necessary and unavoidable compromises later in cell evolution, as we shall see.

There may not be any alternative way for life to begin or to develop than through the reduction and then condensation reactions of organic molecules aided by inorganic ions which require energy and the use of a mutatable reproductive code. Certainly, none is known on Earth. If we look at the molecular machines involved, the principles of their construction and activity appears to have remained virtually unchanged, to this day (see Section 4.7 and Appendix 4C).

The basic outline of autocatalysed linked metabolism, as described in Chapter 4 (see Table 5.3) was originally and to some extent, still is, from CO (or CO_2) by reduction leading to fats, saccharides and with NH_3 to proteins, according to the scheme

$$NH_3$$
$$\downarrow$$
$$\text{amides} \rightleftarrows \text{amino acids} \rightarrow \text{proteins}$$
$$\uparrow$$
$$CO + H_2 \rightarrow HCHO \xrightarrow{H_2} \text{fats}$$
$$\downarrow$$
$$\text{sugars + phosphate} \rightarrow \text{phosphorylated sugars}$$

Some such starts were undoubtedly abiotic but the main paths only became firmly established in reproductive organisms. These pathways are set out in detail in wall charts and in all biochemical textbooks, but they usually omit the inorganic elements essential to all pathways. The primitive pathways to nucleic acids involved the synthesis of phosphorylated sugars and we envisage the bases as starting from HCN and NH_3

$$HCN + NH_3 \longrightarrow \text{purines and pyrimidines}$$

TABLE 5.3

MAJOR PRIMITIVE METABOLIC PATHWAYS OF H, C, N AND O

Pathway	Function	Coenzymes
Glycolysis	Energy production from glucose, small metabolites for other pathways	ATP
Gluconeogenesis	Synthesis of glucose	GTP
Fatty acid synthesis	Synthesis of fatty acid esters for membranes	CoA
Fatty acid degradation	Source of energy and metabolites for other pathways	CoA, NADH
Amino acid synthesis	Addition of –NH_2 group	Pyridoxal
Peptide bond synthesis	Protein synthesis (on ribosomes)	ATP, GTP
Nucleotide condensations	RNA and DNA synthesis	NTP
Reverse citric acid cycle	Incorporation of carbon dioxide	ATP, QH_2, haem
The pentose shunt	Reduction by hydrogen transfer	NADPH, TPP, FAD

The use of HCN is not necessary now (see books on metabolism), but NH_3 is still required today (now from N_2). Immediately we observe that all four major synthetic activities had to be in balance so that the distributions of the basic elements C, H, N, O, P and S and energy were already controlled, that is the metabolome and the proteome, in any given steady state. There are also sets of major degradative steps which can supply energy including the glycolytic pathway (degradation of glucose) and breakdown of amino acids, etc. Finally, the analytical content of free metal ions, the free metallome, was also maintained as we shall see.

We see that even the earliest organisation of a cell *behind a single membrane* (plus a wall) also had a vast feedback/feedforward communication network based on substrates and products, several metal ions and say bound pyrophosphate (MgATP) for energy and phosphate management. Of particular interest is the number of kinases which phosphorylate not just substrates but proteins in order to control activity. A major feature of protein kinases and phosphatases lies in the control over linked gene expression, the cell cycle, and metabolic and mechanical action all linked back to Mg.NTP and energy. The number of such enzymes increases rapidly from tens to hundreds in evolution as deduced from DNA sequences. The extremely critical role of phosphate based on kinetics (or transfer especially) is common to all life. All the small diffusing entities were partly controlled by uptake mechanisms at the membrane requiring NTP so that there was a network of feedback links between uptake, synthetic/degradative activity, energy and expression via DNA and RNA. The controlled networks were probably based in part on fixed equilibrium binding constants for each messenger unit, substrate or metal ion, at exchanging active or allosteric sites of proteins (see Section 3.17 and Appendix 4A).

The above description provides a possible starting background for the description of the beginning of cellular chemotypes, prokaryotes, but even this is less complicated than the only cells for which we have evidence since they have at least two additional groups of more sophisticated chemicals – coenzymes (see Tables 5.3 and 5.4) and certain metal cofactors, which we presume were additions to the most primitive cells. After we have described them, we shall return to the problem of cellular (cytoplasmic) organisation. Note that coenzyme novelty is not in basic pathways but in control of rates and in energy management.

5.4. The Introduction of Coenzymes: Optimalising Basic H, C, N, O, P Distribution

In modern *anaerobic* prokaryotes and in fact in all the prokaryotes we know about, the metabolic paths have changed little from those described above but details have. The essential feature is the flow of C, H, N, O, P and S compounds into and out of cells in an incomplete cycle. The first big evolutionary advance must have been made by introducing coenzymes – both freely mobile, where mobile includes swinging attached arms, and at fixed sites, to aid flow. The known coenzymes are so-called because they all also assist catalysis. This cannot be an

TABLE 5.4

Some Primitive Coenzymes and Cofactors

Coenzyme	Function
NADH, NADPH	Mobile carriers of H^-
Flavin	Fixed carriers of H and e
Pyridoxal phosphate	Transfer of $-NH_2$
Lipoic acid	Thiol equilibration
Biotin	CO_2 activation
Quinone	Mobile carrier of $2H^+$ and 2e (in membranes)
Nucleotide triphosphates	Mobile carrier of phosphate and energy
Glutamine	Mobile carrier of $-NH_2$
Chlorophyll	Light energy capture
Coenzyme M[a]	Mobile carrier of CH_3CO-
Coenzyme A	Mobile carrier of CH_3CO-
Haem (Fe)	Fixed carrier of e
Vitamin B_{12} (Co)	Fixed carrier of CH_3-
Factor F_{430} (Ni)[a]	Fixed carrier of H and CO

[a] Mainly in archaebacteria: the other coenzymes and cofactors occur in many organisms but not all synthesise them.

accidental outcome, but must be the result of a route for a better way to capture, catalyse, and control material and energy transfer in a required progression so as to build a more strongly surviving steady state by faster, more selective but balanced synthesis of the units DNA (RNA), proteins, lipids and sugars and also to catalyse their degradation. In fact they may well have arisen after searching among many types of molecule. Some are listed in Table 5.4, and they are very few in number, even now among millions of kinds of existing organisms, though there are one or two differences between those in bacteria and in archaea. (Those of the bacteria are found in eukaryotes.) The mobile ones carry out facile fragment transfer, that is of H, e, $-NH_2$, $-CH_3$, $-COCH_3$, $-OPO_3^{2-}$, but not O– itself, that is of basic building units, in and between pathways (Table 5.4). Not much of this chemistry has changed in 3.5 billion years, yet nearly all the coenzymes must have arrived before there was competition for resources, which were plentiful initially. Some such as Mg.NTP for energy transfer and control may well have preceded the code itself. Could we devise a better set? (See mankind's operations aimed at energy use and his devices in Chapter 10.) Is it coincidental that they often include the base units of the RNA/DNA code? Proteins that recognise the bases and hence coenzymes must have developed early (see Smith and Munro in Further Reading).

As stated, the mobile organic coenzymes can be thought of as maintaining and controlling a selective balanced distribution of H, C, N and P (energy) between pathways as well as being catalytically active. (Writing P (energy) implies that while pyrophosphate derivatives (MgATP, etc.) transfer phosphate, they also transfer energy.) Some of these coenzymes act in aqueous solution while others act in lipid solutions of membranes. They are connectors between pathways, not to a specific one but always different in synthetic and degradative routes. They are then not just substrates of a selected pathway but are more often simultaneously catalytic aids, carriers, and allosteric factors while they coordinate control by aiding reactions in certain directions (see Section 4.17). Balanced distribution of the basic elements can only be achieved, providing all sites of binding of each individual mobile coenzyme has approximately the same fixed binding constant to proteins of different pathways, which then corresponds approximately to the inverse of the free concentration of that coenzyme (see Section 4.18). For example, MgATP binds to many sites with a constant $K = 10^{3.5}$ M^{-1}, which is close to the inverse of its concentration in all active cells, 10^{-3} M (Table 4.3). The suggestion is that the mobile coenzymes assist in providing a new catalysed transfer, and messenger-based internal *cytoplasmic* steady state close to an *equilibrium pool* for each of transferable H, C, N, and P (energy) units to assist substrates and ions in a particular set of reactions. Similar but separate considerations may well apply to the membrane carriers, quinones. The fixed concentrations and binding constants together with those of substrates and ions ensure that the basic units of synthesis of fats, sugars, amino acids and bases (even of energy capture) act cooperatively but very selectively and evenly throughout all the pathways relative to one another. This basic content of the coenzymes like that of non-metal small molecule units is included in the *metabolome*, which is closely fixed in a given steady state. The balanced pools for binding can be coordinated with a far from equilibrium set of differentiated pathway activities for synthesis by selective negative feedback to the active enzymes by their own products (Section 3.11). Now these coenzymes for the transfer of the above non-metals except for P (energy) and one other case are synthesised from non-metals alone. As mentioned, P (energy) is connected to the metal ion Mg^{2+}, as MgATP, but there is another apparently strange exception – the transfer of atomic oxygen at low redox potential.

Among the transfer and exchange of non-metals, the reactions of atomic oxygen, O, at low potential are unusual in that transport is not required. H$_2$O carries O everywhere but it is not by itself active in O-incorporation into carbon frameworks. It is observed that fixed Mo (W) coenzymes have always been used as catalysts in the oxygen atom transfer from H$_2$O to aldehydes reversibly.

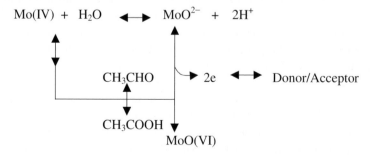

Only metal ions Mo (W), in a certain organic complex later and which can be considered to be a semi-mobile coenzyme (see Section 5.9) are known to catalyse these reactions in low redox potential oxidising conditions. (Note that there is no involvement of high redox potential O-transfer, which requires molecular oxygen or peroxides and their circulation for they were absent in this primitive metabolism. They are employed later in the very different oxygen transfer reactions (from O_2 or H_2O_2) catalysed by $>$SeO, $-$MnO and $-$FeO.) Could it be that the low-potential O-transfer reaction made Mo (W) essential *for all life* no matter how the value of Mo increased later in other cofactors? (Note the redox potentials from Mo(III) to Mo(VI) are all roughly at -0.1 V.) It is quite difficult to transfer O-atoms from most lighter elements, e.g. from $-$COOH, and from NO and SO bonds as we see later. Note that H^+ transfer is similarly transferred by H_2O but does not require activation to react except in redox reactions, see Mn chemistry introduced later.

Reflecting upon this catalysed, non-metal transfer chemistry, we observe that its general purpose remained the building of large molecules, lipids, polysaccharides, proteins and nucleic acids and their degradation yielding energy. The coenzymes have the advantages over the substrates themselves in rapid reaction since they are more catalytically effective due to their aromatic, S or P (or Mo) centres of exchange activity, while most remain small and diffuse rapidly in water.

(It may seem strange that we have left the transfer of sulfur out of this description, but it was available initially as H_2S, which diffuses easily, compare H_2O, and is reactive with metal ions and some organic centres. Sulfur from intermediate states of oxidation of this element, e.g. $S_2O_3^{2-}$, is transferred by molybdenum enzymes. Later, when sulfur became sulfate, a coenzyme (PAPS) was required for its transfer (see aerobes and eukaryotes).)

We next focus on the use of fixed-site cofactors and coenzymes. We note that much of this coenzyme chemistry is now linked to very local two-electron chemistry (H^-, CH_3^-, CH_3CO-, $-NH_2$, O transfer) in enzymes. Additionally, one-electron changes of coenzymes, quinones, flavins and metal ions especially in membranes are used very much in very fast intermediates of twice the *one-electron switches* over considerable electron transfer distances. At certain points, the chains of catalysis revert to a two-electron reaction (see Figure 5.2), and the whole complex linkage of diffusion and carriers is part of energy transduction (see also proton transfer and Williams in Further Reading). There is a variety of additional coenzymes which are fixed and which we believe came later in evolution, and there are the very important metal ion cofactors which are separately considered below.

An additional point of great interest in these energised flow systems, mentioned in Chapter 4, is that similar mobile coenzymes are not used in identical ways. In the case of hydrogen transfer, it is observed that NADH is the carrier of H in degradation reactions (see glycolysis), whereas NADPH functions more frequently in synthesis (see the pentose shunt). Most striking are the different uses of the nucleotide triphosphates, ATP, CTP, UTP and GTP, all associated with Mg^{2+}. Of course, these coenzymes are probably of earlier origin than all others since they are all required in the synthesis of DNA (RNA), in controls and in energy transfer. However, in metabolism we find that each one is used differently. In the prokaryote cytoplasm, MgATP is used in pumping, in glycolysis, and in the reverse citrate cycle. By contrast Mg.GTP is used in gluconeogenesis (the reverse of glycolysis) and protein synthesis. Note that this means that in a cycle

Monomers   Biopolymers

Mg.GTP is used in the forward reaction and MgATP in the reverse adding to the irreversibility of cellular reactions. Later in this book (see Chapter 7), we shall find that the UTP/UDP pair is used in glycogen synthesis and UDP, GDP and CMP are used in glycosylation in vesicles. (Prokaryotes do not carry out these reactions.) Again all the NTP reactions are absolutely dependent on the presence of about 3.0 mM Mg^{2+}. Later, we shall see that with glycosylation, sulfation of sugars occurs outside the cytoplasm where a derivative of MgATP is used, that is Mg.PAPS.

In conclusion, we note that a cell is totally dependent on *external* supply of nutrients and energy and hence the states of coenzymes frequently depend on external conditions. They are also totally dependent on a central DNA code (see below). As coenzyme activity and DNA expression are concentration dependent, the treatment of them belongs with thermodynamic analysis of the whole controlled autocatalytic system.

5.5.   Primitive Metal Reaction Centres

While studying very primitive changes in metabolism of carbon compounds, we showed in Section 4.2 that it is impossible to understand their initial reaction steps without reference to the metal ions in the environment for these metal ions are an essential part of the very basic, original catalysed reactions of N, C, O, H, S and P from environmentally available chemicals. (Perhaps some ions were bound on purely mineral surfaces originally.) The probable concentrations of many of these elements in the cytoplasm of prokaryotes generally is shown in Figure 5.3. While the earliest abiotic chemical systems clearly could not control the handling of metal ions, it is apparent that the earliest known forms of cellular life could do so and had a repertoire of metal-handling and metal-using proteins limited in concentration by feedback control (see Miyakawa *et al.* in Further Reading). Predating the more complicated cofactors such as FeMoco for nitrogen fixation, Moco for low potential oxygen transfer, chlorophylls for light absorption, porphyrins for sulfite and nitrite reactions, vitamin B_{12} and nickel F-430 for H_2/CH_4, all of which require metal insertion reactions (see below) there must have been metal–enzyme sites of reaction, transcription factors, carrier and pumping proteins for at least Mg, Fe, Ni and W (Mo) ions (see Table 5.2). There can be little doubt that *equilibrium* binding in these first biological metal centres was based on donor atoms of the side chains of amino acids of peptides or proteins and that they had to be employed in a discriminating manner. As a proof of this assertion, we can put together the

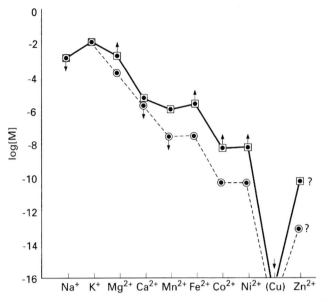

FIG. 5.3. The proposed free metallome profiles of a primitive organism after the synthesis of ring chelates: ⊙, free metallome; ◼, combined or total metallome. Arrows (up) indicate the elevation due to the chelates and arrows (down) indicate pumping out.

principles of equilibrium binding in Appendix A of Chapter 4 and the selectivity of metal binding in Section 2.17 with the observations on the binding sites of metal ions in *prokaryote proteins*. In all cases of known donor groups of prokaryote proteins for binding metal ions, the few that bind Na^+ or K^+ are based largely on neutral O-donors such as carbonyls and ethers; those observed for Mg and externally for Ca (and later largely for Mn) are carboxylates (O-donors), while histidines (N-donors) and carboxylates are used to bind single Fe ions and thiolate S is used for binding Fe, Mo and Ni in clusters. The geometry and bond lengths of special binding groups, cavity shapes and sizes, further aided selectivity including the removal of Ca and Mn (see Fraústo da Silva and Williams in Further Reading). Selectivity was increased too by the use of the higher oxidation state of iron, Fe^{3+}, in mixed-valent (of iron) and mixed-metal clusters, which were held by RS^- and S^{2-} in differently sized clusters and in separate special cavities near protein surfaces, e.g. Fe_2S_2, Fe_3S_4 and Fe_4S_4 (see Figure 5.4(a)). Chains of these Fe/S centres in proteins are the most common, early electron-transfer paths. (The electron does not need a carrier for transfer up to a distance of 20 Å.) Probably nickel was held by similar strong sulfur donors too, e.g. in hydrogenases. The selectivity of binding strengths and stereochemistry are related to model inorganic complex selective formation but are refined, constrained in electronic and stereochemical states, for functional use (see Table 4.10). Even today there are only a very few exceptional examples where small inorganic ligands help to bind the metal ions as in the early Fe(CO)/(CN) centre of prokaryote hydrogenases (Figure 5.4(b)). Such small

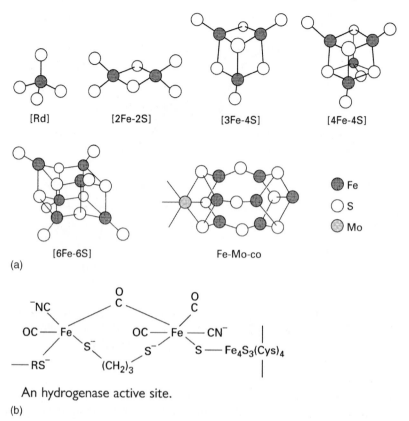

FIG. 5.4(a).   Structures of iron–sulfur clusters found in proteins. (b) An hydrogenase active site. Note the presence of CO and CN⁻ ligands, which could have been present in the primitive anaerobic environment.

ligands could have been available and are still present in the environment in such outpourings as those of the mid-ocean black smokers. Note that these hydrogenases contain nickel but the corresponding enzymes of eukaryotes are Fe-only proteins. Overall, as we have explained earlier, these selected associations of proteins and metal ions in primitive prokaryotes are not so much biologically selected, but the result of inevitable equilibrium organic/inorganic chemistry (see Section 2.17) although they are refined for function. Ion pumping is biologically selected, however, to limit the free ion concentrations which can equilibrate. The full complement of metal ions and their free concentrations, the metallome, are given in Table 4.3 (see Figure 5.3). Little reference will be made until later to other metals such as copper or zinc since these metals would have been very poorly available to prokaryotes in primitive conditions. Much of this metal ion chemistry is present in the cytoplasm and membranes of all cells to this day.

   Although we have described them earlier in general terms, it is well worth stressing again that the concentrations of the free metal ions in these early cells

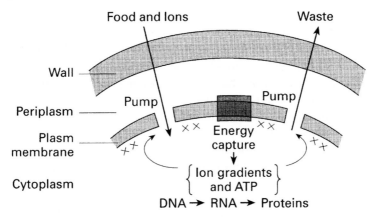

FIG. 5.5. The input and output pumps of cells connected to energy from external sources such as light or unstable chemicals. We show a gap between the inner membrane and the outermost structure which is called the periplasm.

were controlled by the pumps (Figure 5.5), which like their binding proteins were controlled by the transcription factors which expressed them, so that overall enzyme activity incorporating them was kept at a homeostatic level (Section 4.17) while optimalising overall activity and survival. To achieve a homeostatic level of all the activities inside cells the binding strengths of the pump active sites are closely similar to those for metal carriers, transcription factors and certain internal enzymes. To achieve the required selectivity it was also necessary that the production of all these binding proteins are controlled by the levels of the free ions just as the production of enzymes in pathways is related to the availability of substrates (see Figures 3.12 and 3.13). Moreover, to control several networks, the metal ions are in reasonable, rapid, balanced exchange between the networks (see Appendix 4A). There is good reason to believe then that the pumps, the carriers, the transcription factors sites and the active enzyme sites for each metal ion in prokaryotes are governed by *individual chemistry and a fixed selective strength of internal binding different for every metal ion*, (see Appendix 4A). We know this is true for certain metal ions such as Ca^{2+}, Mg^{2+}, Mn^{2+}, K^+ and Fe^{2+}, and it may well be the case for the other metal ions commonly used by the primitive prokaryote as well as modern cells. These *close to equilibrium considerations* are quite different from those of irreversible non-metal chemistry and from the more recent irreversible forms in which metals were inserted later, e.g. in porphyrin derivatives. However, they are similar in character to the exchange of the mobile coenzymes, and substrates common to different pathways (Section 5.4). (Note again that equilibrium constants themselves are not open to evolution.) Along with binding constants, we must also remember the setting of redox potentials of the cell especially by the oxidation state of iron, mainly as Fe^{2+}. This cation especially will be seen later to be a major control factor in protein expression and maturation and in switches between aerobic and anaerobic conditions. There are also important connections between metal ion and substrate metabolism, for example, between Fe^{2+}, H_2, CO, –S–S–/RS⁻ balance, all of which are involved in bacterial homeostasis connected especially to phosphate reactions of histidine kinases, (see Appendix 4B).

In addition to the above-controlled input, some elements were rejected from cells, especially calcium, which has a very important place in evolution. This rejection of calcium from the cytoplasm does not imply that it was of no use to organisms. Calcium was and is used by bacteria in the stabilisation of its outer coat, and this protective use outside cells is maintained throughout evolution. For example, some bacteria produce spores with protective coats of calcium dipicolinate. A further use is associated with activities such as digestion of larger particles, which are only capable of being conducted outside prokaryote cells, not inside them. To this end, bacteria exude inactive digestive proteins to the calcium-containing environment (the sea) where combination of calcium and protein induce the digestive activity. Most interestingly, some of the calcium binding to proteins arose at a structure somewhat resembling a so-called *single EF hand*, but with a binding constant of only 10^3 M^{-1}, sufficient for external binding in the sea (see Morgan *et al.* in Further Reading). We shall see how this protein motif when paired (two EF hands) became one of the high binding constant and a major basis of general signalling in eukaryotes, even to internal digestion. It also became linked to controlled use of O_2 and light (see Sections 7.12 and 8.14) in these cells. In parallel, we shall see that some of the very few internal zinc-binding proteins of prokaryotes had a group of four thiolates (cysteines) as a binding site and this became the zinc site of transcription factors of eukaryotes billions of years later in eukaryotes (see Sections 5.10 and 8.10 and our earlier books in Further Reading).

To handle the import and export of ions and small molecules there are three different types of pumps in the prokaryote membranes which are found in all later organisms. The reversible (free running) F_1F_0 ATP-ases are associated with the production of the proton gradient or potential from ATP or the reverse; the irreversible V-type ATP-ase pumps are similar but act mainly to reject unwanted Na^+ and H^+. Both form molecular machines like that of Figure 4.15. The third is the P-type ATP-ase pumps which pump ions such as Ca^{2+} in exchange for protons, and are more akin to the structure of Figure 4.12. These types of pump act at many different membranes in eukaryotes and also remove poisonous heavy-metal ions. There are also a variety of exchangers, which not only move ions but also small molecules across membranes of all phyla. They are valuable in prokaryotes for the symport of desirable molecules coupled inwardly with the proton or the Na^+ gradient, as both these ions are rejected by prokaryote cells. It is becoming clear that not only are these pumps related proteins of more recent organisms, but that a large variety of properties of cells based on today's proteins can be traced back to the prokaryotes, and that they are based on metal ion or proton gradients.

To conclude this chemical account of the earliest prokaryotes, we can see that there were at least basically two similar anaerobic groups of organisms, archaea and bacteria, which have hardly changed till today in chemical composition, energy capture modes and space occupied but were improved in organisation by the introduction of coenzymes. They remain the basic original chemotypes with limited use of metal ions and are the major anaerobic classes to this day though there have been major advances in evolution of prokaryotes as we describe next and in Chapter 6.

The final anaerobic non-photosynthesising system of chemicals in very primitive lithochemotropic organisms could only be an optimal energy retainer in the sense that it optimalised the energy it could absorb from Earth and degraded it. The next advance in metal ion chemistry was the development of certain porphyrin-based coenzymes which allowed the full use of energy from the sun, and with a special ligand for molybdenum, gave considerable advances in catalysis, with a resultant change of chemotype. Each advance in evolution of living systems as a whole will be seen in the increase of ability to use as many energy sources and as many chemicals as are available and can be incorporated and made to be of functional value. Remember that the thermodynamic objective is the optimal energy capture and degradation mostly in covalent organic chemicals, assisted by essential inorganic ions in the ecosystem. Functional value is seen by cell survival that is strengthening of survival of overall energised chemicals and this will be seen to involve a systematic progression of organisation in organisms, but in our account it will also be seen as a systematic advance in energy flux.

5.6.   Metal/Organic Cofactors

5.6.1.   LIGHT ABSORPTION: CAROTENOIDS AND CHLOROPHYLLS

We consider first the example of the development of metal/organic cofactors for light absorption and the creation of proton gradients since this is so critical to evolution. Light capture drove evolution after a possible start using energised minerals. The discussion on factors that drove the use of light starting from protection from it is deferred to Section 11.13. The early history of light absorption is not certain but there are several possible alternative beginnings (see Des Marais in Further Reading). One is based on unsaturated, long-chain fatty acids without metals – the carotenoids – which leads to modern bacteriorhodopsin in Archaea. Carotenoids are still used in energy transduction in some bacteria and in plants, in control of light capture and in protective devices in plants and in vision in higher organisms. The second is the synthesis of a series of Mg.chlorophylls, in bacteria only and in plants later, leading to modern photosynthesis. The production of carotenoids is easy to explain as a consequence of synthesis of unsaturated fatty acids since such unsaturation is a step in the synthesis of lipids

$$CH_3 \, COOH + CH_3 \, COOH \quad \rightarrow \quad CH_3 \, C = CH \cdot COOH + H_2O$$
$$\underset{\textstyle OH}{\textstyle |}$$

followed by reduction to $CH_3(CH = CH)_n \cdot COOH$. Certain of the desaturases for the synthesis of specific lipids require iron. However, carotenoids capture light only in the wavelength region around 400–500 μm, when they are limited in energy transduction value.

The synthesis of chlorophyll requires a larger number of steps giving first uro-porphyrin (see our books in Further Reading). There are proposed schemes utilis-ing HCN as well as small carbon fragments such that porphyrins could have arisen very early on Earth even without biological intervention. (They may even be abi-otic in origin, but the biotic pathway to them had to be devised.) The uroporphyrin is then modified in cells in different ways to make selectively non-exchanging Mg, Fe, Ni and Co complexes (Figure 5.6). A biological *energised step of selective metal insertion* is often needed for synthesis since the centres of these chelates are not discriminatory in themselves. These syntheses use proteins, for example, a Mg chelatase in the making of chlorophylls for energy capture. We shall consider briefly the nature of such a pathway (similar pathways exist for each of the four metal porphyrins) while we follow the development of the variety of chlorophylls to their final condition in the oxygen-producing plants of today. Here, we make use of the corelations given by Bauer and his colleagues (see Xiong *et al.* in Further Reading). (Several times in this book, we draw particular attention to the interac-tion of light with organisms since it illustrates the nature of the environment/cell ecosystem so well.)

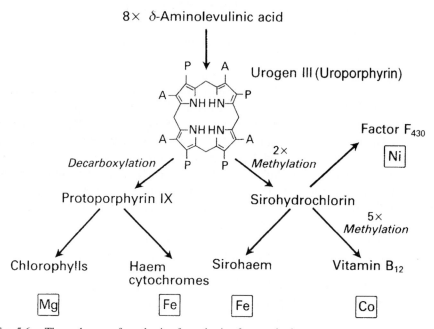

FIG. 5.6.   The pathways of synthesis of porphyrins from a single precursor.

The evolution of photosynthesis in bacteria can be traced using genetic analysis concerning the magnesium chelatase proteins alongside complementary knowledge of the early genes of nitrogen fixation, that is FeMoco formation in nitrogenase, and of cobalt chelatase for synthesis of B_{12}-coenzymes. In essence, these three gene systems lead to metal incorporation in chlorophyll (Mg) and in vitamin B_{12} (Co), both from modifications of urophorphyrin, and of Fe in FeMoco of nitrogenase. (Note that the nickel chelatase for F-430 in archaea, not found in bacteria, can also be used in a parallel way to date synthesis in this family as can the iron chelatase for iron porphyrin synthesis and they are all equally similar early developments from uroporphyrin.) The chelatases resemble one another closely in protein sequence but have differentiated binding centres for the different metal ions, a selectivity described above (see Sections 2.14 and 4.17). In this respect, the proteins resemble certain transcription factors for selective metal uptake and may have been derived from them.

The genetic analysis places the green bacteria as the earliest photosynthesisers using a bacteriochlorophyll. They are of two types, first filamentous and then sulfur bacteria. They cannot produce oxygen from water since they use light of insufficient energy for this process. They are able to use light to energise electrons and then iron and quinone electron-carrier cofactors in making membrane proton gradients for MgATP synthesis while giving hydrogen for energised reduction (see Section 5.12). The main purpose of their light absorption is therefore to make MgATP and to provide reducing equivalents so as to drive synthesis of larger carbon compounds, that is carbon uptake, by the *reversed* citric acid cycle (Figure 5.7). Some of these bacteria can also use H_2S to supply reducing equivalents, where elemental sulfur is a (harmless) waste product. (Note however that as waste solid sulfur could seriously deplete the environment of H_2S in time (see Section 3.8) and the production of O_2 finally removed this gas.) Purple bacteria appeared next and by using a bacteriochlorophyll of a higher potential they can also incorporate carbon using the reversed citric acid cycle but they cannot drive bound hydride and CO_2 to give carbon incorporation directly. The final step brought together these two light-harvesting reactions in combination and acting in sequence (Figure 5.8). By accumulating sufficient energy, the new system was able to oxidise water to oxygen while incorporating CO_2 directly in reduced carbon compounds employing the Calvin cycle utilising MgATP (see Figure 6.4). This occurs in cyanobacteria in which the pigment used is no longer a bacteriochlorophyll but chlorophyll *a* as is used in plants today. The oxidation of water required the novel use of manganese on the outer surface of the membrane to yield O_2 (see Section 5.8 and see our books in Further Reading). The cyanobacteria we know today cannot be very ancient since they not only release O_2, but also have learnt to use it, and have a modest copper content (see Figures 5.8 and 6.10). Note how in Figure 5.8, haem (or later Cu) and Mn have been introduced in addition to the Fe/S centres of the more primitive photosystem I.

Note that the chlorophyll molecules are all not identical in the photosynthesis apparatus. Many molecules in a collecting unit absorb light so that a large volume

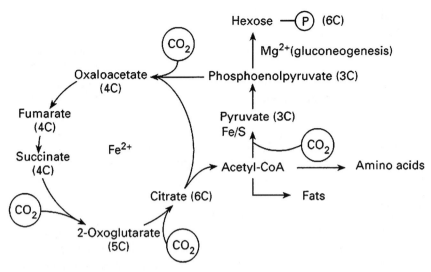

FIG. 5.7. In green sulfur bacteria and in some archaebacteria, a reverse citric acid cycle is used for the assimilation of CO_2. It must be assumed that this was the original function of the citric acid cycle that only secondarily took over the role as a dissimulatory and oxidative process for the degradation of organic matter. A major enzyme here is 2-oxoglutarate: ferredoxin for CO_2 fixation. Note that it, like several other enzymes in the cycle, uses Fe/S proteins. One is the initial so-called complex I which has eight different Fe/S centres of different kinds but no haem (see also other early electron-transfer chains, e.g. in hydrogenases).

of space is utilised to capture energy before it is transferred to special reaction centre molecules. The reaction centres are chlorophyll dimers and have absorption bands at longer wavelength than the other monomer light collectors to allow light to be passed to them. The dimers are also special in that light absorption activates them readily to a di-radical state so that they pass an electron one way and accept an electron from the opposite direction and another source. The increasing effectiveness of these light-capture systems in the production of hydrogen for reduction and MgATP for general use is discussed in Section 5.7. Advances such as these are consistent with the drive towards optimal energy capture and optimal effective use of the available energy and chemical elements but note that much energy is stored relative to reduced carbon compounds in waste O_2 production and then inevitably in the oxidation of the environment and maybe some cellular material with heat loss (see Section 3.8). (If H_2O had been a chemical of low availability its conversion into O_2 would have been limited and caused evolution to come to a halt.)

To achieve the synthesis of all the chlorophylls as magnesium chelates we shall return later to the way the magnesium was inserted while simultaneously the metals iron, cobalt and nickel were prevented from entering the chlorin centre despite the facts that all these metals bind more strongly to chlorin, and they themselves have to be handled specifically and separately to make coenzymes based on them. We note first the general advantages of metal insertion and once again observe the sophistication of the

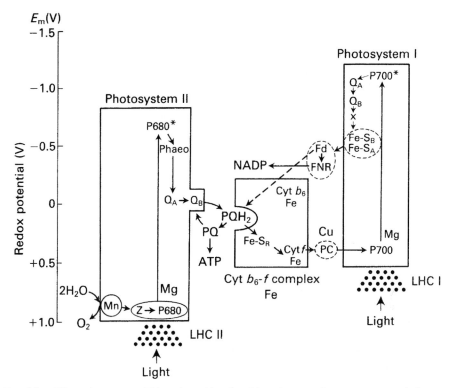

FIG. 5.8.   When photosystem II is activated by absorbing photons, electrons are passed along an electron-acceptor chain and are eventually donated to photosystem I and finally to NAPD$^+$. Photosystem II is responsible for the photolytic dissociation of water and the production of atmospheric oxygen. This pathway is sometimes referred to as the Z scheme because of its zigzag route, as depicted here, but the two arms are in fact remote in space. (Note Plastocyanin (Cu) is an alternative late replacement for an Fe cytochrome complex).

syntheses to achieve an objective, the very selective use of a metal ion utilising its equilibrium chemistry. This mechanism of synthesis is also present in the production of Fe_n/S_n centres and FeMoco so that much energy generation is dependent on these selective metal insertions (see the pathways of oxidative and photophosphorylation).

5.6.2.   PORPHYRIN COFACTORS: THEIR ADVANTAGES AND IRREVERSIBLE METAL INSERTION

The logic of the evolution of insertion can now be considered. Much as the most primitive *selectivity* of the *chemistry* of the uptake process, pumping, carrying and final binding to form a useful enzyme is not an invention by organisms but is a necessary consequence of inevitable equilibrium considerations (see Section 4.17), so the binding of particular metal ions to particular chelatase proteins was similarly selected

TABLE 5.5

New Functions of Metal Porphyrins

Metal Porphyrin	Function
Haem, Fe	Electron transfer in membranes and elsewhere
Chlorophyll, Mg	Light capture and transduction in membranes
Coenzyme B_{12}, Co	Transfer of methyl, rearrangements of substrates
Factor F-430, Ni	Activation of carbon monoxide

as a first step based on the equilibrium chemical strength of the product of free metal ion times the free chelatase concentrations. As a result each of the metal ions, Fe, Co, Ni and Mg, (later even Cu and Zn in special organisms) could be guided into a particular path of the synthesis of the different types of metal porphyrin-related rings (see above and Table 5.5), but the final step of insertion cannot be an equilibrium but must be an irreversible often energised process. (Porphyrin binding itself selects metal ions in the Irving–Williams sequence.) The success of this selective operation depends upon the synthesis of additional proteins each with selective binding to *one given empty porphyrin ring* into which the particular ion has to be inserted. For example, the Mg chelatase recognises the chlorin-binding protein required in the last step to make chlorophyll, an insertion process which here involves energy, and hence is irreversible, and so makes a non-equilibrium combination. The product, chlorophyll, had then to find its way on the carrier protein to a selected centre in a membrane where it is active in a functional unit. Final selection is based on the chemical side chains and shape of each metal porphyrin unit in receptor proteins often enzymes.

The advantages of these porphyrin coenzymes were:

(1)   They are very effective ways of retaining specific metal ions in a non-exchanging site. In effect, each M.porphyrin is a new "element", different from the parent metal ion; compare free Mg^{2+} with chlorophyll, and its organic part is different for each metal ion (see (5) below). Thus a "metal element" becomes like S or P in non-exchanging selectivity similar in a sense to that of organo-metallic chemistry (see Section 2.16). The concentrations of the complexes has then separate controls of synthesis based on novel transcription factors.

(2)   Once formed, the metal coenzymes are more easily selected by proteins than the parent ions due to the idiosyncratic nature of each ring.

(3)   In several cases, the porphyrins (chlorophyll, haem and F-430 but not vitamin B_{12}) are not water-soluble and are often placed in membranes, which is not so easily managed with simple ions.

(4)   The organic structures holding the ions, porphyrins, are themselves quite resistant to decomposition. They are not open to hydrolytic attack but note that they are sensitive to oxidation.

(5) Some of the metal ions, Ni, Co, (Fe), in porphyrins plus proteins can be in low-spin states and so are more effective as binding agents and catalysts. Their spin states and redox potentials are adjustable by binding and cover a very wide range. Some binding centres hold the metal porphyrin on the edge of spin-state change (Section 2.5) so that they can act as allosteric triggers, e.g. later in haemoglobin, and in sensors or transcription factors controlling CO, NO and O_2 chemistry. Binding of the small molecule causes a conformational change, even a cooperative change as in haemoglobin.

(6) The value of these molecules in synthesis and energy capture, photosynthesis and oxidative phosphorylation, together makes the production of organic molecules, which are energy traps, more rapid and hence the total biomass survival is increased. Overall energy retention is also increased. The particular value of Mg^{2+} in chlorin is described in Section 5.7.

(7) The iron porphyrins are much less susceptible to oxidation than the primitive Fe/S units, and unlike Fe/S, Fe/O or Fe/N centres can be used outside cells (see Chapter 6) where we describe the early uses of elements in compartments other than the cytoplasm.

(8) Note that while nickel (in Archaea) and cobalt porphyrins show little or no development and are more or less lost later in evolution, use of iron porphyrins advanced with the use of O_2, NO, SO_3^- and NO^-, that is under new oxidising conditions.

Overall, the specific porphyrins plus the selected metal ion give rise to optimal possibilities of light capture (Mg), electron transfer and spin-state changes (Fe), two-electron, low-spin metal chemistry of carbon and hydrogen (Co, Ni), and later of oxygen, nitrite and sulfite reactions (Fe) (see Table 5.5). All these complexes then lead to increased efficiency of energy and element incorporation into organic compounds and their catalysis, leading us to note how vital are specifically selected metal porphyrins for the further evolution of organisms. Details appear in monographs dedicated to each cofactor and we shall not discuss them specifically any further. Note too that they evolved in the anaerobic evolution of chemotypes in the virtual absence of zinc and copper which do not appear in such chelates, except in eccentric organisms, e.g. a very few photosynthetic (Zn) systems and in the colours of feathers in some birds (Cu).

The above account of selectivity of inorganic plus organic chemistry in synthesis is given rather extensively to stress three points. All the four (Mg, Fe, Co and Ni) porphyrin products came from one source, the synthesis of uroporphyrin. The basis of selection is very different from that in primitive centres which use thermodynamic stability constant selectivity based on different donor atoms for different metal ions. Here, all ion complexes have the same donor atoms, nitrogen, the most constrained being the coordination of Mg^{2+} by five nitrogens exactly as is seen for Fe in haemoglobin. Hence, there also has to be a new control feedback to ensure that the appropriate quantities of each metal cofactor is produced in a balanced way, that is synthesis from uroporphyrin has to be divided based upon requirement for each chelate separately. Additionally, this control during synthesis has to rest with the free metal ion input to the cell from the environment. Here, we see the tight coupling of cell development between material from the environment, metal ions here, but it is also true for the basic non-metal compounds for organic syntheses and metabolism. In a steady state, this control is recognisable by the cell contents – the genome, the proteome, the metabolome and the metallome, the last of which now includes the metal porphyrins. Time and again we stress that solving biochemical pathways without solving the controls over them does not reveal the sophistication of the networks of a cell, which is an autocatalytic system of synthesis and degradation. We have to describe local, concentration-dependent, thermodynamic systems, not just sequences or structures of biopolymers.

The second important point is that the development of these metallo-coenzymes coincides with changes in total metal element composition (see Figure 5.3), of novel energy uptake, of metabolic paths for CO_2 uptake and of organisation of pathways. *Together, they are steps in the evolution of anaerobic prokaryote chemotypes* (see Table 4.2). Note the divergence of increased use of Ni, Co and Fe in Archaea but of mainly Mg, Fe and Mn in photosynthetic bacteria. The major progression is in the greater ability to capture energy, a fundamental feature of our account of the inevitable direction of evolution. (Later in evolution, haem is essential for oxidative in addition to photo-phosphorylation.)

The third point is that the selection of the metals is based on *related proteins with selective different thermodynamic binding constants. Thus, evolution of proteins does not need gross changes of sequences* since this selectivity existed from the beginning of cells in basic functions and is maintained in the long term. Interestingly these, like many other synthesised products, are found in all classes of bacteria. Did they originate in a common ancestor or were there many ancestral exchanges?

5.7.   The Use of Light to Full Advantage

It is very important in the context of the essential advantageous change in energy capture in an ecosystem to see that it is *efficient* considering the production of usable metabolic energy for many purposes, for the synthesis of reduced cell material and in the production of the more oxidised chemicals in the environment, to be used later. Visible light covers a very wide range of the radiation spectrum from the ultraviolet (UV) to the infrared (IR), approximately from 4,000 to 8,000 Å wavelength, i.e. between (UV) and (IR) frequencies. Although of very high energy content, UV radiation of shorter wavelength is dangerous to organisms since it creates free radicals too readily and in fact only a limited amount is present in sunlight. These free radicals, compounds with an electron missing or gained, attack indiscriminately, and destroy many organic molecules such as proteins and nucleic acids. Fortunately, the ozone layer, formed immediately after oxygen was liberated in quantity, came to protect the Earth's surface later from much of this radiation and did so simultaneously with the evolution of eukaryotes, while advancing the rate of generation of heat (see Chapter 7). At the other end of the visible light spectrum, IR radiation stimulates vibrations at considerably lower frequency but before any chemistry takes place these low-frequency forms of IR energy are rapidly dissipated in longer wavelength vibrations, rotations and translations, resulting in a rise of temperature. This temperature rise is not readily directed to specific chemical processes. Visible light up to around 7,000 Å, however, is of sufficient energy to develop two chemical, quite high-energy electronic changes involving bond breaking, to give free electrons, in very selected molecules that absorb in regions of visible light energy. The first is the development of specific (chlorophyll) free radicals, which then give rise to electron/proton transfer in their vicinity, and further possibility of a controlled charge flow in membranes. As indicated in Section 5.6, this reaction chain is based on several metal ion and coenzyme centres. The second (the earlier?) is a light-induced change between molecular isomers, a *cis–trans* geometry change, initiating proton transfer in a membrane. The compounds involved in this case are the carotenoids. Both require a specific molecule in a specific location if they are to be effective. Selectivity arises since few biological organic molecules absorb light above 4,000 and below 8,000 Å, that is, few are coloured and can undergo *cis* → *trans* switches or can give rise to radicals. The fact that these electronic changes are localised in membranes is the reason for their successful use in photosynthesis and vision.

For vision or energy capture without undue risk but high effectiveness, molecules must capture light to cover the whole of the above visible region of radiation to around 7,000 Å. This cannot be achieved with just one type of molecule, so plants, for example, have chlorophylls, which absorb at around 4,000 Å and between 5,000 up to >6,500 Å, and carotenoids that absorb at 4,000–5,000 Å, with other pigments of intermediate unsaturation filling gaps. There is less absorption in the dangerous UV region and very little in the IR regions. The great advantage of *magnesium* chlorophyll is the very high absorption capability around 6,500 Å, much greater than by porphyrins alone and some 10 times stronger than carotenoids. (Note that Mg^{2+} itself in chlorophylls cannot undergo redox changes and is readily available.) For this reason plants look green. However, chlorophyll is not as stable as the carotenoids whose presence is hidden until the cycle of plant life breaks in the autumn. Quite quickly, the green chlorophyll disappears and plants acquire the red and yellow colours of carotenoids before the leaves finally die. The beauty of autumn has a simple chemical explanation. We stress that the process of energy capture from the Sun is highly effective in relation to its energy spectrum. Is it possible to devise a better system? It has existed for at least 2 to 3 billion years with very little modification, yet it developed relatively quickly initially in terms of tens or hundreds of millions of years.

Here we must introduce a complication in the structures of cellular compartments. We have treated the small bacterial cells as one aqueous phase, the cytoplasm, within one containing membrane. Now there are many examples of the membrane inside the wall as a very convoluted structure with "cristae" reaching far from the membrane. Effectively the cristae can be different *kinetic compartments* of the membrane with different protein (enzyme) complements, giving rise to diffusion-controlled separation of reactions. Now by pinching off the membrane "cristae" they isolate, encircle, a fluid different from that of the cytoplasm, within their now separate membranes and solutions, that is they can form vesicles. This point is made here as inspection of many photosynthesising bacteria show them to have just such separate compartments called thylakoids, e.g. in cyanobacteria. The inside of the thylakoid presents an opportunity of short- or long-term storage of molecules or ions, e.g. H^+, separated from the cytoplasm and the environment. Of particular interest in the case of the thylakoid is that it is the site of decrease in pH, proton gain, associated with the absorption of light. Simultaneously, Mg^{2+} ions are rejected to the cell cytoplasm from the thylakoid so as to increase metabolic incorporation of carbon (see Figure 7.1). Proton retention in the earlier bacterial devices uses storage in or across the outer membrane as an uncompensated proton ionic potential. The trick of using a special compartment for retention of material and ions increases through all of evolution and is an effective aid to efficient use of energy. (see Section 6.2 for a description of the special case of *ammonox* bacteria and especially Ca^{2+} gradients in Chapters 7–10.)

We can look upon all these changes as deriving from random mutations, but they are also driven by the fact that the capture of light will cause extra energy flow, and coupling of the flow to the overall cooperative metabolism will lead to increased rate of energy degradation not possible in non-photosynthesising organisms. Remember, however, that light does not penetrate far into the sea, so deep-sea organisms cannot use it.

5.8. Manganese in Cells/Oxygen Evolution

The evolution of manganese chemistry in cells would appear to depend upon its rejection from the cytoplasm relative to the concentration in seawater, which has always been considerable. In this, it may parallel the rejection of the other common, large cations, Ca^{2+} and Na^{2+}, and the anion Cl^-, although it is somewhat smaller than them. Like Ca^{2+}, manganese would then be expected to be bound on the outside membrane (or wall) surfaces. It is in this position that it is found in photosynthesis. The chemistry in photosynthesis involves manganese in a cluster of four Mn atoms, three of which are possibly in a cube completed perhaps with Ca^{2+}, and connected to Cl^- (Figure 5.9, see Rutherford and Boussac, Ferreira *et al.* and Biesiadka *et al.* in Further Reading). The cluster is used to generate O_2 from H_2O. It is doubtful if any other element could carry out this reaction, since it has the accessible oxidation states required (Figure 2.6), but note that its primary function is to provide a route to hydrogen from water using light. In this process O_2 is a waste product (see Figure 5.8).

We shall return to the point that the manganese centre is associated with bicarbonate, Ca^{2+} and Cl^-, which act as control ions for oxygen release. A second bicarbonate ion appears to be bound to the single ferrous atom in part of the

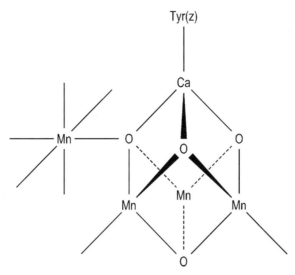

Fɪɢ. 5.9. A possible structure of the Mn/Ca site for the synthesis of oxygen from water. Other structures have been proposed. The site is not very stable and the whole protein holding the metal ions must be replaced frequently. There is a chloride ion nearby (see Rutherford and Boussac and Biesiaka et al., in Further Reading).

photosystem. Manganese is also used later in enzymes, which oxidise S-compounds to sulfate and to some extent in ribonucleotide reductases, in certain catalases and in superoxide dismutases (see Chapter 6).

Interest in the manganese site for oxygen release lies in part with its origin. As stated, manganese as Mn^{2+} was probably rejected from cells together with Ca^{2+}. These two ions have somewhat similar chemistry due to their size and rather low electron affinity. The similarity in chemistry leads to binding to mainly oxygen donor centres. On the outside of the bacterial cell membrane there always would have been sites for Ca^{2+} which could also bind Mn^{2+}. Now the presence of a photosystem close to bound Mn^{2+} could have caused oxidation of Mn^{2+} to Mn^{3+}. The chemical nature of Mn^{3+} is different from that of Mn^{2+} in that it is not lost easily and forms Mn–OH and Mn–O–Mn bridged links readily. Thus it is entirely possible that the observed Mn sites were originally Ca^{2+} sites. The fourth corner of the cubane in Figure 5.9 is possibly still occupied by Ca^{2+} stressing the external nature of the site. It exchanges and can act as a control with the external environment. To this day, the $Mn^{3+}.Mn^{4+}$ resting state of the enzyme can only be made in its protein by photo-energisation of the Mn^{2+} state. A fascinating feature of the protein which holds the Mn ions is that it is so easily damaged by radiation that it has to be replaced once every half hour. (We thank J. Barber and A. Mulkidjanian; see Further Reading for discussion on these points.)

In concluding the section on photosynthesis, this energy transduction mechanism producing oxygen is found in one major group of bacteria, a chemotype with particularly high Mg and Mn. It survives in the light while more primitive chemotypes survive alongside it or in the dark.

5.9. The Molybdenum Cofactor, Moco

There are a series of non-mobile coenzymes based on the so-called isoalloxine rings, which include flavins and pteridines. Their early function was the transfer of hydrogen inside membranes or in complicated enzymes, so we have

$$(\text{mobile}) \; \text{NADH} \longrightarrow \text{Flavine} \longrightarrow \text{Quinone (mobile)} \longrightarrow \text{H}^+$$
$$\downarrow e$$
$$\text{Metal site}$$

where hydroquinones are oxidised by flavines.

Later, they, especially flavin, became involved with oxygen and sulfate activation in enzymes. One of the pteridines is also involved in quite different chemistry. We have already mentioned that oxygen atom transfer is managed by a molybdenum coenzyme, Moco, though presumably molybdenum itself was used at first. (There is even a molybdenum/iron protein in ethylbenzene dehydrogenase in an anaerobe.) In the coenzyme, a special derivative of a pteridine with exposed thiolates (a pteridine dithiolate) is bound to the molybdenum (Figure 5.10). We know that molybdenum has a carrier and is inserted in its coenzymes. The synthesis appears today to be tightly coupled with the introduction of molybdenum into organisms much like the metal ions of the porphyrin coenzymes, Fe_nS_n, and

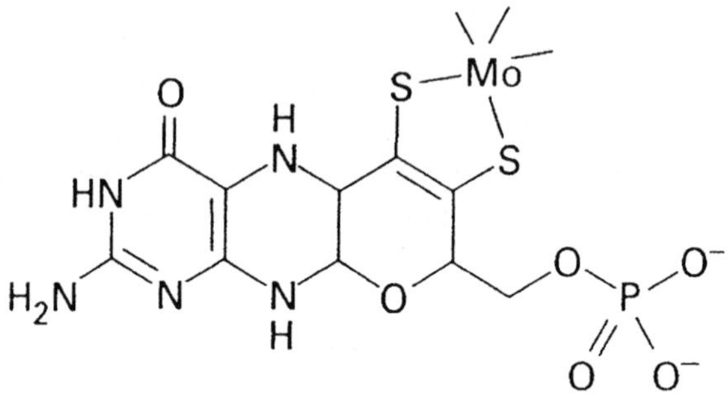

M.Molybdopterin cofactor – Moco

FIG. 5.10.   The formula of one of the mononuclear molybdenum cofactors, Moco. Others have a nucleotide phosphate extension (see references to these elements in Further Reading). In sulfide-rich environments, tungsten replaced molybdenum. In some coenzymes, two pterins are bound to the metal ions.

FeMoco. The selection of the organic cofactor for molybdenum may well concern thermodynamic binding constants for high-valent ions, possibly based upon the particular size of the MoO ion together with the strength of Mo-thiolate binding. The pteridine dithiolene acts as a chelating agent with a very small "bite" (Figure 5.10) and note that Mo(VI) is very small. Few other single metal ions are held by so few protein donor atoms as in this Mo complex, and in some sites there are even 2 mol of the pteridine ligand bound to molybdenum. This may allow exchange of the Mo-pteridine unit, making it a mobile coenzyme. In very primitive organisms, tungsten acted in place of molybdenum. We have already stressed that in very primitive times tungsten could have been more available than molybdenum (Section 1.8), and it could be that the molybdenum chemistry described here was first developed for tungsten. The low values of available molybdenum could well have been due to the presence of H_2S (see also the deep smokers in Chapter 11). Molybdenum then became dominant in slightly more oxidising conditions (see Chapter 6). We give this as one more example of the ability of organisms to devise extremely selective thermodynamic equilibrium modes of binding as well as of activation for special functions characterising in part the chemotype. We can ask again: could any other element besides W replace Mo functionally in this type of reaction? More generally, are elements able to substitute for one another effectively to only a very limited degree? We return to this question when considering the alternative metal ions, Fe, V and Mo, in nitrogenases.

5.10. Early Uses of Zinc, Calcium, Vanadium and Sodium

We consider that the availability of zinc, like that of cobalt and nickel and to a much greater degree of copper, was very limited to anaerobes. However, just as some nickel and cobalt enzymes are known in these organisms, there are some zinc enzymes in them, including certain dehydrogenases and RNA synthetases in which zinc is very tightly bound and non-exchanging (Table 5.6). Clearly, such proteins require expression systems and maybe zinc carriers for synthesis. The use of zinc generally is much less here than we shall uncover in eukaryotes, which are aerobes. We believe that the insolubility of zinc in the presence of sulfide, somewhat more

TABLE 5.6

ZINC IN SOME ANAEROBIC ENZYMES

Enzyme	Function
External proteases	Hydrolysis of proteins
External nucleases	Hydrolysis of nucleotides
Internal synthetases	Synthesis of t-RNA
Dehydrogenases	Conversion of alcohols into aldehydes

soluble in more acid solutions which may well have been the environment of very early forms of life, is the reason for its very modest use before there was oxygen in the environment.

We have described the initial functions of calcium outside cells in Section 5.5. Here we draw attention again to its later function in association with manganese in O_2 production, described in the previous section (see Figure 5.9). We have also left out of this chapter any reference to vanadium as the first functional use of it known to us is in nitrogenase which was probably not required in the earliest organisms (see Section 6.5). The sodium gradient was utilised to cotransport nutrients into cells and this function remains a major use of the Na^+ ion in later organisms.

5.11. Summary of Anaerobic Prokaryote Metabolism

We have now described several kinds of element flow in anaerobic bacteria and their developments:

(1)   Non-metals, C, H, N, O, S and P, in the major irreversible non-equilibrated chemistry, in catalysed flow (organic chemistry) covered in Chapter 4 (see Table 5.3) and equilibrated control systems of substrate and coenzyme bindings

(2)   Metals, mainly Mg and Fe, in simple reversible equilibrium chemistry (complex ion chemistry) also linked to controls

(3)   Specialised chemistry of such elements as Mn, Co, Ni, Se and Mo (W)

(4)   Rejection of Na^+, Cl^- and Ca^{2+} (and Mn^{2+}) to control cytoplasmic ionic solutions, with uptake of K^+, and other elements by pumps or exchangers, mechanical catalysts; the sodium gradient was used to assist nutrient uptake, the gradient itself being driven by the bioenergetic proton gradient

(5)   Metal and non-metal chelatase chemistry leading to irreversible combination as in organic and organo-metallic chemistry but usually taken together with complex ion metal chemistry in (2), of Fe, Co, Ni and Mg (see Table 5.5) and separate from Mo(w) use.

It is the complement of these elements and their compounds and the controlled, different ways of using them with different energy intakes within one simple cell boundary which defines the changing chemotypes present very early in evolution. Points (1)–(5) are totally dependent on energy intake usually from MgATP illustrating that life has always required a large flow of energy and the supply has to be continuous as in any working machinery. Simultaneously, it generates a huge energy store in organic compounds and in the environment, i.e. in an ecosystem in the form of rejected O_2, Na^+, Cl^- and Ca^{2+}. Despite the selective need for energy input, a very important feature of the use of elements in cells is that, as far as we can see, the selection was an inevitable consequence of best chemical practice given the enforced drive to capture and degrade energy utilising element availability in a thermodynamically unstable system, which had, within its limitations, optimal survival strength. The overall selection of elements is particularly revealing as to the fitness of early life. Within the Periodic Table there are the following elements, beside hydrogen, used from the beginning of life of course, in functions in strict anaerobes

									•	C	N	O	•	•
•	•													
(Na)	Mg								•	•	P	S	(Cl)	•
K	(Ca)	•	•	V	•	(Mn)	Fe Co Ni	• Zn	•	•	•	Se	•	•

$$\text{Mo}$$

$$\text{W}$$

where ( ) represents a rejected element and • represents an element not used. The 19 elements are in 13 of the possible 17 useful Groups of the Periodic Table, that is excluding the inert gases. The elements of Groups 3, 4, Cu and Group 13 were hardly available. Hence the table shows that almost immediately cellular life exploited chemistry of all the most available and useful kinds of elements. The use of selectivity of equilibria between metal ions and organic donors is well worth observing and is quite different from the kinetic and gene control over the non-metal compounds. This suggests that the chemotype will change in the elements used not only in novel ways or in new oxidation states, but with one or two additions, e.g. Cu and I, as the environment changed. There is little possibility for the extension of use to many other new elements except by mankind (see Chapters 10 and 11). Finally, we stress again that all the compounds and pathways of non-metals and metals are in a highly effective energy-capture system, which is tightly controlled in its pathways and is autocatalytic.

We look next at energy flow in these anaerobes.

5.12.   Energy Flow in Anaerobes

We have seen in Chapter 3 that the forms energy can take are not all useful for driving uphill *chemical* changes. Light from the Sun, the major source, can be used to generate electron transfer but this is not of much use by itself for vital chemical syntheses such as condensation. However, the flow of electrons in a membrane of a biological cell can be put to use in two ways (Table 5.7). First they, with H^+, can reduce chemicals such as CO_2, CO, N_2, etc. while second, the positive charge generated by electron removal can produce S_n (from H_2S) or O_2 (from H_2O). This flow of negative and positive charges in opposite directions away from an energised centre is managed in a membrane by the assembly of redox centres, reducible to one side oxidisable to the other, in a chain of electron-transfer catalysts (Figures 5.2 and 5.8, and see Williams in Further Reading). The two electron flows become useful in a

TABLE 5.7

PRIMITIVE ENERGY CAPTURE GIVING ATP

Mode	Energy Captured
Breakdown of organic material	Glycolysis $\rightarrow$ ATP
$FeS/H_2S \rightarrow FeS_2 + H$	Charge (e/H$^+$) separation
Light	Charge (e/H$^+$) separation
Charge separation	Formation of ATP
Lightning	NO absorbed and used as oxidant giving charge separation

different way once the electron flow is converted by redox reactions into a proton/hydroxide separation in, on or just outside the membrane. (For an up-to-date description of proton flow see Mulkidjanian *et al.* in Further Reading.) The conversions are managed by redox reactions which are proton-dependent (see Figure 5.2), and which require chemicals, often coenzymes, which convert two one-electron into one two-electron steps, such as quinones, flavins, nicotinamide coenzymes and H_2S/S_n (later oxygen). Note that the process demands at least a pair of compartments so that H^+ has no access to OH^- and reducing chemicals cannot react directly with oxidising chemicals. A membrane separating two aqueous compartments is the preferred device. Even at this stage, the energy in any kind of pH or charge gradient cannot be used directly for driving chemical change such as condensation. A gradient across a cell membrane becomes useful, apart from in exchange uptake, only when it is employed to drive a molecular machine, a (mechanical) device, in which the energy is converted into a bound form of pyrophosphate (MgATP) from P + ADP.Mg (see Figure 4.13), an initial energised condensation, on one side of the membrane. The pyrophosphate (MgATP) is able to drive the absolutely required energised chemistry steps of cellular chemistry, e.g. condensation, since pyrophosphates are drying agents. MgATP is, in fact, the usual energy currency in cells. In this way, a cell comes to contain chemicals energised in three ways relative to the small molecules of the environment (and the S_n (or O_2) rejected by cells). The products are: the *essential irreversibly reduced C/N/O monomers*, the *polymers* formed thereafter by *condensation* of reduced environmental chemicals and the *controlled pumped concentrations* of a variety of chemicals and ions. These steps require other molecular machines (see Figure 4.12(c)). Most of the energy today comes from the sun but the original source of energy flow may have been from Earth's minerals although we do not know for sure what it was. All the time we must remember that the energised chemical system is one of disproportionation of reducing and oxidising equivalents in which one part is internal to cells and the second part is external to life, i.e. in the environment, which we shall see slowly evolves towards a cyclic state.

The proton gradient like the gradients of any other ions, which are established across membranes, as well as the ATP can be used to drive exchange with other ions or to drive uptake of molecules. In cells, uptake driven by H^+ and Na^+ gradients and ATP are common throughout evolution.

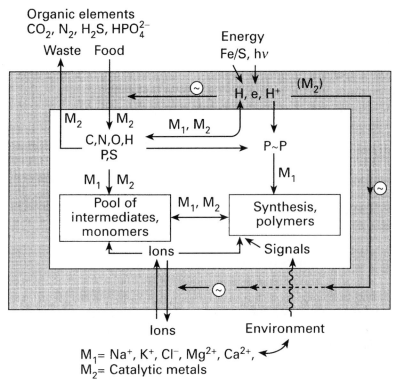

Fig. 5.11. A schematic picture of the roles of different elements in a primitive cell and a low redox potential environmental. O_2 and SO_4^{2-} were introduced later. ~ is a charge or concentration gradient.

At this stage of the present chapter we have looked at elements, energy and small molecules in the cytoplasm (see Figure 5.11), i.e. the metabolome and the metallome and their energetics. The basis of activity rests in the controlled catalysis of their reactions. While we have stressed the roles of metal ions here, we must now turn to the chemistry of the biopolymers, *the proteome and the genome* (see Chapter 4). We must not only see way the organisms developed in time relative to one another as shown by inspection of genes, but also why this series of changes occurred on the basis of thermodynamics.

5.13. The Polymers in Primitive Cells

In Chapters 3 and 4, we referred to the basic reactions leading to the classes of biological polymers DNA (RNA), proteins, saccharides and long-chain hydrocarbons. Each class has a specific functional value. Here we shall not comment further on the special properties these polymers introduce in aiding cell survival including energy capture and essential element incorporation. We draw attention only to these features specific to prokaryotes, which we did not describe before. There are in fact no further discriminating differences in any of the polymers from that in the cytoplasm of all organisms but for the DNA. The DNA of prokaryotes is double-stranded and circular and in the case of bacteria it is unprotected by proteins, histones, but it is so protected in Archaea (see Table 5.1). We return to this feature when we discuss eukaryotes. A point to note is that the circular bacterial DNA reads directly in operons and reading frames with no introns. Mutation in prokaryotes if it is at a rate commensurate with but not shorter than the lifetime is no great disadvantage since the prokaryote has little control over where it is born, lives and dies. The major functions are rapid growth, reproduction and mutation leading to adaptation and to new viable species with a short lifetime. Under starvation conditions it dies or sporulates. The more the DNA is adaptable to circumstances the better.

5.14. Gene Responses in Prokaryotes

Prokaryotes respond to change in the environment in two very different ways. First there is the relatively rapid adaptive response of *latent genes* to a rise in a source of a nutrient, that is to basic supplies of both non-metal and metal ion chemicals, which have been experienced by the cell before but have since that time been virtually absent. This is a particular feature of the controlled link between substrate, DNA, RNA and protein concentrations. The response uses transcription factors, X, proteins which recognise the nutrient, S, bind to DNA and, through their changed conformation to X′, on binding S, cause changes in gene expression. The conformation change is called an *allosteric* switch.

$$DNA \longrightarrow \text{increase in enzymes for S metal}$$
$$X + S \rightleftarrows X'(S)$$

The reaction of X with S must be fast and reversible, close to if not at equilibrium with concentration of S. It can be that there is an intermediate step in which X binds to a protein kinase (a protein which phosphorylates other proteins mostly at histidine residues in bacteria) using phosphate transferred from ATP. It then gives "XP" which is the transcription factor, where concentration of S still decides the extent of phosphorylation. No change occurs in DNA itself. Here equilibrium is avoided as dephosphorylation involves a phosphatase, though changes must be relatively quick since, for example, *cell cycling and division* depend on these steps, which must be completed in minutes. We have noted that such mechanical trigger-proteins as transcription factors are usually based on α-helical backbones common to all manner of such adaptive conformational responses (Section 4.11).

We have seen in Section 4.8 that the major units to which there is fast response are changes of concentrations of C/H/N substrates, H^+, phosphate (energy), thiol of disulfide exchange, Mg^{2+} and Fe^{2+}. There are genetic responses to all these elements and to one or two others such as Mo and Se. These remain most of the basic element internal signalling factors of all cells to the present time, but unlike the prokaryotes, eukaryotes show a greater and greater variety of fast responses to, and gain advantages from, interaction with the environment.

An alternative response is to an entirely new chemical, Y, which may appear in the environment to which the cell has no response in reserve. Here, *mutation or other changes of the DNA* are required so that entirely new proteins, enzymes or rejection pumps, are made. Prokaryote DNA mutates rapidly but we shall have to consider whether the mutation is just a random action on all the DNA or if it is directed, epigenetic (Section 7.13). The full range and nature of "epigenetics" is not known but they are increasingly seen to be extensive (see Caporale in Further Reading). In one sense such organisms through mutations become a population of very many independent species in which variety (of DNA) is an advantage and they can be recognised as different chemotypes. Notice that a prokaryote is not very sensitive to many features of its environment except nutrients and poisons.

At the same time as noting these adaptations of the genes in individual cells we must observe that bacteria can trade genes. This activity is quite different from sexual reproduction in eukaryotes and the trading (*conjugation*) is a kind of gene "grafting". Without conjugation (and mutation), bacterial reproduction would produce identical multiplication – clones. The ability to trade genes is extensive but it does not appear to allow complete merging of chemotypes so that photosynthetic and non-photosynthetic bacteria, for example, are separate.

5.15. Satellite DNA: Plasmids

We have dealt with the possible ways in which DNA could have been formed and could have changed to follow the changes in use of the environment based on (localised) mutations leading to recognition, rejection or use, of new chemicals which were in principle initially toxic to the organism. The way these DNA changes are included in the code can be made simpler if there is not just the one DNA strand. The advantages are similar to those we shall describe of separation of very different reaction paths in more than one compartment (see Chapter 7 onwards). Here however, the "compartment" is due to kinetically isolated reactivity. Prokaryotes in fact often contain plasmids, small separate circular DNA structures, and these structures are known to carry protective genes against unusual organic poisons (drugs) and inorganic poisons (heavy metals and non-metals). A picture of one such plasmid is given in Figure 5.12. Note how the genes may express resistance to a range of heavy, non-essential, elements in Groups 12 to 16 of the Periodic Table. These elements are the very ones which form a large part of the "*b*-group" of metals and have insoluble sulfides (see Section 2.4). The general affinity for thiol side chains is usefully employed in detoxifying from these elements which suggests that many of the metal-binding proteins, transcription factors, carriers and pumps for rejection are

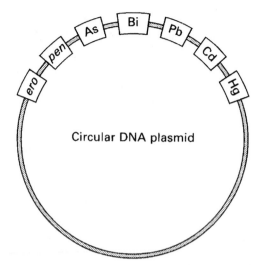

Circular DNA plasmid

FIG. 5.12.   A series of resistance genes built on to a bacterial plasmid. *ero* is erythromycin- and *pen* a penicillin-resistant gene.

derived from one or two precursor genes. (This suggestion carries over later in the value of certain sites for distributing metal ions such as Cu and Zn, for example, in aerobes, see metallothioneins). The question arises as to how such plasmid gene sequences arise. Are they related to existing genes in the main DNA? Are they produced by gene duplication first? Are plasmids protective devices for the main DNA such that they carry resistance to damage? If so, how has selective genetic change arisen in them without disturbing the main DNA? Selective DNA changes could be of extreme value in evolution.

Finally, DNA/RNA must not be viewed in isolation for they have to be synthesised indirectly by autocatalysis and their expression as proteins is complicated.

5.16.   Prokaryote Controls

A further point we have raised but which we stress here concerns the need for control. All the prokaryotes needed to produce a balanced set of many energised chemicals in fixed pathways based ultimately on the same 15–20 elements (Table 5.8) but placed in selected internal, kinetically separated arrangements. This *internal* balance can only be achieved within and between reaction pathways and the genes for the production of the necessary balanced set of structural, pump, and catalytic units (see Chapter 4) by both feedforward and feedback controls (see Figure 4.13). The whole is autocatalytic as we stressed in Chapter 4. Note again that this is managed by messengers especially substrates, phosphate compounds, protons, Mg^{2+}, thiols, Fe^{2+} and coenzymes (Table 5.9) and the special phosphates, including NTP, NDP, NMP

TABLE 5.8

THE ESSENTIAL PRIMITIVE ROLES OF METAL IONS

Metal Ion	Some Roles
Mg^{2+}	Glycolytic pathway (enolase)
	All kinases and NTP reactions[a]
	Signalling (transcription factors)
	DNA/RNA structures
	Light capture
Fe^{2+}	Reverse citric acid cycle
	CO_2 incorporation
	Signalling transcription factors
	Control of protein synthesis
	(deformylation)
	Light capture
W (Mo)	O-atom transfer at low potential
Mn	O_2-release
Ni/Co	H_2, CH_3- metabolism
Na^+, K^+	Osmotic/electrolyte balance
Ca^{2+}	Stabilising membrane and wall, some signalling?

[a] Almost all synthesis pathways.
Note: $K^+/Na^+/Cl^-$ control of osmotic and charge balance while Ca^{2+}, Zn^{2+}, Cu have very little role.

TABLE 5.9

PRIMITIVE MESSENGERS

Messenger	Functional Control Upon
Mobile coenzymes	Distribution of metabolic fragments
	H^-, CH_3, $-COCH_3$, etc.
Nucleotide triphosphates	Distribution of energy
Fe^{2+}, $2RS^-/(RS)_2$	Distribution of electrons
	Redox state balance
Some simple substrates (feedback)	Metabolic products, e.g. glutamine, nucleotide
	bases, amino acids, and upon gene expression
Phosphorylation of proteins,	Gene expression
Mg^{2+}, Fe^{2+} (Mn^{2+})	

and cyclic NMP, in which phosphate exchanges are the most noticeable. Of particular interest are the histidine kinases, which became common in eukaryotes. The network of controls is very intricate so that hydrolytic and condensation controls, which often rests with Mg^{2+} and phosphates, for example, must be connected to oxidation and reduction. A very intriguing example is the use of Fe^{2+}, not only in balancing Fe^{2+}/Fe^{3+} reactions with thiols but also the activation of proteins by deformylation in bacteria and removal of methionine in Archaea, while Fe^{2+} is also central to oxidation/reduction control, connected to phosphate reactions. The concentrations of these messengers are themselves controlled by feedback. There is no escape however from the need for a central control in the whole management and it is clearly so much the better if as much as possible of the control machinery is reproducible through replication of this central management system with its coded DNA, but unfortunately its response time is slow. Here central control means more than DNA itself. It means all the necessary bits and pieces which make the tape of DNA readable and translatable into products, RNA/proteins, and the further units in control of metabolism. *The need is for the transfer of and reception of information to and from the genetic structure* (see Table 5.9) both about the internal condition of metabolism and the external sources of energy and nutrients. We must also be aware that a prokaryote cell is also subject to the gradients across the cell membrane, which inform the cell in part directly to metabolism and there is some suggestion that some bacteria have one or two controls through Ca^{2+} fluxes. They are essential contributors to the whole system since all the information sources must be integrated. As far as calcium ions are concerned a Ca^{2+}-pump and maybe one or two Ca^{2+} carrier proteins are known to exist. There is also a prokaryote sodium pump, but little use of these ions in controls appears until later (see Chapters 7 to 10 and Dominguez in Further Reading). There is little doubt that life, as we know it, depends on *simultaneous* advance of management, central controlled reproduction and the environment with the environment supplying a second form of information, of increasing importance with time.

5.17. Internal Flows and General Movement: Sensing and Searching Chemotaxis

The general description of components, organic and inorganic, and their controls may have given the impression of stationary large molecules within a static organism. In fact it is apparent that there are localised flows in bacteria, which can move considerable units around (see Shapiro and Losick in Further Reading), and the organisms swim. The inner movement is most clearly seen at the time of cell division. At that time the morphology and inner construction of the cell changes dramatically.

However, the general point is that cell internal structure like that of a cloud is in constant flowing motion controlled by field gradients, chemical or electrical, which we see realised in the transient cytoskeleton. What complex relationship this form has to DNA sequences is difficult to know. It is obvious, as discussed in Section 3.10, that a separation of some of the activities in compartments would be advantageous.

The whole body movement of bacterial cells uses flagella motors which drive a peculiar motion of runs and rotations. In this fumbling manner they follow vaguely external gradients of attractants such as sources of food, light or magnetic fields, and they move in the opposite sense in a repellent chemical gradient by chemotaxis.

Thus prokaryote cells have a very limited sense of the environment and a limited search apparatus (see Harold in Further Reading).

5.18. Conclusion: Anaerobic Chemotypes and their Development

Prokaryote evolution from what had to be very simple somewhat random chemical beginnings in protocells has to be seen against the background of securing increasing efficiency of capture of elements, later incorporated into reductive and condensation reactions, and of energy to obtain robust synthesised organic and inorganic chemicals in an enclosed space, a fully fledged cell of considerable survival strength. We stress *the activity was very largely inside the cell*. Most of the activities are then conserved through all evolution (see Table 4.11). The major need was for the optimal catalysed synthesis of particular polymers, lipids, proteins, polysaccharides and nucleic acids by optimal use of the available elements. The work of synthesis demand molecular machines incorporating catalysts. The syntheses could only use as substrates the available non-metal elements that form kinetically rather stable bonds, i.e. H, C, N, O, S and P, and the chemistry of which could be activated, that is not Si or halogens, for example. The reactions had to be both autocatalysed and cooperative selectively as described in Section 3.9 to give the system survival strength. The synthesised biopolymers are considered by us as being the best of all possible compounds for retaining and making the most efficient use of energy in chemicals in an aqueous media at room temperature. If this is so no other non-metal basic form of life exists, on the Earth at least (see Sections 4.5–4.14). Additionally, and to secure this end, energy was used in other machinery to obtain and retain a certain cytoplasmic complement of mainly equally required metal ions which has remained common to all life. The major metal ions in anaerobes were of Mg and Fe with Mo (or W) and Ni for catalytic functions, while Na and K (note also Cl) had ionic strength control roles (see Sections 4.15 and 4.16). We regard the metal in deployment as assisting optimal fitness. The metal ions are present in fixed relative concentrations at equilibrium bindings to organic partners. Several selected non-metal/metal-fixed pathways formed and its proteins were coded for in reproduction. Now simultaneous with this synthesis was the autocatalysed degradation of the original energy (from minerals) or from breakdown of energised molecules from other organisms (food) or even self to give a driven cycle. Such molecules are all unstable and cells accelarate their breakdown. Evolution amongst prokaryotes was then based on greater access to sources of material and their distribution (see coenzymes above), then on greater access to energy, light, and finally to the more resourceful use of all available elements in anaerobic conditions. The changes may be seen as chemically inevitable in the system in the search for optimal energy flux through chemicals whether or not they were achieved through trial and error in DNA changes, which were certainly needed for reproduction. (The increasing variety of the chemotype system we have described is known to be coded, and we know that a coded reproduction is a requirement of what we call a "living" cell, but a code can only exist if something related to it pre-existed.) The overall scheme is (compare Figure 3.11)

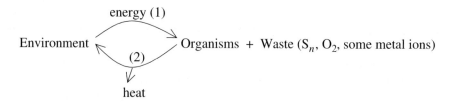

where (1) is synthesis and (2) is degradation.

Note that the prokaryote cells are devices for reproduction at the fastest possible rate to secure survival since they are poorly protected and optimal energy turnover is a driving force. We see evolution mainly in the chemical development of this flow of energy in reactions, with necessary changes in the coded molecules while the whole had to be thoroughly organised (Table 5.10). Note that high density of organisms, high survival and reproduction rates, is part and parcel of high-energy flux as organisms are always active (degrading) machines.

TABLE 5.10

ORGANISATION OF ANAEROBIC PROKARYOTES

1.	Outer cytoplasmic membrane
2.	Wall of well-structured molecules
3.	Information centre, DNA, open to mutation
4.	Positioned uptake/output pumps and some enzymes on or in the membrane
5.	Circulating energy supply, ATP, made from various external sources
6.	Circulating essential ions, substrates and enzymes of pathways (see 10 below)
7.	Coenzymes synthesised later. Note use in atom exchange and control
8.	Development of catalysts, many metal ion dependent, limiting change to particular pathways
9.	Feedback controls from (4) to (8) to (1), (3) and the enzymes of (5) and (6). Note particularly the role of phosphate transfer by kinases and particularly of c-NMPs, NDPs and NTPs
10.	Controlled internal concentrations of substrates, coenzymes and metal ions. In particular, the loosely *bound units* are exchanging at close to equilibrium between particular pathways recognised by binding strength. Equilibria are independent of organism type so that the concentrations are universal in a compartment, e.g. the cytoplasm of all cells. The binding constants for the divalent transition metal ions follow the Irving/Williams series (see Appendix 4.A), they affect maturation of proteins as well as reduction reactions. Mg^{2+} dominates phosphate transfer and Fe^{2+} controls redox reactions, as well as the final step of protein activation (see the metallome)
11.	Some knowledge of the environment through external gradients of food or light and an ability to swim up these gradients. The cells are informed by these gradients, later some families are informed by magnetic and/or gravitational fields
12.	*Later.* Concentric membranes giving rise to two new possible phases in addition to the cytoplasm and its membrane – in particular the periplasmic space (Figure 5.5). The additional thylakoid membrane is another example of a physically isolated compartment (Section 5.7)

We must also look at the waste (pollution) in the above equation since, while there is waste, the material cycle is incomplete and inefficient (see Section 3.8). The cycle is not element neutral. In fact, the waste is not just oxygen after the final adjustments in photosynthesis since death of the organisms will leave debris of energised reduced organic material. The immediate consequence is that not only photosynthesising organisms can increase from the capture of light but the original organisms which probably used the energy of minerals, chemoautotrophs, can now benefit by switching to consuming more reduced debris of energised chemicals, that is, they could become scavengers of waste. The oxygen waste, relative to the synthesised cell material and indeed relative to the original environment, is both a threat and a new source of energy which as we shall see was degraded later to drive aerobic life. Despite the advantage of photo-energy capture in the evolved organisms which synthesised chlorophylls they did not develop and increase alone therefore. In fact, the non-photosynthetic organisms have an advantage in that they do not produce risky oxygen and some came to be able to fix nitrogen from N_2 in the absence of O_2 (see Section 6.4). Note that we shall see later that some cyanobacteria overcame the problem by having a "compartmentalised" cell (see Figure 6.10), to avoid the problem of N_2 fixation in oxygen. There is then a complementary advantage of nitrogen fixers. This pattern of sharing products, best produced separately, is the first form of very loose organisation of many organisms (chemotypes) in a cooperative (synergistic) ecosystem struggling to attain optimal energy capture in chemicals and its more effective distribution. We must look next therefore at the waste, oxygen, and the complications it introduced since oxygen caused many chemical changes in the environment. Although the mutual existence of oxygen, its environmental products, and organic matter generates a huge external/internal energy store relative to organism materials, the fact that oxygen does not enter the cyclic steady state above but it can attack anaerobic organisms means that it was an ever-increasing destructive threat (pollutant) from which all living organisms had either to protect themselves or to find an escape from it or eventually a use for it.

There is another form of "waste" produced by these cells and stored in the potential energy gradients of rejected ions, Na^+, K^+, Ca^{2+}, Cl^- and perhaps Mn^{2+}. The rejection is to the environment and there the ions selectively removed from the cytoplasm are dispersed. While the Na^+ with the H^+ gradient are used in the uptake of nutrients little or no use is made of the gradients of Ca^{2+}, K^+ or Cl^- by prokaryotes. Contrast the use later of these charge gradients by all cells and of proton gradients in bioenergetics. We shall see, when we come to describe eukaryotes, that they not only make more effective use of oxygen and organic waste but they make more and more use of rejected ions. We saw the first such use in the case of Mn^{2+} in the production of hydrogen from water in the photosystem (Figure 5.8). The Mn^{2+} is on the outer surface of the membrane. We may look upon this as achieving "fitness" but it brings disadvantages of a new environment.

Before closing this chapter, we wish to remove any impression which the above description of anaerobic life might have left that the anaerobes are a few small groups of organisms. In fact there is immense variety of these prokaryote organisms, some living under extreme conditions of temperature, salt and pH as well as those living in the environment we experience. They have very great survival strength concomitant with considerable energy degradation, perhaps optimal in the absence of oxidising agents in the environment. The greatest weakness of these anaerobic prokaryotes, apart from the waste they produced, was their inability to recognise features of their environment, except nutrients, or to benefit from the coming of oxygen. We have deliberately restricted our division of the chemotypes of these prokaryote anaerobes into three groups – non-photosynthesising Archaea and bacteria and photosynthesising bacteria. Each division can contain a wide variety of species. Extensive evolution of each prokaryote chemotype is described in the next chapter. This apparently leaves us at this stage with certain chemotypes and an environment but there is another environment and its organisms which we have ignored. Our account is that of a surface ecosystem which is exposed to light and the atmosphere, but there is a dark zone too in which elements have different concentrations, that of the deep sea trenches. We return to it only in Chapter 11 as evolution is different there and occurs in unusual chemotypes, including prokaryotes. The summary of all surface prokaryote evolution in terms of chemotypes is given at the end of Chapter 6, and their general place in the evolving ecosystem is seen in the cover diagram.

Further Reading

BOOKS

Brock, T.D. and Madigan, M.T. (2000). *Biology of Microorganisms* (9th ed.). Prentice-Hall, New Jersey
Caporale, L. (2003). *Darwin in the Genome*, McGraw-Hill, New York
Harold, F. (2001). *The Way of the Cell*. Oxford University Press, Oxford
Knoll, A.H. (2003). Life *on a Young Planet – the First Three Billion Years of Evolution on Earth*. Princeton University Press, Princeton
Margulis, L. and Schwartz, K.V. (1998). *Five Kingdoms* (3rd ed.). Freeman and Co., New York
Morowitz, H.J. (1992). *Beginnings of Cellular Life*. Yale University Press, New York
Pope, M.T., Still, E.R. and Williams, R.J.P. (1980). A Comparison between the chemistry and biochemistry of molybdenum and related elements. In M. Cougwan, (ed.), *Molybdenum and Molybdenum Containing Enzymes*. Pergamon Press, Oxford, Chap. 1
Schlegel, H.G. and Bowlen, D. (eds.) (1999). *Autotrophic Bacteria*. Springer Verlag, Heidelberg
White, D. (1995). *The Physiology and Biochemistry of Prokaryotes*. Oxford University Press, Oxford
Williams, R.J.P. and Fraústo da Silva, J.J.R. (1996). *The Natural Selection of the Chemical Elements – The Environment and Life's Chemistry*. Oxford University Press, Oxford

PAPERS

Abramson, J., Smirnova, I., Kasho, V., Verner, G., Kaback, H.R. and Iwata, S. (2003). Structure and mechanism of the lactose permease of *Escherichia coli*. *Science, 301*, 610–615

Biesiadka, J., Loll, B., Kern, J., Irrgang K/D. and Zouni, A. (2004). Crystal structure of cyanobacteria photosystem II at 3.2Å resolution. *Phys. Chem. Chem. Phys.*, *6*, 4733–4736

Canfield, D.E., Habicht, K.S. and Thamdrup, B. (2000). The Achaean sulfur cycle and the early history of atmospheric oxygen. *Science*, *288*, 658–661

Des Marais, D.J. (2000). When did photosynthesis come to Earth? *Science*, *289*, 1703–1705

Dismukes, G.C., Klimov, V.V., Baranov, S.V., Kozlov, Yu. N., Das Gupta, J. and Tyryshklin, A. (2001). The origin of atmospheric oxygen on Earth: the innovation of oxygenic photosynthesis. *Proc. Natl. Acad. Sci. USA, 98*, 2170–2175

Dominguez, D.C. (2004). Calcium signalling in bacteria. *Mol. Microbiol.*, *54*, 291–297

Ford Doolittle, W. (1999). Phylogenetic classification and the universal tree. *Science*, *284*, 2124–2128

Ferreira, K.N., Iverson, T.M., Maghlaoui, K., Barber, J. and Iwata, S., (2004). Architecture of the photosynthetic oxygen-evolving center. *Science*, *303*, 1831–1838

Forterre, P., Brochier, C. and Phillipe, H. (2002). Evolution of the Archaea. *Theor. Popul. Biol. 61*, 409–422

Graham. D.E., Overbeck, R., Olsen, G.J. and Woese, C.R. (2000). An archaea genomic signature. *Proc. Natl. Acad. Sci. USA*, *97*, 3304–3308

Hagen, W.R. and Avendsen, A.A. (1998). The bioinorganic chemistry of tungsten. *Struct. Bond.*, *90*, 161–192

Huber, C., Eisenreich, W., Hecht, S. and Wachtershauser, G. (2003). A possible primordial peptide cycle. *Science*, *301*, 938–940

Hartman, H. (1998). Photosynthesis and the origin of life. *Origins Life Evol. Biosphere*, *28*, 515–521

Hille, R. (2002). Molybdenum and tungsten in biology. *Trends Biochem. Sci.*, *27*, 360–367

Kletzin, A. and Adams, M.W.W. (1996). Tungsten in biological systems. *FEMS Microbiol. Rev.*, *18*, 5–63

Kashefi, K. (2005). Living hell: life at high temperatures. *The Biochemist*, *27*, 6–10

Leman, L., Orgel, L. and Ghadiri, M.R. (2004). Carbonyl sulfide – mediated prebiotic formation of peptides. *Science*, *306*, 283–286

L'vov, N.P., Nosikov, A.N. and Antipov, A.N. (2002). Tungsten containing enzymes. *Biochemistry (Moscow)*, *67*, 234–239

Miyakawa, S., Yamanashi H., Kobayashi, K, Cleaves, H. J. and Miller, S.L. (2002). Prebiotic systems from CO atmospheres: implications for the origins of life. *Proc. Natl.l Acad. Sci. USA*, *99*, 14628–14631

Morgan, R.O., Martin-Almedina, S., Iglesias, J.M., Gonzalez-Florez, M.I. and Fernandez, M.P. (2004). In special issue on 8th European Symposium on Calcium. *Biochim. Biophys. Acta*, *1742*, 133–140 and other papers therein.

Mulkidjanian, A.Y., Cherepanov, D.A., Herbela, J. and Junge, W. (2005). Proton transfer dynamics at membrane/water interface and mechanism of biological energy conversion. *Biochemistry (Moscow) 70*, 251–256

Norris, V., Grant, S., Freestone, P., Canvin, J., Sheik, F.N., Toth, I., Trinei, M., Modha, K. and Norman, R.I. (1996). Calcium signalling in bacteria. *J. Bacteriol.*, *178*, 3677–3682

Pace, N.R. (1997). A molecular view of microbial diversity and the biosphere. *Science, 276*, 734–740

Rutherford, A.W. and Boussac, A. (2004). Water photolysis in biology. *Science*, *303*, 1782–1784, and see Ferreira, K.N. *et al.* (2004). *Science*, *303*, 1831–1838 and Zouni A. *et al.*, (2001) *Nature*, *409*, 739–743

Shapiro, L. and Losick, R. (1997). Protein localisation and cell fate in bacteria. *Science*, *276*, 712–718

Smith, A.G., and Munro, A.W. (eds.) (2005). Coenzymology: the biochemistry of vitamin biogenesis and cofactor containing enzymes. *Trans. Biochem. Soc.*, *33*, 737–823

Smith, D.C. (2001). Expansion of the marine Archaea. *Science*, *293*, 92–94

Xiong, J., Fisher, W.M., Inoue, K., Nakahara, M. and Bauer, C.A. (2000). Molecular evidence for the early evolution of photosynthesis. *Science*, *289*, 1724–1730

Westheimer, F.H. (1987). Why nature chose phosphate. *Science*, *235*, 1173–1178

Williams, R.J.P. (1961). The functions of chains of catalyst. *J. Theor. Biol. 1*, 1–13

Woese, C.R. (1998). The universal ancestor. *Proc. Natl. Acad. Sci. USA*, *95*, 6854–6859

Woese, C.R. (2002). On the evolution of cells. *Proc. Natl. Acad. Sci. USA*, *99*, 8742–8747

CHAPTER 6

The Evolution of Protoaerobic and Aerobic Prokaryote Chemotypes (Three to Two Billion Years Ago)

6.1. Introduction

Chapter 4 described general biological chemistry and some possible ways in which it could have evolved from very basic beginnings using energised available chemicals in a physically constrained environment, a cell. We avoided too much speculation about the origin of life while we introduced the known components of organisms so that we could proceed to the discussion of the development of the earliest known anaerobic life in Chapter 5. We then indicated how cellular chemistry evolved through improvement in the management of the available energy and materials into groups of chemotypes, the anaerobic prokaryotes. While this system of cells could reach a steady state of flow, as cells were formed and degraded, it inevitably produced waste (Figure 6.1). The inevitability arises from the very nature of the synthesised cellular chemicals best able to retain energy relative to the sources of these elements in the environment. The biomolecules hold elements in reduced states relative to the intermediate oxidation states of the same elements in the environment (Table 6.1 and compare Table 2.11), and the waste was therefore in the form of oxygen or oxidised states of elements C, S, Fe and N for example, in an oxygen-containing atmosphere, and also in reduced cell debris on death. The oxidised states of elements together with the reduced material in cells and debris are a potential new source of energy. It was these compounds, increasing continuously

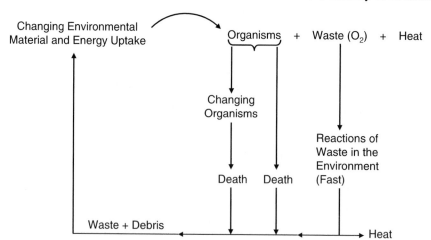

FIG. 6.1   The basic life cycle of organisms. N.B. Organic debris does *NOT* react fast with O_2 of the environment, but H_2S and Fe^{2+} do. Organisms accelerate the organic reaction.

but slowly in the environment, which forced both evolution of anaerobes, firstly of protoaerobes and later of aerobes, and then (as we shall see in Chapters 7–9) compartmentalised eukaryotes in order to utilise this possibility and also to gain a source of the elements for themselves. This process of adaptation of the cellular chemistry in prokaryotes from anaerobes to aerobes is the content of this chapter. Of course the changes took time, but the inorganic oxidation in the environment, not of the debris, was always faster than the changes in the cells. The cellular chemistry, here of prokaryotes, is by its very nature conservative and slow changing, especially the DNA, while the chemistry of the environment generally moves relatively faster through equilibria to successively more oxidised conditions (see Figure 1.14 and Table 6.1). This total evolution of the ecosystem is best seen in a summary diagram where biological waste constantly increases and drives changes of cells, allowing new chemotypes (Figure 6.1). Remember that our thesis is that the whole system, the environment plus the cells, seeks optimal energy capture in chemicals and that the presence of waste causes this process to progress (see Chapter 3). Slowly but surely the whole geological/biological system attempts to build systematically towards an optimal cyclic steady state incorporating all waste with approximately fixed energy sources from the Earth and the Sun. Ultimately, there should be no material loss but irreversible degradation of energy. We must first describe the effects of the *oxidised waste* at the earliest times since anaerobes could certainly adapt to much of it, (see (2) and (3) below). Later, we shall consider aerobes to be those organisms, which began to use oxygen directly (see (4) below). Both steps give new chemotypes, and we stress by employing this description of evolution of groups of organisms that we concentrate on changes of thermodynamic properties of elements in compounds.

TABLE 6.1

SUCCESSIVE OXIDATION OF CHEMICAL ELEMENTS[a]

Oxidation	Potential (volts) pH $= 7$
$HCHO \rightarrow CO_2$	-0.5
$H_2 \rightarrow H^+$ (H_2O)	-0.4
$NH_4^+ \rightarrow N_2$	-0.2
$H_2S \rightarrow S_n$	-0.2
$H_2S \rightarrow SO_4^{2-}$	-0.2
$Mo(IV) \rightarrow Mo(VI)$	-0.1
$Fe^{2+} \rightarrow FeO.OH$	$+0.0$
$CH_4 \rightarrow HCHO$	$+0.2$
$Cu^+ \rightarrow Cu^{2+}$	$+0.3$
$NH_4 \rightarrow NO_3^-$	$+0.5$
$Mn^{2+} \rightarrow MnO_2$	$+0.6$
$HI \rightarrow I_2$	$+0.6$
$H_2O \rightarrow O_2$	$+0.8$

[a] The equation controlling oxidation/reduction change is Oxidised species + ne $\rightleftarrows$ Reduced species. Redox potentials (E) are then related to a standard potential (E_o) by the equation

$$E = E_o + \frac{RT}{n\Im} \ln \frac{[\text{oxidised species}]}{[\text{Reduced species}]}.$$

where T is the absolute temperature, R the gas constant, $\Im$ the Faraday, n the number of electrons in the change, and E_o is the value of the potential when oxidised and reduced forms are equal in concentration. For iron, $n = 1$,

$$E = E_o + \frac{RT}{\Im} \ln \frac{[Fe^{3+}]}{[Fe^{2+}]}$$

and $E_o = +0.7$ V relative to the H^+/H_2 standard potential 0.0 V both at pH $= 0$. Changes of two-fold in any concentration have a small effect on E. Hence, potentials are easily buffered.

Note: The above potentials, E, are for pH 7 at equal concentrations of oxidised and reduced species. These equilibrium values are as important as stability constants and solubility products for an understanding of cellular chemical systems. These are free energy changes in volts, E, and where $n\Im E$ is in kilocalories. The $[Fe^{3+}]/[Fe^{2+}]$ is related to an equilibrium constant, K (see Section 4.17).

The oxidised waste, in two forms, inorganic and organic, as it built up and changed the environment could react back with the prokaryotes in four ways:

(1) It could destroy or poison them. Of course, many anaerobic organisms could and did "hide" in primitive anoxic conditions till today.
(2) It could lead to the generation of new organisms which helped their own survival by degradation of poisons, e.g. by using available catalysts. This

could imply only minor protective adaptation and we could still call the organisms anaerobic. A few new genes would be required.

(3) It could lead to modification (not greatly) of the (cytoplasmic) reaction pathways of existing organisms even introducing some new ones of a mild oxidative kind. This moderate change could lead to a capacity for handling products such as sulfate and Fe^{3+} and, probably later, oxides of nitrogen, when the species might still be called anaerobic or better protoaerobic (see Section 6.2). These chemicals act as mild oxidants of reduced organic material some from debris. Here certain new chemotypes are introduced and a not inconsiderable change of genes would be needed.

(4) It could lead to quite new organisms, which utilised as an energy source the new chemicals in the environment, now including O_2 as well as SO_4^{2-}, Fe^{3+} and nitrogen oxides, many minor oxidised elements and some newly available elements (see Table 1.6). Oxidising agents, including O_2, can react with the reduced organic waste due to the death of organisms and with reduced organic material inside cells. This would require some considerable changes of cell structure, new reaction pathways and large changes of DNA. The new groups of species are called aerobes. These prokaryotes generated a further new group of chemotypes. This catalysed oxidation in organisms is increasingly faster than in the environment.

Some lines of prokaryote development are shown in Table 6.2 with a guide to oxidation/reduction potential ranges in Table 6.3. In all these and further changes the novel chemistry has to be built into the cooperative whole (see Section 3.9). Note again the necessity that the novel features must become part of a controlled autocatalytic restricted set of reaction paths, which become general to any further evolution.

Consider the effect of the waste material. No matter how it came about, those organisms that escaped the damaging features of these novelties or which learnt to use them would be the ones that survived and prospered in the new environment. Feature (3) above obviously provides some gain in energy flow and material capture towards an increase in biomass helped by new catalysts from whence we find that anaerobes evolved to some degree to new chemotypes or protoaerobes. However, there are quite new possibilities under especially feature (4) for increasing biomass, retaining absorbed energy in chemical compounds, provided that the cell systems including DNA and its machinery could adapt considerably. The expectation is that (1) to (4) would occur in a sequence over a very considerable time period lagging behind the changes of the environment, which themselves took long periods of time, about two billion years, due to the slow completion of oxidation of sulfide and ferrous iron. We must be aware that even today the very nature of the Earth has made it possible for many types of related species to exist but in the new different and old environments and in accord with Darwinian views. At the same time we must see that the new types of organisms, i.e. chemotypes, introduced extra ways of degrading energy, effectively giving evolution a direction.

TABLE 6.2

SOME EARLY ANAEROBIC AND PROTOAEROBIC SPECIES

Species	Examples of specific characteristics
1. Bacteroides	Use CO_2 and H_2S reactions
Clostridia	Use CO_2, fix N_2 later
2. Methanogenic bacteria (archaea)	Produce methane as an energy source (or utilise it as a carbon source later)
3. Sulfolobus (archaea)	Use sulfur as an energy source: thermophile and acidophile
4. Rhodobacteria (purple non-sulfur)	Phototrophic (light user) (use reduced carbon as energy source)
5. Chromatia (purple sulfur)	Phototrophic (light user), use H_2S as a source of energy and H_2
6. Azobacter, Rhizobium	Can fix N_2, some are anaerobes
7. Sulfate users (desulfovibrio)	Not strictly anaerobic, use sulfate as an energy source
8. Nitrobacteria	Use oxides of nitrogen, especially note NO, not strictly anaerobic
9. Lactobacillus	Ferment sugars. Do not need iron, use Mn, not strictly anaerobic
10. Fe^{3+} users	Use Fe^{3+} as source of energy, not strictly anaerobic

Note: Groups 1 to 5 were included in Chapter 5, while Groups 6 to 10 are described in this chapter. They are difficult to place strictly in classes even of chemotypes but the time and increase of appearance follows the chemical changes of the environment over at least 3 billion years.

TABLE 6.3

REDOX POTENTIAL RANGE OF ACTIVITY OF SOME PROKARYOTE CHEMOTYPES

Examples of chemotype		Redox potential range (V)	O_2 reactivity
Strict anaerobes	Many (Chapter 5)	-0.5 to 0.0	Poison
Iron Fe^{3+} users	Few	-0.5 to 0.0	Not used
Sulfate users	Many (desulfovibrio)	-0.5 to 0.2 (?)	Detoxification to H_2O
Nitrate users (precursors of mitochondria)	Many (denitrificans)	-0.5 to 0.5	Some can use others detoxify
Aerobes	Very many	-0.5 to 0.8	O_2 requiring

The total rate of energy degradation increased, due to cellular catalysis of the oxidation of C/H containing compounds, especially debris.

During degradation of the cells, including those using photosynthesis, fragments of the biological *energised organic material* would become part of the waste, i.e. debris in the environment. These degradation products would include small molecules such as saccharides, amino acids, fats and bases, as well as biopolymers. As stated in the previous chapter those original anaerobic organisms, non-photosynthetic, as well as the newer oxygen-releasing photosynthetic organisms, always had the possibility to use the material and energy of this organic "waste" in the absence of light. The result is an increase in biomass and better energy retention in all kinds of prokaryotes. However, some organic material was converted into methane, higher hydrocarbons and carbon (coal), which could not be metabolised by anaerobes (and only with difficulty later by aerobes until by mankind's activities very recently) and some would be lost as carbonates due to the unceasing run-off of Mg and Ca into the sea (very slowly cyclic).

6.2.   The Beginning of an Aerobic Environment: Protoaerobic Bacteria

As noted in Chapter 5 the introduction of photosynthesis, generating oxygen from water caused major problems for anaerobes. We cannot date this exactly but biological oxygen generation may have started with some kind of cyanobacteria before or around three billion years ago. This date is made more probable by the first appearance of banded iron formations (see Section 1.10). The problems extended far beyond the effect of oxygen or its partial reduction products H_2O_2 and $O_2^{\cdot-}$. In fact, initially these oxidants were present at only very low levels corresponding to very low O_2 pressures, say from 10^{-6} to 10^{-3} atms for two billion years. At these levels although the chemicals were a menace to all anaerobic cells, since they could oxidise reduced non-metal molecules and metal and non-metal ions in cells, the cells could, by mechanisms we shall describe below, come to protect their cytoplasm by using novel enzymes to reduce O_2, H_2O_2 and $O_2^{\cdot-}$ back to H_2O. Note that this process loses energy as heat without creating biomass. While it helps to increase the rate of energy degradation it is not the most effective way of doing so.

A far greater problem resulted from the slow, but progressive, oxidation of the environment. As noted, environmental chemistry is inorganic and moves towards equilibrium rapidly (see Chapter 2) and, as mentioned above, it will predate any changes conservative organisms can make. Major changes of non-metals in the environment, except in debris, due to oxidation over a period of two billion years were the removal of all sources of carbon except CO_2, the loss of all sources of nitrogen except N_2 and the oxides of nitrogen, and the loss of sources of hydrogen except H_2O since H_2S was converted into H_2SO_4 – also making it difficult to obtain sulfur (see Figure 6.2). Note that the dating of the first biological SO_4^{2-} in Figure 1.10 gives a good idea of the timing of this change. Phosphorus, present as

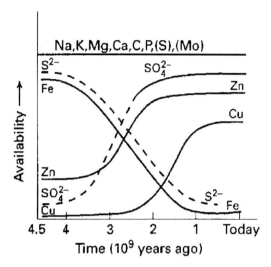

FIG. 6.2   The changing availability of some elements with time and as redox potentials increased due to dioxygen pressure increases. Note that it is solubility as well as standard redox potential that controls availability. NB. $NH_3 \rightarrow N_2$, $S^{2-} \rightarrow SO_4^{2-}$ and $Fe^{2+} \rightarrow Fe^{3+}$ (ppt).

phosphate, silicon as silicic acid and chlorine as chloride were unaffected. Amongst metal ions (see Figure 6.2) free ferrous iron at about 10^{-6} M was replaced slowly by ferric hydroxide with iron, free ferric iron, at the dramatically reduced availability of 10^{-17} M. The process is dated by the formation of mineral iron bands (see Figure 1.10). At the same time, and to this day, the oxidation of mineral sulfides gave ever-increasing sea concentrations of initially "poisonous" free VO_4^{3-}, Co^{2+}, Ni^{2+}, Zn^{2+} and MoO_4^{2-}, followed later by Cu^{2+}, Cd^{2+} and several other heavy elements, while tungsten may well have been only slightly affected in its availability. (The successive oxidation of these metal ions depends on the relative solubility of their sulfides as stated earlier.) Metals not affected were Na^+, K^+, Mg^{2+}, Ca^{2+} (Mn^{2+}?), Al^{3+} and all Group 3 metals, but some heavy elements of Group 12 to 15 became poisons as their sulfides became oxidised. The slow but massive swing in the environment may have caused the extinction of many organisms, forced some to live in anoxic niches and drove the evolution of other organisms (a form of adaptation) into using oxidised states of elements in the order of their progressive oxidation, in new metabolic pathways (Table 6.3). The currents of the seas ensured considerable mixing so that truly anoxic regions are now rare. The primary non-metal needs were to gain access to carbon from CO_2, nitrogen from N_2 and oxides of nitrogen and sulfur from sulfate, while new methods had to be found to obtain Fe^{2+} from Fe^{3+}. Access to hydrogen was already from water (see Section 5.6). Once protection of the cytoplasm had also been secured from oxygen and other oxidising agents, advantage could be gained by utilising these chemicals, and some in energy generation. The sequence was based on treating oxidising chemicals at first as poisons, devising protective activities and lastly adapting them for use

$$\text{New chemical (Poison)} \rightarrow \text{Protection} \rightarrow \text{Use}$$

where use includes the capture of their energy as well as of materials themselves. This important sequence of adaptation is followed to this day. Since these changes took place close to or at equilibrium in the environment there would be long periods in which both SO_4^{2-} and H_2S as well as both Fe^{3+} and Fe^{2+} were roughly equally plentiful (see note to Table 6.1). Hence, for a long while organisms using SO_4^{2-} and/or Fe^{3+} and then NO_3^- could coexist in the same environment with organisms that had not changed and used H_2S and Fe^{2+}. There would have been at first little zinc and virtually no copper present in the environment. Only in the last two to one billion years did the partial pressure of oxygen rise so as to release these metal ions and to help to effectively exclude anaerobes and protoaerobes from the increasingly aerobic regions of the environment (see Figure 1.12). We shall use this change to explain why evolution to complex systems, (Chapters 7–10), was so slow to start and then very rapid some one billion years ago, but in this chapter we are concerned only with the evolution of prokaryotes in more oxidising environments.

We shall examine first the chemistry developed for the protection of the prokaryote cytoplasm from O_2, $O_2^{\cdot-}$ and H_2O_2 and then the first uses of oxidised elements, S, Fe, N and finally the rise of true aerobes.

6.3. Protection of the Cytoplasm of Protoaerobes

We know that prokaryotes developed protection against O_2, $O_2^{\cdot-}$ and H_2O_2 through the evolution of specific catalytic proteins. Oxygen itself was bound and activated and then reduced to water directly by NADH, for example in sulfobacteria, by an iron enzyme. Superoxide was removed by superoxide dismutase, which is often a constitutive iron enzyme in many bacteria but an induced manganese enzyme in other prokaryotes. There is also a Ni^{2+}/Ni^{3+} enzyme in a few bacteria (see Barondeau et al., in Further Reading). Hydrogen peroxide was metabolised by catalase, which could be either a haem or a manganese protein. These protective devices are already found both in "anaerobes" we class as protoaerobic bacteria, which, as stated, are thought to have arisen at very low levels of free oxygen, perhaps considerably before 2.5 or even 3.0×10^9 years ago when Fe^{3+}, oxides of nitrogen and sulfate began to become available as sources of energy.

Protection from any poisonous metal ions liberated from their sulfides by oxidation by O_2 was secured by the use of strong chelating agents in the cytoplasm, most of which are proteins, or small molecules, thiolates, which were connected to exit pumps or to chemical metabolic tricks for metal ion neutralisation (sequestration). The genes that code for these proteins are usually to be found on plasmids in the cytoplasm of the bacterial cells (Section 5.15). Bacteria adapt very quickly to chemical insult but it is not easy to see how this could come about through random mutations everywhere in the parent DNA (see Sections 5.3 and 5.15). Was there some process of directed mutation?

Protection, by itself, though it strengthens survival, is a waste of energy that could be used for new growth increasing survival strength still further. A more effective solution for evolution is to use those waste chemicals, which are initially poisonous, to assist in the complete ecological cycle, that is to increase energy degradation rates. We shall turn to this development after discussing the primary needs for new metabolism of C, N and S compounds.

6.4.   Reduction of Environmental Oxidised Compounds of Non-Metals

If we are correct in thinking that in the most primitive times carbon was readily available as CO, nitrogen as NH_3 and HCN and sulfur as H_2S, also a source of hydrogen, then the oxidation of these molecules by oxygen to CO_2, N_2, oxides of nitrogen and sulfur and H_2O, with increasing O_2, was potentially a slowly increasing disaster for anaerobes. The extent of the disaster is seen in the energy cost of reducing these chemicals to a form suitable for cytoplasmic reactions quite apart from their poisonous nature. It is a reasonable estimate that fixation of carbon from CO_2 costs the energy of the reaction

$$2H_2 + CO_2 \rightarrow HCHO + H_2O$$

$$(\text{contrast } H_2 + CO \rightarrow HCHO)$$

The hydrogen comes ultimately from water or H_2S and requires considerable light or other energy otherwise useful in making Mg.ATP for polymerisation reactions. Again activation of CO_2 by the enzyme ribulose bisphosphate carboxylase (Rubisco, see the Calvin cycle in Figure 6.3, needs 6 ATP molecules, and this or uptake by the reverse citrate cycle (see Figure 5.7) became imperative once CO as a source of carbon was lost. (Note that Rubisco requires magnesium.)

Nitrogen fixation from N_2 is even more energy expensive and requires a new special catalyst. It costs some six ATP molecules for each molecule of NH_3 produced. This underestimates the total cost in that the syntheses of several new proteins are required and these proteins contain three Fe_nS_n centres and the coenzyme FeMoco, of formula $[Fe_7MoS_9 (\text{homocitrate})]$ (see Figures 5.4 and 6.4), which also need syntheses protected from oxygen. Again controls over the reactions are required and these again cost energy. Admittedly once in place these systems catalyse many cycles of carbon and nitrogen fixation.

The handling of sulfate by protoaerobes depends upon the initial energised coupling to adenosine phosphate as APS, since sulfate is difficult to reduce. The reductase is a flavoprotein linked to Fe_nS_n electron transfer centres. Subsequently, released sulfite is reduced by a haem protein (SIR) in which haem is directly bound

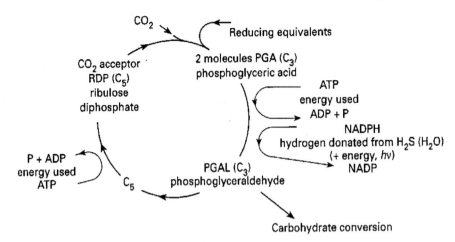

FIG. 6.3   The Calvin cycle or the dark reactions of photosynthesis; see Cooper and also Stryer in Further Reading.

to an Fe_4S_4 centre. SIR uses six electrons taking sulfite to sulfide but it can also reduce nitrate to ammonia. All these reactions may be more recent than the reduction of lower oxidation states of both sulfur and nitrogen by molybdenum enzymes (see Steuber and Kroneck in Further Reading).

   Anaerobic cells could already carry out the reversible reaction

$$X + RCOOH \rightleftarrows XO + RCHO$$

and, as we have shown in Section 5.4, in this reaction the catalyst is based on a molybdenum coenzyme where the O-atom transferred to aldehydes comes from water and MoO_3^{2-} is reoxidised to MoO_4^{2-} in a cycle by a source of hydride. The same catalyst system came to be used in later steps of oxidised sulfur reduction, e.g. of thiosulfate.

   As we mentioned in Chapter 1, we are unsure of the time of the origin of NO in the atmosphere and hence of the beginnings of oxidative processes in membranes associated with nitrogen oxides. One example is provided by Planctomycetes, the anammox bacteria, which carries out the reaction

$$4NH_3 + 6NO \rightarrow 5N_2 + 6H_2O$$

NO can come from NO_3^- or NO_2^-. The fact that they use NH_3 suggests an early origin but they may in fact just scavenge later decay products. The reaction is used to produce ATP and will lead us to a further discussion of the early (?) use of NO, perhaps before O_2 was employed as a major energy source (see Section 6.7, Figure 6.12;

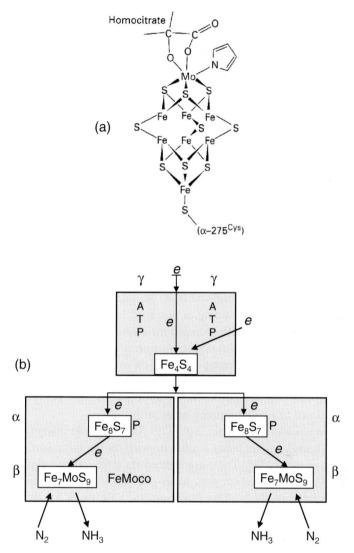

FIG. 6.4 (a) The formula of FeMoco. (b) The essence of the structure of nitrogenase. The Fe protein is in two subunits. The path for electrons is shown but it requires ATP hydrolysis and conformation changes to be activated.

and also Kuypers *et al.* in Further Reading). However, in the modern form in which we can examine them, they have copper enzymes suggesting a more recent origin. Another peculiarity of Planctomycetes is that they appear to have an unusual intracytoplasmic compartment or a huge invagination of the outer membrane, compare cyanobacteria. The membrane in this region and the content of the compartments are of unusual chemicals and contain the NH_3/NO reactions.

It has become apparent from gene analysis that protoaerobes have sensing systems for NO, CO and O_2. All these sensors are haem proteins grouped under the label of NOX (NO or OXygen binding). The NOX of these bacteria is linked to a histidine kinase whereas the NOX of eukaryotes is linked to a guanylate cyclase. The discovery of these bacterial sensors has indicated that NO, more certainly CO, sensing haem proteins may have been the precursors of O_2-sensors and of haemoglobins, and NO and CO signalling proteins in eukaryotes (see Boon and Marletta in Further Reading). These proteins illustrate again the development from detection of and protection from poisons to their use.

In all the above reactions the pre-existing coenzymes are adapted for use, but new proteins and reducing equivalents are needed. Reducing equivalents useful in creating biopolymers are therefore lost. (Note that we know the timescale of the introduction of sulfate into biological chemistry from the geochemical records, see Fig. 1.10.) Probably, later in evolution NO_3^- was reduced to NO_2^- using a molybdenum enzyme and then as stated to NH_3 in a similar way to the reduction of SO_3^{2-}. We stress again the essential value of metal ions in these steps. We may well ask: did the presence of oxidised sulfur compounds and NO_3^- "cause" the localised mutation of the Mo enzyme from its function in the RCOOH $\rightarrow$ RCHO reaction through these anions acting as poisons? In all these cases extra feedback control systems were also needed, hence the energy cost to bacteria is enormous but the alternative was death and simple slower oxidation of reduced debris chemicals in the environment generating heat with an overall loss of energy retention in chemicals. There is no competitive drive since the environment became divided, offered several options, and different chemotypes came into being and could survive in the old and new environments selectively.

Some of the oxidised chemicals, e.g. SO_4^{2-}, Fe^{3+} and especially oxides of nitrogen and even vanadate (see below) could be used on the outer surfaces of cells to produce extra energy (ATP). Their use depended on the oxidation of reduced compounds in cells, or the reaction with hydrogen gas or on oxidation of cell debris taken into the cell directly from the environment. This use of oxidation to generate energy (ATP) would have needed the synthesis of new electron- and proton-transfer enzymes and these chains of enzymes could be the precursors of oxidative phosphorylation, e.g. in NO and O_2 handling (see Figure 6.9 and Wasser et $al.$ in Further Reading). The energy generating reactions are

$$SO_4^{2-} + bound{\bullet}H \rightarrow S_n + H_2O + energy$$

$$NO + bound{\bullet}H \rightarrow NH_3 + H_2O + energy.$$

These reactions add to the complexity of cell activity (see Section 6.9), but the overall gain in biomass was small while much energy was retained in the new environment, e.g. as sulfate, nitrate, and some Fe^{3+}, but much of the iron was deposited and even lost. The cycles of Fe^{3+}, SO_4^{2-} and carbon substrates are shown in Figure 6.5.

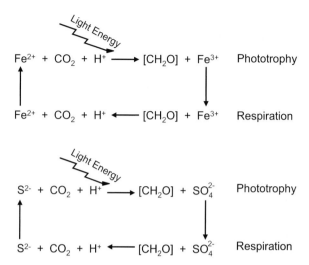

$$Fe^{2+} + CO_2 + H^+ \longrightarrow [CH_2O] + Fe^{3+} \quad \text{Phototrophy}$$

$$Fe^{2+} + CO_2 + H^+ \longleftarrow [CH_2O] + Fe^{3+} \quad \text{Respiration}$$

$$S^{2-} + CO_2 + H^+ \longrightarrow [CH_2O] + SO_4^{2-} \quad \text{Phototrophy}$$

$$S^{2-} + CO_2 + H^+ \longleftarrow [CH_2O] + SO_4^{2-} \quad \text{Respiration}$$

FIG. 6.5   Microbial iron and sulfur cycles that may have dominated biogeochemical cycling before the origin of oxygenic photosynthesis, aerobic respiration and possibly before the use of oxides of nitrogen.

6.5.   The Employment of Metal Ions in Protoaerobes and the Special Cases of Molybdenum and Vanadium

It is important to note that in all the adaptations to new environments there is an intensive involvement of metal ions starting from the initial cases given in Table 5.2. Most of this chemistry is retained in protoaerobes. Remarkably at this stage of evolution and ever since the generation of oxygen from water, (see Section 5.8), both metabolism of N from N_2 (see Figure 6.4) and later from NO_3^- and S from certain anions such as $S_2O_3^{2-}$ depend on *molybdenum* (Table 6.4) or rarely new vanadium enzymes. Later molybdenum became used in the catalysis of the transformation of sulfite to sulfate in aerobes (see Section 5.4). There is no more convincing demonstration of the relevance of inorganic chemistry to life and of the fact that biological systems developed optimal use of the available chemicals in the environment. Again there is no more striking example of selection of elements for chemical fitness. Molybdenum is the best and perhaps, other than tungsten, the only O-transfer metal catalytic agent at low redox potential (see Sections 4.15 and 5.9) with no risk of free radical side reactions, and it is probably the best for transfer of N in its fixation (see Section 6.4). (As we have seen this catalytic power of molybdenum arises from the similar low redox potentials from Mo^{4+} to Mo^{6+} complexes.) *Molybdenum, perhaps with tungsten in the earliest organisms, has always been essential for life* (see Hille and also Kisker *et al.* and also Fraústo da Silva and Williams (2002) in Further Reading). The extension of the uses of iron and magnesium are also clear in the new reactions, while those of copper and zinc came later.

TABLE 6.4

THE MAJOR MOLYBDENUM ENZYMES

Enzyme type	Enzyme families	Examples	Reactions catalysed
Multinuclear M centre	Nitrogenases (Mo, V, Fe)	Mo-nitrogenase	Dinitrogen to ammonia
Mononuclear (pterin-bonded)	I. Xanthine oxidase family (hydroxylases) (15–20 members)	Xanthine oxidase Aldehyde oxidase (W) Formate dehydroferase (W)	Purine or pyrinidine catabolism Aldehyde to acid Formate to CO_2
	II. Sulfite oxidase family (eukaryotic oxotransferases) (2–3 members)	Sulfite oxidase Plant nitrite reductase (assimilatory)	Sulfite to sulfate Nitrate to nitrite
	III. DMSO-reductase family (bacterial oxotransferases) (2 pyranopterins bonded to Mo) (8–10 members)	DMSO reductase Nitrate reduction dissimilatory; terminal respiratory oxidase	DMSO to DMS Nitrate to nitrite
Others		Pyridoxal oxidase Xanthine dehydrogenases Pyrogallol transhydrolase	

Note: W indicates cases in which tungsten may also be used in place of molybdenum.

Another metal, which may well have been used at low oxygen pressure is vanadium, but it is difficult to know when vanadium first became functional in cells. There is no evidence of the early use of vanadium in the transfer of O-atoms from their immediate binding to this atom; compare the value of W and Mo. In fact, VO_2^+ does not exchange the O-atom readily and vanadium only becomes of value much later in a different O-transfer reaction from peroxide for halogenations. This is another example of the increase of atom transfer rate from atoms in later Periods of the Periodic Table. We know that one form of nitrogenase contains vanadium and it is tempting to think that this (or an even rarer all-iron form) was the earlier form. This view is supported by the availability of vanadium relative to molybdenum; MoS_2 is much more insoluble than VS_4, the natural occurring probably primitive sulfide ores of molybdenum and vanadium, respectively. V resembles W more closely in several respects, and while it is known that V is able to form a thiovanadate it, like W, has a somewhat stronger preference for oxygen as a binding group than has molybdenum. Vanadium sulfide in the oxidation state (IV) could equilibrate with the VO^{2+} ion that is favoured with V^{3+} and V^{2+} in reducing aqueous media at neutral pH. (It is in fact the only element in the first transition metal series that has a small dependence of redox potential between several oxidation states, II to IV.) We note that additionally vanadate is a more powerful oxidising agent than molybdate and is used in a parallel fashion to Fe^{3+} and NO_3^- as an external oxidant of hydrogen in some protoaerobic prokaryotes. Cyanobacteria accumulate vanadium for an unknown purpose but it is of interest that these organisms are found associated with shale oil, which contains vanadium porphyrins. (For further use of vanadium later in evolution see Section 8.10.)

There is not much further novel use of metal ions until the oxidation potential increased to give oxygen itself. At this later date in evolution there was already considerable zinc and copper in the environment and, as we shall describe, especially copper became much used in denitrification and O_2 handling, but to see this involvement we have to enter the age of true aerobes.

6.6. The Direct Use of Oxygen: Aerobes

As the partial pressure of molecular oxygen increased further some 2×10^9 years ago then O_2 itself became a very useful cellular source of energy in the oxidation of reduced materials, e.g. of debris. These materials included not only sources of hydrogen from organic carbon compounds but also the reduced compounds of nitrogen and sulfur. *The new organisms are true aerobes.* Looking at all oxidative processes, and remembering that cellular chemistry is essentially reductive, we observe that initial oxidations are of environmental chemicals. Evolution then generated cellular oxidation using these oxidised chemicals of the environment ultimately as energy aids to reductive growth, by oxidation of some of their own reduced chemicals and those of their debris faster than the same steps in the environment. (*Note.* It does not appear that archaea were very successful in making these changes to an aerobic life style. They neither produce oxygen nor do they use it as successfully as bacteria, much though they come to use it in oxidation of methane. Was this because their DNA was too protected? They are however apparently impor- tant in the evolution of aerobic eukaryotes (see Figure 6.11 and Chapter 7.)

If oxygen was used in steady-state catalytic burning of organic matter in organ- isms at the same rate as it was produced, to no advantage evolution would have been stopped at this steady-state limit. We could then write a cyclic steady state in which there is no waste but the state is far from optimum in energy retention in chemicals and would have a slower rate of energy degradation than could be achieved by further evolution (see Chapters 7–10) since the possibility to create additional cells would have been lost. Instead, as much as possible oxygen reaction with excess C/H/O compounds over that directly necessary for growth was directed by new organisms to produce ATP, which helped to produce all cellular materials and then extra cells. The sequence of reactions for incorporating CO_2 became reversed from the cycle of anaerobes, that is it became the *forward oxidising cit- rate cycle* (Figure 6.6) and it produced CO_2 while CO_2 itself was fixed by other pathways, especially the photosynthetic route. More energy from light is turned to heat in the new total environment/life ecosystem but that from direct simple burn- ing organic chemicals outside or inside cells is removed. The cells in steady state in fact increase the total rate of oxygen and carbohydrate metabolism.

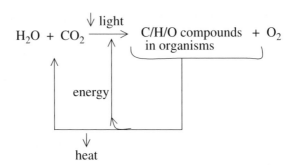

This means that while the energy of light must initially produce C/H/O compounds and oxygen, reaction of these products could be used to optimalise increased steady- state growth of cells by employing these oxidative reactions in what is essentially an internal fuel cell to drive the many condensation reactions leading to biopoly- mers. The total trapped energy in organic compounds could in principle increase until the O_2 level reached a new cyclic steady state in which cells burn sponta- neously (see the Gaia hypothesis in the Appendix to Chapter 11), but observe that we do not see life as "purposefully" driving the environment (as in Gaia) but that life adapts to previous environmental change. We do not consider that the optimal cyclic oxidation/reduction system, optimal rate of heat production, has been reached by these aerobic cells for reasons given in Chapters 7 and 11, and maybe it cannot come about. Now the oxidations not only gave extra energy for organic syntheses but they provided energy and chemicals for the uptake of all the essential non-metal elements and essential metals such as Fe required to maintain a cell, as well as

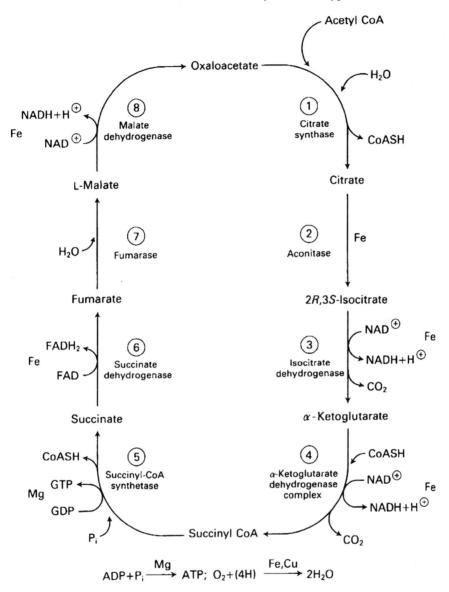

FIG. 6.6   The modern version of the tricarboxylic acid cycle (Kreb's cycle).

energy for pumping out poisons. It may be helpful to summarise the following chemical aspects of non-metal chemistry of the aerobic prokaryote evolution.

(1) Oxygen + inorganic elements led to oxidised chemicals such as sulfate, nitrate and Fe^{3+}, which became part of the environment as in the modern sea. Any such reaction results in a loss of some stored energy over that which would have

existed if O_2 had remained unreacted. Now this oxidation also deprived the primitive organisms of the initial source of the elements S, N and Fe (see (3)). Without life this store would have been only slowly degraded, so new energy-requiring reductions were devised. To some degree SO_4^{2-}, oxides of nitrogen and Fe^{3+} (and perhaps some VO_4^{3-}) also became part of the biological energy supply for synthesis of the required polymers before O_2 was available. New enzymes evolved to utilise these chemicals by protoaerobes, new chemotypes.

(2) A system of stored waste, O_2 + organic C/H/O chemicals (debris) as they increased further and even in the presence of possible environmental catalysts, provides a further long-lasting store of chemical energy in the environment. The C/H/O chemicals may be converted into very inert compounds such as oil, gas and coal that are deposited in the environment when they are best protected by being stored deep in the Earth away from atmospheric oxygen. They contain trapped energy relative to reaction with O_2 to give CO_2. However, this storage deprives the cells of elements in a non-cyclic (steady) state of flow and reduces energy degradation rate. (The fossil fuel reserves, e.g. methane, did become somewhat available and used by some aerobic bacteria and are extensively used today by mankind (see Section 10.6).

(3) Oxygen and other oxidised compounds came to be used to produce useful energy by aerobic cells in the form of pyrophosphate (ATP) at the expense of burning internally some C/H/O energised molecules such as sugars or organic debris taken in from dead cells (Figure 6.7). These rates of oxidation are now faster than the reactions in the environment. The ATP is produced by chains of catalysts in the membrane that yield an intermediate proton gradient

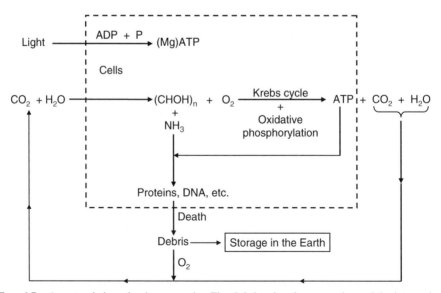

FIG. 6.7   An extended version incorporating Fig. 6.6 showing the energetics and the interaction across the cell membrane between the cell and the environment.

(see Figure 6.8 and Figure 5.2). The reactions also give rise to chemical intermediates for syntheses. Note the use of metal ions (now with Cu) in *helical*
proteins in the membranes (see Section 4.7). There is in this case some heat
production so that some energy is lost but the total energy, which can be
stored for a while in a kinetically safe way in cells, is increased in the full
cyclic steady state of synthesised proteins, nucleotides, fats, sugars and of
ionic gradients, and which are protected from oxygen. The result is increased
cell growth and replication. *The total energised chemicals and their energy
degradation in all cells may then increase* towards a final state that has a high
overall rate of converting light into heat in a material cycle. These organisms
are new aerobic chemotypes. The implication is still that oxygen will rise
until it begins to destroy the very system at the same rate as it makes it unless

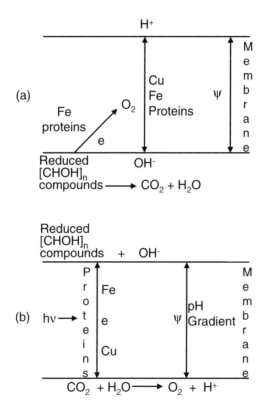

Fig. 6.8   (a) In respiration the dioxygen and reduced carbon compounds $[CHOH]_n$ are used to
create, across a membrane, a pH and potential gradient while remaking CO_2 and H_2O. Electrons
and protons are moved within the membrane using Fe and Cu protein 'wires'. The Fe and Cu
proteins are strikingly similar to those used in (b) photo energy transduction (see Figures 5.2 and
5.8). Note that a gradient of any ion such as of Na^+ can be used in the above manner but the versatility of hydrogen chemistry also allows coupling to metabolism. It is very important to
observe the helical nature of the vast majority of proteins in membranes which allows mobility—
see all pumps and energy transducing proteins.

other factors intervene so that a final state is not reached. One extra factor is the continuous rate of buffering of oxygen by materials released from the heavily reduced underlying mantle and crust of Earth.[*]

(4) Oxygen released a series of heavy metals from their sulfides and this could have caused serious poisoning of cells, but some cells developed protective pumping outwards and later use of these metals. They gave rise to an ongoing very long term, billions of years, evolution (see footnote [*]). Aerobes especially gained some access to such metal ions as those of zinc and copper (see Section 6.7). Again they needed new proteins to handle and utilise them. We must be aware that some released metals may have no use and must be avoided or rejected at a cost of energy, e.g. Pb^{2+}, Hg^{2+}.

(5) Skilful protected metabolism of O_2 (or of $O_2^{.-}$ or H_2O_2) in a cell leads to quite novel organic chemicals and they can be used in a variety of ways which will become more obvious when we describe eukaryotes. Such reactions can increase biomass (hence energy retention) and protective devices. We then find special new hydroxylating and epoxide synthesising enzymes in aerobes including the versatile simple ferrous proteins and the iron P-450 cytochromes. They must be employed so that their reactions are isolated from general cytoplasmic metabolism, examples of kinetic confinement (see Section 6.8). Note the increasing use of iron itself as well as of iron porphyrins and some use of copper (see, for example, Murrell and Dalton in Further Reading).

(6) Other non-metals became oxidised, such as selenium to selenate, and this element became a detoxifying agent, used in destroying peroxides, e.g. in glutathione peroxidase, as well as a hydrogen transfer centre.

(7) Eventually, it became possible to oxidise halides so as to halogenate organic aromatic compounds (see redox potentials in Table 6.1). The introduction of organic chlorides came very late in evolution mainly in bacteria, algae and fungi and they provide protective chemicals for these organisms, especially fungi (see Section 6.8).

We now see that in aerobes the chemotypes manage organic and inorganic chemistry and energy in new ways that were not possible earlier, but they remain autocatalytic and controlled. (In an overall sense the advance can be looked upon as an increase in energy degradation catalysed by cells.) We might believe that evolution would have been completed at this stage but of course we know that the greatest developments were still to come in eukaryotes and we must show why this too was a systematic advance dependent on oxidation. First, we have to see in more detail

---

[*] Under (3) and (4) there remains a huge store in Earth of reduced metals and their compounds. The ultimate limit of oxidation is then their conversion to oxidised products. It is a question of the rates of these non-biological reactions as well as those of organisms as to the ultimate fate of the whole ecosystem. Any very large sudden up-thrust of reducing minerals would set the geological/biological system back billions of years, only for the whole process to continue again (see Chapter 11).

the handling of (i) metals in addition to that of non-metals, as just described, and of (ii) the communication network to organise oxidative chemistry together with the necessary reductive chemistry of the cytoplasm in aerobic prokaryotes which had to advance also in a systematic way.

6.7. The Handling of Metals by Aerobes

We consider the handling of iron in an oxygen atmosphere first. A major problem for aerobes was the loss of ferrous ions from the environment, yet there was increased demand for it in cells. (Examples are given in Table 6.5). A striking development, often based on novel oxidative metabolism in organic synthesis, was the production of small scavenger molecules (siderophores) for picking up iron(III) from the environment and delivering it to the cell membrane where it was reduced and passed into the cell. The production of the scavengers was regulated at the DNA level based on transcription factors, ferric uptake response (protein) (FUR), and for-mate/nitrate reduction regulator (protein) (FNR), which may have existed previously to this use as a general regulator based on Fe^{2+} in anaerobes (see Section 5.16). FUR and FNR, when not bound to Fe^{2+}, trigger via the DNA the production of all the proteins and small molecules needed to make the scavenging system for Fe^{3+}. The syntheses are switched off when bound Fe^{2+} and supply is sufficient. Fe^{2+} + FUR and FNR also switch on synthesis of many Fe^{2+} requiring enzymes (see Table 6.5), perhaps all the citric acid cycle for gaining energy from molecular oxygen. The FNR redox regulation is connected to the kinases, Arc ABC, that is to aerobic respiratory control, and other regulators, Reg AD, so that free iron concentration is linked into virtually the whole cell responses to the oxidation conditions in the environment. It is also observed that an sRNA is involved in control here (see Massé and Gottesman in Further Reading). The balance of the Fe^{2+}/Fe^{3+} couple in complexes is perhaps the major adaptive switch of the aerobic/anaerobic facultative change. The essence of the system is found in all aerobic organisms including mankind (see Williams and Fraústo da Silva in Further Reading). Iron is

TABLE 6.5

SOME INTERNAL METALLO-OXIDASES IN AEROBIC PROKARYOTES

Enzyme	Reaction centre
Methane oxidase-type	Fe–O–Fe (later Cu)
Ribonucleotide reductase	Fe–O–Fe (Mn also)
Penicillin synthetase-type	Fe
Cytochrome oxidases	Haem (copper)
Ascorbate oxidase	Haem (later Cu)
Cytochrome P-450	Haem

Note: Most of the above enzymes reduce O_2 directly in H_2O, in contrast with one-electron extracellular oxidases which produce radicals.

an absolutely essential element for life and controls very many processes. Transcription factors related to FNR and later to iron hydroxylating enzymes are also used as signals of O_2 partial pressure. Remember here that the equilibrium binding constant for Fe^{2+}, about 10^7 M^{-1} in the cytoplasm, has to be maintained unchanged from anaerobic conditions and links enzymes, pumps and transcription systems. The management of the more common metal ions, e.g., Mg^{2+}, K^+ and Na^+ used by anaerobes also, did not change. We draw attention again to the use of minerals in protection and sensing (see Baeuerlein in Further Reading).

As mentioned before several heavier metal ions were released from their sulfides as the oxygen tension increased further towards 10^{-3} atm. This process is ongoing to this day and is made more active by mankind's industry. The major ions involved were and are V(V), Co^{2+}, Ni^{2+}, Cu^{2+}, Zn^{2+}, Cd^{2+}, Mo(VI), and minor ones are Sn^{2+}, Hg^{2+} and Pb^{2+}. These are the metals which have the most insoluble sulfides. It would appear that anaerobic cells had already devised ways of obtaining a sufficiency of Co and Ni and a little V, Zn and Mo (or previously W) before oxygen was liberated in quantity. (We have stated that this makes it likely that early life formed in acidic media where the sulfides are more soluble.) All new release of metal ions and especially those of Ni^{2+}, Cu^{2+}, Zn^{2+}, Cd^{2+} and of course Sn^{2+}, Hg^{2+} and Pb^{2+} represented a threat especially to all early life as they could compete with both Fe^{2+} and Mg^{2+} in the cytoplasm. We shall see in Chapter 7 how this threat was overcome and later used in eukaryotes. Aerobic prokaryotes too can manage increased Cu^{2+}, and Zn^{2+} at low concentrations (see Section 6.10) and use them to some degree (see Table 6.8) (and even Cd^{2+} in one case; see Lane and Morel in Further Reading). The protection from all these metal ions required new carriers, pumps and transcription factors as well as novel proteins as enzymes, (see Finney in Further Reading). The elements Sn, Hg and Pb remain a threat to all life in all chemical forms but bacteria may well adapt and find a value for them; they are already protected from these metal ions (see Section 5.15).

In some ways it is surprising that aerobic bacteria have not made more use of zinc, internally, and calcium generally, especially in controls since we know they present no redox threat and we shall see that their uses increase dramatically in eukaryotes. The aerobic bacteria do have genetic connections for controlling zinc (e.g. the transcription factor ZUR and ZntR genes) but its use is not extensive. The absence of full use of Ca and Zn may well be due to the limited space and the fast time of the bacterial cell metabolism and life cycle.

One potentially damaging metal element already mentioned, which eventually came to be of considerable value in aerobic bacteria was copper. Particularly intriguing is the development of its use in denitrification and the value of these reactions in energy transduction (Figure 6.9). We can best see this development by starting analysis from the way anaerobic bacteria developed the use of proton gradients, derived from oxidation of carbon compounds or other reactions not using oxygen, e.g. driven by sulfur metabolism, to give ATP (see Figure 5.2) and how this demand for reducing equivalents, also needed in synthesis, led to the introduction of the use of H_2O with the liberation of O_2. The generation of oxygen would probably have led to the

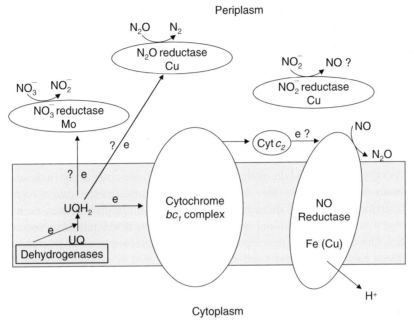

FIG. 6.9   The catalysts for denitrification. Nitrate is reduced by a molybdenum enzyme while nitrite and oxides of nitrogen are reduced today mainly by copper enzymes. However, there are alternatives, probably earlier iron enzymes. The electron transfer bc_1 complex is common to that in oxidative phosphorylation and similar to the bf complex of photosynthesis, while cytochrome c_2 is to be compared with cytochrome c of oxidative phosphorylation. These four processes are linked in energy capture via proton (H^+) gradients; see Figure 6.8(a) and (b) and the lower parts of Fig. 6.9 which show separately the active site of the all iron NO-reductase, and the active site of cytochrome oxidase (O_2^- reductase).

oxidation of ammonia before the full rise of oxygen itself and this could have given rise to a quite powerful oxidant, NO, also obtainable from nitrate, nitrite or directly from the atmosphere. (Note that the redox potentials in Table 6.1 show that NO_3^- would form before a considerable rise in O_2.) Thus, organisms probably had a source of considerable energy before dioxygen became fully available. At this stage of evolution NO reduction would have become coupled to the production of proton gradients (see Figure 6.9) using, as is observed today in certain protoaerobes, an all iron reaction centre initially together with the earlier NADH oxidase, Fe_nS_n, enzymes (see Figure 6.12). The only other metal ion employed in early organisms in oxidation reactions is Mn. Later we find NO reductase in a series of enzymes in which the NO-reaction centre and electron-transfer centres contain copper, now resembling the cytochrome oxidase enzymes for energy transduction using oxygen. Copper also became used in enzymes for N_2O and NO_2^- reduction and we observe it also in photosystems in place of a cytochrome. The simplest explanation for all these uses of copper is that the earlier requirement for Fe^{2+} iron was jeopardised seriously by its precipitation as Fe^{3+} outside the cells. Several of the iron functions were therefore economically switched to the use of the increasingly available copper. Copper in fact can be more readily used, especially externally, at high redox potentials than iron, and copper oxidases became involved even in iron uptake. There are parallels with the exchange of tungsten and possibly vanadium by molybdenum or *vice versa* according to the absence or presence of sulfide, and the replacement of cobalt (B_{12}) by zinc as it became more available, as we shall see later. These changes not only allow continuation in changed circumstances, they also make possible new efficient reactions in energy and material trapping. (Note how evolution follows environmental change in all these cases, that is, as oxidation potential in the environment rises.) The use of zinc and copper increased dramatically later in eukaryotes, especially in those which are multi-cellular due to the development of compartments (see Chapter 8). To understand this lack of more extensive development in prokaryotes, even today, we need to look at other limited features of prokaryote development in Sections 6.8 and 6.9. Observe in these developments how the concentrations of free and bound new metal ions are built into a self-regulating catalysed system by feedback limited by internal equilibria, as for the old ones.

6.8. Cytoplasmic and Membrane Organisation of Proteins

To appreciate one advance in aerobic prokaryotes we must return to the general ways of organising chemical reactions within a compartment in addition to those due to physical barrier constraints and/or by field gradients, as described in Chapter 3. Undoubtedly, both of these play a part in any cellular organism. We explained in earlier chapters how reaction pathways could be isolated from one another also through the development of highly selective catalysts which bind only to their selected substrates. Many sequences of chemical change such as glycolysis use a series of enzymes, which are limited to the catalysis of particular members of the sequence in a given way, which is a form of isolation. However, these substrates and products are free in the cytoplasm. We have drawn attention to the fact that the advent of oxygen included the possibility that some of its reactions could be equally well managed and isolated by even more selective "protected" enzymes in the cytoplasm. Hence, reactions of oxygen here such as those involving cytochrome P-450,

$$H_2 + O_2 + RH \rightarrow ROH + H_2O, \text{ hydroxylation}$$

$$H_2 + O_2 + R \rightarrow RO + H_2O, \text{ epoxidation}$$

must be protected. In the above reactions reduction is achieved by an electron donor system and H^+

$$(XH_2 \rightarrow 2e + 2H^+ + X) + O_2 + RH \rightarrow ROH + H_2O$$

in which one oxygen of O_2 becomes water before the second oxygen gives ROH. Note that the enzymes are devised so that their reactions occur in fixed timed sequences: they are mini, "isolated organisations". In aerobic prokaryotes these enzymes are used to oxidise many unreactive hydrocarbons in synthesis and in removing poisons. The hydroxylating enzymes, now including simpler iron enzymes, gain importance throughout evolution irrespective of the site of activity (see Hewitson *et al.* in Further Reading). Obviously it is safer if such enzymes are physically separated in a compartment such as in a membrane or completely outside the cytoplasm. (This further development of enzyme organisation is most clearly seen in eukaryotes, Chapters 7 and 8, but note the periplasm of bacteria in Section 6.9.)

Quite another way of isolating components, especially elements, is storage in precipitates. A simple but extremely valuable case is seen in the storage of iron in ferritin, $FeO(OH)$, or in magnetite, Fe_3O_4, in aerobes. These stores are effective buffers of total iron and are seen in all later organisms where, as in aerobic prokaryotes, they also help to buffer free iron, Fe^{2+}, in the cytoplasm. Remember that iron is relatively inaccessible in an aerobic environment and so this element must be handled and conserved preventing its loss. Other elements stored selectively are copper and zinc, now found in complexed forms, i.e. soluble *buffer* molecules, which again were increasingly developed in evolution opposite increasing use of these elements. The low free concentrations of elements such as copper and zinc meant that carriers for their distribution, often mistakenly called "chaperones", were required.

From this account of the protoaerobic and aerobic prokaryotes it is clear that the change of the environment from a moderately reducing to an oxidising environment with the need to maintain the reductive metabolism of the cell cytoplasm placed a burden on what was possible for these small cells. We describe another way round these problems other than protected cytoplasmic reactions before we discuss the extra problems of controlling the new activities.

6.9. The Need for Extra Compartments

As already stated, all prokaryotes faced a problem with the introduction of oxidation reactions since one of their major compartments had to be reducing in nature. Due to oxidation of the environment their cytoplasm, as noted before, also faced the problem of new metal ions dangerously competitive with internal Mg^{2+} and Fe^{2+} functions. Yet, as explained, it is useful for cells to use oxidation to gain extra energy and to use certain novel metal enzymes to assist in these reactions. Considerable risks to basic processes are also present, for example reduction of N_2 to NH_3, which must be cytoplasmic, is extremely sensitive to oxygen. (In fact today cells that produce O_2 or use it do not generally fix N_2 in the same compartment.) We find then that many bacteria (and other organisms) that use oxygen do not carry out protein synthesis without an external supply of directly usable nitrogen. Quite interestingly we also find bacteria using oxidised products such as sulfate, ferric ions or nitrate *as distinct species*. Effectively these bacteria are all mutually beneficial, separate

FIG. 6.10   Cyanobacteria in a chain of photosynthesising cells, A, which produce oxygen from light and water, with a heterocyst "cell" B which is anaerobic and fixes nitrogen gas.

chemical "compartments". Note again the highly convoluted membranes of many bacteria, which could help to localise reactions in effectively isolated compartments and which become real physically separate compartments in eukaryotes, but the limited size of a bacterial cell and the constraint of the walls are severe restrictions on more extensive development of such internal spaces. (A most revealing early case of "compartmental" activity now found in one organism is that of cyanobacteria (Fig. 6.10). They are in fact the first organisms to show *cell differentiation* (almost), where one type of "cell", better called a heterocyst, with the same DNA as all the others in the same organism performs functions different from its neighbours in the cytoplasm. (A second somewhat uncertain group of bacteria which quite probably have a separate compartment are the planctomycetes; see Section 6.4.) There was however one general way in which prokaryotes could develop genuine compartmental activity, which was to use the space outside the cytoplasm but inside the wall, that is in the periplasmic space.

6.10.   The Periplasmic Space and Oxidative Metabolism

When we introduced the idea of chemotypes we included the possible classification of the types the novel users of space to carry out their chemistry. The ability of aerobic prokaryotes to use a concentric compartment, the periplasm between outer and inner membranes (see Figure 6.11) gave the advantage that in it they could handle safely much oxidative chemistry working at a higher redox potential, even that of O_2 itself, than is usual in the cytoplasm. (Note that O_2 is not such a menace as $O_2^{\cdot-}$ or H_2O_2, which are destructive and must be removed quickly.) This periplasmic compartment also allowed the use of the metal ions that were damaging to the cytoplasm. For example, the machinery for production of oxygen by photosynthetic bacteria that uses *manganese* for O_2 production in the cytoplasmic membrane places this metal ion so that it faces the periplasm, as does the *copper/haem* a_3 oxygen site used in oxidative metabolism by cytochrome oxidase in all aerobes. In the periplasm or facing it there are also many of the enzymes catalysing sulfate and denitrifying reactions, for example, of nitrate. Here *molybdenum* is often used in high oxidation state reactions of NO_3^- but a novel metal, that bacteria also employ in this compartment, is *copper*. (Probably released late from its sulfide less than two billion years ago.) In the denitrifying steps of $NO_2^- \rightarrow NH_3$, several *copper* enzymes are found in the periplasm of nitro-bacteria but none are found in the cytoplasm (Table 6.6 and Figure 6.9). Most high potential electron-transfer *iron cytochromes c,*

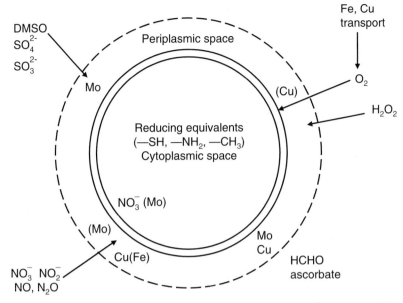

F$_{IG}$. 6.11 The distribution in periplasmic space of the major molybdenum (and copper) enzymes except nitrogenase. Note the types of substrate.

TABLE 6.6

S$_{OME}$ C$_{OPPER}$ P$_{ROTEINS}$ $_{IN}$ U$_{NICELLULAR}$ A$_{EROBES}$

Nitric oxide reductase (P)	Dinitrification (Cu)
Nitrous oxide reductase (P)	Dinitrification (Cu)
Ascorbate oxidase (P)	Redox buffer (Cu)
Cytochrome oxidase (PM)	Energy capture from O_2 (Cu)
Copper ATPase pumps (PM)	Export of copper

Note: P periplasmic; PM periplasmic face of outer membrane.

where haem is cross-linked to the protein, are also confined to the periplasm. Remember that simple ferrous ion enzymes could not be used outside the cell cytoplasm since Fe^{2+} does not bind strongly enough to avoid oxidation and precipitation. We regard the separation of these high potential substrates and their enzymes from the cytoplasm as a necessary (inevitable) development for continued evolution. Note that there is little if any synthesis by condensation in the periplasm since no ATP is available there so that an export system of proteins had to be devised. However –S–S– bridge formation, which is due to oxidation, is common to protect proteins. Thus, the use of particular series of reactions in the periplasm defines a new set of chemotypes. Although the periplasm performs an extremely useful function, its use has complicated the feedback control system necessary for cooperative management of the whole cell. One such control is that of the free and bound metal ion concentrations where control is restricted by equilibrium considerations as discussed in Sections 4.18 to 4.20.

6.11. Novel Forms of Control and Organisation: New Genetic Features of Aerobes

The aerobic bacteria developed even more complicated controls of necessity. They had to maintain homeostasis in a cytoplasm very similar to that in anaerobes, but in a more difficult environment, and the periplasmic space had to be managed. Oxidised chemicals in the environment, which led first to protection from and then use of them by enzymes, required genetic additions (see below), but also needed new feedback controls. For example, sensors are observed, i.e. transcription factors, in some bacteria for molecules such as O_2, CO and NO and for facultative switches and all these sensors are similar to those seen later in eukaryotes. CO was no longer a useful chemical since it blocked O_2 reactions (but see Section 8.8). There are also regulatory systems for other new non-metal compounds and for metal ions, for example nitrate, sulfate and copper, and for the new uptake systems for iron. The new aerobic prokaryotes were by these means increasingly informed about the chemical environment and had the ability to respond appropriately, but there does not appear to be extensive use of Ca^{2+} gradients (see Section 5.16 and Table 8.22). As we have stressed in Chapter 4, increasing demands on the ability of a central single organisation to handle new and additional processes is best met by decentralisation even of the genome. As described in Section 6.9, this need finds expression in different chemotypes carrying out different metabolic tasks and through consumption of different debris. Mutual gain results especially if the organisms can live in mixed populations. This very loose organisation of interactive species acting in separate related parts of space is effective but while the individual chemotypes limit their DNA just to cover special environmental situations there is no direct control of all activity. A striking example is the separation of photosynthetic from non-photosynthetic aerobic bacteria leading forward to the separation of plants and animals (with fungi). An alternative solution is for cells to increase the numbers of internal compartments and then the number of connected differentiated cells to isolate specialisation within separate volumes but under one central overall control in one organism. However, this requires an increasing size of the DNA and of cell size. Thus, the overall process of a successive small appropriate improvement in organisation and effective energy retention in prokaryote cells with increased capacity to use the environment had but little ability to exploit it fully. Notice that information to the genetic structure or metabolism still concerns largely chemicals useful as nutrients or poisonous items to be avoided, and is not used to manipulate the cell structure or to any great degree the cell motion.

Turning to the additional metabolism that needs extra controls there was little change in basic prokaryote reductive metabolism in the cytoplasm from anaerobes to aerobes but as stated above there were added several oxidative reactions and machineries. These additions required new proteins and hence the development of the genetic code. Some changes are found in the main, not in satellite DNA. Gene duplication plus gene modification was one possibility for small alterations (see the changes of uses of molybdenum proteins described in Sections 5.9 and 6.4). We can see this also in the development of some of the oxidative phosphorylation proteins using O_2, for example, in the reversal of the citrate cycle and enzymes for dinitrification (see Fig. 6.9), which are related to earlier photosynthetic proteins. Thus, parts of the two pathways of electrons and protons have striking similarities (see Figures 6.8 and 6.9). For example, the NADH reactions, leading in both cases to the reaction of the cycling of quinones, require a series of very similar Fe_n/S_n proteins. This series of Fe_nS_n proteins may even have been derived from simpler hydrogenases, which are related to the genes for assembly of Fe/S centres. The new haem bc_1 (oxidative phophosphorylation) and the old bf (photo-phosphorylation) series of electron-transfer complexes are also similar while the modern use of a copper protein in photosynthesis, plastocyanin, and the two copper centre of cytochrome oxidase resemble one another (Figure 6.12). Part of the new DNA could have arisen therefore through local DNA changes (mutation) due to the attack of oxidising agents on valuable proteins preceded by gene duplication and transfer (see Section 7.13). As mentioned before, a very interesting development that has its corresponding change

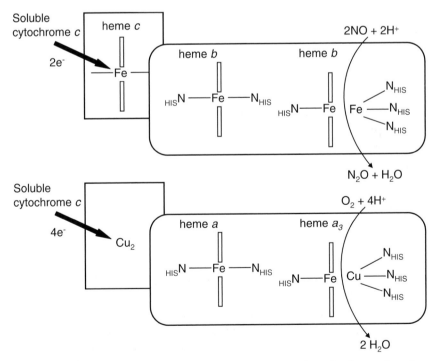

FIG. 6.12   A comparison between the NO- and the O_2-based oxidases for the creation of a proton gradient.

in genes is that from an all *iron* nitric oxide reductase (see Section 6.7) to a *copper-*dependent system where there is a close relationship to all the cytochrome oxidase proteins, (Fig. 6.12; see Wasser *et al.* in Further Reading). However, a more difficult problem involves the genetic control over the periplasm. It is a long-standing puzzle as to how organisational developments such as transport to and from new compartments came to be introduced in genetic changes and we shall return to this problem in Section 7.13. We do not wish to analyse genetic advances in any detail in this book but prefer to refer to the increasing collection of as yet incomplete data about changes in DNA with time. By comparison within and between the approximately 250 bacterial and 25 archaea DNA sequences now known (2005) we can often see how a given class of protein has developed from aerobic beginnings. For example, it is readily observed that the signatures of say multi-thiolate zinc or copper-binding sites arise in a related way. We leave this species by species analysis to others, as it is not our main area of interest. The additional comparison with eukaryote sequences is however a very instructive way of analysing evolution but it has to be used with care since it is not only the *sequence but also the extent of expression as proteins* that is the critical feature in a cell system.

In the satellite plasmid DNA of bacteria there are many genes clustered together, which are regulated in expression and used in detoxification but not in essential metabolism. (Parallel genetic additions are used to combat modern drugs, e.g. penicillin; see Section 5.15.) A very intriguing different control, the beginnings of which are not clear, is found in *E. coli*. The regulation of expression of genes involved with iron metabolism is by an sRNA, a special short piece of mRNA. Much further work is needed here (see Massé and Gotternman in Further Reading).

6.12. Summary of Prokaryote Development

In concluding the two chapters on prokaryotes we note again the major evolutionary developments, taking all prokaryote chemotypes together as part of the total ecosystem of all living organisms, i.e. one part of the total ecosystem, Sun/Earth/Organisms, before eukaryotes appeared and they are very much in existence today. They are listed in Table 6.7. We note the following features stressing the use of the elements.

(1) Initially some 15 to 20 elements, that is those *available in the original anaerobic environment*, gave the general observed features of all cells (see Table 4.11).

(2) Step by step introduction of more effective ways of handling cytoplasmic chemistry, using *coenzymes and special metal cofactors*, e.g. haem (Fe), vitamin B_{12}(Co), F-430(Ni), chlorophyll(Mg) and Moco(Mo).

(3) Gain of access to light (use of chlorophyll) and inadvertent production of O_2, use of manganese, creating an increasingly aerobic environment.

(4) Carefully controlled reactions where oxygen was removed from the cytoplasm taking O_2 directly to H_2O so that little of the dangerous $O_2^{\cdot-}$ and/or H_2O_2 are produced, using Fe.

(5) Protection from $O_2^{\cdot-}$ and H_2O_2 by destructive enzymes, involving Fe or Mn.

(6) Adjustment of bacteria following the environmental changes initiated by O_2 especially, for example, new nitrogen and carbon dioxide metabolism which frequently took place in particular cells, chemotypes and their species. Novel metabolic paths avoided dangerous intermediate products from nitrate and sulfate on the way to H_2S and NH_3, using Mo and Fe (see Table 6.8).

(7) Protection of the cytoplasm from damaging metal ions, for example, of increasing Zn and Cu, released by oxidation of sulfides, which was managed by internal carrier chelation or chemical modification and transfer to exit pumps.

TABLE 6.7

THE PROKARYOTE CHEMOTYPES

Prokaryote	Chemotype
Archaea	Special membrane lipids, coenzymes and uses of nickel (several specialist types see under bacteria) Extremophiles of various kinds. Use of sulfate and limited use of light and O_2
Bacteria	Different membrane lipids and coenzymes
	(a) Strict anaerobes non-photosynthesising
	(b) Sulfate, ferric iron, and later nitrate users: protoaerobes
	(c) Photosynthesisers: extra Mg, release O_2
	(d) Aerobes utilising Cu, O_2 reactions giving new organic products

TABLE 6.8

A POSSIBLE PROGRESSION OF PROKARYOTES

1. Earliest anaerobic non-photosynthetic autrotrophs including archaea and bacteria
2. Development of co-enzymes and porphyrins
3. Purple photosynthetic bacteria using the sulfide to sulfur reaction
4. Green photosynthetic bacteria (various)
5. (3) + (4) to give photosynthetic bacteria producing oxygen from water
6(a) Sulfobacteria using $SO_4^{2-} \rightarrow S^{2-}$ metabolism
6(b) Bacteria using $Fe^{3+} \rightarrow Fe^{2+}$ metabolism
7. Nitrobacteria using oxidised ammonia, for example N_2 and nitrate
8. Aerobic bacteria; full use of oxygen and oxidised chemicals both non-photosynthetic and photosynthetic; some use of newly available zinc and copper

Different groups of chemotypes in this book are represented by 1 + 2; 3 to 7; 8

Note: Some of the above reactions supply energy as well as materials.

(8) Retention of much of the cytoplasmic chemistry while novel chemistry was largely placed in the periplasm of bacteria or at least on the membrane surface facing it, e.g. the management of sulfate, ferric ion or oxidised nitrogen, often involving Mo, as a source of energy.

Much of (1) to (8) was managed in what we have called protoaerobes, which was followed on increase of aerobic conditions by:

(9) Use of protected oxidation of substrates in the cell cytoplasm, e.g. by cytochrome P-450 (Fe).

(10) The major change in aerobes was the direct use of oxygen together with excess reducing equivalents (debris) to gain energy with the reformation of $CO_2 + H_2O$. Hence in different prokaryotes, photosynthesisers and non-photosynthesisers, the cycle of material and energy degradation exists:

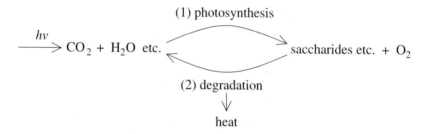

$$\xrightarrow{\ hv\ } CO_2 + H_2O \text{ etc.}$$

(1) photosynthesis

saccharides etc. $+ O_2$

(2) degradation

$\downarrow$

heat

We shall see that the photosynthesis becomes isolated in "plant" chemotypes using derivatives of photosynthesising bacteria, chloroplasts, while degradation will be found in "plants" (no light), "fungi" and "animals" using derivatives of non-photosynthesising bacteria, mitochondria. These are cases of symbiosis.

Note that internal degradation is also an essential part of nitrogen incorporation and that oxidation reactions initially only in the environment ((3) above) became increasingly used by cells ((8) to (10) above) to drive internal reduction overall.

(11) Oxygen became useful for the preparation of new, partially oxidised organic chemicals, which acted as protective agents.

(12) Use of newly available elements, copper especially, on the cell's periplasm side of the membrane, with high redox potential substrates, O_2, and nitrogen oxides. (Note that there is no corresponding increase in the use of increasingly available Ni or Co.)

(13) The use of copper required, as for other metals, the synthesis of novel proteins and controls over them to meet the requirements of equilibria (see Section 4.18).

Notice that rejected and "poisonous" elements increased in the environment and are then increasingly used in cells.

We now add points concerned with the general development of aerobes.

(14) There is not so much competition between organisms following these developments as there is specialisation since new secondary energy and chemical sources are best employed in different compartments, here largely different isolated cells, chemotypes. (Use was made of debris by cells from other organisms.) Separation of anaerobes, plant- and animal-like aerobes, including different chemotypes, where their coexistence and cooperativity is more notable than competition, was an essential evolutionary step towards a cyclic state of the whole ecosystem.

(15) Increased information to the cells through internal sensors and transcription factors. Cells, therefore, became somewhat more aware of the changed environment and more able to respond to needs and risks presented by new resources for growth. This is a major theme of organisational improvement throughout evolution.

(16) Additions were obviously needed in the genetic complement, and some of them arose through gene transfer making species difficult to define in the wild.

All these features and activities are controlled in each cell so that each is part of a total organisation of internal genes, proteins, metabolism, metal ions, etc., and, to some degree, the environment are linked and limited in autocatalytic pathways. The major drives in all the changes are simultaneously greater energy intake, greater biomass, greater degradation rate of organic chemicals and a resultant greater rate of heat, thermal entropy production. However, a major failure was the unavoidable poor use of space, with a rather inflexible membrane and low level of total organisation leaving the possibility for larger organisms to appear in evolution. In fact the prokaryotes were not well informed about the environment since there was little more than chemical sensing, with chemotasis, of small molecule or ions for nutrition or of poisons. The wall prevents sensing of objects by touch, prevents uptake of large particles, while general sensing and searching at a distance followed by directed movement is poor. Finally the prokaryotes were unable to digest large molecules and particles of food except externally. Table 6.9 summarises some of the advantages and disadvantages of these organisms.

Before proceeding it is worth noting that the prokaryotes established the fundamental features of life in the coordinated production of DNA, RNA, proteins, saccharides and lipids, mainly from H, C, N, O, S and P and in the functional use of many inorganic elements including Na, K, Mg, Ca, Mn, Fe, Co, Ni, Cu, Zn, Mo, Cl and Se. In this book we deliberately stress this role of the available elements in the environment especially as they change with time in evolution in advance of that in cells. In evolution many of the functions of these molecules and ions remain only modified not fundamentally changed. While the uses of metal ions were restricted by equilibria (Sections 4.18–4.20), the synthesis of porphyrins allowed effectively novel elements,

TABLE 6.9

THE ADVANTAGES AND DISADVANTAGES OF PROKARYOTES

	Advantage	Disadvantage
Genetic structure	Bare in bacteria, cyclic, easily and quickly reproduced fast response time (contrast archaea) Mini-circle plasmids	Vulnerable as unprotected Difficult to expand greatly
Membrane	Strong protection due to wall (bacteria) or by type of lipid (archaea)	Inflexible, not able expand/contract but note invaginations
Wall	Rigid (bacteria), strong in defense. May absorb minerals, e.g. SiO_2	Prevents flexibility and limits sensing
Uptake of material including minerals and digestion	Ability to seek food by swimming in a gradient	Unable to digest large particles. Has to send digestive enzymes outside. No compartments so cannot make shaped minerals
Ability to absorb energy from the environment, light	Ability to seek light by swimming	Small light-sensitive volume. Initial difficulty handling oxygen generated by light
Response to environment change	Fast mutation Easy gene transfer use of plasmids	No compartments except periplasm to keep incompatible reactions separate in space
Use of inorganic ions	Catalysis of organic reactions	Many metal ions are not fully used Ca, Na, K, Mn, and see heavier elements, Zn, Cu

forms of Mg, Fe, Co and Ni, to be used not restricted by equilibria, but also kinetically controlled by feedback to uptake. The prokaryotes not only discovered the necessary components but they discovered the management of the coded, autocatalytic, feedback network, which was essential for survival. It is the linking together of internal pathways based on external conditions and supplies and DNA expression, which is a decisive feature. Moreover, they generate together the developments seen in Chapters 5 and 6 of flexibility and openness to variation, seen in chemotypes with multiples of species, relatively quickly adapted to environmental change. Correlations of internal parts of organisms is insufficient information for an understanding of evolution. We stress again the strong link to the inorganic environment especially through metal ions, their selection, limited by availability, and the matching of their potential to their uses. This is not to take away from the basic nature and the bulk necessary for use of kinetic stability of the combinations of a few non-metal elements in compounds of high (the most?) functional value, i.e. of reduced states of C/H/N/O/S, but to complement it.

From this account of prokaryotes it must be clear that the switches from anaerobic to protoaerobic and then to aerobic chemotypes amongst prokaryotes was not a gain in intrinsic cellular efficiency in the processes of fast reproduction and survival except in the context of the environmental changes. The efficiency of anaerobes in an anaerobic environment exceeds that of any of the later prokaryotes in that the later organisms all have to use much energy to reduce the sources of oxidised required elements, C, N, S for example, to the required states in small substrates and for subsequent condensation leading on to biopolymers. The anaerobes "discovered" the basic system of energised autocatalytic coupled pathways. It is not by chance but because of necessity that protoaerobes and aerobes evolved *in the novel environment, created by photosynthesising, oxygen-releasing anaerobes* themselves, since it was only through novel protection helped by novel energy capture that they could exist. Clearly the combination of anaerobes, protoaerobes and aerobes each in several chemotypes formed in a variety of environments and from thence an ecosystem of somewhat dependent chemotypes emerged which could utilise one another's debris (see Silver in Further Reading). Under oxidising conditions there evolved later more effective forms of life, chemotypes with many compartments, that is the eukaryotes. They are more effective in both the uses of novel non-metal elements and metal ions and in energy use (see the following chapters) but they also have disadvantages. The system strives to be element neutral in a cycle turning to use pollutants.

Despite the reservations about their efficiency as individual cells, we have to stress that prokaryotes as a whole were and are extremely successful (see Whitman *et al.* in Further Reading). In closing the account of them we must not forget their rapid reproduction and their easy adaptation to a changing environment. This evolution

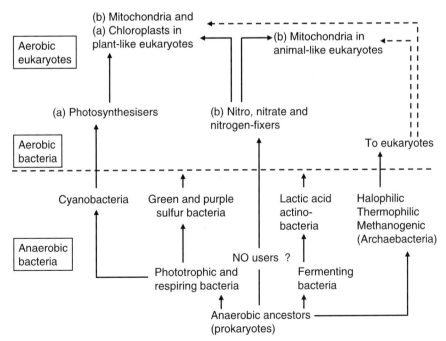

FIG. 6.13   The development of aerobic eukaryotic organisms from the earliest anaerobic ancestors via aerobic bacteria, all single cells.

(see Figure 6.12) has the overall effect of a gradual increase in the rate of degradation of energy not just directly by the internal degradation of synthetic products but in the oxidation of debris to generate a cycle of material. The faster the total amount of material is synthesised and then returned to the environment, or cycled internally, and the more useable the chemicals which are synthesised the greater is the survival strength. Increased survival strength is the greatest overall factor that drives individual cell evolution, but underlying it is the increased rate of the transformation of light to heat. However, we stress that the need for prokaryotes to reject excess of certain elements, such as Na, K, Ca, Cl and a whole range of poisonous heavier metals, and some non-metals including oxygen, without being able to use them fully to assist further growth is a waste of external/internal material and energy storage. When we turn to eukaryotes we shall see a greater degree of efficiency but at a cost – making eukaryotes together with prokaryotes the inheritors of the total ecosystem. It is the survival of this ever-developing system that we observe against the ever-increasing rate of degradation of energy of light or in chemical reactions, see the cover picture.

Note again that throughout the two chapters on prokaryotes the directional path of evolution has been analysed through inorganic rather than organic chemicals.

Further Reading

BOOKS

Baeuerlein, E. (ed.) (2004). *Biomineralization: From Biology to Biotechnology and Medical Application*. Wiley – VCH. Weinheim, Germany.

Cooper, G.M. (2003). *The Cell: A Molecular Approach*. Sinauer Associates, Inc., Sunderland, MA.

Cracraft, J. and Donoghue, J. (eds.) (2004). *Assembling the Tree of Life*. Oxford University Press, New York

Fraústo da Silva, J.J.R. and Williams, R.J.P. (2001). *The Biological Chemistry of the Elements – The Inorganic Chemistry of Life* (2nd edn). Oxford University Press, Oxford

Harold, F.M. (2003). *The Way of the Cell*. Oxford University Press Inc., New York

Murrel, J.C. and Dalton, H. (eds.) (1992). *Methane and Methane Users*. Plenum Press, New York

Purves, W.K., Orians, G.H. and Craig-Haller, H. (1998). *Life—The Science of Biology* (5th edn). Sinauer Associates/W.H. Freeman, San Francisco

Silver, S. (2006). *A Bug's Eye View of the Periodic Table*. Elsevier, Amsterdam

Stryer, L. (2002). *Biochemistry*, 5th edn. W.H. Freeman, New York

White, D. (1995). *The Physiology and Biochemistry of Prokaryotes*. Oxford University Press, Oxford

Williams, R.J.P. and Fraústo da Silva, J.J.R. (1996). *The Natural Selection of the Chemical Elements—The Environment and Life's Chemistry*. Oxford University Press, Oxford

Williams, R.J.P. and Fraústo da Silva, J.J.R. (1999). *Bringing Chemistry to Life—From Matter to Man*. Oxford University Press, Oxford

PAPERS

Anbar, A.D. and Knoll, A.M. (2002). Proterozoic ocean chemistry and evolution: a bioinorganic bridge. *Science, 297*, 1137–1142

Arnold, G.L., Anbar, A.D., Barling, J. and Lyons, T.W. (2004). Molybdenum isotope evidence for widespread anoxia in mid-proterozoic oceans. *Science, 304*, 87–90

Auffinger, P., Hays, F.S. Westhoff, E. and Shing Ho, P. (2004). Halogen bonds in biological molecules. *Proc. Natl. Acad. Sci. USA*, *101*, 16789–16794

Barondeau, D.P., Kassman, C.J., Bruns, C.K., Tainer, J.A. and Getzoff, E.D. (2004). Nickel superoxide dismutase structure and function. *Biochemistry*, *43*, 8038–8047

Boon, E.M. and Marletta, M.A. (2005). Ligand specificity of H-NOX domains. *J. Inorg. Biochem.*, *99*, 892–902

Burgess, B.K. and Lowe, D.L. (1996). Mechanisms of molybdenum nitrogenase. *Chem. Rev.*, *96*, 2983–3011

Einsle, O., Tezcan, F.A., Andrade, S.L.A., Schmidt, B., Yoshida, M., Howard, J.B. and Rees, D.C. (2002). Nitrogenase MoFe-protein at 1.16 Å resolution: a central ligand in the FeMo-cofactor. *Science*, *297*, 1696–1700, see also J. Kim and D.C. Rees (1992). *Nature*, *360*, 553

Finney, L.A. and O'Halloran, T.V. (2003). Transition metals speciation in the cells; insights from the chemistry of metal ion receptors. *Science*, *300*, 931–936

Green, M.T., Dawson, J.H. and Gray, H.B. (2004). Oxoiron (IV) in chloroperoxidase compound II is basic: implications for P450 chemistry. *Science*, *304*, 1653–1656

Hallam, S.J., Putnam, N., Preston, C.M., Detter, J.C., Rocksar, D., Richardson, P.M. and Delong, E.F. (2004). Reverse methanogenesis: testing the hypothesis with environmental genomics. *Science*, *305*, 1457–1462

Hewitson, K.S., Granatino, N., Welford, R.W.D. and Schofield, C.J. (2005). Oxidation by 2-oxoglutarate oxygenases: non heme iron synthesis. *Phil. Trans Roy. Soc. A363*, 807–829.

Hille, R. (1999). Molybdenum enzymes. *Essays in Biochemistry*, *34*, 125–137

Holm, R.H. (1990). The biological relevant oxygen atom transfer of molybdenum. *Coordin. Chem. Rev.*, *100*, 183–222

Kim, H.J., Graham, D.W., DiSpirito, A.A., Alterman, M.A., Galeva, N., Larive, C.K., Asunskis, D. and Sherwood, P.M.A. (2004). Methanobactin, a copper acquisition compound for methane oxidising bacteria. *Science*, *305*, 1612–1615

Kisker, C., Schindelin, H. and Rees, D.C. (1997). Molybdenum cofactor – containing enzymes: structure and mechanisms. *Annu. Rev. of Biochem.*, *66*, 233–267

Kuypers, M. *et al.* (2003). Anaerobic ammonia oxidation by anammox bacteria in the Black Sea. *Nature*, *422*, 608–611

Lane, T.W. and Morel, F.M.M. (2000). A biological function for cadmium in marine diatoms. *Proc. Natl. Acad. Sci. USA*, *97*, 4627–4631

Lyalikova, N.N. and Yurkova, N.A. (1992). Role of microorganisms in vanadium concentration and dispersion. *Geomicrobiol. J.*, *10*, 15–26

Martin, W. and Russell, M.J. (2003). On the origins of cells: a hypothesis for the evolutionary transitions from abiotic geochemistry to chemoautotrophic prokaryotes, and from prokaryotes to nucleated cells. *Phil. Trans. R. Soc. London, B 358*, 59–85

Massé E. and Gottesman, S. (2002). A small RNA regulates the expression of genes involved in iron metabolism in *Escherichia coli. Proc. Natl. Acad. Sci.*, *99*, 4620–4625

McEvoy, J.P. and Brudvig, G.W. (2004). Structure-based mechanisms of photosynthetic water oxidation. *Phys. Chem. Chem. Phys.*, *6*, 4754–4763

Mc Master, J. and Enemark, J.H. (1998). The active sites of molybdenum and tungsten enzymes. *Curr. Opin. Chem. Biol.*, *2*, 201–207

Moura, I. and Moura, J.J.G. (2001). Structural aspects of dinitrifying enzymes. *Curr. Opin. Chem. Biol.*, *5*, 168–175

Peña, M.M.D., Lee, J. and Thiele, D.J. (1999). A delicate balance homeostatic control of copper uptake and distribution. *J. Nut.*, *129*, 1251–1260

Que Jr, L. and Watanabe, Y. (2001). Oxygen pathways: oxo, peroxo and superoxo. *Science*, *292*, 651–653

Rajogopalan, K.V. (1998). Molybdenum: an essential trace element. *Adv. Inorg. Chem.*, *35*, 103–177

Russell, M.J. (2003). The importance of being alkaline. *Science, 302*, 580–581; and see also Hanczyc *et al.*, *Science*, *302*, 618

Silver, S. (1998). Genes for all metals. *J. Ind. Microbiol. Biotechnol.*, *20*, 1–12

Smith, B.E. (2002). Nitrogenase reveals its inner secrets. *Science*, *297*, 1654–1655

Steuber, J. and Kroneck, P.M.H. (1998). Anaerobic dissimilatory sulfate reduction. *Inorg. Chim. Acta.*, *52*, 275–276

Wasser, I.M., de Vries, S., Moënne-Loccoz, P., Schröeder, I. and Karlin, K.D. (2002). Nitric oxide in biological denitrification. *Chem. Rev.*, *102*, 1201–1234

Whitman, W.B., Coleman, D.C. and Wiebe, W.J. (2004). Prokaryotes: the unseen majority. *Proc. Natl. Acad. Sci. USA*, *95*, 6578–6583

Williams, R.J.P. and Fraústo da Silva, J.J.R. (2002). The involvement of molybdenum in life. *Biochem. Bioph. Res. Comm.*, *292*, 293–299

Yocum, C.F. and Pecoraro, V.C. (1999). Recent advances in understanding the biological chemistry of manganese. *Curr. Opin. Chem. Biol.*, *3*, 182–187

CHAPTER 7

Unicellular Eukaryotes Chemotypes (About One and a Half Billion Years Ago?)

7.1. Introduction

Eukaryotes, which probably emerged some 2.0 to 1.5 billion years after prokaryotes, (Table 7.1; see also Table 1.9) are commonly defined by the fact that they are large cells, have a recognisable nucleus with a nuclear membrane and also internal filaments (Figure 7.1). However, there is recent evidence that a few bacteria may have a similar nucleus (see Pennisi in Further Reading) and some have filaments (see Kürner *et al.* in Further Reading), but this is certainly not the case for the vast majority of prokaryotes. We shall therefore use the general finding that the eukaryote cell has a nucleus, several compartments and permanent filaments together with other features, such as novel uses of chemical elements, e.g. calcium, to distinguish this group of organisms as defined chemotypes (see below). The two types of cell, prokaryotes and eukaryotes, have evolved subsequently, that is since some 1.5 billion years ago, in an increasingly, often symbiotic, association to today. The eukaryotes have produced in this period a huge variety of unicellular and multicellular species within different chemotypes. In this chapter after describing the unicellular variety we shall discuss how they originated and then study the nature of

TABLE 7.1

THE GENERAL PROPOSAL OF EVOLUTION OF EUKARYOTES

Date (million years ago)	Organisms	Events	Atmospheric oxygen (~%)
3800	Prokaryote chemoautotrophs	Origin of life	0
3500–3000	Prokaryote heterotrophs; precursors of cyanobacteria. Stromatolites. Sulfur bacteria	Beginning of photosynthesis	Traces
2100	Filamentous spirally curled organisms, (Grypania)	Major land masses; shallow seas, Iron deposits, BIFs	0.1%
2000	Cyanobacteria tolerant to O_2	Sterols in bitumen (fossil organisms)	0.2%
1700	Spheromorph Acritarchs, primitive unicellular eukaryotes	Atmosphere oxidising Endosymbiosis. Aerobic respiration	0.3%
1200	Red algae and metaphytes	Large cells. Endosymbiosis. Aerobic respiration. Meiosis. Genetic recombination	0.5%
1000–550	Various primitive multi-cellular eukaryotes in precambrian fossils, some mineralized. Green algae dominant. Early land plants	Fossils and tracks. Oxygen and ozone accumulating	1–4%
450–present	Full flourishing multicellular eukaryotes; land living organisms	Ozone layer completed. Crust movements more pronounced. Super continents formed. Ocean basins altered	10–21%

their chemistry before describing the multicellular variety in Chapter 8. Given the timing of their evolution it would appear that unicellular eukaryotes (sometimes called protista) evolved through chemistry forced by the insult anaerobic prokaryote life suffered due to its own abstraction of hydrogen from water using photosynthesis, a reaction that released the poisonous (to prokaryote anaerobic life) molecular oxygen, O_2. It is probable that these eukaryotes evolved shortly after

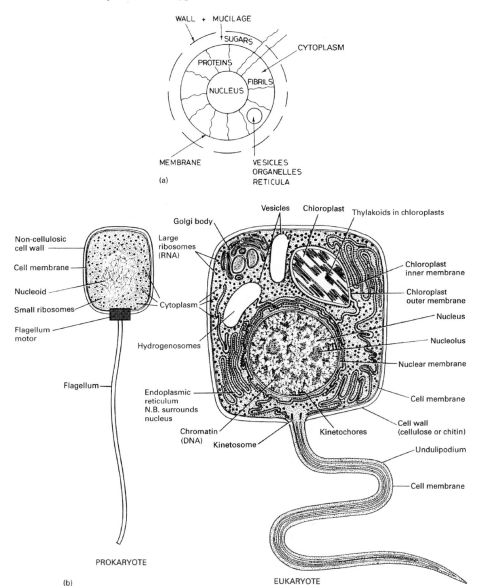

FIG. 7.1.   (a) A schematic diagram of a eukaryote. (b) Typical prokaryotic and eukaryotic cells, based on electron microscopy. Not every prokaryote or eukaryote has every feature shown here. (From Margulis and Schwartz (1998). *Five Kingdoms.* Copyright W. H. Freeman and Company, New York. Reproduced with permission.)

fully aerobic prokaryotes appeared, as described in Chapter 6, as more effective organisms in some way but with disadvantages (see below). Oxygen, with the products of its reactions, e.g. metal ions from sulfides, was the first major pollutant for life in the environment due to life itself (see Chapter 11 for the description of later pollutants), thus forcing evolution. We can date the rise of O_2 through the changes of the isotopes of sulfur and the successive appearances of banded iron formations (see Chapter 1). The evidence, apart from fossil dating, that oxygen was a primary, if not the only necessary, new chemical for eukaryote evolution is twofold: (a) the membranes of eukaryotes differ from those of prokaryotes in that they almost always contain cholesterol, an oxidation product of squalene chemistry involving oxygen reactions that uses a pterin coenzyme catalyst (see later in this section for odd exceptions) and (b) they almost invariably have many compartments, including the organelles, for example, mitochondria, which generate energy using oxygen. The demand to cope with and then make use of the increasingly available chemistry of oxygen, while maintaining the necessity of central reductive chemistry of the cytoplasm, *increased the number of compartments essential for efficient energy and material management in a single cell* (see Section 3.10). The limited size and structure of the earlier prokaryotes, anaerobic or aerobic, left little space for such development in one cell, while all eukaryotes are large cells, 10 to 100 μm, allowing several compartments. Their large size and a new strong but flexible membrane enabled them to dispense with a limiting cell wall and to digest large molecules, particles and even bacteria. We conclude that this together with the greater number of compartments of eukaryotes (see Figure 7.1), and hence, as we shall show, greater energy and chemical effectiveness (see Section 7.8), allowed the eukaryotes to evolve and survive alongside aerobic prokaryotes in an oxygen atmosphere despite slower reproductive and adaptive rates and hence a larger possibility of being attacked. (Remember prokaryote survival is also helped by diversification in different chemotypes as shown in Chapter 6.) We shall see that eukaryotes have considerable necessary internal organisational complexity, which also assisted their survival strength (Table 7.2). We shall also note that they were able to sense and respond more rapidly to the environment. An increasing feature of evolution is that the link with the use of the environment increases with information from the environment all the way up to mankind. This increase in complexity had to be matched by increase in genetic complement and feedback generally.

[Before proceeding, we should note that there are some apparently "anaerobic" parasitic eukaryotes such as *Trichomonas* and *Giardia*, which do not have mitochondria. Characteristically they have hydrogenosomes, for deriving energy from H_2 reactions, which are related to mitochondria (see Embley *et al.* in Further Reading). It is possible in fact that these organisms were aerobic eukaryotes, which reverted to an anaerobic state with degraded mitochondria inside them (see Knight in Further Reading). There remains a further puzzle since their "eukaryote" membranes do not contain cholesterol. It is known, however, that these apparent anaerobes must have felt the presence of some oxygen since they do have eukaryote-type protective enzymes against superoxide. Possibly these eukaryotes should be called protoaerobic (see Section 6.1). From here on we shall assume that all eukaryotes are aerobes existing in a partial pressure of O_2 greater than 10^{-3} times that of today].

Now while in the preceding chapters we have been able to go directly to the chemical contents of given chemotypes to illustrate features of their distinctive

TABLE 7.2

NOVEL FEATURES OF EUKARYOTES

1.	The DNA is linear but wound up into nucleosomes and then chromosomes. It is protected by a membrane in a nucleus and by many proteins, e.g. histones, and terminated by telomeres. Archaea have histones, a few bacteria may have a nucleus
2.	Introns are found but they are absent in bacteria; archaea have introns
3.	Increase in number of translated genes, exons, as proteins; from 4000 to 6000
4.	Their membranes are flexible although they contain cholesterol, which gives them strength
5.	The membrane and the nucleus as well as other structures are supported by cross-connecting filaments, that is new proteins, tubulins and actomyosins
6.	The size of a cell increased more than 10-fold in radius relative to prokaryotes
7.	Cells have several types of internal enclosed compartments. Some are of weaving reticula including the endoplasmic reticulum and the Golgi body. Glycosylation occurs outside the cytoplasm in vesicles and there is also storage of ions such as calcium in vesicles. Others compartments are derived from trapped bacteria and are called organelles. The two well-known are mitochondria, for oxygen metabolism, and chloroplasts, for capture of light and oxygen production. (A few bacteria have modest compartments, thylakoids.) The compartments often have unusual pH values and other concentrations
8.	Novel external/internal sensing and signalling systems, often based on calcium gradients, and novel internal signalling, often due to organic phosphates. Histidine kinase, common in bacteria, are often absent in eukaryotes
9.	Novel protective systems leading to "immune" responses
10.	Protein synthesis is initiated by methionine as in archaea and not by formyl as in bacteria. Removal of these end-groups necessary for protein maturation is catalysed by a *ferrous* peptidase in both cases. The Fe(II) is relatively weakly bound and is a control element of protein maturation
11.	It is quite probable that intro-processing giving untranslated RNAi generates a novel control system. s-RNA control may be more general
12.	In addition to iron, magnesium and the other metals common in prokaryotes the eukaryotes make increasing use of zinc and copper, for example in superoxide dismutase, relatively little used by even aerobic bacteria and less use of nickel, for example in hydrogenases, which are now, Fe enzymes
13.	Eukaryote cells have larger ribosomes
14.	Cell walls, where they exist, are made from cellulose or chitin, in contrast to prokaryotes
15.	Eukaryotes have novel modes of direct body movement and swimming, based on sensors
16.	The mode of reproduction differs in that eukaryotes use sexual combination

nature, in this chapter, we have to describe the general features of different eukaryote cells first due to their compartment differences (Section 7.2). Thereafter, we shall analyse the common characteristic features of the chemotypes of all the major groups of unicellular eukaryotes in the order of (1) the division of compartments within the organisms and their possible origins (Section 7.3); (2) the changes in energy capture (Section 7.4); (3) the chemical purposes of compartments (Section 7.5) and (4) the reproduction, growth and form of the cells (Section 7.6), before we turn to the overall organisation of chemical activity. Chemically, there are several similarities with aerobic prokaryotes but with certain factors separated and enhanced.

7.2. Plant, Animal and Fungal Eukaryotes and Interactions between them

The unicellular eukaryotes include species in the three "advanced" major group-ings of chemotypes–plant-, animal- and fungal-like (see Table 7.3). They are sep-arated by the differences in metabolism, for example light-dependent plants have two types of organelles, chloroplasts and mitochondria, while animals and fungi have only mitochondria. Animals again are readily distinguished from *stationary fungi* in that most fungi have hyphae, running filaments, which can be multi-nucle-ated, not multi-cellular; but many plant cells, such as various algae, are also multi-nucleated. Unicellular animals are almost invariably *mobile* and motile and have single-nucleus cells. We stress that all types of single-cell eukaryote have vesicles as well as nuclei and organelles. We shall often use information from yeast, a fun-gus, as a reference organism since it has been so well studied. Table 7.3 gives examples of other types of single-cell eukaryotes related to plants and animals. We have discussed previously the reasons for this diversity of chemotypes initially seen in aerobic prokaryotes (see Section 6.13).

Many unicellular eukaryotes are free-living cells, but may form huge local com-munities, which are especially beneficial to the homeostasis of the ocean/atmos-phere carbon cycle, e.g. coccoliths. Many others are not free-living, but are extremely valuable in symbiotic relationship with multi-cellular plants and animals. Unfortunately, some unicellular eukaryotes are the causes of disease, for example *Trypanosoma*, which are animals and cause sleeping sickness in humans (see Section 8.9 for parallel diseases of plants). These facts are reminders that while we consider that the whole ecosystem works to one general purpose (Section 4.4), this does not exclude the obvious feature that within its overall associations we can see diseases inflicted on one species by another or competition between similar species. Many bacteria are also causes of serious eukaryote diseases. Even so at the end of

TABLE 7.3

SOME SINGLE-CELL EUKARYOTES

Plant-like	Animal-like	Fungi-like
Euglena	Zooflagellates, e.g. Trypanosomes	Some yeasts Sporozoa
Dinoflagellates	Amoeba	Slime molds
(Coccoliths[a])	Foraminiferae	Plasmodia
Acritarchs (algae and plankton)	Radiolaria	
	Ciliates Acantharia[a]	

[a] Some of these species can carry out photosynthesis, but they may do so by engulfing symbiotic algae. It has been suggested that Grypania found in iron sediments could be the earliest eukaryotes.

Chapter 8 on multi-cellular eukaryotes, we shall show that the total system of different eukaryotes and prokaryotes, chemotypes, is an efficient cooperative one, related to the drive towards a steady-state ecosystem of optimal energy retention and overall rate of energy degradation. Do not forget that the ecosystem involves the energy input from the Earth and the Sun and the whole geochemical activity of the surface of the Earth, which is not totally favourable to organisms. To increase awareness of the geochemical link with living organisms, especially eukaryotes, Section 7.13 describes the minerals produced by organisms, much of which are now part of geological formations, e.g. carbonates, silica and phosphates, and are only weakly energised chemicals if they are energised at all. We shall see that mineralisation is very common amongst several chemotypes of eukaryotes.

7.3. Connections between Eukaryotes, their Compartments and Prokaryotes

Let us consider the possible physical/chemical connections between eukaryotes (Table 7.2) and different prokaryotes (see Table 6.10), which might explain their origin and that of their compartments (Figure 7.2). We know from DNA/RNA/protein sequences that there is some connectivity to both archaea and bacteria, but it appears

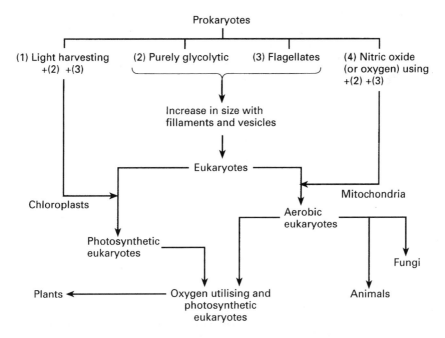

Fig. 7.2. A possible scheme to show the interdependence of the origin of many eukaryotes upon genetic crossing and combination with early prokaryote species.

to be necessary to assume the involvement of a further type of early organism (some suggest a virus; see Martin and Russell in Further Reading). The eukaryote DNA itself indicates that genes from archaea dominate regulation of genetic machinery while regulatory genes for metabolism come from bacteria. There are in fact some eukaryote cells, which are apparently derived from archaea alone, called *Archaezoa*. They have a nucleus but no other compartments and might have been the earliest eukaryotes, but there is disagreement in the literature on this point. Also, as in archaea, all eukaryote DNA is protected by histone-like proteins and has introns. Many other features of more conventional multi-compartment eukaryotes may have been derived from one or the other class of prokaryote. The internal filamentous features common to eukaryotes are not totally unexpected, since some filaments are necessary in cell division of all prokaryotes, though they are neither a permanent feature nor are they made from the same proteins (see Kürner *et al.* in Further Reading). The fact that there are convoluted membranes such as the endoplasmic reticulum in almost all eukaryotes is also not a great surprise in that prokaryotes often have long weaving invaginations of the outer membrane or possibly layers of membranes. A glance at the large lamella structures of thylakoids or the cristae of mitochondria (Figures 7.1 and 7.3) gives a quick reminder that the bacteria, from which these organelles are thought to have been derived, do not necessarily have simple internal membrane structures, though much of their cells usually have simple closed often cylindrical shapes. Note the special case of the separate thylakoid in cyanobacteria (Figure 6.10) and all plant chloroplasts and the peculiar case of anammox bacteria (see van Niftrick *et al.* in Further Reading and Section 6.4). It is possible that all internal eukaryote vesicles derived their membranes from invagination of the cytoplasmic membrane of bacteria since archaea have a different set of lipids. We can use this general enfolding of the eukaryote membrane as an explanation of the origin of the formation of the nuclear membrane too, but again there are different views, including those which arise from the presence of a "nucleus" in some bacteria (see Pennisi in Further Reading). Prokaryotes may also have external filaments, flagella, which act to propel the cells along gradients in a rough and ready way, and such external filaments have parallels amongst eukaryotes but now made from different proteins and the propulsion is well directed. Prokaryotes have sensing devices for food and poisons (perhaps to a limited degree linked to messengers) and even for magnetic and gravitational fields, but while such general sensors are developed in eukaryotes, these multi-compartment organisms also have quite extensive novel directional sensing and response systems (see below). Unfortunately, the only species available for comparative study with modern eukaryotes are modern prokaryotes, which are quite evolved, and hence the early link to the eukaryotes of yesterday is not readily discernible. It is not quite clear even how different eukaryotes came to have different organelles derived from bacteria, mitochondria and chloroplasts, which are probably from specific prokaryotes (Rickettsia and Cyanobacteria, respectively), since to describe the origin of these organelles we have to assume the nature of bacterial life some 1.5×10^9 years ago at about the time the first eukaryotes developed. All we can say is that these organelles have a

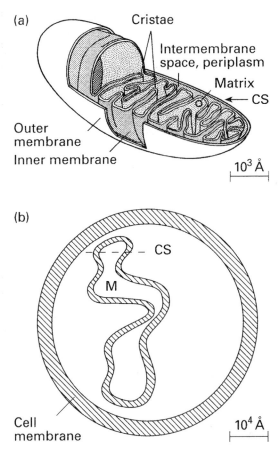

FiG. 7.3.   (a) Old impression of mitochondria (M) based on cross-section (CS) shown also in (b) a modern view of this weaving larger organelle in a cell. See Perkins *et al.* (1997) *J. Struct. Biol. 119*, 260–272.

very clear relationship to particular prokaryotes alive today. We do know however how prokaryotes can be taken up into eukaryotes today. Eukaryotes are large cells with flexible membranes and have the ability to ingest large particles, including prokaryotes, by enfolding them in the cytoplasmic membrane, invagination, so that they enter the cell in vesicles (Figure 7.4). This process is called endocytosis, or in particular cases phagocytosis, which has a great advantage in the sensing and capture of already energised food. Inside these cellular vesicles, fused to lysosomes (other vesicles), the food is digested, thereby keeping the process of digestion away from the cytoplasm. These vesicles, much like vacuoles in plants, are often acidic.

It is clear that no matter how eukaryotes evolved, their physical/chemical features indicate that a step by step evolution of such cells could have given an ever

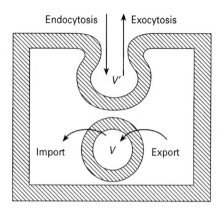

Fig. 7.4. Vesicle uptake and release mechanisms: V, vesicle, V', vesicle coalesced with membrane.

more complicated series of viable multi-compartment chemotypes. We need not seek for a sudden one-off complete transformation of prokaryotes to eukaryotes, but the gap in our direct knowledge of eukaryote evolution is very large despite the above possible links, which we shall see again in several groups of proteins. We shall turn to the connections between prokaryotes and eukaryotes in the use of chemical elements and biopolymers in Section 7.8. We must be aware that every time new compartments are added new organisation is required, including DNA changes.

Some organisms, notably parasites, have "organelles", which appear to be relics of chloroplasts but have no photosynthetic capacity. They are called *apicoplasts* but are probably not the origin of chloroplasts. Also note that a feature of symbiosis is that genes that are no longer required can be *selectively* lost as was probably the case in the example of loss from bacteria on becoming mitochondria (mentioned above).

If cells of eukaryotes are much slower to reproduce and adapt, why did they evolve at all and so successfully? Why did they continue to evolve; yet their predecessor prokaryotes also evolved (as aerobes); and both they and early and late eukaryotes are also extant in considerable numbers today? Clearly a large number of types of organisms, chemotypes, co-exist. Why? We shall provide answers to these questions in terms of the different efficiencies of different chemotypes within the ecosystem, including mutual chemical dependences.

7.4.   The Organelles of Eukaryotes

The development of eukaryotes has a major novel feature: their largest energy sources are in two organelles. They all have mitochondria and plants have additionally chloroplasts, which together are the primary sources of cell energy (see Figure 7.3). The proteins of mitochondria especially and some of chloroplasts are not all coded in the organelle DNA, less than 10% of those of mitochondria, but have most of their genes in the main cytoplasmic DNA. This illustrates again the ease of gene transfer across membranes effectively between species during evolution (see Section 5.1 and Andersson *et al.* in Further Reading). The very existence of DNA transfer to the central nucleus must remind us again of the possibility that eukaryotes evolved from many types of prokaryotes, and not just from two or three. The *Chromista*, a group of diatoms, have DNA traceable to four different organisms – their own as a green algae, from red algae, from purple bacteria and from cyanobacteria, many concerned with their organelles. History based on few facts is very uncertain and we have few facts as yet as to the origins of eukaryotes. As a consequence of the loss of DNA from organelles, the need for protein transfer to organelles, especially mitochondria, requires very active protein import pumps apart from import/export pumps for energy, ions and metabolites. Another feature of mitochondria (and chloroplasts) is that they can multiply independently from the central cell DNA in a given cell. They are flexible and are often, perhaps most frequently, fused together into huge multi-nucleate organelles resembling the cytoplasmic reticulum (see below) crisscrossing the whole cell (see Figure 7.3). Here and there they contact other compartments, forming a kinetically localised network. Since the eukaryote cells are large, they and their organelles must be expected to be differentially active in different local regions so that kinetic "compartments" within cells arise (see Section 3.10).

It may seem strange that the two major energy-producing reaction pathways in mitochondria and chloroplasts are not incorporated completely into the eukaryote genome, but are incorporated effectively in degraded symbionts, derived from prokaryotes. In a sense, when we compare a prokaryote with a eukaryote, we are in fact comparing a single organism with a complicated ecosystem of *cooperative* organisms in one organism. We shall find this association again and again – remember that all plants are effectively an essential resource for all animals. This takes us back to the benefits and needs of a multi-compartmentalised system related to symbiosis and/or to cooperative existence of separate chemotypes (see Sections 3.10, 5.15 and 6.12). The mitochondria also utilise somewhat dangerous free radical oxidative reactions. In fact, they degrade fatty acids and carboxylic acids quite separately from the synthesis of these compounds in the cytoplasm, it is easily seen that there is an advantage in placing all these reactions in separate compartments; compare the periplasm of prokaryotes. The chloroplasts are involved in the generation of oxygen, which is again open to the production of side-products damaging to DNA.

Another feature of mitochondria is that they handle the syntheses of both iron/sulfur and iron haem units. It appears as if handling metal ions, including that of iron, is risky in the cytoplasm. In fact, other metal ions are pumped out of cells or into vesicles and in some cases into mitochondria; – especially calcium.

Now the chloroplast is more complicated than the mitochondrion in that it synthesises pentose sugars from CO_2 as well as making ATP (see the pentose shunt in Section 4.5) and it has retained more DNA. In these organelles there is an inner compartment with its own membrane, the thylakoid, which control proton flow and transferable hydrogen and oxygen production. The bioenergetic gradient across the membrane is a pH gradient not largely a potential (ψ) gradient as in mitochondria indicative of the fact that mitochondrial energy transduction derives from the outer membrane of bacteria while photosynthetic bacteria have compartments, the thylakoids. The outer membrane of cells cannot use a large pH gradient since the pH gradient in bacteria (related to mitochondria) would include the environment, the sea! The chloroplast then has a compartment for synthesis, incorporation of CO_2, outside the thylakoid, called the stroma, which itself has a double membrane in keeping with the structure of bacteria and mitochondria. The double membrane holds the periplasm, sometimes called the lumen, as is normal. The increase in the number of compartments of the chloroplast is small but protective of the cytoplasm which cannot tolerate pH change.

So far, the four clear cut advantages of eukaryotes over prokaryotes are (1) their ability to digest prokaryotes (and large particles) in a compartment, separating dangerous hydrolysis of proteins in acid solutions and relieving the need for some synthesis from strictly environmental chemicals; (2) the incorporation of separate organelles for energy production keeping certain types of reaction, degradative and synthesis, out of the cytoplasm; (3) a nucleus for protection of DNA keeping DNA/RNA reactions separate and (4) special ways of handling metal ions in compartments. However, we must always remember the disadvantages of slower reproduction and less adaptability when compared with prokaryotes. Increasing organisation of activities in different parts of space always has advantages and disadvantages.

7.5.   The Uses of Other Compartments: Further Separate Activities

It is interesting to note the uses of other physically separate compartments, now vesicles without DNA, in all eukaryotes (Table 7.4 and Figure 7.5). We find that the major compartment is the large weaving endoplasmic reticulum connected to a central Golgi apparatus. The whole unit is able to synthesise both the large saccharides, to sulfate them and to glycosylate proteins. The sulfate is formed from sulfite by a molybdenum enzyme under some conditions (see Table 6.4). The oxidising conditions in these compartments are favourable too to —S—S— bond cross-linking as opposed to the RSH conditions in the cytoplasm. These are valuable new synthetic products for use on the outside of cells since the products of the endoplasmic reticulum metabolism are usually not kept there, though some are, but are exported using exocytosis (see Figure 7.4) to form external layers including minerals which they support. We can neither miss the striking beauty of the shells of coccoliths ($CaCO_3$), diatoms (SiO_2) and acantharia ($SrSO_4$) constructed in

TABLE 7.4

COMPARTMENTS AND REACTIONS IN AEROBIC EUKARYOTES

Compartment	Reactions	Elements
Cytoplasm	Synthesis of DNA, RNA proteins	Mg, Zn, K
	Glycolysis	Mg
Mitochondria	Oxidative phosphorylation	Fe, (Cu), (Ca), (Mg)
	Citric acid cycle	
Chloroplasts	Photophosphorylation	Mg (chlorin), Fe(Cu)
	Carbon assimilation	Mg
Golgi apparatus	Glycosylation	Mn
	Sulphation	
Endoplasmic reticulum	$2RS^- \rightarrow R–S–S–R$	(?) (Ca)
Vacuoles	Urea hydrolysis	Ni, (Mn)
Peroxisomes	Oxidative degradation	Fe, Mn

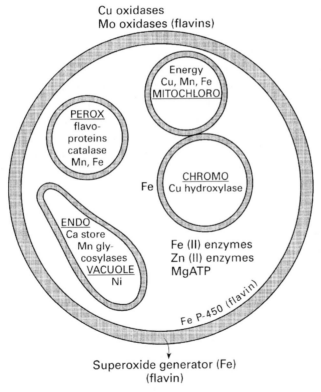

Cu oxidases
Mo oxidases (flavins)

Energy
Cu, Mn, Fe
MITOCHLORO

PEROX
flavo-
proteins
catalase
Mn, Fe

CHROMO
Fe Cu hydroxylase

ENDO
Ca store
Mn gly-
cosylases
VACUOLE
Ni

Fe (II) enzymes
Zn (II) enzymes
MgATP

Fe P-450 (flavin)

Superoxide generator (Fe)
(flavin)

FIG. 7.5. A schematic indication of some of the different membrane separated compartments in an advanced cell. PEROX is a peroxisome; MITOCHLORO is either a mitochondrion or a chloroplast; CHROMO is a vesicle of, say, the chromaffin granule; ENDO is a reticulum, e.g. the endoplasmic reticulum. Other compartments are lysosomes, vacuoles, calcisomes and so on. Localised metal concentrations are shown. The figure is of a transverse section. To appreciate a cell fully it is necessary to have serial plane sections in parallel along the z-direction.

vesicles nor their apparently "purposeful" design for protection. The endoplasmic reticulum is also used for storage of calcium, and other vesicles, the vacuoles, are used to store or remove heavy metal ions, the value of which is seen in Section 7.12. All such features seem to appear suddenly in the record of life. As we observed before, the absence of well-defined precursors is often a grave problem in evolutionary connectivity.

There are other compartments, called peroxysomes, in which peroxide is produced to destroy unwanted bacteria or large particles. They remain a part of the defense mechanisms of later eukaryotes. This is another example of the advantage of the eukaryote compartments – keeping the cytoplasm as free as possible from risk while introducing new, especially high redox potential, chemistry, compare the advantages of the periplasm of aerobic bacteria.

As stressed earlier, the vesicle and organelle membranes are in themselves extra compartments, as is the cytoplasmic membrane containing pumps and enzymes as well as detectors/receptors. It is important to observe that the vesicle membrane pumps face in the inverse direction to those of the outer membrane, so that vesicles take up free Ca^{2+}, Cl^-, Na^+ and Mn^{2+}, often protons, and later excess free Ni^{2+} and Zn^{2+}, for example. Of special interest are the so-called V-type ATP-ases, some of which pump protons. Thus the interior of vesicles can be of much lower pH than the cytoplasm, as low as pH 2. This pH hydrolyses many proteins and hence is helpful in digestion. The V-type ATP-ases that pump ions are strongly "irreversible" in contrast with the F_1F_0-type ATP-ases of organelle energy transduction (see Appendix 4C). The compartments often have therefore a very novel inorganic content giving rise in part to the novelty of eukaryote chemotypes. Each also have a controlled protein and modified protein content, so that there has to be designed trafficking of proteins directed selectively to many compartments. Now all these compartments have to be managed and shortly we turn to communication using messengers between them and the cytoplasmic components and especially DNA and RNA. We shall see that these compartments aid response to external events, providing a means of protection and of selection of environments. A general property of all the eukaryote membranes appears to be flexibility but use of them requires mechanical devices and energy often linked to the internal filaments.

Before we describe the chemistry of the compartments involved, note that like prokaryotes, a number of oxidative enzymes are found in the cytoplasm but they do not release damaging chemicals (see Section 6.10). We also observed that such kinds of kinetic "compartments" are not enclosed by physical limitations such as membranes. We have also mentioned that increased size itself makes for kinetic "compartments" if diffusion is restricted. In this section, we see many additional advantages of eukaryotes from those given in Section 7.4. How deceptive it can be to use just the DNA, the all-embracing proteome, metabolome or metallome in discussing evolution without the recognition of the thermodynamic importance of compartments and their concentrations? These data could be useful both here and in simpler studies of single-compartment bacteria even in the analysis of species but not much information is available.

7.6.   Reproduction, Growth and Form

A further set of complications are involved in the simple description of eukaryotes. They reproduce sexually by a combination of male and female gametes (half-cells); they grow and change form and their form may be mobile, especially that of the more animal-like, as animals are motile in a directed, sensed, way. We cannot describe the advantages and disadvantages of these features here, but the reader should be aware of the huge diversity of patterns of growth, of form and of behaviour of unicellular eukaryotes. We can give them morphological labels and analyse their genotype, but we are far from being able to discuss their chemotypes today except in the broadest groupings, plants, fungi and animals. In fact, simple analytical data of element content are not available for most cell compartments. However, given our stress on flowing systems, we can note some additional general features of these single-cell eukaryotes that appear during growth. In many, if not in all, of the adult cells there is cytoplasmic directed flow for the distribution of chemicals. This flow occurs over or through the organisation of larger units, filaments and vesicles, but all these units can be re-arranged to increase efficiency under a given circumstance. The state of energisation of mitochondria and thylakoids clearly alters their shape within the cell. We stress that as chemotypes the plant-like organisms are individually very efficient in capturing light while fungal- and animal-like organisms are very efficient in digestion, both increase efficiency by circulation of material as well as by readjustment of form, external and internal, opposite circumstances. Digestion requires capture of food, and here motility of the whole cell or of parts can be aids as we see in the way an amoeba can catch other organisms. We deal with these topics in somewhat more detail in Section 8.6.

An interesting problem is how does the final internal organisation and external form it arise? There is an undoubted contribution from the genetic information expressed as proteins. However, as we have seen in Chapter 3, form is dependent on physico-chemical fields, the environment, and transfer of energy and information between all the different flows of material and energy to which a eukaryote but not a prokaryote cell is sensitive. A considerable part of spontaneous formation of dynamic structure lies in this interaction with the environment, i.e. its information content and the metabolism of organisms which does not involve genetic (DNA) response of necessity. Another part concerns the degree of flexibility of internal self-organisation, for example, of proteins once produced. This question of the involvement of factors other than those directly genetic has exercised scientists for many generations and is not yet solved (see Thompson D. and also Harold F. and references therein in Further Reading).

Finally, no matter how the organisation arose, it has to be related to the restricted and controlled pathways, new and old, and they must be autocatalytic since their active species, proteins, are made from their own substrates or ions. There is a unity of purpose in the cell to aid survival.

We now turn to the major purpose in this book – the description of these organisms in chemical terms, their chemotypes.

7.7.   The Threat of Dioxygen : The Chemistry of Protection

Looking again at the evolution of eukaryotes and its link to the increasing concentration of molecular oxygen, we note that the threat from this gas was to the more vulnerable reactants in the cytoplasm as in earlier prokaryote cells. Outstanding amongst these are the Fe/S, pterin and flavin containing enzymes inherited from prokaryotes. As we have stated, protection is the first way forward to assist survival and, as in aerobic prokaryotes, eukaryotes have devices for nullifying the greatest danger from O_2 in its reduced forms, O_2^- and H_2O_2. A very interesting fact is that while prokaryotes remove O_2^- using an iron or a manganese superoxide dismutase and both mitochondria and chloroplast still do so, the cytoplasm of all eukaryotes has a superoxide dismutase containing *copper* and *zinc*. Now divalent iron and manganese enzymes dissociate readily, but since divalent copper and zinc enzymes do not (see Irving-Williams series in Section 2.20), the latter are to be preferred in this function. Zinc and copper binding also stabilises protein folds, and so bound forms of these metal ions are of low risk. Note that as we have stressed, the metals, zinc and copper, are late additions to the environment, requiring the oxidation of their sulfides by oxygen to increase their availability, so that these enzymes indicate again the late arrival of eukaryotes and the increased complexity of their chemistry. By contrast, prokaryotes and eukaryotes have similar systems for the removal of H_2O_2 by iron-containing catalases. We draw attention to the development of H_2O_2 into a very useful protective chemical in vesicles (see peroxysomes above), especially in plants, and superoxide is also employed in this manner here and there in animals and plants. These are clearly examples of the development from a poison to a useful reagent. In all the above we see again a further advantage of eukaryotes in keeping oxidation reactions away from the cytoplasm generally, only allowing them there in protected enzymes, while using them extensively in vesicles and outside cells.

The vulnerability of Fe/S, flavin and *pterin* enzymes to molecular oxygen did not just lead directly to a flavin Fe/S based way of removing this threat, but it also led, as stated above, to use of oxygen, for example, in the production of cholesterol (Figure 7.6), by a *pterin* protein. This appears to be a vital contributing step to evolution. In this oxygen chemistry, we see again the progression where the earliest step was the removal of O_2 via NADH and flavins or pterins:

$$poison \rightarrow protection \rightarrow use$$

We have observed this sequence earlier and will do so later, and it probably applies to the uses of both zinc and copper, in all aerobes, and to calcium generally (Section 7.11), and even to sodium and chloride (see Chapter 9), and it may even apply to the impact of light (see Chapter 11).

The difficulty of handling oxygen appears to have led on the other hand to another case of cooperation in metabolism but now in true symbiosis. The vast majority of eukaryotes cannot metabolise N_2 and rely on bacteria for a conversion

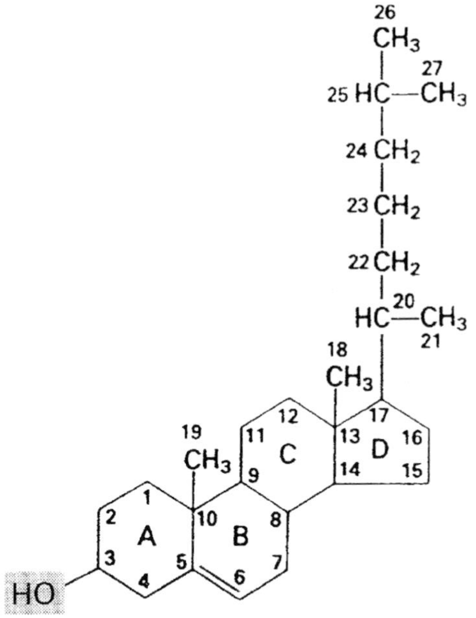

Fig. 7.6. Formula of cholesterol, note requirement for hydroxylation.

of N_2 into a usable chemical. Why was the gene construct for N_2-metabolism not passed on to all eukaryotes? Is there a desirable situation for some reactions to be removed from O_2 itself? Note that N_2 reduction requires the simultaneous release of hydrogen. This is, we consider, another example of the advantage of compartments; here seen as separate symbiotic organisms to develop organization, best able to capture energy in chemicals.

Of course, the eukaryotes inherited from prokaryotes many ways of handling essential organic synthesis, but unfortunately the old starting materials such as HCN, CO, NH_3 and H_2S became poisons in that they were inhibitors of O_2 uses. Again, as we saw was the case for aerobic prokaryotes, the ever-increasing presence of inorganic elements such as cobalt, nickel, copper and zinc, liberated from sulfides, gave rise to a most difficult problem since if allowed to enter in high concentration they could compete with Mg and Fe in the cytoplasm. Since in the bound state these two metals found new uses mainly in new compartments, methods of transport and insertion of dangerous metals without exposing the cell to risk were required. Carrier, pump and buffer proteins were synthesised for dangerous metal ions such as Ca^{2+}, Cu^{2+}, Zn^{2+} and even in part for Fe^{2+} as in aerobic prokaryotes, so as to maintain very low free ion concentrations. The binding of nickel was reduced as is seen in the change in hydrogenases from enzymes using mainly Ni/Fe in prokaryotes to Fe/Fe in eukaryotes. We shall see in the next chapter that the value of both Ni and Co became of still less use in multi-cellular eukaryotes. In some cases, especially in plants, these elements are rejected via vesicle storage. Before discussing further uses, we are now ready to describe the content of the elements in the compartments of these single-cell eukaryotes.

7.8. Additional Distributions of Elements in Unicellular Eukaryote Compartments: the Eukaryote Metallome and the Advantages of Compartmentalised Oxygen Metabolism

The distribution of elements in *single-cell* non-photosynthetic eukaryotes is probably best seen in terms of the well-defined compartments of yeast. The central cytoplasmic compartment containing the nucleus has many *free* element concentrations, only somewhat different from those in all known aerobic prokaryotes (Figure 7.7). (The nuclear membrane is a poor barrier to small molecules and ions and so we include the nucleus with the cytoplasm.) We do not believe in fact that the free cytoplasmic values of Mg^{2+}, Mn^{2+}, Fe^{2+}, Ca^{2+}, and possibly Zn^{2+}, have changed greatly throughout evolution. As stressed already there are limitations since free Mg^{2+} and Fe^{2+} are essential for the maintenance of the primary synthetic routes of all cells, and changes in other free metal ions could well have imposed

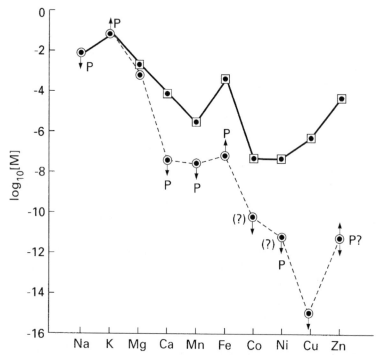

FIG. 7.7.   The proposed free metallome profile of the cytoplasm of the aerobic cell ⊙ and of the combined metallome of the organism ▣. P represents pump. Compare Figs. 5.3 and 4.2.

serious risk and the nature of chemical equilibria is a decisive influence here (see Chapters 2 and 4). Rejection of many *free* metal ions, e.g. most Cu^{2+}, Ni^{2+}, Co^{2+} and Cd^{2+}, and scavenging for some, e.g. Fe^{3+} to maintain Fe^{2+}, which was lost due to oxidation, are common activities of the cytoplasm of all these eukaryotes. Free Mn^{2+} and Ca^{2+} were rejected to even lower levels than in prokaryotes perhaps to protect further the essential biopolymers and phosphates. In the case of Ca^{2+}, this rejection but with storage in vesicles has allowed its use as a messenger in eukaryotes (see Section 7.12), while Mn^{2+} is stored and used in glycosylation in the endoplasmic reticulum. The free copper may have risen somewhat in the cytoplasm of all aerobes, to say 10^{-15} M while Zn^{2+} is maintained at around 10^{-11} M. On the other hand, the values for the *bound* elements *in the cytoplasm* changed considerably and we observe less bound nickel and cobalt—almost no cobalt in many later plant cells (Section 8.8)—and quite considerable values of bound zinc but very little of even bound copper. Note that zinc cannot be involved in oxidation/reduction reactions and is therefore relatively safe in the cytoplasm at low free concentration, $\sim 10^{-11}$ M and in bound forms. Zinc became a central cytoplasmic element but did not increase in organelles. It is especially associated with a new group of transcriptional proteins, zinc fingers. This use of zinc, apart from being applied as a cross-linking and stabilising agent, is linked to the long life of eukaryotes, which gives time for zinc to exchange and hence it could become an internal messenger connecting to and probably coordinating the action of many transcription factors and groups of messengers during growth; see also zinc in multi-cellular organisms. It is then interesting to compare the faster coordinating action of exchanging Mg^{2+} and Fe^{2+} maintained in many cytoplasmic functions much as seen in prokaryotes. Zinc is also used extensively in the same form, zinc fingers, in other new proteins in the cytoplasm (see Section 8.11). The increasingly extensive use of bound zinc *in the cytoplasm* is then a further indication that eukaryote evolution followed oxygen increase and its consequences in the geo-environment. This does not mean that other elements already present in the cytoplasm could not be of value in some new functions, e.g. bound ferrous iron and selenium in various carefully managed "isolated" oxidative catalyses (see Section 8.11). As mentioned earlier, there is also a small but important use of bound copper, with zinc, in the cytoplasm in superoxide dismutase (see above), so that for this and other uses copper too needed transcription factors and carriers. Copper became most useful in other compartments and externally in the later multi-cellular eukaryotes (see Chapter 8). As we have made clear, copper released after zinc into the environment, due to the greater insolubility of its sulfide (Figure 1.8), which requires a higher oxygen concentration to dissolve it, undoubtedly made for its later extensive use.

We now turn to the known or estimate concentrations of elements in the other compartments of eukaryotes that show a different distribution of elements, free and in proteins, from that in the cytoplasm. Most of these compartments, vesicles such as the endoplasmic reticulum, operate at a higher redox potential than the cytoplasm and several at a lower, more acidic, pH. Some of the element distributions there, and their functions are given in Table 7.5. Note, as mentioned before, that

TABLE 7.5

NEW ELEMENT BIOCHEMISTRY AFTER THE ADVENT OF DIOXYGEN

Element	Biochemistry
Copper	Most oxidases outside higher cells, connective tissue finalization, production of some hormones, dioxygen carrier, N/O metabolism
Molybdenum	Two electron reactions outside cells, NO_3^-, SO_4^{2-}, aldehyde metabolism
Manganese	Higher oxidation state reactions in vesicles, organelles and outside cells, lignin oxidation (note especially plants); O_2 production
Nickel	Virtually disappears from higher organisms
Vanadium	New haloperoxidases outside cells
Calcium	Calmodulin systems, signalling; general value outside cells
Zinc	Zinc fingers connect to hormones produced by oxidative metabolism
Selenium	Detoxification from peroxides, de-iodination?
Halogens	New carbon–halogen chemistry, poisons, hormones
Iron	Vast range of especially membrane-bound and/or vesicular oxidases; peroxidases for the production of hydroxylated and halogenated secondary metabolites; dioxygen carrier and store

free Mn, Ca, Na, Cl are usually high in these vesicles where both oxidation, glyco-sylation, –S–S– bonding and sulfation can occur. Membrane pumps drive these elements into vesicles but out of the cytoplasm. Ca^{2+}, apart from its storage as a messenger (see below), then controls much protein folding there. The elements that are not found in vesicles of *unicellular* eukaryotes to any large degree are K, Mg, non-haem Fe (e.g. Fe_nS_n), Zn, Mo and Se, but some plant cells accumulate danger-ous metals such as Co and Ni in special vesicles for protection, vacuoles. Turning to organelles, we have already noted that energy transduction that generates or uses oxygen is found in them. Their membranes are very highly concentrated in bound iron and copper (and magnesium and manganese in chloroplasts). The cytoplasm of the organelles is a second cytoplasm, and their periplasmic space (they are bac-terial in origin) is very similar to the periplasm of aerobic bacteria. As stated, inter-nally, these organelles are also the sites of fatty acid oxidation and of the Krebs cycle reactions (mitochondria) and of CO_2 assimilation (chloroplasts), where in all the cases Fe proteins are required. It is the inner compartment of mitochondria and the outer compartment of chloroplasts which are used for these reactions. Especially of interest is the appearance of bound copper in vesicles generally, and in the periplasm of organelles and in the outer surfaces of their membranes. As stated in Section 6.9, once oxygen levels reached about 10% of present day values, ferrous iron, relatively weakly bound except when protected in haem, could only be used in a protected reducing medium, the cytoplasm; so the handling of strong oxidising agents such as O_2 and the oxides of nitrogen outside the cells became more and more the function of copper enzymes. Copper proteins also became a general part of electron transfer and in reactions in as well as in the organelles. In peroxyzomes, we find the haem enzymes, peroxidase and catalase in the tightly bound *oxidised ferric* state. All the proteins of these metal related functions are coded in the nucleus of the cytoplasm, so there are both issues of controlled production and of transfer of proteins introduced into the genes of DNA. How did this arise? Note that the element content outside the cytoplasm is largely responsible for the novel chemical content of eukaryotes, in part defining their chemotype.

In conclusion, the differential distribution of elements in different compartments is very marked but is little known. Analysis is essential for detailed thermodynamic knowledge of these cells.

7.9.   The Proteome and the Metabolome

We turn now to the more conventional topic of analytical biochemistry, that of the organic molecular rather than the inorganic contents. We must remember that the proteome and the metabolome like the metallome are extensive properties, that is concentration-dependent, unlike DNA, and any discussion of their significance is not linked only to sequences and structures, but also to system, thermodynamic properties including compartmental containment, concentration and temperature. In eukaryotes, there are individual compartments very different in protein and metabolite types and concentrations, but usually not of RNA content, from those in prokaryotes, though much of their cytoplasmic content may be similar. In Section 7.14 we will discuss changes in RNA activities, and of genetic structure. In many respects, the protein content of the cytoplasm of the eukaryote is not much changed from that of the prokaryotes since the basic pathways have to remain. It was this fact that allowed us to treat all organisms together in Chapter 4. We have already described the compartmental separation of the energy-producing pathways of mitochondria, including the citrate cycle, its peculiar membrane content and the problem of the origin of its proteins and DNA. Most of its proteins require import and hence special transport systems are needed. The chloroplast poses even more distinctive features since it includes the reactions of carbon dioxide incorporation as well as the transduction of energy from light. Clearly, the protein and metabolite concentrations of the organelles are particular to them as is the content of the endoplasmic reticulum with the special metabolic features described above. Many of the new proteins, listed in Table 7.6, have special functions, particularly, as mentioned before, zinc fingers. Much of the total increase here and later in evolution is due to gene duplication, which allows variants of one kind of protein or enzyme to be produced, e.g. of related enzymes, iso-enzymes. The several iso-enzymes generate a variety of metabolites often on the basis of the simpler central themes of prokaryote syntheses. Through these alterations the basic metabolome has been changed very little in its major constituents, but the total metabolome is now compartmentalised and there are various modifications. Hence, it is necessary to import many small molecules to the mitochondria and the chloroplasts amongst all other compartments. The major import mechanism uses the proton or the sodium gradient so that much energy is expended in transport. At the same time, ATP from the

TABLE 7.6

SOME NOVEL PROTEINS IN SINGLE CELL EUKARYOTES

Protein	Function
Myosin	Contractile protein (Ca^{2+})
Actin	Structure/contractile protein
Tubulin	Vesicle guidance filament
Zinc fingers[a]	DNA-transcription factor
Calmodulin[a]	Ca^{2+} signalling
Calcineurin	Activator of transcription (Ca^{2+}, Zn^{2+}, Fe^{2+})
Superoxide dismutase (Cu,Zn)	Catalyst for superoxide dismutation
Squalene oxidase	Cholesterol synthesis
Tyrosine kinases[a]	Phosphorylation signal
Extra serine/threonine kinases[a]	Phosphorylation signal

[a] As eukaryotes increase in complexity (see Chapter 8), the varieties of these and other types of protein, e.g. kinases, also increase considerably as seen in DNA sequences. They are associated with controls.

organelles is continuously exported and ADP plus P_i are imported. In the cytoplasm only a few new metabolites in small quantities are found. Many became more important in multi-cellular organisms as messengers and we defer discussion of them. Most obvious then are the novel dispositions of protein and metabolite concentrations in compartments.

In order to place proteins outside the cytoplasm, their synthesis starts in ribosomes on the cytoplasmic side of the compartment membrane (Figure 7.8). Then as the protein is processed, the thread of the protein is passed through the membrane to the compartment where it is folded and used. The protein can be processed from the vesicle by moving it to the outside of the cell (Figure 7.8). Minerals, with their organic matrices formed in certain vesicles, can also be transferred to the outside of cells. Of course, all these protein-involved processes in vesicles have to be controlled by new messages to and from the centre of organisation, the DNA. Messenger development was found to be an essential part of all eukaryote evolution connected to information transfer. The proteome and the metabolome are now extremely complicated, being related also to these messenger concentrations in different compartments. It is these complications, vast number of organic chemicals, which makes it difficult to see change as systematic.

We see immediately that the trafficking of many new ions, small molecules, proteins and even large units in eukaryote compartments, as is also the case of several earlier compounds, is an example of controlled directed diffusion. The controls are more and more connected to the environment as we shall see when we focus on message systems in Section 7.11. A central feature of evolution is the extension of organisation to larger and larger volumes, so that slowly the surface geosphere and the biosphere are interacting to become one thermodynamic whole – the ultimate

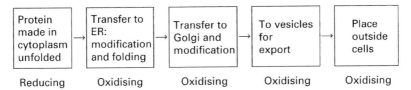

Protein made in cytoplasm unfolded	→	Transfer to ER: modification and folding	→	Transfer to Golgi and modification	→	To vesicles for export	→	Place outside cells
Reducing		Oxidising		Oxidising		Oxidising		Oxidising

FIG. 7.8.   The exporting steps for a protein from the cytoplasm to the exterior.

aim of a steady state of optimal energy retention and degradation. Before we discuss these topics, we wish to draw attention to the special part of the proteome concerned with metal ions since the role of essential elements is a dominant theme in this book.

7.10.   The Proteins for Metal Ions in Eukaryotes

Before we conclude the functions of the elements and the proteome, there is a second general feature of eukaryote cells, much of which evolved from that of the prokaryotes – the types of metal-binding protein. The general supposition is that the number of folds are limited and certainly the number of metal-binding sites for any one metal ion is closely limited (see Section 4.15). We find that there are some general rules for protein-binding centres of metal ions and their geometry, mentioned only in brief in Chapters 5 and 6.

(1)   Non-haem, non-Fe_nS_n proteins bind iron, Fe^{2+}, in an approximate octahedral site using carboxylate and histidine ligands. The protein has largely a β-sheet fold.

(2) Fe_nS_n proteins has each Fe in a tetrahedral site linked to a single cysteine residue. The protein has a β-sheet fold. The Fe is usually a combination of Fe^{2+} and Fe^{3+}.

(3)   Fe in haem is bound in a mostly α-helical protein usually by histidine but sometimes by methionine, cysteine or tyrosine. Several such proteins are to be found in membranes.

(4)   Cu is bound by a β-sheet protein (enzyme) in 3 or 4 coordination using histidines or/and cysteines. Cu^+ is bound by thiolates in storage proteins.

(5)   Zn is bound in 4/5 coordination in two major ways:
   (a)   a tetrahedron of thiolates with or without histidines in a β-sheet
   (b)   a 4/5 coordinate site of histidines and carboxylates.

(6)   Mn is bound by longer bonds than those binding Fe and Mg to carboxylates and one or two histidines in 6/7 coordination.

(7)   Ca is bound usually in 7 coordination with carboxylate and neutral O-donors in *helical* proteins.

(8) Mg is bound in a 6 coordinate site by four or less protein O-donors and several water molecules often in mixed α/β proteins in machinery. (Note the extraordinary binding of Mg^{2+} to five-nitrogen donors in chlorophyll proteins.)

(9) K is bound by up to eight neutral O-donors in DNA, rarely in proteins.

Now, although not quite universally true, these descriptions are in general accord with equilibrium thermodynamic and ligand field expectation (see Chapter 2), and have to be maintained throughout evolution as the binding of many metal centres is in fast or relatively fast exchange. Equilibrium is an unavoidable thermodynamic control over site selection and is universal and independent of life. The possible sites for selective binding are then very limited, generating one almost fixed metallome (Figure 7.7) in a compartment, and these sites are close to optimal functional value (see Section 2.20 and Appendix 4A). It is not too surprising that the folds are also maintained since they give *the site required stability and additionally certain properties desirable for a particular function*, for example, more rigidity for electron transfer and enzyme activation of substrates, and therefore β-sheets, more flexibility for triggers and machinery, and therefore α-helices (see Sections 4.11–4.13). A given fold can be used to create sites for different metal ions using site-selective differences and there is the possibility of following evolution of, for example, transcription factors, carriers, enzymes (β-lactamase fold, for example, as iron sites were modified later for say zinc functions) or more obviously Mo and W were exchanged in closely similar sites. This is a metal-ion "mutation". It is a general truth that the binding proteins and their selected metal ions are strikingly matched to generate optimal effectiveness. Can we then trace protein development from prokaryotes to eukaryotes as function changes? Interestingly we observe the following:

(1) Maintenance of similar zinc proteins in proteases and peptidases for digestion in both cell types and then to hydrolytic management of connective tissue in later eukaryotes.

(2) Extended use but similar coordination chemistry of haem proteins from reductive to oxidative reactions.

(3) The earlier switch of the molybdenum cofactor from carboxylate to sulfate and nitrate reduction.

(4) The EF-hand construct of calcium-binding proteins in cellulases (prokaryotes), but only single hands, becomes extended to triggers (eukaryotes) in twin hands.

(5) The tetrahedron of zinc thiolates is found in enzymes in both cell types but is most frequently seen in transcription factors, zinc fingers of eukaryotes.

Other connections no doubt will be found as DNA sequences are explored. A further feature of some proteins are that they are used to retain, carry, and buffer metal ions in compartments ready for exchange, for example, metallothionein for Cu and Zn in the cytoplasm and calcium-binding proteins in both the cytoplasm and the reticulum. There are relatives of these proteins in prokaryotes.

Now the creation of new protein centres, for example, for copper, is then of great interest since it is difficult to see how they were derived except in a few cases such as the copper metallothioneins that could well be modified tetra-thiolate binding sites for Zn^{2+} that became suitable for Cu^+ (see Calderone et al. in Further Reading). Did the very presence of copper in the environment force the evolution of modified proteins? If so, how did this happen? The general suggestion is by random mutation but could not this activity be localised within the DNA? An obvious problem is the radical change in the superoxide dismutases from prokaryotes (Fe, Mn) to eukaryotes (Cu, Zn). Finally, the selected proteins and the metal ions are found in cells again in fixed concentrations linked to equilibrium binding constants (see Appendix 4A). Many metal proteins are catalysts, all of which must be carefully regulated and controlled and form part of the autocatalytic system of the cell (see Section 3.9).

If we now look away from metalloproteins there are some obvious further connections between prokaryotes and eukaryotes but there are also other novelties such as the arrival of the proteins actin, myosin and tubulin associated with MgATP and mechanical activities. How did they arise?

7.11.   Messengers in Single-Cell Eukaryotes

As mentioned above, the development of compartments in large cells also brought the special difficulties of longer growth and life periods and of their control. Survival of long-life organisms needs much better protection, which requires response to the environment and, with the possession of many compartments, also requires extensive new internal controls. Apart from the protection already described to maintain low concentrations of oxygen, or from its partially reduced products $O_2^{\cdot-}$ and H_2O_2, and from many metal ions in the cytoplasm, a second protection was to develop sensing of the environment so as to be able to take avoiding action away from any foreign substance if there was a threat. A threat was also met by the release of poisons from vesicles. Alternatively, the cell could adjust its shape to an object or external condition if there was an advantage in doing so. For example, certain green algae have an "eye-spot" and direct swimming to the light using channel rhodopsins (see Nagel et al. in Further Reading). Thus, eukaryotes, on the one hand, became able to change shape or even to move away from or repel competitors, and on the other hand, to adjust or move towards and capture large objects of food. They could also recognise surfaces in the environment. All this organisation of activities requires novel sensing and messengers between the flexible exposed membranes and internal metabolic paths including those in vesicles and to and from many other activities such as membranes pumps, and often eventually to DNA, but the last is a slow response. The sensing apparatus that is needed is more usefully described when it developed further (see Chapter 8).

The novel external messenger system, which appeared first, was, we believe, in the form of Ca^{2+} ion fluxes from the outside of the cytoplasm to the inside, often coupled to release of membrane-bound sugar phosphates, as well as to and from vesicles and organelles in response to environmental events (Table 7.7 and Section 8.11). The flow in was initiated by sensors (receptors) and there followed, first amplification by extra release of Ca^{2+} from vesicle stores, and then immediately after response, outward Ca^{2+} pumping. We believe that zinc too acted increasingly in transcription factors in the coordination of slower novel protein syntheses, in adaptive responses during growth for example. In other words, while the prokaryote metabolic coordination inside the cytoplasm due to Fe^{2+} and Mg^{2+} and small organic molecules including phosphates such as NTP, NDP and c-NMP remained and increased in complexity, e.g. in numbers of kinases, at least two new external and compartmental (Ca^{2+}) and internal (Zn^{2+}) information carrying systems evolved. Of course the old and the new had to be linked together. We shall describe in Chapter 8 other additions especially, of organic chemical messenger controls as they are of greater importance in multicellular organisms. The choice of the use of free calcium and zinc as messengers have already been explained as they, together with Mg^{2-}, are the only available divalent ions that cannot cause damage in the cytoplasm through catalytic reactions involving oxygen since they do not participate in redox processes. The advantage of the calcium ions is that the temporary actions of calcium are faster than those of Zn^{2+}, Mg^{2+} and Fe^{2+}, as they bind quite strongly (not too strongly) and a large external/internal gradient of calcium existed. The Ca^{2+} ions can also be removed quickly, and they are therefore capable of generating a quick effective pulse (see Figure 2.7). Their actions are rarely connected to genes while those of Zn^{2+} are much slower and their actions, as well as those of Mg^{2+} and Fe^{2+} ions, are often so connected. In the case of zinc the on/off binding rates are often so slow that they, as messengers, are linked to genetic processes in development not required in prokaryotes (see Section 8.6).

TABLE 7.7

THE CHANGES INTRODUCED FOR CALCIUM SIGNALLING

Change	Location
1. Compartmental concentration of Ca^{2+}	Endoplasmic reticulum (chloroplast and mitochondria). Other vesicles, Golgi
2. Calcium channels	Membranes (cell membranes and vesicle membranes)
3. Calcium receptors	Membranes of vesicles. Many enzymes (kinases). Filaments for contraction (calmodulins, annexins, troponin)
4. Calcium pumps	Membranes (driven by ATP)
5. Links to phosphate signals	Kinases and phosphatases with attached Ca^{2+} trigger proteins
6. Calcium stores	Calsequestrin in reticula
7. Nucleus (separation) (calcium concentration?)	Calcium signals to the nucleus from the surrounding endoplasmic reticulum (Figure 12.5)

The extra advantages to eukaryotes in addition to those described in Sections 7.4–7.6 are then flexibility of response, chemical and mechanical to many kinds of external events, with different rate constants and controls giving protection and advantage in searching the environment for food and other advantages using sensing devices in membranes. Before proceeding further we give a note concerning calcium chemistry involved in many of these responses, since the development of calcium chemistry in evolution is a major consideration in the increase of all organisation. The involvement of calcium in evolution stresses the need to consider inorganic chemical systems, and not just organic molecules or their sequences, in an effort to understand systematic development of organisation in contact with the environment.

Note that it is possible to use proton or sodium ion gradients in signalling and a possible example is provided by the reactions of the proteins of the eye-spot of green algae (see Nagel *et al.* in Further Reading).

7.12. The Crucial Nature of the Calcium Ion

The use of the calcium ion is ideal for its purpose as a messenger in that a gradient of some 10^4, inside close to 10^{-7} M, outside 10^{-3} M, was established at the very beginning of prokaryote life of necessity. [There was little or no need for the use of Ca^{2+} as a messenger in prokaryotes since their short life made reproduction dominant over the complexity of response to environmental challenges.] Free Ca^{2+} concentrations in vesicles are often 10^{-3} M, and much is stored there in rapid exchange so that Ca^{2+} can be released from these vesicles as an amplification of input from outside. Now the rates of calcium binding to a target are very fast, say 10^{-9} s. So for a binding constant of 10^6 M^{-1} the off-rate from the target is 10^{-3} s, a millisecond, which is long enough and the binding of Ca^{2+} is strong (energetic) enough to bring about a protein conformation change inside cells. *This rate, a millisecond, then became a fundamental restriction on all eukaryote biological transformation due to the properties of the calcium ion and proteins.* The new proteins in question, e.g. calmodulins (see Table 7.6), are α-helical and such proteins have a high internal mobility and switch conformation readily (see Sections 4.11 to 4.14 and Carafoli in Further Reading). These properties of Ca^{2+} ions should be contrasted with those of other fast exchange ions. Ions such as Na^+, K^+ and Cl^- do not bind well enough to cause conformation changes, even though they can flow rapidly down gradients and exchange rapidly. We shall find that Na^+, K^+ and Cl^- gradients are used in *cooperative* electrical, that is bulk physical not chemical, ion triggering much later in evolution (Section 9.3). Of the ions that could cause conformation changes, only Mg^{2+} is in high enough concentration and binds strongly enough, but it undergoes rather slow exchange and has too small a gradient across a membrane to be of much use as a signal. Again it was already fully used in high concentration as a catalyst and connectivity control in the cytoplasm. (It may be used to switch on slowly certain synthetic activity in the cytoplasm of plant cells when chloroplasts are illuminated and Mg^{2+} is forced out of the thylakoids as H^+ increases there.)

We indicate how Ca^{2+} became connected to many parts of the cell in Figure 7.9, where Ca^{2+} changes are also linked to energy generation in mitochondria and chloroplasts. Note additionally the interlocking of Ca^{2+} with phosphate messengers such as c-AMP, GTP (G-proteins) ATP (phosphorylation), phosphatases and organic phosphates derived from membrane lipids, i.e. virtually the whole variety of the huge internal phosphate messenger system and to Mg^{2+} via Mg NTP (see Section 8.14). There is also a link to oxidative signal systems and therefore iron, Fe^{2+}, in, for example, NO release.

An important point to note here and elsewhere in the description of cell activity is that the particular nature of calcium biochemistry, including the availability of the element and its necessary rejection from the prokaryote cell, when linked to stimulated input and interaction with specific internal proteins of selected properties, made it *uniquely suitable for the function as an elementary ionic fast in/out messenger*. It was then capable of signalling to cell changes once cell size and organisation increased beyond the elementary level of a cell with one small, rapidly

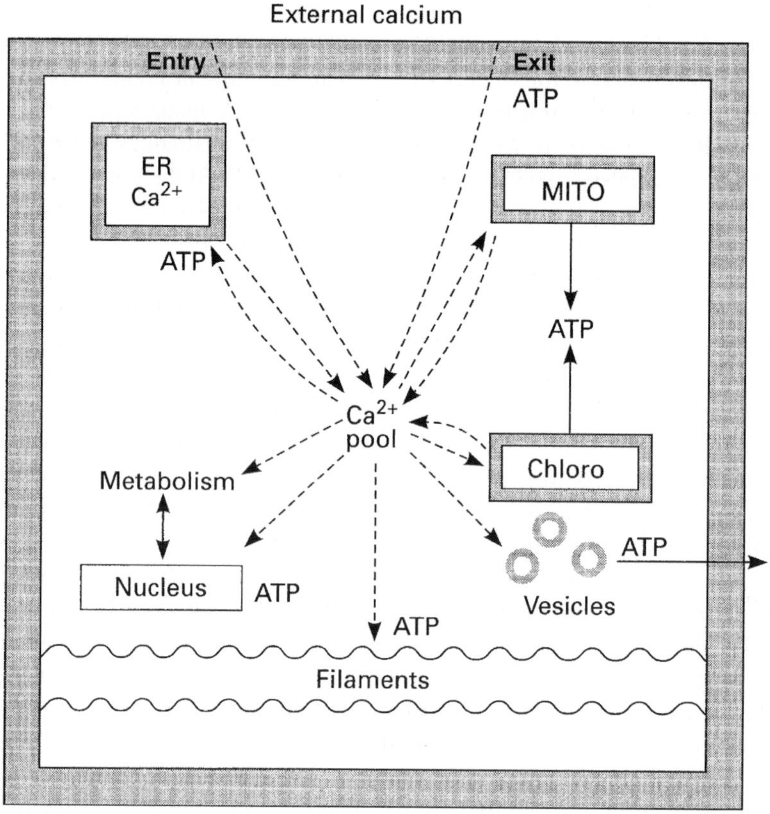

FIG. 7.9.   The calcium circuit in an active transient state. All the activated proteins have very similar binding constants for calcium. (ATP is really MgATP.)

reproduced, internal compartment (see Table 7.7). No other element has the same inherent and environmental properties. The genetic machinery of eukaryotes had to discover the value of this calcium chemistry and to code the proteins involved with it. It is not just a matter of random mutation but of opportunity meeting necessity as this was an inevitable advance if optimal energy capture and use in chemistry was to be secured within organisation of a large, environment-sensitive organism. The increasing dominance of calcium as a messenger in all eukaryotes will be seen as evolution advances further so that some hundreds of proteins respond to calcium input in very advanced animals (Section 8.11). But how did the genetic code develop this function, absent in prokaryotes? Once developed, gene duplication could allow variation and specific functioning.

While we have concentrated attention on the calcium communication network, note that it does not interfere with the immensely important internal signalling since Ca^{2+} is rejected quickly. If it remained high the cell would and does undergo apoptosis – cell death. The Ca^{2+} outward pump is of early origin, it had to be present in prokaryotes, while we note the later origin of the outward copper (zinc) pumps, since the early environment contained little of these elements. All the metal ion pumps can be seen in evolution of the genes (see Figure 9.3). The use of these elements later stresses also the advantages of individual elements both early and late in evolution to provide uniquely associated novel properties, indicating the inevitable link to availability in the environment.

Of the great advantages of eukaryotes, the foremost is the development of structured internal compartments and filaments, but of equal importance is the ability to inform the dynamics of the membranes of these vesicles and filaments, and hence to use them in changing structure and motion in response to external conditions using sensors (receptors). The response mainly employs Ca^{2+} signalling to directed motion, in contrast to that of prokaryotes. Increase in such information transfer is an essential feature of evolution flows within cells and between the environment and cells as cellular chemistry became increasingly interlocked. This immediate response is of every increasing importance in evolution (see muscle reactions in Chapter 8). However, we shall see that regularly and rapidly repeated Ca^{2+} pulse can cause genetic change of expression.

Finally, note that eukaryote chemotypes have as a general feature the increase inside the cell vesicles of elements, here calcium, previously confined to the outside of prokaryotes, but these increases are different in different vesicles and organelles and in different organisms separating cells into different chemotypes. This is also seen in their minerals. There is as yet far too little quantitative analysis of calcium or indeed of elements generally to allow us to build a full picture of chemotypes together with genotypes (see Table 8.22).

We must also observe that the rejection of Ca^{2+} and of other elements, like oxygen, to aid survival produced an energy store. Although, unlike oxygen, this store could not be used to drive metabolism, it is *energised gradients*, here of calcium, which are central to information transfer from the environment and then of fast action, independent of DNA.

7.13.   Minerals in Unicellular Plants and Animals and their Deposition

Fungi do not in general have associated minerals, but both plants and animals may have. The major difference appears to be that single-cell animals use calcium salts, mainly carbonates, for their shells, as in the foraminiferae, while plants utilise silica as in diatoms (see Table 8.12). The minerals are placed outside the cytoplasm. The difference is continued throughout multicellular species of animals and plants as we shall see later. Thus the cells of the different eukaryote chemotypes use elements *outside* cells somewhat differently and in many cases this mineralisation occurs initially in specialised vesicles before export. Calcium, silica, phosphate and iron have to be pumped into cells and then into vesicles for this purpose, even in the sea. The formation of compartments greatly assisted the progress of mineralisation, which is now in precisely defined, species-specific, form, e.g. in coccoliths and radiolaria. The morphology of these mineral shells has long been used in typing organisms. For recent references see Mann and also Veis in Further Reading.

Deposits of these shells of unicellular eukaryotes form sludge at the bottom of the Earth's oceans and then appear on land as minerals sometimes after compression and further processing. Worldwide there are huge deposits of carbonates due to algae, foraminiferae and coccoliths, of phosphates from some animals and of silica due to radiolaria. Examples are siliceous earth in the Sahara desert and chalk, $CaCO_3$, of coral islands and the white cliffs of Dover (see Section 1.9). The immense mineralisation of carbon as carbonate is a very important part of the whole ecosystem since it is a way of retaining carbon on Earth in a geochemical slowly available form. Apart from mineral deposits from organisms we must remember the huge deposits of carbon not only in coal, oil and gas, but also in organic compounds in all soils. These are primary forms of retention of the elements carbon and hydrogen as well as nitrogen to a lesser degree. (Note that hydrogen is largely retained as liquid water and with bound nitrogen and carbon in organisms.) There are many other fossil deposits that we will discuss later and they with the carbonates can be used in the determination of dates of evolution using isotopes. An example is the Burgess Shale deposits of minerals about 500×10^6 years old, but there are also younger fossils. Even the history of our climate is studied using these minerals. When we turn to mankind's use of elements, we shall see that some of these biological mineral deposits are used in buildings and others in fertilisers re-entering the element cycles. Finally, note that these elements, Ca and Si, form a special link through solubility and complex ion formation between the organisms and the environment. Their extensive use also creates new chemotypes and, like all other evolutionary changes, requires the expansion of the genetic apparatus for their management.

7.14.   Gene Development in Eukaryotes

Deliberately we have kept the discussion of genes towards the end of the Chapter. We know that much of evolution can be followed through gene sequences, but this discussion often appears as if it is an analysis of random events which dominates the selection of the "fittest" *species*. This leaves the impression that there is no rational explanation of the general development of life and to the limits of biological evolution towards ecological fitness. Our stance does not question that this description of *the random origin of species is correct*, we believe it is, but we consider that the species-embracing chemotypes and their divisions, which include very large groups of species in well-separated classes of organisms, have developed differently in an inevitable logical sequence forced by equilibrium thermodynamic environmental, and largely kinetically controlled life chemistry. We repeat that the principle is that a chemical system, the Earth, exposed to energy, mainly from the sun, and able to use or enclose space in an organised manner will produce novel chemicals and eventually organisms made into different chemotypes, starting with prokaryote species. We shall not discuss and probably we cannot know if this had a chance but an inevitable beginning. Some of these organisms on capturing light and through their chemistry produced oxygen and forced environmental change. In turn, and with time, these new environmental chemicals forced further changes on the chemotypes of organisms, including the advantages associated with multi-compartment single-cell eukaryotes. Now we wish to know how these changes could impose themselves on the genetic structure for they must do so if evolution is to be sustained. We do not believe that what was in effect an insult due to environmental change could have been met by an entirely random response of the genetic material generally. There is a good hint from prokaryote genetics of a different possibility (Sections 5.14 and 5.15), that is exposure of these organisms to a range of poisonous metals and offensive chemicals resulted in the main response of genes being found in the *localised mini circular DNA, plasmids, not in the main DNA*. This implies that a region of DNA can change locally, although such localised mutation may still be to some degree random. We shall see later that in higher eukaryote organisms only the genes related to the immune system, not the whole genetic structure, respond to foreign agents that could lead to many diseases. Already in Section 6.11, we have proposed that local DNA change in aerobes was brought about by the effect of damaging chemicals on selected proteins, which had to be relieved not just by protection but by the production of new protective, then useful, proteins. How is this possible in eukaryotes? Before approaching this question we note the differences in the genetic apparatus between prokaryotes and eukaryotes.

DNA in prokaryotes is circular, is continuously being copied and it has a very rapid rate of mutation – a way of sustaining life of these species. In eukaryotes it is observed to be linear, also double-stranded, with protected sections in chromosomes, where the DNA is wrapped around a group of proteins, histones and held in compact form, nucleosomes (Figure 7.10). Moreover, it is now composed of inactive regions (introns) and actively expressed regions (exons). The nucleosomes disks are packed into rings to form a solenoid. The whole is held together by Mg^{2+} (and K^+) ions. What is the advantage of these changes? Certainly one is the reduced vulnerability from the condition of an unprotected DNA, which ensures a

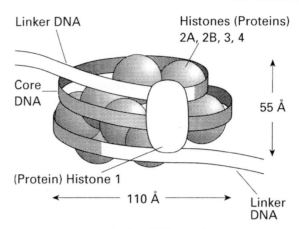

Linker DNA

Histones (Proteins)
2A, 2B, 3, 4

Core
DNA

55 Å

(Protein) Histone 1

← 110 Å →

Linker
DNA

FIG. 7.10.   The nucleosome structure that has 147 base pairs.

lower overall mutation rate, but how then did evolution proceed at all? Here we remember that in transcription to RNA, eukaryote DNA also becomes single-stranded and unwrapped and, therefore, more vulnerable at localities and within it there are "hot spots", sequences of DNA known to be vulnerable to chemical attack. There is then a mechanism for local DNA response to an attack from a novel chemical. The novel chemical, however, does not necessarily attack a section of the DNA directly, but may well damage particular proteins belonging to a small part of the whole cell organisation related to just such a section. The response to this upset to homeostasis, due to the pressure of intensive feedback, has to be an increased production of the very proteins damaged or blocked. These selected proteins must be regenerated by increased synthesis of the corresponding messenger RNA, which is related to a *special piece of DNA*. This piece of DNA alone then becomes more frequently single-stranded and more exposed to chance mutation during the synthesis of its RNA as in prokaryotes. Hence, a local region of the DNA is open to greater mutational or other change due to a damaging insult to particular proteins and effectively this region can be mutated more intensely in order to provide a remedy to an insult to the organisation. Darwinian natural selection "of the fittest" may then act to favour mutations in a local region, giving organisms better adapted to the changed environment and/or better able to reproduce in it. We consider this to be a type of *epigenetic environmental mechanism*. Since the original gene is still required we have to consider that gene duplication occurs simultaneously or earlier due to insult. We realise that this explanation has a touch of Lamarkian as well as Darwinian inheritance, but we do not believe it is out of line with much of recent thinking in genetics (see Caporale and also Jablonka and Lamb in Further Reading). Now and then such mutations in groups could cause a change, unlike that from one species to another generating a cell with a new cooperative activity, a new chemotype.

In this context we must see the value of introns, non-translated sections of DNA, largely introduced into the genome in eukaryotes. One advantage is that sections of proteins separated by introns can be shuffled and so it is possible to get variety from one DNA sequence. Another possible advantage is that short lengths of intron, "interfering" RNA, RNA_i and s-RNA, are released into the cytoplasm and may well act as feedback messengers by binding to DNA – a novel set of controls – or as protection of the intron (see Section 4.10). In all activities it is observed in fact that introns have a low rate of mutation. These possibilities are under intensive investigation (see Linder; Matzke; Wickelgren; and Freeland and Hurst in Further Reading). There are also ways of using RNA and proteins to transfer sections of DNA from chromosome to chromosome, called transposons. All such rearrangements are known to be selective and dependent upon environmental factors. Once again this illustrates that changes are not random within the whole coded molecule. Note that unlike the linear intensive information in the DNA sequence the RNAs are expressed in concentration dependent terms, that is they have extensive properties of their own and such expression like that of proteins has conventional thermodynamic dependencies. The relation of the proteome to the DNA and of both to the environment is in need of great experimental activity. The difficulty is in seeing how quite new chemotypes arose relatively quickly at periods after the first two billion years. One answer is that the environment changed more quickly introducing many initially poisonous chemicals and they caused evolution, including DNA mutation and change rate to increase and to change of organisms.

We must also see that chromosomes, separate lengths of DNA, have an advantage over a single (circular) DNA. As organisms grew more complex the number of genes increased. There is an obvious problem in the handling of single, very long strands of bases, which is overcome by breaking it down to several lengths. As in all "compartmentalisations", it is necessary to have coordination, and hence a more structured DNA environment, for growth reproduction and generally for transcription to RNA (then translated into proteins) related to particular cell compartments. The requirement is for mechanical coordination at division and message coordination during growth. What then coordinates chromosomes?

Now the inorganic chemistry of the DNA has also changed. Apart from the association with zinc transcription factors the termini of the duplex DNA, the telomeres, are held in a particular structure by incorporated potassium ions. The lengths of the telomeres are related to the number of times that DNA can be reproduced, hence to cell death, *apoptosis*, since in each cell cycle telomeres shorten! We return to the problem of DNA changes and epigenetics in Section 11.13. Here we draw attention to the fact that certain genes were *deleted* from higher organisms as they became dependent on lower organisms for substances vital to both, as discussed in the next section. How are genes *selectively* deleted?

Finally, reproduction in all eukaryotes is dependent on sexual interaction (see Margulis (1995) in Further Reading). The implication is twofold: that the DNA is at all times less exposed to horizontal gene transfer that is from species to species, which is common in bacteria; but the DNA is more open to recombination as a source of variation for more effective adaption amongst similar species. The great advantage of sexual reproduction is that it assists the sorting in a population of systematically strengthened survival strength in a breeding species. The evolution of sexual reproduction is a major feature of the eukaryotes but how did it develop?

7.15. Mutual Dependence of Eukaryotes and Prokaryotes

Looking at Table 7.8 it is readily seen that in terms of fitness eukaryotes and prokaryotes each have advantages and disadvantages so that *the best solution for total optimal energy capture is coexistence assisted by cooperation not competition.* If life could have started from a situation where there had been an ample supply of sufficiently reduced basic small molecule materials and energy and no reduction but only molecular combination was required, it may well have been that anaerobic prokaryotes would have remained exclusively successful. They are still in abundance on the Earth. There would have been no need to reject oxidised materials, and increase in the number of compartments would have had little advantage since the observed chemically necessary syntheses (see Chapter 4) are simply

TABLE 7.8

COMPARISON OF SINGLE CELL EUKARYOTES AND PROKARYOTES

Advantage of Eukaryotes

1. Flexible membrane
2. Large size allowing uptake of particles including prokaryotes using (1)
3. Digestion in internal compartments, phagocytosis or endocytosis
4. Vesicular compartment control over export, exocytosis, of enzymes, poisons and constructed minerals
5. Separate management of radical photo- and oxygen chemistry in organelles and of peroxide chemistry in peroxysomes to decrease risks during energy uptake. Glycosylation of proteins in the endoplasmic reticulum for protection
6. Protected DNA by histones, etc
7. The beginnings of sophisticated sensing devices
8. Signalling to inside to give notice of environment, amplified by stores in vesicles, especially Ca^{2+} messengers
9. Internal filaments to allow management of vesicles and the membranes
10. Increased external protective layers, including minerals
11. Sexual reproduction with protection from horizontal gene transfer

Disadvantages of Eukaryotes

1. Unprotected membrane of some cells, compare walls (bacterial) and rigid lipids (archaea)
2. Slow response of DNA, slow mutation rate; slow adaptation on exposure to changed environment with new nutrient
3. Compartments introduce complexity of controls limiting ability to do everything in one organism
4. Slow reproduction, increased risk of exposure to damage

cooperative and additive with little chance of destructive interference. Under the real circumstances of the primitive Earth, reduced carbon materials could only be made in cells from the moderately higher level of oxidation of the then existing compounds of the environment and oxidised materials and potentially poisonous ions were rejected. Initially, this could have been sulfur from H_2S (essential for reduction), which was of limited availability, and ions such as Na^+, Cl^-, Mn^{2+}, and Ca^{2+}. The later rejection of oxygen, to get the abundant source of hydrogen from water and to maintain sulfur in circulation, was imperative to avoid reducing the extent of cytoplasmic chemistry of any primitive cell. Oxygen release gave rise to waste as various oxidised elements in the environment (pollutants), where waste relative to reduced cell material is a new energy store. Useful re-employment of waste is best managed to avoid interference between oxidation and reduction pathways for any given cell if it is compartmentalised, that is as seen in eukaryotes. At the same time heavy (poisonous) metals had to be rejected, but also used in compartments or in screened reactions. Also consider the fact that prokaryotes create debris of large particles (even living or dead whole cells), which could only be digested by external hydrolysis by prokaryotes, while novel large cells, eukaryotes, have the extra advantages that they can ingest and digest internally such particles, especially effectively in acidified compartments. Finally, eukaryotes could carry special new syntheses, such as –S–S– bond formation and glycolytic attachments again in compartments. Each compartment could have a special quota of catalysts including metals. However, as explained the larger cells of eukaryotes use compartments at a price – slower rates of reproduction due to complexity of organisation and slower adaptability. The difficulties of organisation are relieved somewhat by removing the need to produce all essential basic chemicals and by more extensive controls. Returning to the need for extra controls, eukaryotes required increase in information transfer involving now particularly the rejected calcium. Relief from the complications of synthesis was also to be found in utilising products from the digestion of prokaryotes. The eukaryote can then do without genes for many products needed by both prokaryotes and eukaryotes, and which have to be made from very simple environmental chemicals. The free prokaryotes became general sources of energised chemicals and the suppliers of many small molecules and ions, for example, mineral elements and coenzymes, some of which are called vitamins, to eukaryotes, while the prokaryote derived organelles inside eukaryotes generated much of the energy of all eukaryotes! This naturally led to extensive symbiosis or closer combined existence of free prokaryotes and eukaryotes, not so much competition, as the prokaryote population also lived off extra debris from (dead) eukaryotes, and eukaryotes utilised prokaryotes as above. We shall see that this process of evolution, mutual cooperation and the increase in organisation are general novel features in eukaryotes, leading together to a more successful exploitation of the environment and greater energy retention in chemicals, that is development of the ecosystem. An example is seen in the huge coral banks all over the Earth, which are a symbiotic combination of animals and algae separating photosynthesising activity from scavenging. The total outcome is not contradictory to the fact that when resources are limited, competition may come into play especially between similar organisms and note that eukaryotes consume prokaryotes yet prokaryotes came to live inside eukaryotes. The competition can assist the drive to greater overall effectiveness of the whole ecosystem in energy degradation. Population genetics then describes the way balance is achieved in the ecosystem.

Now even while unicellular eukaryotes developed they faced ever-increasing concentrations of especially zinc and copper as sulfides became increasingly oxidised to soluble sulfates. These two elements together with further controlled oxidation of the environment became particularly instrumental in assisting the further advance of the handling of chemicals outside the cytoplasm. We must remember that highly organised compartmentalised systems are able to utilise energy in diverse chemicals more effectively than those which are poorly organised, and especially so if the latter act as sources of simple chemicals. We stress as we have done at the end of Chapters 5 and 6 that it is the environmental change, more highly oxidised conditions including liberation of metal ions, which generated, even enforced, evolution along one direction towards greater energy retention, through greater possible varieties of chemistry, of necessity in more and more organised compartments. We must not be surprised that this creates the possibility of larger and larger organisms that can reach out into and "enclose" the environment. This is not to say that the role of prokaryotes diminished, it may well have increased considerably, for example by assisting large eukaryotes, but the total effectiveness of energy retention in organisms as a whole increased with the coming of larger and larger aerobic eukaryotes. We note that single-cell small eukaryotes also continued to evolve as did prokaryotes. The eukaryotes remain of little or no consequence in the anaerobic zones. (For a summary of the evolution of eukaryotes as chemotypes see the end of Chapter 9). In summary, the role of the changing inorganic elements in the environment played a large part in the evolution of an ever-more effective ecosystem of increasing rate of energy degradation, rate of thermal entropy production by a variety of chemotypes. Meanwhile we observe that there is only one set of fundamental thermodynamic features in all organisms as set out in Table 4.11. There is a drive to increase energy flux utilising all available elements while becoming element neutral, that is utilising all waste, while moving towards a complete cycle (see Chapter 3). We show the rise of the eukaryotes within the cone of ecological evolution in the figure on the cover of this book.

Further Reading

BOOKS

Caporale, L. (2003). *Darwin in the Genome.* McGraw-Hill, New York
Carafoli, E. and Klee, C. (1999). *Calcium as Cellular Regulator.* Oxford University Press, New York
Corning, P.A. (2003). *Nature's Magic: Synergy in Evolution and the Fate of Mankind.* Cambridge University Press, Cambridge
Harold, F. (2001). *The Way of the Cell.* Oxford University Press, New York
Jablonka, E. and Lamb, M. (1995). *Epigenetic Inheritance and Evolution.* Oxford University Press, New York

Japp, J. (1994). *Evolution by Association – A History of Symbiosis*. Oxford University Press, Oxford
Mann, S. (2001). *Biomineralization: Principles and Concepts in Bioinorganic Materials Chemistry*. Oxford University Press, Oxford
Margulis, L. and Sagan, D. (1995). *What is Life?* Simon and Schuster, New York
Margulis, L. and Schwartz, K.V. (1998). *Five Kingdoms* (3rd ed.). Freeman and Co, New York; W.H. Freeman and Company, New York
Margulis, L. (1998). *Symbiotic Planet*. Basic Books, New York
Maynard-Smith, J. and Szathmáry, E. (2000). *The Origins of Life – From the Birth of Life to the Origins of Language*. Oxford University Press, Oxford.
Postlethwhait, J.H. and Hopson, J.L. (1992). *The Nature of Life*. McGraw-Hill Publishing Company, New York
Stanley, S.M. (2002). *Earth System History* (3rd printing). W.H. Freeman and Company, New York
Thompson, D.W. (1961). *On Growth and Form* (abridged edition) J.T. Bonner (ed.); Cambridge University Press, Cambridge (reprinted 1988)
Turner, B.M. (2001). *Chromatin and Gene Regulation – Molecular Mechanisms in Epigenetics*. Blackwell Science, Oxford

PAPERS

Andersson, S.G.E., Karlberg, O., Canbäck, B. and Kurland, G.G. (2003). On the origins of mitochondria: a genomic perspective. *Philos. Trans. Royal Soc. B. 358*, 165–177
Calderone, V., Dolderer, B., Hartmann, H.-J., Echner, H., Luchinat, C., Del Bianco, C., Mangani, S. and Weser, U. (2005). The crystal structure of yeast copper thionein. *Proc. Natl. Acad. Sci. USA, 102*, 51–56
Dacks, J. and Ford-Doolittle, W. (2002). Novel syntaxin gene sequences from Giardia, Trypanosoma and algae: implications for the ancient evolution of the eukaryotic endomembrane system. *J. Cell Sci., 115*, 1635–1642
Embley, T.M., and Hirt, R.P. (1998). Early branching eukaryotes? *Curr. Opinion Genet. Dev., 8*, 624–629
Embley, T.M., van der Giezen, M., Horner, D.S., Dyal, P.L. and Foster, P. (2003). Mitochondria and hydrogenosomes are two forms of the same fundamental organelle. *Philos. Trans.: Biol. Sci., 358*, 191–203
Falkowski, P.G., Katz, M.E., Knoll, A.H., Quigg, A., Raven, J.A., Schofield, O. and Taylor, F.J.R. (2004). The evolution of modern eukaryote phytoplankton. *Science, 305*, 254–260
Freeland, S.J. and Hurst, L.D. (2004). Evolution encoded. *Sci. Am., 290*, 56–63
Hartman, H. and Fedorov, A. (2002). The origin of the eukaryotic cell: a genomic investigation. *Proc. Natl. Sci. USA, 99*, 1420–1425
Keeling, P. (1998). A kingdom's progress: Archaezoa and the origin of eukaryotes. *BioEssays, 20*, 87–95
Knight, J. (2004). Giardia – Not so special after all? *Nature, 429*, 236–237
Knoll, A.H. (1992). The early evolution of eukaryotes: a geological perspective. *Science, 256*, 622–627
Knoll, A.H. (1999). A new molecular window on early life. *Science, 285*, 1025–1026
Kurner, J., Frangakis, A.S. and Baumeister, W. (2005). Cryo-electron tomography reveals the cytoskeletal structure of *Sprioplasma melliferum*. *Science, 307*, 436–438
Linder, P. (2004). The life of RNA with proteins. *Science, 304*, 694–695
Margulis, L., Delan, M.F. and Guerrero, R. (2000). The chimeric eukaryote. *Proc. Natl. Acad. Sci. USA, 97*, 6924–6929
Martin, W. and Russell, M.J. (2003). On the origin of cells. *Philos. Trans. Royal Soc. London, B 358*, 59–85 (and references therein)
Matzke, M. and Matzke, A.J.M. (2003). RNA extends its reach. *Science, 301*, 1060–1061
Nagel, G., Ollig, D., Fuhrmann, M., Kateriya, S., Musti, A.M., Bamburg, E. and Hegeman, P. (2002). Channel rhodopsin-1, a light-gated proton channel in green algae. *Science, 296*, 2395–2398
Pennisi, E. (2004). The birth of the nucleus. *Science, 305*, 766–768

Perkins, G.A., Reuken, C., Martone, M.E., Young, S.J. and Frey, T. (1997). Electron tomography of neuronal mitochondria three dimensional structure and organisation of crystae and membrane contacts. *J. Struct. Biol.*, *119*, 260–272

Roper, A.J. (1999). Reconstructing early events in eukaryotic evolution. *Am. Nat.*, *154*, S146–S163

van Niftrick, L.A., Fuerst, J.A., Sinninghe Damste, J.S., Kuenen, J.G., Jetten, M.G.M. and Srauss, M. (2004). The anammoxone: An introcytoplasmic compartment in anammox bacteria. *FEBS Microbiol. Lett.*, *233*, 7–13

Vellai, T. and Vida, G. (1999). The origin of eukaryotes: the difference between prokaryotic and eukaryotic cells. *Proc. R. Soc. London, B 266*, 1571–1577

Veis, A. (2005). A window on biomineralisation. *Science*, *307*, 1419–1420

Vogel, G. (1997). Searching for living relics of the cell's early days. *Science*, *277*, 1604

Waller, R.F., Keeling, P.J., van Dooren, G.G. and McFadden, G.I. (2003). Comment on a "green algal apicoplast ancestor". *Science*, *301*, 49a

Wickelgren, I. (2003). Spinning junk into gold. *Science*, *300*, 1646–1649

Williams, R.J.P. and Fraústo da Silva, J.J.R. (2004). The trinity of life: the genome, the proteome and the chemical elements. *J. Chem. Edu.*, *81*, 738–749

Woese, C.R. (2002). On the evolution of cells. *Proc. Natl. Acad. Soc. USA*, *99*, 8742–8743

CHAPTER 8

Multi-Cellular Eukaryote Chemotypes (From One Billion Years Ago)

8.1. Introduction

Many of our readers will be chemists unfamiliar with the later steps of the evolution of complex organisms, which led to the vast range of multi-cellular living forms that we see around us, all dependent on aerobic conditions. We have therefore divided this chapter into three parts. Part 8A will give a simple introduction to their characteristics for those who know little of such organisms. We then show in Parts 8B and 8C how their evolution was dependent on functions of newly available elements and of new states of those already available elements, and on the more effective use of energy. In this way, elements and energy in these organisms contribute to the whole of the ecosystem's development towards optimal energy flux through chemicals. Part 8B will describe the element and energy content of whole organisms while Part 8C will describe those of the well-defined separate organs in these organisms. This division is made since although it was relatively simple to explain the slow changes in unicellular prokaryotes and eukaryotes, the changes and increases of multi-cellular eukaryotes of different but related chemotypes over a relatively very short period of evolution were much more complicated within compartments requiring more elaborate description. It is even difficult to place in time the initial appearances of these organisms, so we have estimated one of around 1 billion years ago though very clear evidence only exists for animals from nearly 500 million years later, when a vast range of complex fossils were deposited (Table 8.1 and Figure 8.1). Note that just before this period there was a considerable rise in O_2 and a change of composition of the sea over some half a billion years. We believe that this environmental change induced the biological evolution of these new chemotypes, while it also helped to simultaneously advance prokaryotes and unicellular eukaryotes, but to a lesser degree. The description of plant and of fungal physical evolution, two of the three major chemotypes, is somewhat easier to give than that of animals as different functions in separate organs in one organism became organised. There is not a vast range of organs and senses in plants and most structure is visually obvious (see Section 8.2). Similarly, multi-cellular fungi derived from yeast do not present great difficulty in visualising their evolution (see Section 8.3). Multi-cellular animal evolution is not so readily described since there are invertebrates and vertebrates of all manner of shapes and sizes and internal structures (Figure 8.1), which appeared relatively very rapidly. We shall therefore give first an introduction to multi-cellular plants and fungi. Here and indeed throughout this book we are not concerned with the distinctions between different species, but only with those between chemotypes or major physical/chemical divisions within them, as defined in Chapter 4. Our objective is to see the physically obvious evolution in terms of novel chemical sophistication, in kind, in space and in organisation, and in the absorbed energy increase in them. It is likely to be impossible to give a reason for the appearance of so many species when there are about one million insect species

TABLE 8.1

ApproxImate Origin Time of the Major Multi-Cellular Plant and Animal Groups[*]

Organism Group	Time of Origin
Marine plants (algae)	>800 million years ago
Marine invertebrates	>570 millions years ago
Fish	505 million years ago
Land plants	438 million years ago
Amphibians	408 million years ago
Reptiles (birds)	320 million years ago
Mammals	208 million years ago
Flowering plants (angiosperms)	140 million years ago
Homo sapiens	<100 thousand years ago

[*] See also Table 1.8.

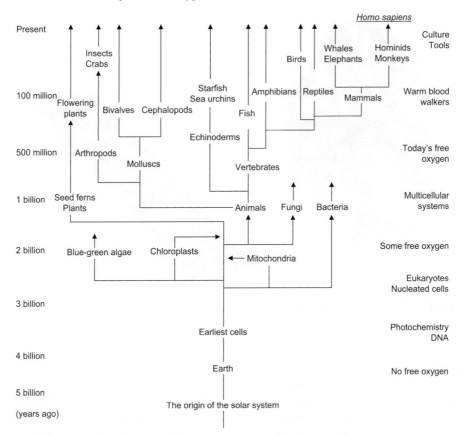

FIG. 8.1   A general scheme of cellular evolution leading towards organisms of today indicated by arrows at the top of the figure.

alone today. We shall describe the fully formed physical nature of the organisms before we discuss their growth from single cells, which is a separate, strikingly novel feature of multicellular organisms (see Section 8.6).

8A.   The Morphological Nature of Multi-Cellular Eukaryotes

8.2.   The Evolution of Multi-Cellular Plants

The step from a multitude of unicellular to an even greater diversity of multi-cellular plant organisms took place first in the sea among algae, e.g. the groups of *Ulva* and *Spirogyra* (see Plate) though many large algae are multi-nucleate

a)   b)

PLATE   Some examples of green algae: (a) *Ulva lactuca*; (b) *Spyrogyra*

rather than multi-cellular. Further development took the form of an increase in the number of different compartments, organs, e.g. root and leaf, each of which contain large numbers of cells of the same kind. Somewhat strangely, it is obvious in the sea and on land today that even very small localities of a few square metres house very many different multi-cellular plant species, which are roughly equal in survival strength, i.e. in their capacity to take up nutrients and to absorb light. Of course, the variety of protective devices of multi-cellular plants, bark, thorns, poisons, etc., help survival, as does the means of dispersal of seed, but it might have been thought that the complications of multi-cellular plants would have greatly limited the number of species. In fact the striking feature is that the coexistence of and cooperativity between many plant species, not competition, appears to dominate, e.g. only some chemotype groups are associated with N_2 fixation. Even the introduction of "better" ways of generating progeny such as flowering meets the observation that earlier spore producers, such as ferns, horsetails and mosses, abound together with the flowering plants. It may be argued that each has a special environmental niche but that would be very hard to prove. These observations make the general point that we must accept, that this multitude of plant species, and likewise those of fungi and animals, arose by chance (see Section 4.18) and, as long as each one had roughly the same survival strength in a given environment, it has come to stay even over a billion years in some cases. As a group, all multi-cellular plants are superior in energy capture to organisms preceding them (Figure 8.2) but they have slow, complex development and reproduction so that many single-cell eukaryote plants and photosynthetic

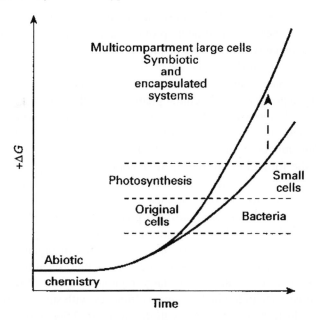

FIG. 8.2 The increase in absorbed energy per cell. Fast reproductive and adaptive rates could be maintained for advanced bacteria but not for the larger cells and certainly not for eukaryotes. This gave rise to a balanced set of organisms.

bacteria (which may have appeared up to 2.5 billion years earlier) have survived, evolved and thrived to live alongside multi-cellular plants. We turn to the detailed reason for this in Section 8.19. It seems that once a particular chemotype appears, it will be able to survive even if some species of it become extinct. Of course, aids to survival of organisms of lesser complexity are faster reproduction and greater adaptability. We shall therefore treat the multi-cellular plant coverage of Earth as developing in relationship with more primitive and other advanced kinds of organisms as a whole in order to show the general direction of evolution. We maintain the overall stance that in an evolved ecosystem the more complex organisms will utilise supplies from the more simple and primitive, and vice versa. Synergism is of the essence of evolution of systems (see Margulis and also Corning in Further Reading). We shall insist of course that chemotypes themselves, not their many species, arose of necessity with a chemical direction, and not by random selection (see the ecological cone on the cover of this book).

Plants developed from photosynthetic bacteria. As we have explained in Chapters 5 and 6, bacteria could evolve on the top surface of the Earth by increasing their ability to capture light (energy) and by obtaining and using more effectively 15–20 elements from seawater, later the seabed and land, and with three or four from the atmosphere, while utilising novel chemistry. The need to adapt to oxygen later forced the development of compartments already seen in "differentiated" cyanobacteria (see Section 6.9) but most clearly in unicellular eukaryote plants. As we have argued, these compartments provide more efficient, organised ways of carrying out separately a multitude of protected, different chemical activities. In what follows, we explain that once the presence of oxygen allowed the possibility of *multi-compartment* single eukaryote plant cells (see Chapter 7), it was even more efficient to go on to multi-cellular plants especially since the increasing presence of more and more oxidised chemicals in the environment can be used to create integral, helpful, separate parts of these new constructs, which frequently developed into tubular form (Figure 8.3). Intermediate steps in organisation, as stated, were the formation of controlled "colonies" of cells as seen in the green algae, *Volvox* (Figure 8.4). The further step to organs, groups of more closely identical, differentiated cells, is a move to increased overall efficiency in growth in the sea with more defined structure and with greater utilisation of space in the air, waters and underlying mud often requiring increased size. The increase in efficiency arises from the effectiveness of types of cells in organs, e.g. roots and leaves, for their separate functions in these particular separated environments together with their cooperativity over large regions of space. (Note here that we must not confuse such individual organism efficiency with survival strength, which includes reproduction, and also not with the overall efficiency of the ecosystem.) However, the real strength of this advance was not to come until they inhabited sedimentary land, linking again the geochemical and biochemical evolution. It was the earlier erosion of the land by rain, which brought about finely divided fertile minerals and allowed primitive plants to live and die there and thus form even more

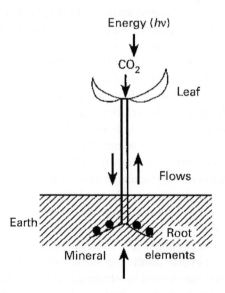

FIG. 8.3   The basic plant structure receives energy from the sun and carbon from the air. All other elements (15–25) come from the Earth as inorganic minerals or from bacteria ($N_2 \rightarrow NH_3$), shown as filled circles.

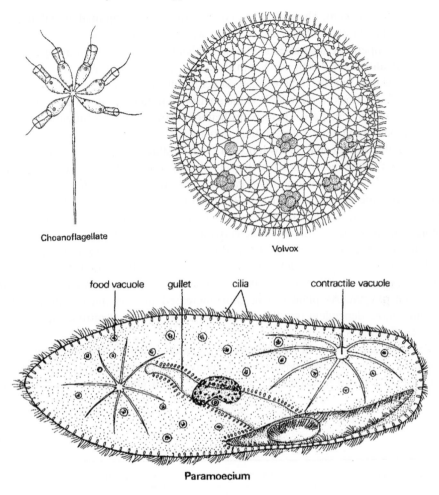

Choanoflagellate

Volvox

Paramoecium

FIG. 8.4 *Volvox* shows different types of association of cells into loose organisation. *Choanoflagellates* are organisms with connected parts while *Paramoecium* is a highly organised, ciliated animal protozoan.

fertile soils at the river and sea edges. Once having invaded the land, the evolution of new multi-cellular plants is a development of increasingly better physical devices for the capture of energy, light, while turning CO_2 and H_2O largely into saccharides, such as cellulose, often in huge structures. The resultant growth was towards an optimal height and an absorbing spread-out surface of leaves limited by geological factors and hours of light for energy capture. Increased or changed mineral element availability from the Earth was utilised via specialised, extensive roots. In order to maintain the organs in a unity, the plants developed not only physical structures, stems, but internal fluids in them carrying material and energy and distributing them. The final physiological limitation on optimal growth in trees is purely the ability to obtain and to pump water and nutrients to about a hundred metres. Temperature also affects growth, so that tree lines are seen in high mountain areas and as the North and South Poles are approached. It appears that extensive grasslands survive where other larger plants fail due to their inability to survive fluctuations in climate. It is the environment that decides the form of surviving plants in fair part. There is no better illustration of the limitations of the environment upon gene potential than in plant life.

Given the present state of evolution of plants (Table 8.2) we believe that treated in isolation from other organisms, they are already tending to meet the *steady-state condition of optimum energy (light) capture and use in chemical flows*. Of course, plants continue to pump out waste oxygen that is not utilised fully by them (see the next sections), and which, relative to reduce materials, is a part of the total energy storage in unstable chemicals in the ecosystem. Another unstable store of energy lies in the ion gradients of Na^+, Cl^- and Ca^{2+}, also not much used by plants. The debris from plants are widely distributed, are not easily re-used by them since plants are static and they form a part of the reduced materials in an energy store opposite oxygen. As plant organisation increased in size and complexity they became more dependent on other species, e.g. for sources of nitrogen and minerals (see Section 8.3).

We shall see that the development of these multi-cellular from single-cell structures of plants, together with other developments, were only possible through the use of a changed chemical environment, which allowed the required novel

TABLE 8.2

EVOLUTION OF PLANTS

Types and Nature	Examples	Approximate Number of Species
Algae		
Rhodophyta	Red algae, many fan like or filamentous	4,000
Phaeophyta	Brown algae; includes giant kelp and *Sargassum*	1,500
Chlorophyta	Green algae; includes *Volvox, Ulva* and *Chlamydomona*	7,000
Non-vascular plants		
Bryophyta	Mosses, liverworts, hornworts	24,000
Vascular plants		
a) Seedless plants Ferns and allies	Ferns, horsetails, club mosses	13,000
b) Seed plants		
Gymnosperms	Conifers, cycads, ginkgos, gnetae	700
Angiosperms		
Monocots	Lily, corn, onion, palm	50,000
Dicots	Apple, bean, rose, daisy, etc.	235,000

chemistry, mostly outside the cell cytoplasm, and led to organisation by quite new messengers, the synthesis of which also depended in fair part on the new chemicals from the environment (see Table 8.24). Remember that the unavoidable drive, as we see it, is towards optimal retention of energy in chemicals, that is of new chemotypes working with previous ones in an energised, largely cyclic ecosystem. Plants absorb energy efficiently but fail to complete a material cycle or an efficient degradation of energy. Their weaknesses are those of all eukaryotes, especially the multi-cellular variety, in their slow reproduction rate, their almost negligible adaptability and their complexity. They are not competitors for smaller unicellular eukaryotes or prokaryotes but to a large degree cooperative with them.

8.3. The Evolution of Multi-Cellular Fungi

Yeast cells, which we have described in Section 7.2, are unicellular but are related to larger fungi generally, mushrooms for example, which appear to be multi-cellular and have internal structures and distribution systems. They probably developed at the same time as multi-cellular plants; both are immobile. In fact these fungal "cells" are often fused, so that they are multi-nucleated rather than multi-cellular. Such fused "cells", organised by effective, internally controlled diffusion, not physical diffusion across membranes, are seen as branched filaments called hyphae. The head of a mushroom is in fact a huge bundle of cross-linked filamentous hyphae and there do not seem to be many different types of cell in it. All fungi are either dependent on plants or sometimes on animals for food and energy (C/H/O/N compounds). In association with roots of plants in particular, they frequently form mutually beneficial combinations since they are very good at scavenging elements from minerals, which they make available to plant roots in exchange for carbon compounds, a form of symbiosis. In particular, many trees depend on fungi for the supply of ions from the soil. A second example of good scavenger systems of fungi are the lichens on rocks, which are a combination of a fungus and a photosynthetic (plant) bacterium growing symbiotically. Note that through this association some fungi are sometimes very dangerous to plants, especially trees – a good example is the honey-fungus. Any observer will note how frequently fungi are associated with rotting wood (of trees) and their ability to use this source of nutrients can become an attack on live plants. Ignoring this "evil" that comes from their selfish, apparently "good intentions", fungi are a major set of organisms in the recycling of dead (rotting) plant debris by aerobic metabolism. In this way they contribute to the recycling of waste, organic chemicals and oxygen, with degradation of energy (heat production) as well as increasing the retention of the energy taken up by plants by supplying them with minerals. They are then an important part of the cyclic mineral/organic chemistry of the ecosystem, but notice that they too required the changed environment to make their essential structures and chemicals (see Part 8B). Fungi are rarely important for animals and some are noxious and may cause disease, e.g. plasmodia cause malaria. Mankind, however, uses some fungi for food and to obtain antibiotics from others, such as penicillin. Finally, fungi, like plants, can only scavenge very locally.

8.4.   The Evolution of Multi-Cellular Animals

The further step in evolution of organisation is a very logical one of improving the recycling of "waste" chemicals by other forms of life – the multi-cellular animal eukaryotes, which became purposefully mobile. Food for animals eventually or immediately comes from higher plants, which developed first, and indirectly from them by consuming other animals. As sufficient food does not come to them easily at all they need senses in order to direct searching and catching together with a set of organs, muscles, to move rapidly. They are scavengers *par excellence* and all animals therefore need a digestive system able to consume as huge a variety of sizes and kinds of food as possible. It may be that the modes of digestion for different chemicals, e.g. celluloses and bulk protein, mean that once again different animals manage best only one digestive style. Thus, ruminants eat grasses while the cat family eats animals. In effect they are different chemotypes for the reasons of classification already given. Immediately, it is seen that to achieve these activities in a coordinated animal, the whole animal organism has to be of some size and highly organised, much more than a plant, and in fact the more organised the better. Hence, their evolution is one of a succession of increasing numbers of organs for improving and organising the above activities including mobility and also ways of passing information between organs; this information includes as much knowledge of the environment as possible. Of course, this helps them to escape from predators, other animals, and in the capture of yet others for food. We shall show in Section 8.9 that all this development was made possible by utilising chemical changes of cell surroundings, but intake of food had also to be increasingly controlled in internal and extracellular fluids so as to distribute it between cells.

We can believe that multi-cellular invertebrate animals evolved from colonies of cells as seen in corals, for example. The development is believed to have started with single-cell choanoflagellates (see Figure 8.4) and has now reached the stage of coordinated human communities (see Chapters 9 and 10). It has a huge potential for variation in species within chemotypes. Perhaps the first true organs to develop were digestive, separate from motility organs, and the first animals, not motile, may well have been just a set of cells in a rough cusp, which could be activated to catch, trap and digest food from waters, which flowed passed them. We see this in the cavities of early invertebrates, sponges and medusa (see Figure 8.5), which existed 500×10^6 years ago. They required only the senses of touch and smell (taste), already found in more primitive species (Sections 5.17 and 7.11). "Muscles" for catching food developed very quickly so that the (stationary) animals could actively pump water containing food through themselves while it was digested. Chemical sensing of the water content was much like that in earlier eukaryotes. About 500 million years ago, invertebrates evolved into more sophisticated pumping systems, that is muscular,

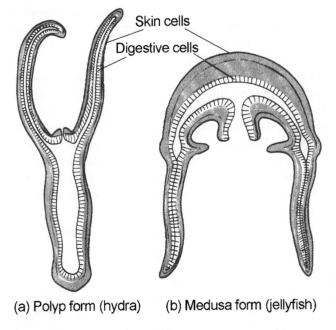

Skin cells

Digestive cells

(a) Polyp form (hydra)   (b) Medusa form (jellyfish)

FIG. 8.5   The polyp and medusa body forms characteristic of phylum Cnidaria are structurally similar. (a) The polyp form as seen in *Hydra*. (b) The medusa form is basically an upside-down polyp. The mouths are a primitive muscle system. Note the separation of organs.

worm-like forms, which were now bodily mobile with advanced senses so that they could search using additional skeletal muscles. Usually these animals have tubular form but there are unusual examples, e.g. in the "black smokers" (see Chapter 11). The muscles were associated directly with two types of internal coordinating messengers for response to sensing, including new and better uses of light (eyes), of smell (noses) and of sound (ears). Probably the first such internal messengers were certain chemical compounds, some already present in unicellular eukaryotes, and we know that others, many called hormones or transmitters, were present in such animals as early worms. The second kind of message system, the appearance of a nerve system, is not found in the earliest invertebrates, but is seen first in *Aplysia* (sea slugs) and some flat worms, where it is only a network without real central coordination. The first signs of central (brain) coordination of muscles are in the nematode worms, which have linking nerves in a ring near to the head where sense organs are also located. These developments of senses and muscles allowed directed movement in the search for food and avoidance of danger. At this time, further increasing awareness of the environment arose with the development of more sophisticated eyes and hearing equipment, special organs connected via the brain to an extensive series of different muscle types controlling separately posture and locomotion. We shall describe later the chemistry of the evolution of the complicated nerves and brain in

Chapters 9 and 10 while in this chapter we concentrate on *cell–cell structures and their organic chemical communication* and the very simple nerve networks between senses and muscles. At the same time a complicated series of organs became involved in intake, synthesis, distribution of material and waste excretion so as to supply suitable material with energy to the whole body and remove excess chemicals. Probably to protect and strengthen the structures, the invertebrates developed external shells but it is only with the arrival of vertebrates, animals with bones, that great internal structural strength with mobility evolved (see Figure 8.6 and Table 8.3).

We need to point additionally to the evolution of special organs, glands, which have stored chemical message systems to help coordinate activities and are associated with nerves and the brain (Figure 8.6 and Table 8.3). Glands are advantageous in organisation as chemical substations connected to different remote parts of the

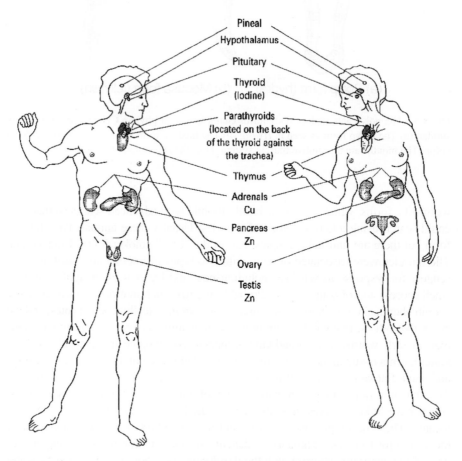

Pineal
Hypothalamus
Pituitary
Thyroid
(Iodine)
Parathyroids
(located on the back
of the thyroid against
the trachea)
Thymus
Adrenals
Cu
Pancreas
Zn
Ovary
Testis
Zn

FIG. 8.6   The muscular structure and endocrine glands of man with some specially required elements indicated. (Adapted from Purves *et al.*, Life – the Science of Biology (5th ed.). Associates, Inc. and W.H. Freeman, Sunderland, MA.) Note that organs such as liver are rich in other elements, for example, iron. The large part of the body is linked by internal bones.

TABLE 8.3

THE MAJOR ORGANS OF ANIMALS

Organ	Function
Muscles (heart, etc.)	Digestion, fluid circulation, motion, mechanical work
Stomach/intestine	Major zones of digestion
Liver	Major zone of synthesis
Kidney (bowel)	Major external rejection ion/water balance
Reproductive organs	Reproduction
Glands	Chemical controls
Brain (Chapter 9)	Electrical control
Sense organs	Environmental detectors

TABLE 8.4

NOVEL FLOW IN MULTI-CELLULAR ORGANISMS

Flow	Example of Function
General extracellular fluid of animals	Bloodstream containing special cells, erythrocytes Lymph containing special cells for protection
Cerebrospinal fluid of animals	Brain and spinal cord chemical communication in animals
Plant plasma	Movement of chemicals up and down stems and note phloem and xylem

Note: A major problem for an aberrant growth, a cancer, is the need for blood flow to be redirected to it.

organism and giving rise to different, not very fast, chemical responses. This division of instructions with different time constants through different chemical transmission (and feedback) mechanisms is central to a highly organised system of many activities. It illustrates once again the efficiency benefit of multi-compartments, which could only develop with a systematic evolution in size, but one should be aware of the disadvantages of such complexity. Note that all the separate compartments are linked by internal fluids (see Table 8.4). Most recently, organisation in animals introduced control of body temperature, which fixed reaction rates (see Chapter 2).

The general thesis we maintain is that increase in diversity of functional demands, the functions being only made possible by utilising new environmental chemicals (see below), is met with division of activity in different compartments, repeatedly for good organisation reasons. For example, muscles are a large part of digestion, posture, circulation of fluids and motion while other organs, liver for example, control synthesis and metabolism generally. The glands are the main source of the slow chemical information transfer while nerves are the source of the fast electrolytic, physical information transfer. All these messages are transmitted outside, that is between cells, but had to connect to the already present communication networks internal to cells. This complicated physical evolution of animals cannot be described further in this book but can be found in detail in biology textbooks (see Further Reading).

Here, we must reflect on the remarkable and rapid development of the three major groups of chemotypes, plants, fungi and animals as cooperating organic chemical synthesisers in a newly evolved environment. Though separated in type as are the parent unicellular organisms, they all developed multi-cellularity together in time, which is surely indicative of *the enforcing underlying environmental chemical direction of change* and we shall see this commonality later in this chapter. Given these three chemotypes, we observe an increasing ability to absorb light energy and to utilise, synthesise, and then recycle the chemical elements in an efficient way. All three are parts of the dynamic flow of energy, intake, store and degradation, in very similar chemicals in all three, creating with the environment (which they help to create) a strong approach to optimal energy throughput from light to heat. We need to see how the novel chemical environment allowed the development of this efficient ecosystem but first we need to note some extra features within each of the three major chemotype divisions. Notice that the development of all these organisms requires changes in DNA and morphology but that the increased ability of animal organisms to respond to events in the environment via metabolic responses quickly does not require the intervention of genetic (DNA) operations (see below). The reader should see that step by step the molecular machinery of prokaryotes has become the macro-machinery of multi-cellular organisms and finally the external machinery of mankind with a lessening of genetic controls. Finally, we repeat the closing sentence of Section 8.2. Large multi-cellular eukaryotes, now animals, are not competitors of unicellular eukaryotes or prokaryotes but are cooperative with them in the total ecosystem.

8.5.   Diversity within the Major Chemotypes

We must not leave the impression that all plants, fungi and all animals within each of these three separate broad groups of multi-cellular chemotypes have identical systems of organs. It is evident that there are many different kinds within each group and there is some chemical variation. For example, there are the plants with broad leaves and those with grass-like structures, the difference between them being noticeable in their uptake of the element silicon. The uptake of silicon goes back in evolution to unicellular eukaryotes and among early plants to horsetails, spore producers. There are different groups of animals, as we have mentioned already, some of which specialise in consuming plants (sheep) and they require symbiotic bacteria, using nickel, to digest cellulose, while others (cat family) live mainly off other animals, and cannot digest cellulose and have little use for nickel. Again some animals are invertebrates, often with a calcium carbonate shell, while others are vertebrates with a varying calcium phosphate bone content. In this book we will not attempt to describe or analyse this complexity within the three major groups of chemotypes since compartmental chemical analysis of them is incomplete. We also observe organisms of very different size and shape, morphology, within one chemotype. In Section 3.7, we showed that size and shape were decided by the total internal system, limited in its material and energy flow by external boundary conditions, that is including fields (e.g. see clouds in Section 3.4). These same limitations apply to the growth of both unicellular and multi-cellular eukaryotes, but their sizes and shapes are now increasingly controlled *by the external factors,* e.g. the temperature zone as well as by *internal and genetic information.* Each cell in multi-cellular organisms is controlled, or stressed, by its neighbours so that *the cell/cell environment is informative* as well as is its DNA and the external environment of the organism. Under stress, the body shape, even of humans, can be managed, but this is more obviously true of trees in the prevailing climate and winds but for all these organisms the internal information from DNA and feedback systems still provide the major control. We refer readers to books on genetics on the one hand and to the outstanding book by Thompson "On Shape and Form" (see Further Reading) on the other. We shall observe a novel effect of the environment on characteristics, those of behaviour, when we describe humans in Chapter 10. (As an aside we have to note that many animals consume others, as well as plants, but overall this "competitive" activity has no noticeable effect on the number of chemotypes.)

8.6. Growth of Plants and Animals from Single Cells

In the previous chapters, we have not had to consider growth in any detail but in the description of multi-cellular organisms it is a dominant feature since they start from a single cell, which is followed by differentiation of cell types. We can only touch on this topic giving a very elementary description and then a way of thinking about growth, which allows us to connect it to our major theme – the development of the use of chemicals and energy within evolution as the separation of oxidised and reduced chemicals increased with increasing use of space and organisation.

A plant grows from a single fertilised cell surrounded by food stores in a seed or a spore. (Fungal growth is somewhat similar and we shall not describe it.) The growth is easily seen as the seeds or spores send out differentiated cells in the shoot and the root before the first green leaves (cotyledons) appear. As cells multiply, connective tissue, communication, and differentiation develop with a special layer of cells near the surface, which gives rise to both a dead "skin" or bark and a circulation system. The processes are coordinated and linked to light, temperature and mineral uptake by chemical messengers (see Section 8.13). The growing plant is "informed", therefore, by both external conditions and by the genetic machinery involving DNA. Further growth is largely in a pattern of few compartments before flower and seed or spore formation provides a renewed cycle. The pattern of development is largely, tubularly *linear* up and down with periodic *planar* spread to gain access to requirements, especially light. Note that leaves have many purposes – light capture, photosynthesis, transpiration and poison rejection, while roots collect mineral elements, act as fixed supports and contact essential bacteria and fungi. The resultant combination of particular *concentrations* of organic and inorganic chemicals, adjusted with time, is essential for growth and reproduction. The control of these processes by chemicals, hormones, is quite new in evolution, and while plants and fungi use hormonal messages they have few sense organs and virtually no extremely fast responses using transmitters. Because of the changing chemical composition during growth the chemotypes will be described later only by major differences in the chemical content or organisation in adult plants. A possibility of growth of plants and fungi is cloning, requiring dedifferentiation either completely or in a limited way to allow new organisms to develop from say a leaf cell, for example. All differentiated cells arise through the changing environment with growth.

Animal growth from a single cell is quite different in that, starting from one cell, *three-dimensional* doubling of cells leads to a roughly spherical body which infolds so that the organs, later very many, split away from one another internally (see Figure 8.6 for example) and externally at various stages. Differentiated growth is then tightly regulated by novel, three-dimensional cell–cell contacts through which physical and chemical messages are exchanged. An essential feature is cell migration, not seen in plants, which has a risk later in animal life, that of metastasis leading to cancer. There is no nerve system in any animal until quite late in the embryo of higher animals when tubular shape has developed but messengers are present at a very early stage. The messengers connect and coordinate growth of the dispersed organs via extracellular fluids and in later animals through circulating liquid in veins and arteries (Table 8.4). Again, as for plants, growth demands connective tissue which is constantly modified by synthesis and breakdown. The growth in embryo is very much dependent on genetic factors, food supply and element balance even *in utero*. For example, among minerals, zinc deficiency in early and juvenile stages leads to stunted growth. Examples in human life are the requirements for manganese in milk production and for zinc in the transitions at puberty. The final form that is attained is three dimensional due in part to the demand of muscle and bone structures, the actions of which are coordinated. In the case of a higher animal, the older it becomes after birth the more its *behaviour* becomes related to the external experiences relayed quickly from its senses to storage devices, later in evolution via nerves and in the brain, which are largely activated after birth. Hence the system of an animal, especially *Homo sapiens*, becomes *better informed from outside as it grows*. This type of chemotype has a variable chemical composition until the adult stage (see Chapters 9 and 10). Note that unlike plants, cloning of animal cells is difficult and may not be really fully realisable.

While considering these issues it is worthwhile to have in mind the quite amazing transformations that occur both early and late in the life of certain animals. One only has to examine the life of many an insect from egg to grub to chrysalis to adult butterfly to be puzzled as to what chemical mechanism could bring about the gross changes in morphology.

We cannot give a wider description of the diversity of multi-cellular eukaryote growth patterns in this book, so we refer the reader once again to general books on biological sciences (see Further Reading). We now turn our attention to the changes in chemical element content in the environment and in those organisms, which we consider made the above developments possible. In this book we are deliberately stressing the link between life and the environment with reference to the roles of the (inorganic) elements. Now there is a new difficulty since the organs we refer to above are based on *differentiated cells* in which, although each cell has the same DNA it is differently expressed and has somewhat different chemistry leading to different chemical element content in each organ and even in different cells within an organ. This differentiation includes the selective uptake or rejection of elements especially in the vesicle compartments of each cell, already described in single-cell eukaryotes. Hence, we shall look first at the chemistry of organisms as a whole in the adult stage but we must also examine later the chemistry in different cells belonging to different organs or at parts of organs or even at cells at different times in a life cycle, including both their cytoplasmic and vesicle contents. We shall first give the major compositional chemicals of the whole adult organisms in Part 8B of this chapter, and then some impression of the different chemistry in different cell types, while we include considerations of the organisation of them, in Part 8C, so that the reader can see the diversity of chemical changes in multi-cellular life from organisms based on one cell. At the end of Part 8B of this chapter, we hope to have shown that, of necessity, novel chemical element distribution in living organisms is related to functional development and the whole is dependent upon the advance in chemical composition of the environment.

(Finally, we must always be aware that evolution can go backwards if a unicellular organism finds a home in an environment, say within a multi-cellular organism. For example, an aerobic unicellular eukaryote can find a niche in a multi-cellular organism which is of low oxygen exposure when this cell may revert towards an anaerobic condition. In this connection, we have mentioned both symbiotic and parasitic organisms in previous chapters. How can an organism lose genes selectively?)

8B. The General Chemical Changes in the Ecosystem Some One Billion Years Ago

8.7. The Chemical Changes of the Environment

As in Chapter 7 on multi-compartment single cells, we have had to delay discussion of the chemistry of multi-cellular organisms until we had described their compartments, and their organs. What we wish to demonstrate next in this chapter is that all the major features of multi-cellular development described above required newly available chemistry while we leave differences between organs until Part 8C. We shall accept that, multi-cellular organisms evolved first in the sea at some time not too far removed from one billion years ago, while such organisms arrived on land some 400 to 500 million years later (see Table 8.1). One billion years ago, oxygen in the atmosphere had risen to some 5% of the present level and then it rose more rapidly over a further few hundred million years to present day levels of 21% (Figures 8.7 and 1.12) with some variation up to 35%. Thus, there was a general rise in the energy storage between the oxidised environment and the reduced organism chemicals in living systems and as the second became debris on death there was a greater possibility for living organisms to obtain more energy. As explained in Section 1.9, before this time period, the concentrations of free ferrous iron and sulfide had already been lost to a very large degree as $Fe(OH)_3$ and sulfate, respectively. (It would appear that it took from some 3 to 2 billion years to remove most of this ferrous iron and sulfide from the sea; see Section 1.4.) After this time, the rises in vanadium, nickel, cobalt and particularly of zinc and then of copper would have been rapid as their insoluble sulfides are much less buffering than that of the plentiful and soluble iron sulfide. At the same time many other elements became available in new oxidised states (see Section 1.12). This relatively fast, late change of the element concentrations coincidentally with the faster

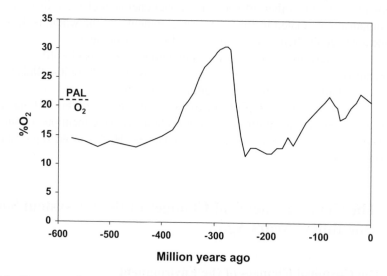

FIG. 8.7   The recent changes of atmospheric oxygen. This graph shows how the amount of oxygen in the atmosphere (expressed as a percentage of its present-day value) has evolved with time. Note that the atmosphere contained essentially no oxygen until about 2 billion years ago and then rose rapidly for the next 1.5 billion years but with fluctuations have moved towards the 21% of today. (Adapted from Berner, see Further Reading.)

production of free oxygen, due to increasing plant life, should be seen to be biologically causative and entirely in keeping with the quite *parallel* and subsequent pattern of evolution of chemicals in all organisms. In effect, we consider in each chapter including this one, that while the development of unicellular then multi-cellular eukaryotes was most dependent on these late changes of available elements it was slowed due to the conservative nature of DNA and cell chemistry. The cause of change rested not simply directly upon the element changes, of course, but upon the opportunity they provided to increase energy retention also in previously used and new chemicals, while utilising increased size and organisation of cells.

As the oxygen partial pressure increased above one-tenth of its present level the effect of the incoming UV radiation from the Sun upon it created an increasing ozone layer. This layer then became an approximately cyclic steady state $3O_2 \rightleftharpoons 2O_3$, some 500 million years ago (see Chapter 3). It may well be that this layer so protected the land that it aided the ever-developing invasion of land by living organisms. It certainly helped evolution there. Remember that UV radiation is not an effective source of energy for life and in fact it is damaging, especially to molecules like DNA.

8.8. Chemical Changes in Whole Multi-Cellular Organisms

We now look at the changes of element content and concentrations of organisms against the above background. As stated above we shall divide the discussion looking first at all cells of all multi-cellular organisms together as if they all were of one kind before we look at the element contents and concentrations of organs. The elements concerned are taken up by plants and fungi and then incorporated into animals (Table 8.5). When we look at the average analytical concentration of the free elements *in the cytoplasm* of all the multi-cellular eukaryote cells, their free metal-lome, we observe little change from that in other aerobes (Section 7.8) or even anaerobes (see Figure 7.7). As stated, this is due to the absolute demand for conditions suitable for syntheses of limited kinds of essential biopolymers. However, there is a large increase in several *bound* elements, that is in the bound cytoplasmic metallome (see Figure 8.8, Tables 8.6, 8.8–8.11). Especially interesting is the increase in the *cytoplasmic* content of zinc in a variety of proteins and iron in the multiplication of some oxidases. There is also a great variety of iso-enzymes which contain other previously well-used elements as well as these two. Turning to their vesicles next and then the extracellular fluids (Table 8.6), we find that there are some vesicles which contain high free zinc and many more others than were seen in earlier single-cell eukaryotes which have high free calcium. There are also clearly considerable amounts of these elements and copper bound by increased numbers of proteins for these and other elements in many types of vesicle. There are, for example, some special vesicles, with considerable amounts of bound manganese, or haem or copper, all within selective proteins, while some tissues have quite a high content of nickel in vesicles (storage of poison), but there is now very little cobalt

TABLE 8.5

ELEMENTS TAKEN INTO BY PLANTS AND FUNGI AND THEN ABSORBED BY ANIMALS

Element	Source	Absorbed Form
Non-mineral elements		
Carbon (C)	Atmosphere	CO_2
Oxygen (O)	Atmosphere	CO_2, O_2
Hydrogen (H)	Soil	H_2O
Nitrogen (N)	Soil	NH_4^+ and NO_2^-
Nitrogen (N)	Bacteria	RNH_2
Mineral nutrients		
Macronutrients		
Sodium (Na)	Soil	Na^+
Phosphorus (P)	Soil	$H_2PO_4^-$
Potassium (K)	Soil	K^+
Sulphur (S)	Soil	SO_4^{2-}
Calcium (Ca)	Soil	Ca^{2+}
Magnesium (Mg)	Soil	Mg^{2+}
Silicon (Si)	Soil	$Si(OH)_4$
Micronutrients		
Iron (Fe)	Soil	Fe^{3+} complex
Chlorine (Cl)	Soil	Cl^-
Manganese (Mn)	Soil	Mn^{2+}
Boron (B)*	Soil	$H_2BO_3^-$ and HBO_3^{2-}
Nickel (Ni)	Soil	Ni^{2+} complex
Cobalt (Co)	Soil	Co^{2+} complex
Zinc (Zn)	Soil	Zn^{2+} complex
Copper (Cu)	Soil	Cu^{2+} complex
Molybdenum (Mo)	Soil	MoO_4^{2-} complex

* Not essential for animals.

TABLE 8.6

GENERAL FEATURES OF FREE AND BOUND ELEMENTS DISTRIBUTION IN MULTI-CELLULAR ORGANISMS

Elements not Concentrated, i.e. Evenly Distributed	Elements Concentrated in Cytoplasm of Cells	Elements Concentrated Outside Cells or in Vesicles
Mg^{2+*}, H, O	K^+, HPO_4^{2-}, Fe, Co, Zn, C, N, P, S, Se (in organic molecules)	Na^+, Cl^-, Ca^{2+}, Cu, Zn, Mn, Ni, Si, (Mo), (V)

*The reference is to ions of elements where charges are shown, i.e. not to magnesium in chlorophyll. Most elements are concentrated in combined forms.

anywhere in cells. The changes in vesicle content contribute considerably to the new chemotypes. Outside the cells are the extracellular fluids essential for element distribution and communication between cells and organs. These fluids and their element contents are quite new and are in a few, quite separate compartments.

TABLE 8.7

Extracellular Fluids (mM)

	Na$^+$	K$^+$	Ca^{2+}	Mg^{2+}	Cl$^-$
Seawater	470	10	10.2	54	550
Typical freshwater	0.6	0.01	2.0	0.2	0.5
Marine invertebrates	450	10	10–20	10–50	530
Freshwater invertebrates	10–100	0.5–4	10	0.5–5	10–150
Terrestrial animals	10–150	5–30	4–15	1–40	20–150

In general, their composition, which is homeostatically controlled in concentration, certainly in higher plant sap, fungal fluids and animal blood, is as shown in Table 8.7. The reason for the high Na$^+$, Ca^{2+} and Cl$^-$ concentrations in these organisms, which were absent in single-cell organisms, is now seen to be that the extracellular fluid carries the cell "environment", and in a sense it is, in this respect, the necessary equivalent of the sea, creating an immediate independence of cell metabolism on their truly external environment but a dependence on their internally held "external" environment. The extracellular fluid also contains many sulfated and oxidised side chains of saccharides and proteins allowing new Ca^{2+} binding, and novel carrier proteins for elements as well as a series of metallo-enzymes. Finally, there are the extracellular elements accumulated in minerals (see Table 8.13), which are largely confined to a few combinations of calcium salts that are insoluble as shells or bones in animals and silica spherules in plant.

Thus, the *total* free metallome in the organisms (Figure 8.8) has changed quantitatively and so has the total bound metallome (see Fraústo da Silva and Williams in Further Reading). This and the changed organisation of chemistry permits us to state that new groups of chemotypes appeared. The changes had to be backed up by DNA and genetic machinery change too (see below). There is no doubt that throughout the chemotypes, changes of many chemicals of multi-cellular organisms are directly connected with the somewhat earlier changes of element content and concentration in the environment, but equally important is the indirect consequence of the novel environmental elements upon the ability to generate new functions in multi-cellular organisms. We have seen parallel but less extensive evolution in both prokaryotes (Chapter 6), and single-cell eukaryotes (Chapter 7).

8.9. Novel Proteins Associated with Multi-Cellular Organisms

Before we draw attention to new functions of elements we must ask if there has been any basic developments of proteins, particularly those associated with metal ions. In Section 4.11, we described their very basic fold structures while in Sections 5.5 and 7.10, we referred to the features including the folds of metal-bound

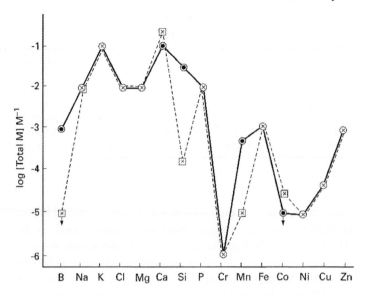

proteins in the previously described anaerobic and aerobic chemotypes. In the multi-cellular organisms, there are the proteins of cytoplasm and the vesicles, both of which were present in unicellular eukaryotes, but there is further gene duplication and gene splicing in protein production. In the cytoplasm, there do not appear to be many new types of fold but greatly extended application of known folds in novel proteins (see, for example, Tables 8.8–8.11). However, outside cells and in the vesicles, there are major changes due to oxidation of protein side chains, formation of –S–S– bridges and protein glycosylation. Oxidation, particularly hydroxylation of proline and lysine in collagens, of aspartate in fibrillins, and of carboxylation of glutamate in blood-clotting proteins, has allowed the development, especially in animals of novel structures. In the case of collagen, this is the triple-helix strand, in the case of fibrillins the stretched structure of so-called EGF domains, which is Ca^{2+} binding, and in the case of γ-carboxyglutamate a novel kind of Ca^{2+}-binding site fold for external activation of enzyme functions. Glycosylation, often with sulfation of sugar side chains, of these and other proteins, helps folds to be extended and located between filaments in multi-protein open meshes again in the extracellular fluids of all chemotypes. Many of these glycosylated proteins are also associated with calcium ions. These new structures and certain other new extracellular proteins which have little fixed fold, "random" structures, allow the properties associated with dynamic properties of the whole organism. The understanding of their functions is not so much related to chemical as to physical properties such as elasticity, ability to withstand direct stress and

TABLE 8.8

SOME ZINC PROTEIN CLASSES

Oxido-reductase	Dehydrogenases	All phyla
	Cytochrome oxidase	
Transferases	Aspartate trans-carbamoylase	All phyla
	Kinases	All phyla
	RNA polymerases	Bacteria
	DNA transferases	Very few
Hydrolyases	Metallo-proteases	Mainly eukaryote
	Endonucleases	All phyla
	Glycosylases	All phyla
Ligases	t-RNA synthetase	All phyla
	DNA ligase	
	Carbonic anhydrase	
Transcription factors	Zinc fingers	Eukaryotes
	LIM domains	Eukaryotes
Zinc storage	Metallothioneins	Eukaryotes

Source: From Messerschmidt, A., Bode, W. and Cygler, M. (2004). *Handbook of Metallo-Proteins* Vol. 3, John Wiley & Sons, Chichester, UK.

strain. Note that many of these developments occur in the extensive endoplasmic reticulum or other vesicles before exocytosis. We stress that they all originate from increase in oxidation and the use of particularly a novel element copper, in this extracellular oxidation.

In plants and fungi, the extracellular matrices are more generally of polysaccharides, although they are present in animals too. The polysaccharides are modified celluloses, etc. and are often cross-linked by calcium ions.

8.10.   New Functional Uses of Elements: General Outline

As already indicated, the qualitative presence and general uses of elements in the multiple-cell eukaryote changed little from those introduced into the single eukaryote cells, but the variety and extent of use of several bound metal ions and of one or two non-metals increased greatly, especially in animals. We need to see the functional value of these changes especially of zinc, copper, haem iron, selenium, manganese and calcium. In the case of *zinc*, there are general extensive uses in hydrolysis outside the cell cytoplasm, and a very noticeable change is its much-increased employment in transcription factors and internal control of enzymes in the cytoplasm (see Table 8.8). In higher animals, there are over 100 zinc proteases

TABLE 8.9

SOME COPPER PROTEINS AND ENZYMES: LOCALISATION AND FUNCTION

Protein or Enzyme	Localisation	Function
Cytochrome oxidase	External face of mitochondrial membrane	Reduction of O_2 to H_2O
Laccase, tyrosinase	Exracellular	Oxidation of phenols (reduction of O_2 to H_2O)
Caeruloplasmin	Extracellular (blood plasma)	Oxidation of Fe(II) to Fe(III) (reduction of O_2 to H_2O)
Haemocyanin	Extracellular (blood plasma)	Transport of O_2
Lysine oxidase	Extracellular	'Cross-linking' of collagen (reduction of O_2)
Ascorbate oxidase	Extracellular	Oxidation of ascorbate (reduction of O_2 to H_2O)
Galactose oxidase	Extracellular	Oxidation of primary alcohols to aldehydes in sugars (reduction of O_2 to H_2O_2)
Amine oxidase	Extracellular	Removal of amines and diamines
Blue proteins	Membranes (high potential) thylakoid vesicles	Electron-transfer (Many different kinds)
Superoxide dismutase	Cytosol	Superoxide dismutation (eukaryotes)
Nitritc reductase	Extracellular	Reduction of NO_2^- to NO
Nitrous oxide reductase	Extracellular	Reduction of N_2O to N_2
Metallothionein	Cytosol	Copper(I) storage
ACE-1 (MAC)	Cytosol	Transcription factor
Dopamine mono-oxygenase	Vesicular	Hydroxylation of Dopa
Coproporphyrin decarboxylase	Extracellular	Production of Proto-porphyrin IX
Ethylene receptor	Membrane	Hormone signalling
Methane oxidase	Membrane	Oxidation to methanol
Terminal glycine oxidases	Vesicular	Production of signal peptides
CP-x type ATP-ase	Membrane	Copper pump
ATx-1 (Lys 7)	Cytosol	Copper transfer

and hundreds of the so-called zinc-finger transcription and other factors in both animals and plants. Between 1 and 5% of DNA expression is now controlled by these slow-exchanging zinc proteins. Very many of these controls are related to changes in growth, metamorphosis and puberty adjustments, that is in organ-to-organ communication. The very different external zinc proteases are used to break-down extracellular tissues to allow growth as well as to digest food, especially in animals (Table 8.8). At the same time, the general use of *copper* and *haem iron* greatly increased in catalysing oxidative protein cross-linking and the synthesis of signalling messenger molecules. Copper enzymes developed especially in selected

TABLE 8.10

SOME IRON HYDROXYLASES

Organism	Hydroxylation Involvement
General	DNA repair
Bacteria	Penicillin synthesis
Plant	Ethylene synthesis
Plant	Gibberellin synthesis
Animals	Procollagen synthesis
Animals	Factor of hypoxia induction

See Hewitson *et al*. in Further Reading for details of these enzymes.

TABLE 8.11

SOME CLASSES OF CALCIUM PROTEINS

Protein	Location and Function
Calmodulin*	Cytoplasm, trigger of kinases, etc.
Calcineurin*	Cytoplasm, trigger of phosphatases
Annexins	Internal associated with lipids, trigger
C-2 domains	Part of several membrane-link enzymes
S-100*	Internal and external: buffer, messenger, trigger
EGF-domains	External growth factor but general protein assembly control e.g. fibrillin
GLA-domains	External, associated with bone
Cadherins	Cell–cell adhesion
Calsequestin	Calcium store in reticula
ATP-ases	Calcium pumps

* EF-hand proteins.

regions of animal (Table 8.9), but note the association of copper with the action of the plant hormone ethylene. The oxidation reactions take place outside cells or in vesicles, while in plants and insects, a parallel development of oxidation is seen but mostly through the use of *haem enzymes*. (Plant multi-cellularity developed before that in animals when maybe copper was less available but we have noted in Chapter 7 that haem oxidases would also create a greater risk of free-radical reactions in animals if used.) We describe these developments in detail in Section 8.14. Protection from external damage to plants uses haem peroxidases and these and other haem (myelo)-peroxidases are employed either in vesicles or extracellular fluids in plants and animals, respectively, in immune responses. At the same time, there was development of internal iron hydroxylases, based on haem or on simple iron but not Fe/S sites (Table 8.10).

Protection from unwanted side products of oxygen reactions uses the cytoplasmic Cu/Zn superoxide dismutase and vesicular haem catalases as in all eukaryotes as already mentioned, but there is much development in multi-cellular organisms, especially animals, of *selenium* in the enzyme, glutathione peroxidase, which also protects from oxidative damage. The switch to oxidative conditions allowed the change of the use of selenium from a reductive synthesis site in anaerobes to an oxidative, protective one in aerobes. It also allowed the adjustment of the involved number of *iodine* atoms inserted by peroxidases in phenolic rings, e.g. in the animal hormone, thyroxine. The similar redox potentials of half-cell couples of iodine and selenium is suggestive in that they may have allowed their roughly simultaneous combined use. In the synthesis of external proteins the *manganese* function is much increased in protein glycosylation in the Golgi vesicles, where a special Mn^{2+} pump into the endoplasmic reticulum developed, but also especially in chloroplasts for the now extensive photosynthesis in plants. In fungi, the lignin-degrading enzymes use the new oxidising power to oxidise Mn^{2+} to Mn^{3+} and organic free radicals as intermediates. The fundamental functional value of metals such as free and bound *non-haem Fe/S iron, molybdenum* and *magnesium*, though extended, did not change greatly but it is noticeable that *nickel* and *cobalt* (vitamin B_{12}) are now of decreased use. Higher plants do not use cobalt in vitamin B_{12} and in fact Zn enzymes have replaced those containing this coenzyme, in methylation for example, and the cobalt complex has become a vitamin in later animals. It is very intriguing to observe the continuous increase in use of zinc and copper against the diminishing use of nickel and cobalt with the change to more aerobic conditions and as eukaryotes developed in complexity. Nickel and cobalt are of prime use in group transfer of H and carbon fragments ($-CH_3$ and $-COCH_3$) in prokaryotes (and even in chemistry) (see Sections 2.20 and 5.3) but when the environment became oxidised, this type of chemistry was no longer required directly. (Care has to be taken here since cobalt and nickel are still essential for higher organisms as these elements are required by symbiotic anaerobic organisms necessary for the life of the multi-cellular organism.) The transfer of oxygen as OH and O became increasingly important, functions not generally manageable by nickel or cobalt, and we note the use, greatly increased, of Fe, Fe haem, Mo, Se and Cu, in catalysis. As stated, a grave problem, however, was the low availability of iron, and multi-cellular organisms had to develop further special methods for the controlled uptake and distribution of this element. Particularly the iron carrier protein, transferrin, is observed in many extracellular fluids (see our previous books in Further Reading).

In Section 6.5, we have drawn attention to the value of *vanadium*, especially under weakly oxidising conditions in prokaryote organisms for nitrogen fixation. Here, we wish to stress further the use of this element under stronger oxidising circumstances. Through its higher oxidation states, vanadium as vanadate could be used in a number of reactions of high redox potential in multi-cellular organisms. A very good example is the oxidation of halides, especially in fungi, to give halogenated organic compounds involving bromide and chloride. The reactions here involve the use of H_2O_2 as a substrate. Vanadium readily forms a comparatively stable, oxidation state V (five), peroxide $V(O_2^{2-})$ complex and it is this complex and not a redox change of the vanadium, which is the attacking agent, in contrast to the haem-bearing halogenases. The reactions occur in vesicles or at least outside the cytoplasm. The accumulation of vanadium here and in other organisms, such as in tunicates (ascidians), allows us to classify the organisms as special chemotypes under aerobic conditions. The value of vanadium to the tunicates is uncertain but observe that vanadate itself competes with phosphate (compare VO_4^{3-} and PO_4^{3-}) and halogenated products too may well be considerable hazards. Vanadium chemistry in organisms needs further study since its changes, like the chemistries of other oxidisable metal elements, is closely linked to environmental evolution (see Sigel and Sigel, Tracey and Crans, and Fraústo da Silva and Williams in Further Reading).

Finally, signalling using especially *calcium* is increased greatly relative to its functions in single-cell eukaryotes and with it the variety of calcium proteins (see Table 8.11), some with novel oxidised side chains. In Chapter 9, we shall show a novel value of *sodium* and *potassium* in signalling introduced later in the nerves of higher animals. Among non-metals we note that *chloride* too became much used in nerve messages also in higher animals. Like the new functional use of calcium the value of these three elements is totally dependent on the novel extracellular fluid compartments (Table 8.5).

Certain higher plants take advantage of the availability of calcium and silicon in external waters and make extensive precipitates of calcium oxalate, for example, in nettle hairs, and hydrated silica in many grasses and some more primitive plants (Table 8.12). The evolution of silica in flowering plants is thought to be a relatively modern development (see Krögger *et al.* and Hodgson *et al.* in Further Reading). The exo-skeletons of animals are usually of calcium carbonate or calcium phosphates since these salts are close to saturation in seawater and now in extracellular fluids. The large shells of molluscs, for example, are all protective of the organic matter of organisms. Endo-skeletons are almost invariably of calcium hydroxyphosphates which have remarkable, continuously changing growth patterns due to their piezo-electric proton conductivity based on local stress and st rain: they are living tissues unlike shells. These minerals are supported by some novel proteins with oxidised side chains (see also Section 8.14).

What is apparent from this large list of novel and increased uses of inorganic elements is that the evolution of multi-cellular organisms could not have occurred without the changes in element availability in the presence of a considerable level of oxygen in the atmosphere. We repeat that the conclusion is simple – the geochemical changes created the possibility of the generation of new chemicals which in turn created the possibility of the new structures and activities. Uptake and retention of energy increased (in plants) and was used for increased growth in novel ways in parts of space not open to unicellular organisms, and then led to increased energy degradation rate, especially in animals, as a new steady state of the total life cycle is approached. This is in keeping with the general hypothesis that energised steady states will grow towards an optimal energy retention and thermal degradation in cycling chemicals. Novel chemicals arising from waste, initially poisons oxidised minerals for example, allow novel kinds of growth while earlier systems are modified and sustained.

TABLE 8.12

THE MAIN INORGANIC SOLIDS IN BIOLOGICAL SYSTEMS

Cation	Anion	Formula	Crystal	Occurrence	Function
Calcium	Carbonate	$CaCO_3$	Calcite Aragonite Vaterite	Widespread in animals and plants	Exoskeleton Gravity, Ca store Eye lens
	Phosphate	$Ca_{10}(PO_4)_5(OH)_2$	Hydroxyapatite	Shells, some bacteria, bones and teeth	Skeletal, Ca store (piezoelectric)
	Oxalate	$Ca(COO)_2 \cdot H_2O$	Whewellite	⎰Insect eggs ⎱Vertebrate stones ⎰Abundant in plants	Deterrent Cytoskeleton Ca store
		$Ca(COO)_2 \cdot 2H_2O$	Weddellite		
	Sulphate	$CaSO_4 \cdot 2H_2O$	Gypsum	Coelenterate statocysts	Gravity S store, Ca store
Iron	Oxide	Fe_3O_4	Magnetite	Bacteria–chitons teeth	Magnetic device
		$FeO(OH)$	Ferritin	Widespread	Iron store
Silicon	Oxide	SiO_2	Amorphous (opatine)	Sponges, protozoa abundant in plants	Skeletal, deterrent
Magnesium	Carbonate	$MgCO_3$	Magnesite	Reef corals	Skeletal

Before concluding this section, we draw attention to the fact that much of the above account of our knowledge of the changes in the bound element content including the metallome is not due to direct analysis but to the knowledge of DNA sequences. There are at least 30 eukaryote and 200 bacterial gene sequences already determined. Many of the ("presumed") metallo-proteins are deduced from amino acid sequences characteristic of known binding selectivity (see Section 7.10). The DNA evidence shows that the development of the numbers of related Mg^{2+}, Ca^{2+}, Zn^{2+} and Cu^{2+} proteins increased very rapidly with the evolution of multi-cellular eukaryotes but with relatively few completely novel protein structures and major functional changes. Through DNA analysis there are now well over 1000 known zinc and copper proteins in animals and over 500 Mg-dependent kinases in humans alone. What is lacking is thorough analysis and knowledge of location and functional use.

8C.   The Use of Elements in Compartments and in Signalling

8.11.   Growth and Differentiation

We wish to treat next some of the developments in multi-cellular eukaryotes in more detail in particular processes and often in separate organs in different chemotypes while illustrating further our view that the evolution of these organisms depended on the chemical evolution of the environment. There were five general developments of multi-cellular organism which were (1) a controlled pattern of growth from single cells; (2) differentiated cells in organs; (3) the synthesis of connective tissue, holding cells and organs together in fixed positions; (4) the functions of special chemical extracellular fluids held in the limiting whole body already mentioned; and (5) the synthesis of completely novel set of organic messengers and message centres for information exchange linking internal development and later homeostasis of cells and organs via external fluids but also connecting inner metabolism to external environmental events through senses to a much greater degree. We reserve a separate chapter for the discussion of the special late appearance of nerve connections and the brain in animals, since this added greatly to the interaction with the environment and use of novel chemistry and organisation and therefore its evolution defined further new chemotypes (see Chapter 9). An "overview" of the developments is given in Tables 8.22–8.24 at the end of this chapter. We are seeking in this Part 8C a view of the controlled autocatalytic parts of the organism, that is within organs, rather than of the whole as above. The organs, for example muscles, are added to the earlier molecular machines.

It is generally considered that the pattern of growth with differentiation in all multi-cellular organisms is dependent in the first instance on cell–cell contacts after the original cell divides in a roughly symmetrical pattern to the eight-cell stage. The cells hereafter become subject to different gradients of products as chemical contents and concentrations of internal and extracellular fluids change and lower symmetry of cell contacts are created by further cell doubling. The gradients affect the further activity and the positioning of cells, which are then effectively semi-permanently differentiated, that is different cells using the same DNA have limited and different gene expression. Hence the cell's interior environment affects DNA expression and DNA expression relates to the environment cooperatively. As the positioning evolves, the gradients change and the whole organism develops both in morphology and in localised chemical cell (organ) activities (see Figures 8.3–8.6). Notice that the genetic structure is now more or less permanently subjugated to the cell's environment. Of course, the structures can only be managed by controls of activity inside the cells in addition to controls imposed by the surrounding fluids, contained in the whole organism, since internal synthesis of proteins for outside use is required. Let us look in turn at the chemical nature of differentiated cells, the messengers in the adult organisms and the cross-linked external structures. At the present time there is inadequate knowledge of the detailed chemistry of their developmental stages for us to be able to discuss much more than organs of adult cells in outline. Once the adult stage is reached, gene suppression and expression is maintained locally even when cells are replaced. It would appear to be extremely difficult to predict such kinetically controlled and maintained differential development from the genes themselves. As an aside notice that differentiated cells in one organism can be likened to a collective of cooperating different cells of many organisms.

We stress first that the separation of activities in different cell types in organs, differentiation, required quite new idiosyncratic *chemical* control of expression and activity of proteins both during growth and in homeostasis in different organs. It is intriguing that much of the control of expression is based on novel limitation of DNA transcription. Every cell in a given organ loses its general autonomy but it has to maintain many basic activities of synthesis, degradation and energy transduction even though their rates are now controlled differently. The limitation to particular expression rests with new particular transcription factors responsive to local space conditions and reception of concentrations of *hormonal-like messengers,* sometimes called *morphogens,* for selective long-term activity while local metabolic activity is also controlled by new *transmitters* for specified short-term fast responses. The "hormones" are largely responsible for the relatively slow, required production of the cell's proteins and do change in concentration with age. In this book, we wish to draw attention to the dependence of such features as differentiation on the new possibilities of use of a changed chemical external environment. We note immediately, as stated above, that many of the new transcription factors are *zinc fingers*, *zinc-ring* proteins, or *zinc domain* (homeobox) receptor proteins, which provide binding sites for the long-life hormones and can only exchange zinc slowly and are different in different organs. Muscles do not respond to steroid hormones in the same way as do sexual organs. The correspondence between several hormonal and zinc exchange rates indicates that zinc links internally in anyone cell the different hormonal messages from outside cells using different zinc receptors (transcription factors) in different cells. Zinc is a master hormone. The differentiated expression is of a wide variety of other metal proteins too and we note especially the different concentrations even of whole organelles and vesicles such that cells are often even of different colour associated with mitochondrial and chloroplast expression of iron, magnesium or copper proteins. Further element analysis of differentiated organs is given in physiology texts. A striking example is the difference between muscle types and their different reliance on mitochondrial, oxidative (red) or anaerobic (white) metabolism, that is their haem iron (and copper) content, and in the difference between green (chloroplast) leaves and white roots of plants in their Mg and Mn contents. Very intriguingly, the protein expression in the muscles of advanced animals is controlled by their different connection to nerves. By switching nerve connections and making repeated activation by Ca^{2+} pulses, muscles can be switched in type between slow, posture, muscles and fast more active muscles. We must remember that control of major as opposed to such minor differentiation is flexible early in all life, and remains so for plants, such that root cells go green in the light, and cloning is even possible though difficult and uncertain from differentiated animal cells. The dependence on the environment and external messengers and contacts is striking and reflects the interactive nature of the code expression and the environment.

It must be recognised that the development and maintenance of differentiation is a topic of immense interest and importance to which we can only draw attention here while noting element involvement.

8.12. The Production of Chemical Messengers between Cells in Organs

We turn now to the production and reception of the differentiated communication network necessary for coordinated activity. A primitive example of external messengers affecting the behaviour of unicellular cells is the development of the slime mould when short of food. Here the messenger molecule is c-AMP, previously an internal bacterial (phosphate) messenger (Table 5.9). The dispersal of this molecule outside single mould cells causes them all to form a shaped upright fruiting body which then ejects spores, its way of scattering dormant cells to distant places, where food could be more plentiful. Note that as elsewhere the formation around cells of a chemical gradient, here of c-AMP, is a way of creating an information source in the environment. The equally important reception of this information is not initially by the genes, the DNA and the production of new proteins, but by the metabolism of the slime mould cells so that they all swim towards the centre. Note how such a gradient of c-AMP is basic to the development of *form* of this organism. This type of *chemotaxis* using external messengers may be common to many colony-forming single-cell organisms in search of food and development. It is amusing to note that addition of the thyroid hormone to the environment of the axotl drives it through metamorphosis from a tadpole-like to a frog-like condition.

Now, when we come to consider modes of information exchange by messengers between cells in multi-cellular organisms we have to recognise that they must be chemically different from those already in use in the cytoplasm of all cells and from Ca^{2+} utilised by all eukaryote cells to give information about the nature of the immediate *external environment* of a given cell which is now the extracellular fluid with a fixed Ca^{2+} concentration. These signalling systems remain essential for the multi-cellular organisms but have to be coupled to several novel chemical messengers for long-range, cell–cell communication, in which a selective message is sent out from one particular cell or organ to another in the extracellular fluids. The choice of these messengers was limited to organic molecules since all simple inorganic ions which can perform message functions – diffuse and then bind – have been used earlier in evolution mostly internally, e.g. organic phosphates, iron, and magnesium and so on, and there are many cell types which must receive different messengers (see Tables 8.13–8.15). (Note however that Na^+, K^+ and Cl^-, which do not bind, have not been used and their employment awaits the coming if nerve messages; see Chapter 9.)

Release of a variety of specific organic messenger molecules was then the inevitable extension to meet the need for signals to a particular cell type or organ, which had specific receptors and would respond in a particular way. In effect, they provided a novel way of *information* transmission and reception and for maintaining differentiation. As stated, two devices were possible, both involving diffusion of the messenger molecules *in the extracellular fluid* and assisted by circulation of that fluid, up and down in a plant, and round and round in an animal. The two types of messengers are: (i) hormones (morphogens) for long-term management and (ii) transmitters which could cause a rapid response in metabolism through coupling to pre-existing local outside/inside message systems. The obvious coupling connection is via the Ca^{2+} fluxes and activation of organic phosphates, some released from membranes often Mg^{2+}-linked, e.g. in kinase cascades (see Figure 8.9). There is then a novel network of interacting signals with specific effects in different cells and organs and with different time patterns both in the developing and the fully developed organism. First we look at the syntheses, reception and destruction of these new organic messengers, different in different organs, noting the dependence of many on the newly available environmental inorganic chemicals, especially Zn and Cu, and on the greater availability of oxygen.

The hormonal messenger molecules for slow response must have a long life and hence must be difficult to destroy; in fact they usually circulate in the external fluids at sub- or low-triggering levels but these levels are also adjusted in time leading to periodic growth behaviour, e.g. in metamorphism. Many are released from special organs, glands, in particular parts of the organism. The responses are timed by internal changes in cells with age where times of change vary from days to many years. These slow-acting messengers in animals are molecules such as steroids, long-chain acids (retinoic acids) and thyroxine (Table 8.13) with parallels to those in plants (Table 8.14). Synthesis of many of them depends on *in-cell oxidation* by protected iron-dependent enzymic processes, e.g. P-450 cytochromes, peroxidases and non-haem iron hydroxylases in the donor cells. A particular case of great importance in plants and animals is nitric oxide, NO, while most hormones are different in the two chemotypes. The hormonal molecules (not NO) pass through membranes and act directly on *zinc-finger transcription* factors to produce proteins being usually destroyed later by oxidases based on haem, such as other cytochrome P-450 enzymes in the cytoplasm of the receiver cell. Note again the

TABLE 8.13

SOME CLASSES OF ANIMAL HORMONES IN MAMMALS

Hormone Class	Gland	Enzymes
Many peptides	Stomach, duodenum, pancreas	Synthesis, Cu oxidases
Often amidated	pituitary, pineal, etc.	degradation, Zn peptidases
Sterols	Adrenal cortex, testes, ovaries	Synthesis and degradation
		Haem Fe, P-450 enzymes
Adrenaline	Adrenal medulla	Fe and Cu oxidases
		Cu oxidases in degradation
Thyroxine	Thyroid	Fe peroxidases
		Se removal of iodine
*Histamine	Many cell types	Many synthesised by Fe enzymes
Prostaglandins		
NO, CO		

*Not released from glands; chemicals acting more quickly than hormones, but like them consequentially. Sometimes all the above set of messenger molecules are described under the endocrine system. Some are released by the peripheral sense receptors. There are many other possible hormones such as glucose and several of the simplest hormones are related to bacterial sensors e.g. NO and some ions (see Chapter 6).

TABLE 8.14

PLANT HORMONES*

Hormone	Associated With
Ethylene	Induction of growth (Cu-receptor)
Auxins	Derivative of tryptophan – e.g. indole acetic acid. Controls H^+/K^+ balance and growth
Gibberellins	Growth, complex 5-ring hydroxylated alicycle. Requires iron containing enzyme for synthesis
Abscisic acid	Growth inhibitor largely linear unsaturated aliphatic compounds. Hydroxylated by iron enzyme
Zeatin	Derivative of purine. There are other similar purines controlling cell activities
Cytokinins	Peptides controlling cell activities

* There are no new messengers which act as cell-membrane transmitters in plants except cytokinins; calcium is more widely used than in unicellular organisms but much less so than in animals. Plants respond to light via phosphorylation and changes from dormancy requires the change of cell calcium. Response times >1 s. There are several other sensors which are sometimes described as hormones, e.g. glucose and NO.

use of the powerful intermediate oxene FeO in these enzymes and remember the different, earlier less powerful use of MoO in oxygen transfer (Section 5.9) whereas intermediate oxidising power uses >SeO for protection. Such new oxidising entities could not arise early in evolution and note similar reactions linked to copper arrived later, e.g. the ethylene hormones of plants. It is very likely that iron and selenium, for oxidation, and zinc binding were made possible by environment changes before copper was available. We have stressed earlier that these hormones (morphogens) and zinc are linked in a totally interactive exchange system which affects growth and metamorphism. Perhaps, as remarked earlier, zinc is an internal hormone aid in that it links many hormones to DNA. Copper is rarely involved with hormones but note the role of ethylene in plants and adrenaline in animals.

Fast messages, transmitter (Table 8.15) systems, are mainly to be found *in animals* for the obvious reasons. The simplest new method of synthesising a transmitter messenger *in the concentrated solutions* needed for *this fast signalling* was to use the very vesicles in which the messenger molecules were stored and from which they were then released. As a consequence we find in animal-cell vesicles a variety of stored aromatic and peptide molecules synthesised, often internally cross-linked (–S –S–) and hydroxylated, e.g. adrenaline (again) and novel amidated peptides. There are complications in that much of synthesis demands energy only available to a large extent in the cytoplasm. Hence, initial precursor synthesis is in the cytoplasm, using energy and substrates available there, and from where the molecule is transferred to a vesicle by a pump before it is modified usually oxidatively. In keeping with the already described development of evolutionary oxidation reactions using haem iron, or iron hydroxylases (see Section 7.10)

TABLE 8.15

SOME SMALL INFORMATION TRANSMITTERS

Simple Ions	Complicated Ions[*]	Molecules (Transmitters)
Na^+, K^+	c-AMP (P)$^-$	Acetylcholine (+)
Ca^{2+}	c-GMP (P)$^-$	Glutamate (2 −)
Cl^-	ATP (P)$^{4-}$	Adrenaline (+)
Zn^{2+}		8-OH Tryptamine (+)
Fe^{2+}		NO (CO)
		Peptides
Mg^{2+}		Amino acids (+ −)
		Prostaglandins

[*] c-AMP, cyclic adenosine monophosphate; c-GMP, cyclic guanosine monophosphate; ATP, adenosine triphosphate. (P)$^-$ indicates the charge on the molecule. Note also membrane-derived phosphates, e.g, IP_3.

these enzymes are used in steps on the cytoplasmic side of the vesicle membrane, but *copper* enzymes are then used in vesicles or outside cells as in single-cell eukaryotes to complete synthesis. We observe that copper enzymes are more greatly employed in transmitter synthesis than in that of hormones that came earlier. Examples are given in Table 8.16. We see here and in other examples that copper dependence is more noticeable in animals than in plants. This in-vesicle synthesis is organ-selective and the presence of messenger systems of a given kind is indicative of a particular concentration of an element, here copper. The transmitter molecules before or after release must not diffuse through membranes so that their release from vesicles is prompted by a *calcium pulse initiated by an external sensory event which generates their exocytosis* (see also nerve transmission). Their later excitory action is localised on selective receptors on the outside of the membrane of receiving cells, not at the genes. Now, since many of these messengers are required for relatively fast responses to threats and opportunities while they can only act at the membrane of a sensitive cell, a secondary fast signal has to be propagated through the membrane that connects to signalling systems in the cytoplasm (see Section 7.11). We observe that the links are from a second direct input of calcium, from extracellular fluid to the cytoplasm of the receptor cell, and a liberation there of sugar phosphates from membrane lipids or/and a cascade of G-protein receptors (using Mg.GTP) and/or other Mg.kinases. Both of the latter may then activate an extra influx of Ca^{2+} internally from vesicles (Figure 8.9) or even activate Fe, Zn phosphatases, e.g. calcineurin. Note the large rise in the number and type of Ca^{2+}-activated changes especially in animal cells, as organism complexity increases (see Table 8.17) and in the differential cell distribution of calcium (in vesicles) and in calcium-binding proteins, especially linking to an expression to Mg^{2+}.ATP kinases, now very numerous. The messengers generate quick response via special controls of metabolism without new protein production, that is, there is no immediate connection to DNA although repeated waves of stimulus can act as

TABLE 8.16

Organic Messengers Produced by Oxidation

Messenger	Production	Reception	Destruction
NO	Arginine oxidation (haem)	G-protein	? Haem oxidation
Steroids	Cholesterol oxidation (haem)	Zn fingers	Haem oxidation
Amidated peptides	Cu oxides	(Ca^{2+} release)	Zn peptidases
Adrenaline	Fe/Cu oxidase	(Ca^{2+} release)	Cu enzyme?
δ-OH tryptamine	Fe/Cu oxidase	(Ca^{2+} release)	Cu enzyme?
Thyroxine	Haem (Fe) peroxidase	Zn finger? ([*])	Se enzyme
Retinoic acid	Retinol (vitamin A) oxidation	Zn finger?([*])	?

([*]) In the nuclear receptor super family of transcriptional receptors.

transcription stimulators. Some internal Ca^{2+} responses are also linked to Fe^{2+} messenger system in cells, e.g. NO signalling. Much signalling therefore needs special proteins inside the cell leading to control over metabolism and separately to DNA expression. In particular, the cascade of kinases and phosphatases has increased greatly in the multi-cellular eukaryotes. We draw attention to these cascades by reference to Ray, Adler and Gough (see Further Reading). We are, of course, deliberately stressing the role of elements in this book, here Fe, Cu, Mg, Ca and P, so that the connection to the environment is kept in mind.

A fast organic messenger molecule has to be degraded very shortly after release from binding at a receptor to allow rapid, repeated response and is therefore free for any length of time only at extremely low fixed concentration. The degradation of the fast-acting peptides is achieved by the action of *zinc* peptidases while that of fast-acting aromatic amines is by amine oxidases, some of which are *copper* enzymes; both occurring outside the cell.

Internally, cell management also has to prevent new differentiation of cells once they are permanently associated with particular organs. In animals, loss of such control can lead to aberrant cancerous cells, for example. A very important zinc protein for such control is called P-53 and minor mutations affecting its zinc-binding site appear to allow cancer to develop (see Campisi and also Marx in Further Reading). There are also zinc domains associated with maintaining cell structure, such as the so-called LIM homeobox domains of the cytoskeleton. In both cases, certain calcium proteins have also been suggested as having a role.

To summarise the way in which the increase of oxidation using new elements and novel states of previously used inorganic elements, created in environmental conditions, enabled multi-cellular organisms to evolve new selective organic communication systems between differentiated cells, we note: (i) the increase in the cytoplasm of selected cells of *iron* oxidases, e.g. for steroids, thyroxine, retinoic acid and prostaglandins production; (ii) the increased use of oxidation outside selected cells and in vesicles by *copper* enzymes, e.g. adrenaline and hydroxylated peptides production; (iii) the involvement of oxidation of halides, e.g. in thyroxine synthesis as well as of even carbon and nitrogen (e.g. oxides CO and NO) employing *haem iron* enzymes in cells; (iv) the value of selected *zinc* receptors to monitor oxidised products and (v) the use of *zinc* peptidases to destroyed oxidised peptides. The developments apply to all multi-cellular chemotypes, but differentially to different cell types and organs. Only in animals is there a necessity for a vast range of transmitters. Oxygen increase together with that of the external availability of some metal ions in protein complexes (Table 8.18) was clearly essential for this evolution. We stress again that zinc and copper were released increasingly into the environment by oxidation of their sulfides only around one billion years ago. It is the degree of uptake of these elements related to oxidation, which led to

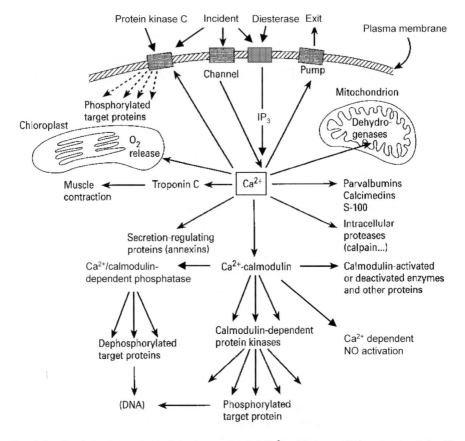

FIG. 8.9   Condensed overview of the interaction of Ca^{2+} with intracellular proteins (after S. Forsen). The extracellular systems are equally important. The connections between the Ca^{2+} input and linked signalling from the membrane are shown.

TABLE 8.17

CALCIUM MESSAGE-DEPENDENT EFFECTS IN CELLS

Effect (System)	Proteins
Muscle contraction	Troponin
Energy availability	Mitochondrial dehydrogenases
Oxygen formation	Chloroplast Mn protein
Membrane activities	IP$_3$-receptor
	Certain kinases (calmodulin)
	C-2 domains
Apoptosis	Calpain
Water soluble hormone release	Vesicle trafficking proteins
Transmitter release	
Flow of vesicles	Tubulin related proteins
Fertilisation	Many
Morphogenesis	Many
Differentiation	Many
Cell division	?
Genetic connection	Phosphatases (calcineurin)
	Kinases
Nerves and brain	Many
General metabolism	c-AMP, c-GMP cyclases
Nitroxide synthetase	NO generation, muscle relaxation

the major analytical distinction as chemotypes. We stress the general parallel changes in the use of some elements and proteins from unicellular to multi-cellular eukaryotes in Table 8.18.

Now a differentiated cell and the organ in which it resides have to be positioned by fixed connective tissue within the contained whole organism. The way in which the organs are arranged within the whole body of the organ, and the way this body comes to fixed size with an outer skin of cells and connective tissue is as yet little understood. The biggest developments of novel chemistry here were of connective tissue and the containment of body fluids around them.

8.13. Connective Tissues

The whole of a multi-cellular organism is contained by outer cell layers, which are described in biology texts, and maintained by connective tissue. Connective tissue is a novel, external biopolymer structure of multi-cellular organisms found within their new extracellular, circulating fluid compartments (see Section 8.9). As mentioned there, the main connective tissues, covalently cross-linked structures, are: (1) those of plants, celluloses (polysaccharides), often cross-linked by lignin; (2) those of lower animals and insects, mixed cross-linked polysaccharides and

TABLE 8.18

METALLO-PROTEIN SETS IN TWO ORGANISMS

Protein Set	Yeast	Worm (*C. elegans*)	Metal
Nuclear hormone receptor (Zn)*	0	270	Zn
Binuclear GAL cluster (Zn)*	54	0	Zn
Metallo-proteases	0	94	Zn
Na+ channels*	0	28	Na
Mg^{2+} adhesion*	4	43	Mg
Calmodulin-like proteins*	4	36	Ca
K+ Channels (voltage gated)*	1	68	K
EGF, Ca^{2+}-binding cysteine-rich repeats*	0	135	Ca
Kinases (tyrosine)	15	63	Mg
Cytochrome P-450	3	73	Fe

* Absent in bacteria. Note that values, generally larger, are also available for man and *Arabidopsis*
Source: After Chervitz, S.A. *et al.* (1998). *Science*, 282, 2022–2027, see also Deotyarenko, K.N. and Kullkova, T.A. (2001). *Biochem. Soc. Trans.*, 29, 139–147.

some proteins, chitins; and (3) those of higher animals, the main one here being collagen: all cross-linked by oxidation. All three polymers are thoroughly described elsewhere (see Stryer and other biochemical texts in Further Reading). The major features of the polymer chains are that they require enzymes to cross-link them so as to gain rigidity, but these rigid structures have to be cut at intervals by yet other enzymes to allow change of position of cells, not random migration, and growth. Of these activities, the basic linear protein synthesis has to be cytoplasmic while polysaccharides are made in vesicles. The necessary oxidative cross-linking chemistry once these linear polymers are put outside cells, or in vesicles before ejection, makes the connective tissue, often further stabilised by ionic calcium bridges and –S–S– links, a firm base for cell–cell organisation giving the organism a fixed shape after development. At the same time, this extracellular matrix had to allow diffusion in the extracellular fluids to give access for food and messages to cells. The open matrices are generally synthesised as *polysaccharides*, often carboxylated and sulfated, in which their anion side chains repel one another, and bulky, ill-fitting, glycoproteins. (Note that sulfate is not used, though it is reduced, in the cytoplasm due to its easy reduction and was not of great value earlier in evolution.)

We note that the first multi-cellular, cross-link connected cells in organisms were very probably in sea plants. As stated, the particular chemical development of connective tissue here is that of cross-linked polysaccharides. It appears that the oxidative covalent requirement is for a small molecule, HO• from H_2O_2 derived from oxygen and often an atomic ion, Mn^{3+}, generated at *haem peroxidases* to carry out the cross-linking reactions usually involving radical phenolic oxidation (Table 8.19). Manganese was always available outside cells but previously not free in an oxidised state. This chemistry could develop rapidly as soon as oxygen became sufficiently available, but as we have reminded the reader before, it is

TABLE 8.19

SOME CROSS-LINKING PROCESSES

Reaction	Catalyst
Hydroxylation of proline	Iron proline hydroxylase (animals)
Hydroxylation of lysine	Iron lysine hydroxylase (animals)
Oxidative coupling of collagen	Copper lysine oxidase (animals)
Cross-linking of chitin	Copper tyrosinase (animals)
Incorporation of phenols (lignin)	Manganese and iron peroxidases (plants)
Incorporation of calcium (saccharides)	(plants)
Disulfide bridges	(general)

somewhat risky due to associated random free-radical chemistry. In a plant, such chemistry is not so dangerous since plants unlike higher animals do not have circulating cells, so that a disease such as cancer does not metastasise in them. The use of copper enzymes, e.g. laccases and phenol oxidases, in these oxidations, not very usual, in plants is probably a late addition to the oxidative enzymic complement. It is a general feature of plants that they are richer in manganese but poorer in copper than animals. Cross-linking in fungi is probably similar to that in plants.

The cross-linking in animal tissues is quite different and its major features are distinctive in different groups of animals. As mentioned already, chitin is much used in insects and other invertebrates while collagen is dominant in vertebrates. Chitin is cross-linked by *copper* enzymes using phenols as agents. By way of contrast, collagen is cross-linked by oxidative changes of protein side chain lysines to aldehydes and its prolines are often hydroxylated to form additional H-bond cross-linking. Again aspartate is oxidised to hydroxy-aspartate to give a calcium-binding site (see fibrin). Many of these oxidative actions are brought about initially, while the protein is in the cell, by *iron* enzymes but the final steps of oxidative cross-linking are managed by *copper* enzymes outside the cell (Table 8.9). Referring back to their mechanism we see that copper enzymes do not produce dangerous free radicals and are more numerous in animals which came later. (In the event of accidental internal production of peroxides there are detoxifying selenium enzymes present in most tissues of animals.)

It is necessary to remember that as well as organic cross-links, elements such as boron, silicon and calcium cross-link all the major external proteins and saccharides even in the walls of prokaryotes. Many of the cross-linking binding sites are of oxidised side chains of biopolymers. As described in Section 8.10, certain of these elements form mineral deposits but now these minerals are frequently found inside the multi-cellular organisms. Here, we see a great difference between the chemo-types of plants and animals. The acidity of the extracellular fluids of plants differs from the neutral fluid of animals. It is not possible to precipitate calcium carbonates (shells) or phosphates (bones) in plants due to the weak acid character of these anions (see Table 8.12). Plants therefore precipitate silica and calcium oxalate and indeed some precipitate strontium or even barium sulfate. The gross analytical element content clearly distinguishes the different chemotypes. A further example of the influence of oxidation outside cells which provides a stabilising influence on extracellular structures is the formation of –S–S– cross-links which are stable there while in the cytoplasm they are at best in equilibrium with 2RSH.

Now the need to cut connective tissue for organism growth requires ways of breaking bonds in the polysaccharides and proteins. For this purpose the major classes of enzyme in plants are *Ca-dependent* cellulases, and *haem-dependent* ligninases sometimes with Mn^{3+}, while in fungi, insects and some lower animals, *haem-dependent* chitinases break down polysaccharides. In animals, *zinc* metallo-proteases are used for degrading proteins, particularly collagen (but see Table 8.20). Note too that chitin and lignin are usually laid down on the protective surface of plants, fungi and insects, and become "dead", permanent structures, while animals, especially vertebrates, use collagen in internal extracellular fluids and skin which, like bone, remain "alive" and are constantly remade. Many of these calcium and zinc enzymes are related to extracellular enzymes for digestion found even in aerobic bacteria. Note that zinc proteases released by cancer cells destroy connective tissue and allow aberrant growth and migration.

In concluding this section, we stress again the novel dependence of the extracellular connective structures on chemistry, especially that of copper and iron using oxygen, and zinc proteins for hydrolysis, which did not and could not have taken place before more than one billion years ago. They arose mainly after the development of unicellular eukaryotes, and were dependent on additional environmental change. Even several external uses of calcium depend upon new oxidation of the side chains of proteins.

Finally, we have to observe that all this chemistry of differentiation, of communication and of connective tissue formation has to be connected to changes in genetic content, that is ultimately of DNA. The DNA expresses all the novel proteins of synthesis, oxidation and hydrolysis involving the metals Fe, Cu, Zn, Mn

TABLE 8.20

METAL ENZYMES OF THE EXTERNAL MATRIX

Enzyme	Metal	Activity
		Hydrolysis of
Elastase	Zn	Elastins
Collagenase	Zn	Collagens
Stromelysin	Zn	Proteoglycans
Gelatinase	Zn	Gelatins
		Oxidation
Lysine oxidase	Cu	Cross-links collagen and elastin
Laccase	Cu	Cross-links via phenols
Phenol oxidase	Cu	Cross-links via phenols
Chitinase	Mn/Fe	Destroys chitin

and Ca in particular but we must not forget the novel involvement of the non-metal halogens (iodine in thyroxine), selenium in Fx2 compounds and sulfur (as sulfate). It is the expression of these proteins which affects the chemotype and it is linked to the environment. How did many new genes evolve as a response to environmental change and how were some genes selectively lost, e.g. for cobalt metabolism? Before we try to answer these questions we wish to stress again the very important role of the metal elements and their availability from the environment. We do so here by further reference to the calcium ion that is deeply involved in all the sequential advances in evolution of eukaryotes.

8.14. A Further Note on Calcium

As mentioned previously, through the analysis of DNA from many organisms or by direct examination we can follow the successive increases in the numbers of Fe, Cu, Zn, Mn, Se, etc. proteins in the chemotypes from prokaryotes onwards. The chemotype is continuously evolving and increasing in diversity of inorganic content in evolution in line with other developments such as of compartments and organisation only possibly useful and required after oxidising conditions arose. A striking example uncovered by genetic analysis is the increasing functional significance of calcium. We have seen how single prokaryote, weak-binding external Ca^{2+} sites, an EF hand, evolved into internal strong-binding twin EF hands and how oxidation of side chains of external proteins gave rise to novel external Ca^{2+} functions in eukaryotes (see Morgan et al. and also Williams in Further Reading; Table 8.21). In Section 7.12, we stressed the role of this element in the evolution of single-cell eukaryotes and contrasted that with the virtual absence of internal use in prokaryotes. Its importance is vastly increased again in multi-cellular organisms through transmitter sensing (Figure 8.9, Tables 8.17 and 8.18), and in stabilising cross-links. We know of several types of temporary binding site and function of its proteins in cells but it is the protein diversity associated with differentiated cells as well as generally in organisms, which is so striking. There are at least 500 different calcium proteins in tens of different classes in all advanced animals (see Table 8.21). Calcium became central to signalling, coupling external changes around cells to the cytoplasm and vesicles, to properties of the extracellular fluids, to connective tissue, to minerals, and we shall see later that its functions are also central to nerves and the brain. It became advantageous to hold calcium concentration fixed in the extracellular fluids so as to refine signalling and in higher animals this control is quite remarkably good especially in animals with temperature control. There is then a hormonal control on Ca^{2+} homeostasis acting on epithelial cells.

Now we have noted in the last chapters how strengthened calcium signalling employed storage of calcium in various vesicles, which act as a capacitance. The advantage of using these stores rather than the cells' external environment, here the extracellular fluid, is twofold. The message signals so amplified can progress (as waves through cells) not dying out on leaving the surface, and the recovery of

TABLE 8.21

DISTRIBUTION OF DIFFERENT CA²⁺-BINDING PROTEIN MOTIFS IN ORGANISMS

	Binding Proteins					
	Excalibur	EF-hand	C-2	Annexins	Calrecticulum	S-100
Archaea	—	6*	—	—	—	—
Bacteria	17	68*	—	—	—	—
Yeasts	—	38	27	1	4	—
Fungi	—	116	51	4	6	—
Plants	—	499	242	45	40	—
Animals	—	2540	762	160	69	107

* Single hands only (see Chapter 5).
Note: The table is based on DNA sequences available in 2004. The activities of the proteins are unknown in most cases.
Source: From Morgan *et al.* (2004). *Biochem. Biophys. Acta, 1742*, 133–140.

these stores by pumping calcium back into vesicles and not out of the cell removes the loss of these ions to circulating fluids. This release and recovery in local vesicles is a common way of avoiding waste of energy and extends from protons in thylakoids of bacteria, to calcium in all eukaryotes, to that of Na^+, K^+ and Cl^- in extracellular fluids, which is of unique value in nerve action (Chapter 9). The homeostasis of these vesicles and extracellular compartments is another efficiency feature. In Chapter 7, we recorded how the calcium signal was connected to energy stimulation in mitochondria and chloroplasts and this use is extended in multi-cellular organisms. The universality of Ca^{2+} as a messenger cannot be overstressed.

As a specific illustration of the different functioning of calcium we turn to light switches in plants and animals in the next section. Note we have deliberately used the discussion of the use of light previously as an example of how the environment became increasingly and directly but differently linked to chemotypes (see Chapter 5). This linkage is a dominant theme in our description of ecosystem evolution and it is here that calcium-ion fluxes are almost universally involved. We observe too how these calcium messages almost always interact inside cells with phosphate messages frequently linked to magnesium. These elements, Mg and P (with Fe^{2+}), are the most critical inside the cytoplasm, but we stress again that free not bound calcium, is central to information going from the outside to the inside of *all eukaryote* cells, not between cells with one or two exceptions (e.g. see S-100 proteins, see Morgan *et al.* in Further Reading).

8.15. Light Switches in Plants and Animals

This section is introduced to stress the difference in two major chemotypes, plants and animals, based on the reception of light, that is use of senses in different ways rather underplayed in the above, but which is very obvious in that plants have very different responses to light from those of animals. These responses are quite separate from the use of light in energy transduction in plants (also dependent on calcium) (Section 5.6), and in both chemotypes light affects growth and protection. The growth response to light by plants is via phototroponin, a system of proteins consisting of a light-sensitive protein, called LOV, which switches on reactions for growth. The light is absorbed by a flavin cofactor (FMN), which on activation by light forms a covalent link with a protein cysteine –SH. The changes of the flavin force a conformational change on the protein helices, which then transmit change to a phosphorylation system. The phosphoproteins produced are transcription factors in themselves or promote secondary activity such as inward flux of *calcium*. The calcium triggers metabolism in the cell and the two, Ca^{2+} and Mg^{2+} phosphate metabolism, then trigger growth. It is important to note that the whole system of plant growth and development, e.g. flowering, are environmentally dominated and that a factor of the environment, light or heat, does not just help to form the organism but also informs it. Overall, light regulates plant life in the slow way a hormone acts. Moreover, the light does also have a diurnal rhythm forced upon plant functioning.

One major obvious light switch in animals is quite different since it must be fast. It is a matter of sending a message from the eye to the central nerve organisation, the brain, to promote action. Hence, as we shall see, this fast response, like that to other senses in higher animals, depends on a series of transmitter messages. The light initiates a carotenoid *cis/trans* switch (see Section 5.6), followed by a reaction of *magnesium* guanidine phosphates and a *calcium* flux to give nerve pulses. The whole visual system does not require protein synthesis but, via responsive nerves from the brain, leads to immediate Ca^{2+} stimulated metabolic and muscular response. Hence, the difference between a plant and an animal in response to light illustrates the major difference between a slow hormonal (here in plant) and a fast transmitter (only in animal) activation. (Repetition of a visual image in animals results in a memory involving the storage of transmitters or ionic charge and synaptic change in patterns in the brain (see Chapter 9) but there is again no need for immediate protein synthesis as the required proteins are also stored at the synapse.) There are other effects of light in animals since they too have a diurnal rhythm, which is controlled by the pineal gland, yet this gland cannot sense light directly as it is hidden in the brain! Finally, light is of value in animals in the control of calcium in the formation of bone through the production of the hormone related to vitamin D, without which animal growth is stilted.

8.16. The Protection Systems of Plants and Animals

As organisms grew in size, the time for development and their lifetimes became much longer. This lifestyle, if it was to survive, demanded increased protection from the adverse effects of the environment and from other species and even from its own kind. Oxidation and the new catalysts for oxidising reactions were added to the attack on unwanted chemicals and invading species (see Section 8.10). In plants, the protection is provided largely by sealing off areas of damage or infection by the synthesis of extracellular biopolymers cross-linked by iron peroxidase reactions. The consequent healed wound is seen as a heavy browning of the surface or even a bulge on the surface – e.g. a gall. The peroxide attack is limited by catalase removal of peroxide. The control of peroxide action may be due to the organic compound salicylate (note here the use of salicylates in human medicine under the name of aspirin). In Section 8.10, we drew attention to the different use of peroxide by fungi in the synthesis of halogenated compounds as protective poisons.

In animals, there is a protective system similar to that in plants and fungi in which a peroxidase is used to attack an invading organism trapped within a vesicle. The attacking agent may be chloride, oxidised to hypochloride, ClO^-. Note that such oxidation requires very high redox potential catalysts, here haem, only available later in evolutionary time. There is a risk in this procedure of the accidental liberation of peroxides and superoxide. As stated in Section 8.10, peroxides are removed in the cytoplasm by a selenium-containing enzyme as well as by catalase and to remove superoxide there is also a Cu/Zn superoxide dismutase. These protective systems can only operate under conditions which produce oxygen. Additionally, animals devised a protection from especially UV radiation by synthesising melanin in the outer skin using oxidation of aromatic molecules with *copper* enzymes. Animals also developed a particular protection called the immune system. It is too complex to be described here, but it has one feature of importance in the context of this book – the production of antibodies (proteins) which recognise an object deleterious to the organism and alert the cells able to counteract it. The genetics of antibody production indicate that it is only the *local parts of the DNA in the immune cells* which are responsible for production of antibodies, and upgrading them by (localised) mutation in the DNA. The cell environment, therefore, has a direct *localised* effect on the genetic structure and protein production. The similarity of this mechanism to that we have discussed for directed DNA mutation in evolution must not be overlooked. The question arises as to whether any such mechanism can be inherited. We return to the problem in the next chapters. We must enquire here into the link between all the above changes and the changes in genetic structure.

8.17.   Changes in Genetic Structure

In Section 7.14, we have seen the major changes in the genetic structure of single-cell eukaryotes from that in prokaryotes. The nature of the genetic structure in multi-cellular organisms is not changed further except in its size and in the development of expression. Thus, the number of chromosomes has increased to 46 in humans. There is also an increase in both introns and exons, the latter related, though not closely, to the increase in proteins in these organisms. Many of the increases in the expressed parts of the DNA arise from gene duplication, or increased numbers of introns which allow splicing into novel varieties of proteins.

In sexual reproduction, there is also a high probability of recombination so as to keep variety and adaptability, as noted in Chapter 7. Thus, we see numerous examples in the expansion of the numbers of genes for similar proteins, for example linked to Ca^{2+}, Cu^{2+}, haem, Fe^{2+} and Zn^{2+}. Some 5% of genes are linked to zinc fingers and there are some 200 Fe^{2+} iron regulator proteins. Now the increase in genetic content does not relate directly to what we see as the complexity of organisms. The expressed size of the genes of humans is little different from that in a plant such as *Arabidopsis*. Moreover, many plants have several more chromosomes. There are fish with very large quantities of "unused" DNA compared with those of other animals, and some with almost none. This suggests that perhaps the major change in organisms is not just in the content of the DNA or other parts of the gene structure (as organising centres) but lies in the relationship to the information from the environment interacting with metabolism and then on some genetic controls (see Chapter 9). Insofar as splicing, for example, is dependent on the environment then the environment affects gene expression. We explore this point of view further in Chapter 11. As we have mentioned in Section 7.14, an alternative explanation is that the bits of transcribed DNA, that is of intron-derived RNA not translated form a pool of small interfering RNAs (RNA_i), which are themselves involved in novel control (see Couzin, and also Mattick in Further Reading). Note again that if this is the case the roles of Mg^{2+} and K^+ associated with RNA increase in importance and of course the RNAs have concentration dependences. There has to be an explanation of why so many species, even chemotypes, arose when oxygen entered the atmosphere in quantity (see Section 11.10). Was it an indirect effect of environmental change via controls?

A different problem concerns the condensation of chromatin beyond the stage of nucleosomes into compact stacks. In these stacks the DNA is silent and it is known that the silencing depends on the presence of various DNA-binding proteins and their chemical modification. All of this complicated apparatus is part of the differentiation system but it is not known if any small molecules or ions are also involved (see Mohd-Sarip and Verrijzev in Further Reading).

Notice too that several essential genes found in simpler organisms have been lost in multi-cellular organisms. These organisms are therefore more dependent upon other organisms than are the unicellular eukaryotes. Thus, the animals are hosts to symbiotic bacteria and lower eukaryotes, which produce vital chemicals for them. How did this gene loss come about unless the genes "knew" of the supplies from outside? It would be strange if there was random losses of genes! We expand on this theme in Chapters 10 and 11.

8.18.   Degradation Activity and Apoptosis

Deliberate degradation of biopolymers, even of cells, is very developed in multicellular organisms. We have already stressed the need for it externally in the constant remodelling of organisms during growth. There is also the obvious need for external digestion of food and for protective degradation of foreign cells and biopolymers. In all cases, similar zinc proteases and peptidases, often activated by calcium, are released from organs of animals for digestion. There is also a set of nucleases (Ca/Zn) and saccharases (Ca) produced for similar use, which, acting all together can give extra protection from foreign bodies. Internally, protein degradation is managed initially by a process called ubiquitinisation in which a group of proteins aids unfolding of proteins opening them to hydrolytic attack in a special particle, a proteasome. The ubiquitin group includes a *zinc* protein.

Systematic cell degradation and death, apoptosis, on the other hand, is thought to be necessary to avoid cell damage accumulating so as to cause incorrect differentiation. The process is very complicated involving activation of a number of destructive enzymes where once more increase in cell calcium often initiates the hydrolyses using special internal calcium-dependent enzymes, calpains (see Demaurex in Further Reading).

8.19.   Conclusion

The advantages of the evolution of multi-cellular plants are listed in Table 8.22 and those of multi-cellular animals are brought together in Table 8.23 with the link to human beings in Chapters 9 and 10. It is very important to recognise that these organisms and unicellular eukaryotes only apparently dominate the biosphere due to their visibility. In fact, prokaryotes and unicellular eukaryotes are far more numerous than the multi-cellular eukaryotes. Moreover, the latest organisms to evolve are dependent upon those previously present for supplies of basic elements and many compounds (Table 8.24). The reason for the existence of large plants is

TABLE 8.22

THE EVOLUTIONARY GAINS OF MULTI-CELLULAR PLANTS

Gains	Organ
Optimal light absorption	Leaves
CO_2 uptake	
Optimal energy conservation to	All cells through distribution via veins,
C/H/O compounds	stems, branches, trunks
Commensurate H_2O and mineral uptake	Roots (+fungi)
NH_3 uptake from N_2	Roots + bacteria
NO_3^- uptake	Roots (+bacteria)
Improved rejection of poisons	O_2 release by leaves
	Shedding of leaves release of Al, Ni, Co
Development of hormonal control	Leaf and root exchange control
Improved message system	Growth connected to light. Ca^{2+} triggers
	leaves
Improved reproduction	Seeds replace spores (flowers)

TABLE 8.23

THE EVOLUTIONARY GAINS OF MULTI-CELLULAR ANIMALS

Gains	Organ (Elements)
Fast observations	Nerves (Na, K, Cl, Ca)
Fast reactions (motile)	Muscles (Ca message)
Controlled growth	Glands (hormones)
Food consumption	Digestive systems (Zn, Ca)
Element distribution	Blood (Fe(O$_2$), all elements)
Element control	Extracellular fluid (Table 8.7)
	Liver (Fe, etc.)
	Kidney (Na, K, etc.)
Internal minerals	Bone (Ca, P)
Memory	Brain (see Chapter 9)

TABLE 8.24

EXAMPLES OF EXTERNAL SUPPLIES TO MULTI-CELLULAR ORGANISMS

Multi-Cellular Organism	Supply
Plants	Nitrogen from bacteria
	Minerals from fungi
Animals (humans)	Vitamins i.e. many coenzymes in food
	Minerals in food
	Essential amino acids in food
	Essential saccharides in food
Animals (ruminants)	Carbon from cellulose (bacteria)

then that they can take advantage of their organised ability to capture and use available energy (light) and materials by occupying previously unoccupied space, while that of large animals is their searching and scavenging abilities. They then increase synthesis and energy degradation rate, respectively. Note, for example, how higher organisms increase energy turnover in that "bacterial" organelles for degradation (e.g. mitochondria) are linked with bacterial syntheses (e.g. chloroplasts) in a unity in the ecosystem through plants and animals. The two increase even together in plants but in a sense plants and animals are only large-scale bacteria degrading light to heat. However, their life and organisation is very complicated and is of considerable length and hence they are not competitors of free unicellular organisms of less complexity, faster reproduction rate and easier adaptation. The possibility of their existence with prokaryotes was closely dependent on the rise of oxygen content of the atmosphere and the consequent change in the available elements from the environment. A feature we must stress is that with them the ecosystem as a whole is moving towards an all-inclusive, element neutral cycle. We shall not look further at this cooperative structure of the living part of the ecosystem until we consider it together with the role of the environment in the evolution of the whole ecosystem in Sections 9.11–9.13 and again in Chapter 11. These sections come after we have examined the nature and chemistry of a further group of chemotypes – animals with a nervous system and a brain, which are the topic of Chapter 9, and finally of man, in Chapter 10 But keep in mind the continuous evolution from molecular cell to synchronous cell macro-machinery as cells cooperate.

The features of all the changes in the ecosystem which we have now outlined are continuous and unavoidable in a thermodynamic system, which seeks to become thoroughly efficient and cyclic. We must see that evolution is a systematic change of the ecosystem through adjustment within chemotypes that are cooperative. Here, chemotypes refer to organisms making greater use of available elements with as little waste as possible. The direction of change is due to the inevitable increase in oxygen which, with its products, becomes one half of the energy storage in the cycles we describe. Its increase is systematic as the total trapped hydrogen and carbon in life increase. It is the oxygen build-up that requires the systematic change in use of elements in the organisms. This change requires better use of space, albeit divided space and greater localised organisation. These parameters of change, components, concentrations, space (with divisions) and communication cannot be seen by examining sequences or morphologies. They indicate that the linear trunk and main branches of the cellular evolutionary tree are inevitable thermodynamic consequences, the drive to increase the rate of thermal entropy production, given the initial starting system. They occur with the three dimensionally continous evolution of chemicals in the environment to give an evolving evolutionary cone, see the cover of this book. This directional change is associated with increase of light and material capture and is in accordance with the survival of fitness of the ecosystem and that of organisms. The change does not contradict survival of the randomly produced fittest species in selected environments. We shall see how mankind's activities are a considerable further leap in this direction but with qualifications, since suddenly a species becomes a completing novel chemotype, guided by new features.

Further Reading

Books

Corning, P.A. (2003). *Nature's Magic*. Cambridge University Press, Cambridge

Fraústo Da Silva J.J.R. and Williams, R.J.P. (2001). *The Biological Chemistry of the Elements*. Oxford University Press, Oxford

Mann, S. (2001). *Biomineralization: Principles and Concepts in Bioinorganic Materials Chemistry*. Oxford University Press, Oxford

Margulis, L. (1998). *Symbiotic Planet*. Basic Books, New York

Postlethwait, J.H. and Hopson, J.L. (1989). *The Nature of Life*. McGraw-Hill, New York

Purves, W.K., Orians, G.H., Heller, H.C. and Sadava, D. (1998). *Life – the Science of Biology* (5th ed.). Associates, Inc. and W.H. Freeman, Sunderland, MA

Sapp, J. (1994). *Evolution by Association*. Oxford University Press, New York

Sigel, H. and Sigel, A. (eds.) (1995). *Metal Ions in Biological Systems, Vol. 31- Vanadium and its Role in Life*. Marcel Dekker, New York

Southwood, R. (2003). *The Story of Life*. Oxford University Press, Oxford
Stryer, L. (1995). *Biochemistry* (4th ed.). W.H. Freeman and Company, New York
Thompson, D'Arcy, M. (1988). *On Growth and Form*. Cambridge University Press, Cambridge (reprinted from the 1961 edition)
Tracey, A.S. and Crans, D.C. (eds.) (1998). *Vanadium Compounds – Chemistry, Biochemistry and Therapeutic Applications*. Am. Chem. Soc. Symposium Series, Vol. 711, American Chemical Society, Washington, DC.
Villee, C.A., Solomon, E.P., Martin, C.E., Martin, D.W., Berg, R.B. and Davis, P.W. (1989). *Biology*, (2nd ed.). Saunders College Publishing, Philadelphia
Williams, R.J.P. (2004). Calcium, biological fitness of. In *Encyclopedia of Biological Chemistry* Vol. 1. Elsevier, Amsterdam, pp. 270–273

PAPERS

Berner, R.A. (2003). The long-term carbon cycle, fossil fuels and atmospheric composition. *Nature, 426*, 323–326
Campisi, J. (2004). Fragile fugue: p53 in aging, cancer and IGF signalling. *Nat. Med., 10*, 231–232
Coghlan, A. (1998). Sensitive flower. *New Sci., 26*, 24–28
Couzin, J. (2002). Small RNAs make a big splash. *Science, 298*, 2296–2297
Demaurex, N. and Distelhorst, C. (2003). Apoptosis – the calcium connection. *Science, 300*, 65–67
Forsen, S. and Kordel, J. (1993). The molecular anatomy of a calcium-binding protein. *Acc. Chem. Res. 26*, 7–18
Hewitson, K.S., McNeill, L.A., Elkins, J.M. and Schofield, C.J. (2003). The role of iron and 2-oxoglutarate in signalling. *Biochem. Soc. Trans., 31*, 510–515
Hildebrand, M., Volcani, B.E., Grassmann, W. and Schroeder, J.L. (1997). A gene family of silicon transporters, *Nature, 385*, 688–689
Hodgson, M.J., White, P.J., Mead, A. and Broadley, M.R. (2005). Phylogenetic variation of the silicon composition of plants. *Ann. Botany, 96*, 1027–1046.
Kröger, N., Lorenz, S., Brunner, E. and Sumper, M. (2002). Self-assembly of highly phosphorylated sillafins and their function in biologic morphogenesis. *Science, 298*, 584–586
Marx, J. (2004). Inflammation and cancer: the link goes stronger. *Science, 306*, 966–968
Mattick, J.S. (2004). The hidden genetic program of complex organisms. *Sci. Am., 291*, 31–37
Meyerowitz, E.M., (2002). Plants compared with animals: the broadest comparative study of development. *Science, 295*, 1482–1485
Mohd-Sarip, A. and Verrijzev, C.P. (2004). A higher order of silence. *Science, 306*, 484–485 and related articles referenced there
Morgan, R.O., Martin-Almedina, S., Iglesias, J.M., Gonzalez-Florezm M.I. and Fernandez, M.P. (2004). In special issue on 8th European Symposium on Calcium. *Biochim. Biophys. Acta, 1742*, 133–140 and other papers therein
Pennisi, E. (2003). Symbiosis – fast friends, sworn enemies. *Science, 302*, 774–775
Pennisi, E. (2004). DNA reveals diatom's complexity. *Science, 306*, 31 and see also pp. 79–86
Pennisi, E. (2004). The secret life of fungi. *Science, 304*, 1620–1622
Perry, C.C. (2000). Biosilicification – the role of the organic matrix. *J. Biol. Inorg. Chem., 5*, 537–550
Ray, L.B. Adler, E.M. and Gough, N.R. (2004). Common signalling themes. *Science, 306*, 1505 and the related articles referenced there
Redman, R.S., Sherman, K.B., Stout, R.G., Rodriguez, R.J. and Henson, J.M. (2002). Thermotolerance generated by plant/fungal symbiosis. *Science, 298*, 1581
Schofield, C.J. and Ratcliffe, P.J. (2004). Oxygen sensing by HIF hydrogenase. *Nat. Rev. Mol. Cell Biol., 5*, 343–354
Thauer, R.K. (2001). Nickel to the fore. *Science, 293*, 1264–1265

CHAPTER 9

The Evolution of Chemotypes with Nerves and a Brain (0.5 Billion Years Ago to Today)

9.1. Introduction

At the end of Chapter 8, we have emphasised our major theme that the whole co-operative development of organisms from prokaryotes to higher animals and plants within an evolving environment had one drive – the optimal uptake and use of energy moving towards a final cyclic steady-state flow system involving all those chemical elements, which will be finally available on Earth, and in this drive there is a movement towards functional optimalisation of chemical element use *in particular compounds*. Because all the chemicals produced are unstable (in bond energy and/or gradient), it is not only the drive to optimal rate of energy absorption but also of optimal degradation of chemicals that is achieved in a fully cyclic system, and the activity therefore corresponds with an optimal rate of thermal entropy production by the system on reaching the final steady state; see Section 3.6. We have seen how the cellular organisation with its environment, that is the ecosystem on Earth, moved towards this objective developing from energy a greater and greater separation of reduced from oxidised chemicals in a larger and larger number of organised compartments. (Note that the final objective may be unattainable as we discuss in Chapter 11.) The cellular organised activity was based, at first, mainly upon the internal DNA/RNA code, with appropriate feedback molecular machinery to create material and information for controlled internal overall development, homeostasis with fast repair and reproduction. We saw that later certain more complicated organisms were increasingly informed by the external field gradients, structures and the chemical composition of the environment. The organisms as a whole changed in an unavoidable general systematic way so that they could manage the changes of the environment which preceded them and which took place in an inevitable oxidative manner. Since we concentrate upon the chemistry and in particular the chemical element changes, we gave the name chemotypes to organisms at different evolutionary stages. The increased complexity of the total necessary set of chemical operations in later organisms made reliance on one another, especially earlier organisms, that is between chemotypes of organisms, unavoidable to gain efficiency. The later organisms are slower in reproduction but have organisational advantages that allow coexistent and cooperative survival with the earlier more numerous life forms. The final objective can only be achieved if all organisms within the environment work together in an ecosystem to retain energy to as great a degree as possible without further environmental waste, even though some, mostly similar, organisms, may compete with and even feed on one another. The retention of energy is not in chemicals of organisms alone but it is also with respect to those of the environment. As stated, there is no discernible management of this whole ecosystem except the overall drive of all the chemicals together to capture and degrade energy, hence increasing thermal entropy production. Each organism is tightly managed, but our concern with individuals or species (until man arrived) is slight since they only made contributions to the survival strength of the whole, that is within and between chemotypes or very large groups of species. We shall see in the chapters 10 and 11 how new management has been introduced by one species, humans, a chemotype in its own right, using additional external information to that sensed by earlier organisms and this is leading to mankind's attempted new organisation of the whole ecosystem in the hidden belief that his understanding will lead to an optimal sustainable ecosystem. In other words, one species has become a novel chemotype with strong ideas concerning its own (even individual) importance and it is now attempting domination of the ecosystem, environment plus organisms. We summarise in this chapter the way this situation has arisen; chemical details of the rise of this dominant species are given in Chapters 10 and 11. It arises through the practice and improvement of scavenging and utilising the environment material and energy. It is on organisms which can use the environment better that evolution depends.

Biological chemistry is overwhelmingly that of some 15–20 elements, which must be obtained and utilised using the flow of available energy. The main energy flow is that of sunlight managed by plants in the production of carbohydrates and oxygen. Plants, as we have seen in Chapter 8, cannot gain easy access to all the elements even for their own lifestyle, nor can they scavenge their own debris. Consequently, plant life had to be supported by better collectors and scavengers to complete the biological cycle. It was in part the complexity of higher plants that led to the additional need of reducing the complications of synthesis. We have seen that bacteria and fungi help to supply elements and some metabolites to plants and with animals they are also major scavengers. As plant life has developed to what may be close to an optimal condition of light capture, animal and fungal life has increased to keep pace with the needs of plants and the opportunities of debris consumption. Molecular scavenging machines were complemented with macro-, organ, machines increasingly, especially in larger and larger mobile animals. Chapter 8 described this increase, particularly of the multi-cellular animal as a consumer of the excess carbohydrate of plants, using it, with the oxygen, also released by plants, to get energy for the creation of a larger total biomass. The animals took in a much larger percentage than did plants of elements other than C/H/O. Now the demand of an overall good scavenger system can be best met by the development of a mobile organism with senses. As mentioned in Chapter 8, this required the further development of an organised centre in animals with generalised connections to all its parts. At first, these connections were made by chemical messages, but later animals acquired a variety of senses, which stimulate particularly muscle response by a quite new mechanism via nerves. To begin the description of this development we shall describe the senses, which clearly have to be coordinated with action via nerves if response is to be fast. We shall then describe the nature of the new connection through nerves to a new organising centre for response, the brain, which allowed the necessary fast transfer and storage of information from the senses and finally led to an ability to scavenge and use the environment very effectively. The organisms are new chemotypes with nerves and a brain. In turn, this has led to behaviour independent of DNA and to quite new possibilities for evolution to be seen in the activities of mankind (see Chapters 10 and 11).

9.2. Senses

Before proceeding with the nature of the nerve message cells, we must stress what it is that the senses of the primitive and then the advanced organisms can detect or sense (Table 9.1). We have already described the advances from prokaryotes to multi-cellular animals in Chapters 5–8. The original purpose of the communication network of these simple multi-cellular animals (and plants), as we have noted, was to have control over *internal* response to opportunity in the environment, that is to coordinate energy and metabolism in addition to controlled slow development and growth. They used calcium and some organic messengers in external fluids for these moderately quick responses as we have described. As organisms grew in size and had to scavenge, there was the need for more rapid coordination of muscle movement with information from the external environment using sensors and long-range internal connections. The initial responses of eukaryote organisms were dependent on what could be sensed by touch and chemical gradients (food), smell, and may have included the detection of gravitational, electric and magnetic fields. Many early animals also sensed temperature (radiant heat) with and without touching solid objects, and could detect light and possibly sound. Now the information, which an improved sensor system can possibly use, apart from the detection of

TABLE 9.1

POSSIBLE ORDER OF DEVELOPMENT OF SENSES

Sense	Period of Development (Organism)	Source
Chemical smell/taste	Very early (Prokaryote)	Food Chemical environment
Physical fields gravity, magnetic electric	Very early (Prokaryote) later balance in the "ear"	Earth's properties
Contact touch	Two billion years ago (single-cell eukaryote)	Physical environment Ca^{2+} messages
Light	< Two billion years ago eye spot (single-cell eukaryotes)	General environment (Sun) radiation waves
Sight colour discrimination	5×10^8 years ago eyes multi-cellular Organisms	Long-range environment discrimination radiation waves
Sound hearing	5×10^9 years ago ears language (50,000 years ago)	Air motion other organisms Sound waves (air)
Machines	200 years ago radiation	Machines to senses

these features of the environment lies in *a discriminatory sensing to cover separated inputs* of a wide range of frequencies of both electromagnetic (light and heat) and sound. Seeing and hearing in animals are clearly more sophisticated than the more primitive senses, since they have detectors that distinguish these frequency and intensity differences quite sharply. The environment contains both advantageous and disadvantageous conditions, which these radiations reveal in detail. Today, in higher animals, the detectors in the receivers (eyes and ears) are extremely discriminating and sensitive to them. As we have stated, the development of these senses over a long period of evolution is well documented in biological texts (see Further Reading). The upshot is that the information from the environment from radiation of all kinds has increased gradually and remarkably. (It has been increased further by mankind's devices; see Section 10.8.) The acknowledgement of receipt of information at first was only by an automatic *fast cooperative* muscle response achieved, as we shall see, by nerve connections, which added speed to the slower chemical message systems described in Chapter 8. However, at some stage in evolution, the automatic response was complemented by storage of the environment information, no matter how received, in a nerve network, the brain, located near the sense organs in the head. The senses plus the nervous system then generated a store of information not about internal chemistry but about the general surroundings of the organism. The degree of exchange between the brain and the genetic structure is still unknown, but the genetic structure remains responsible only for initial and in part long-term slow internal management. As the reader knows fully well, it is the brain store that plays an equal but very different part from that of genetic controls in higher animal behaviour, especially in the adult human. Like each previous development, the nervous system and then the brain arose from the quite novel use of a group of chemical elements. Note that this use could only come about subsequent to the evolution of a large multi-cellular, motile structure and it is not about short-range *chemical communication* but *fast long-range physical networking* linked by chemical signals to bulk organs. It is not useful for plants but extremely valuable for animals. From our definition, those animals which use elements in such a novel way in nerve compartments are a new chemotype, animals with a nervous system and a brain. Is it the last of all possible developments of internal organisation using materials synthesised inside organisms? Notice that slowly organisms became proactive in the environment.

9.3.   The Development of Nerves

We have observed that animals required fast responses with multiple connectors to many sense and muscle organs. One special fast messenger we have already discussed, which was used in a general way and connected to all kinds of inputs from the environment and outputs of individual cells of unicellular and multi-cellular organisms, was the calcium ion. As we have explained, the use comes from the fact that all cells produce gradients of fast-diffusing, activating calcium ions, low in the cytoplasm and high outside; Major use of these gradients is as immediate cell messengers to alert the cells to the environment. Calcium was incorporated into the external/internal messenger systems for most if not all exchange of information at single-cell membranes of multi- as well as single-cell eukaryotes. The message is received internally at proteins by binding a very small number of these ions. Since cells also have a physical gradient of charge across the membrane due to these ions, bulk calcium ion flow over a large cell area can also be activating as a physical current causing a depolarisation of a part or the whole of enclosing membrane of, say, a muscle. In fact the relatively slow potential pulses of the heart, a multi-cellular machine, in animals utilise calcium currents to activate concerted heart contraction at around one to a tenth of a second. The activity demands calcium input/calcium receipt and calcium rejection in repeated membrane electrical pulses coupled to energy. (The heart is a cooperative pump for the general circulation of fluids and is the equivalent of a massive multimolecular pump in membranes.) While the calcium message enters heart (and other) muscle cells through this transmission as an electrical current, it simultaneously, through relatively strong binding in the interior of the cell, activates temporary specific chemistry before fast removal. The fastest possible environment to cell transmission, about 1 ms, is to some degree inhibited by this need to invoke cell, muscle, action by binding. Clearly, what is desirable for a quicker response of a large organism for faster long-range communication is a charge carrier stored in a gradient, which does not bind, and hence does not cause chemical but only field change during transmission to the point of required action. Any chemical activity is a time and energy waste until the message reaches its action site. In the response, only a local part of a whole body needs to be informed from the senses, which have been activated through specific connections with the site to be made useful via an effective line of connections in an electrolytic system analogous to an electrical wire. These are the demands of a highly mobile large organism, where sense organs are far from points of action and which can be both predator and prey. (It is easy to appreciate the way the whole system works by reference to mankind's electronic circuits.)

The observed system for this fast coordinated action is in fact the only one open to biological evolution – the use of the gradients of the very weakly binding ions generated of necessity by all cells from the beginning of life, that is of Na^+, K^+ and Cl^- (see Section 4.15(d)). Any transfer of such ions across membranes causes fast electrical potential, *physical*, change but no *chemical* change (see Figs. 9.1 and 9.2). Use of this long-range communication based usually on Na^+/K^+ gradients needed cells that conduct over great distances at exchange speeds $>10^3$ per second. We see this development in nerves, long thin tubes extended over metres, which are only possible in a multi-cell organism with already formed connective tissue, and hence shape. The evolution of this message system also depended upon the ability

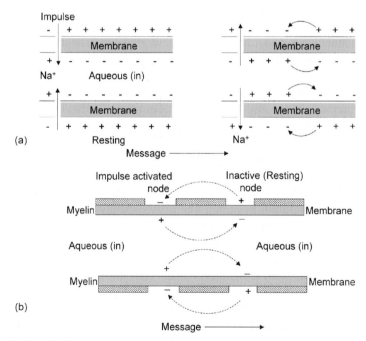

FIG. 9.1   (a) The distribution and flow of sodium ions carrying a message in an axon of a neuron (see Fig. 9.2). The neuron in (a) is not myelinated while in (b) it is, and this enables faster transmission in higher animals. Note K^+ flow is in the opposed sense to that of Na^+.

(a)

(b)

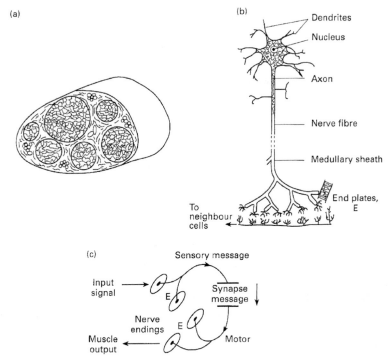

Dendrites

Nucleus

Axon

Nerve fibre

Medullary sheath

End plates,
E

To
neighbour
cells

(c)

Sensory message

Input
signal

E

Synapse
message

Nerve
endings   E

Muscle
output

Motor

FIG. 9.2   A general impression of the construction of nerve cells (a) cross-section of a nerve trunk, (b) a neuron, (c) sensory and motor nerve connection via a synapse.

to recover quickly from the Na$^+$/K$^+$ change by fast pumping out just as the Ca^{2+} message system is dependent upon somewhat slower exiting of the Ca^{2+} ions. (The energy used in transmission is considerable.) The evolutionary linkages between some of the known pumps are shown in Fig. 9.3. Note how the early type I ATP-ase pump eliminates excess heavy metals while the later type II pump removes excess of simpler ions starting from Ca^{2+}. The Na$^+$/K$^+$ ATP-ase pump is of very recent origin. However, the nerve message had to produce a response at termini, that is, for example at the locomotive organs, muscles, nerve relay junctions or any other organs that are capable of being activated. Therefore, there had to be at the nerve endings, the so-called synapses, an accessory system of messenger units, which could bring about action. The obvious choice was influx of calcium ions, which, stimulated by an Na$^+$/K$^+$ nerve current locally at synapses, could flow in pulses, into and then out of a cell. As stated, Ca^{2+} ions cause contraction and chemical changes directly in say muscle cells by binding. Additionally, organic ions, transmitters such as acetylcholine, adrenaline, etc. can be released from vesicles from within the receiving cell synapse by the calcium pulse to communicate selectively to adjacent cells. The organic messengers can stimulate in turn the selected adjacent cells at receptors causing calcium input across the second cell's membrane.

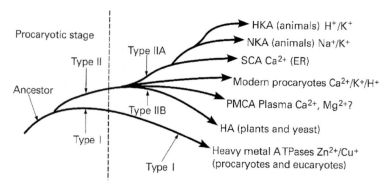

FIG. 9.3   A possible evolutionary connection between the P-type ATP-ase pumps. All are Mg^{2+}-dependent. (After J.V. Møller *et al.* (1996) *Biochimica and Biophysica Acta*, *1286*, 1–51.)

The calcium influx can then initiate a further Na^+/K^+ message flow, making an amplified conduction cable to more distant cells. We can see, with hindsight again, that this use of elementary chemistry, conduction by chemically innocuous ions – and electrolytic physical currents-connected to previously devised flux of Ca^{2+} and chemical binding agents was inevitable, given the pressures to increase scavenging. Once the demands of osmotic and electrical neutrality of the first prokaryote cells forced energy to be used to create physical gradients of these ions, Na^+, K^+ and Cl^- (see Section 4.15(4)), evolution was almost bound to use them sooner or later in messages as organisation increased in size. It waited upon organisational need in *large scavengers* for them to be used coordinatively, since in single cells this kind of communication had no advantage. The choice of calcium and then of organic transmitters at synaptic terminals was also virtually inevitable since they must be small (for fast diffusion), charged or polar so as to be retained in vesicles or held outside cells, bind relatively strongly and must not be substrates of the main metabolism (Table 9.2). Note that the order of the use of messengers in evolution follows the order of organisational complexity: (a) single small cells, prokaryotes, Mg^{2+}, Fe^{2+} and internal organic transcription factors especially phosphates; (b) larger single eukaryote cells with added calcium internal/external messengers connecting to extra-internal organic factors, especially phosphates; (c) multi-cell eukaryotes with added organic molecules external to and going between cells but stored in the organism, as well as the above Ca^{2+} and internal transmitters, and finally (d) all of these messengers were combined together with Na/K ionic transmission in advanced animals. Some of the organic transmitters are largely used in the brain, e.g. glutamate, but they are also employed elsewhere and their receptors can be traced back to bacteria. There are three more possible stages for still larger organisations – the uses of completely external sound, electron and radiation transmission – and in chapter we shall see how and when they were introduced. How else can evolution proceed given that it starts as a "simple" chemical cell system except by this gradual increase in types of message systems as organisation developed

TABLE 9.2

MESSENGER COMPOUNDS BETWEEN CELLS IN THE BRAIN

Excitatory (+)/inhibitory (–)
  Glutamate (+)
  Glycine (–)
  GABA (γ-aminobutyric acid) (–)
  Acetylcholine
  ATP (+)

Noradrenaline (norepinephrine)
Dopamine
Serotonin
Inorganic signals
  Na^+, K^+, Ca^{2+}, (Zn^{2+}), Cl^-
Nitric oxide
Carbon monoxide

Peptides
Neuropeptides (large number > 20)
Substance P
Cholecystokinin
Corticotrophin-releasing factor
Melatonin

---

using more and more space? The space was occupied by larger and larger units connected by bulk communication and mechanical devices. Also note how critical deployment of information carriers (as well as of information itself in say DNA) is to the control of cells in evolution. Surely the development is systematic with the whole development of chemotypes generated or allowed by environmental change (see Williams (2003) in Further Reading).

By its very nature this complicated detection and action system enables animals to sense the environment, move and act appropriately. Such a system involved several separate senses and it is not surprising to note that in scavenging animals there is different development in different species. For example, a nematode worm in the soil needs only a local sense of touch; many animals, e.g. dogs use smell strongly; birds need the sense of sight more than that of smell. It is interesting to speculate which came first in the form of a fully developed organ for a given sense, but we must remember that even unicellular bacteria have two or three sensing devices for environmental features. As mentioned above, a multitude of sensors needs a fast coordinating centre to be effective, and this cannot be based on synthesis such as that of the slow DNA-based response. A bundle of nerves can be made ideal for this purpose, as we show next, but, before we do so, observe that energy is now spent in physical activity and electrolytic transmission to a much greater degree than seen previously.

9.4. The Brain

Given a set of electrical nerve connections between parts of the body, the need for joint action was for equally fast cooperation between them and the cells or organs involved needed to have their subsequent chemically transmitted messages coordinated. Then there developed a central organisation of nerves closely associated with sensing apparatus in the head of animals. As stated, the first example is the nematode worm. The connections between muscles for whole body movement in it are via these long parallel tubes, nerves, running to a ring of linked nerve termini, i.e. synapses, in the head. The effect is to coordinate all the muscle actions of the body automatically and rapidly by synchronous response initiated at the head but independent of genes themselves. The whole body has become a macro-machine developed from cooperative molecular machines.

As the brain developed, it became encased in a separate extra-cellular fluid, i.e. the cerebrospinal fluid (CSF), to improve conduction for transmission. The encasing encloses the nervous system down the spine as well as in the head and its inorganic/protein composition is novel (see Table 9.5). The major electrolyte balance of Na^+, K^+, Cl^-, Mg^{2+} and Ca^{2+} is much as in blood plasma but the trace element concentrations carried by proteins and of proteins themselves are extremely low. Notice that the use of the ions and molecules in nerve conduction in extra-cellular fluids conserves energy and material as they are no longer rejected to the environment, see also the function of the kidney and of the thylakoid in chloroplasts. Clearly, the nerve cells are protected from very many adverse factors including heavy metal elements. Note how the content of the free heavier metal ions from Mn to Zn is reduced in steps from the modern sea to blood plasma to CSF (see Table 9.4). Much though these elements in bound states are valuable as free ions, they are poisonous and are rejected by cells to low internal values. There is some suggestion that a second set of brain cells, glia, supply the nerves with the required metal ions and nutrients but their functions generally are uncertain (see Berry *et al.*, Fields, and also Miller *et al.* in Further Reading).

As this central organisation grew it developed a novel feature. The effect of a nerve message to a bundle of central nerves in the brain came to bring about *specific cell growth* and *storage of chemicals* in them (see Thompson in Further Reading). These two features, nerve growth and storage with communication between them reflected experience of the environment (a memory). Rather than being a centre of automatic responses the nerves in the brain became increasingly a store of images, an information database in a memory, at nerve termini, synapses, related somewhat differently but cooperatively to the different senses as they evolved (see Fig. 9.4) (see Kandel and also Ullan *et al.* in Further Reading). "Bits" of information here have a novel feature – intensity within considerable 3D structure. Quite differently from DNA and a computer code a bit is now not an all or nothing store but a (variable) concentration at a synapse and as a result it is much more prone to variation even error (and in the end perhaps to what we call imagination and even creativity?). Concentration storage gives memory a variable intensity factor dependent upon the size of and the concentration in the store, but a store can and does decay. Nerve cells also are cross connected by extensive neuron axons (see Fig. 9.2) so that while having a local intensity storage (memory) they have an interlocked three-dimensional character. A very interesting feature of such a system is that to retain a memory there has to develop, to grow, cell–cell connections. In other words, the synapses must make contacts which locally have a coded representation of an observed object, an image, and which can be recalled on demand. We must see that this development of memory in the brain is a novelty, which is distinct from that in DNA and unrelated to it, though protein synthesis capability is still DNA dependent of course. Certain synthesised proteins are held at the synapses in vesicles to enable fast growth of contacts from sense observations without novel synthesis. Apart from the intensity factor which gives a huge potential for variation, the number of synapses is at least 10^5 times larger than the number of bases in human DNA. The brain also puts the environment in immediate fast connection with the organisms in line with the development of evolution, generally which shows an increasing connection between the two without involvement of genes. Of course, this connection is not inherited in any obvious way, yet there are observations suggesting that such examples as migratory maps used by animals are inherited. Maps represent the environment and are connected to small, local single organism units not species! They are almost individual since, for example, every bird has its own nest. At present we do not understand fully the link between directly inherited DNA and such external information – if there is one. (Note also that the information in the brain can be inherited in the sense that it can be passed on by learning from a previous generation.) It has to be noted that abilities have appeared very rapidly in animals as if discovering use over generations can be imposed on genes. If true, how does this come about? Let us consider the way in which the brain has evolved. Before doing so, we note again that much energy is constantly spent in the continuous activity of the brain cells that has nothing to do with the previous energy flow in the body of animals.

An important point is that the brain retains instructions introduced into it as a memory so that the dictates of DNA can be of less importance in behaviour. The decision-making process is now extremely complex.

9.5. The Physical Evolution of the Brain

As in previous chapters, to appreciate the use of chemical elements in a new structure, here the brain, the nature of its compartments must be described. The evolution of the brain from the nematode can be traced all the way up to man in its quickly changing construction. No doubt there are lots of possible different arrangements of the zones of the brain but even so a line of "ascending" organisation can be drawn over a few hundred million years (Fig. 9.4 and Table 9.3). Particular features not stressed so far are the mutual development of the brain nerve and glial cells in zones, the possible functions of the zones (Table 9.4), the position

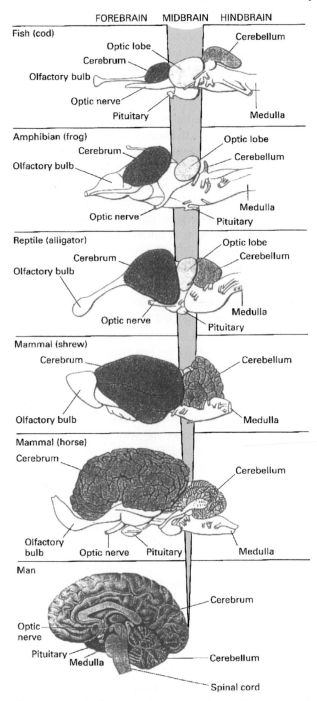

FIG. 9.4   Evolutionary change in the zones of vertebrate brains. Note that the figures are related here to the cerebrum cortex so that in man the hindbrain appears twisted downwards to allow for human upright stance.

TABLE 9.3

POSSIBLE EVOLUTION OF THE PHYSICAL PROPERTIES OF THE BRAIN

Species	Brain Characteristics
(1) Nematode (>500×10^6 years)	Simple nerve net from body to head region. No senses except touch and general chemotaxis. Ring (of nerve cells) brain
(2) Amphioxus (~500×10^6 years)	As for (1) with an eye spot. Some coordination of sight with the body. No evidence of olfactory or hearing senses. Little or no telencephalox or cerebral cortex. Has forebrain, pineal and hypothalamus like systems
(3) Jawless fish (450×10^6 years)	As for (2) with olfactory system and a cerebellum for maintaining posture while viewing. Very small cortex and telencephalox
(4) Vertebrates with jaws (400×10^6 years)	As for (3) with ever-more complex states of awareness. Development of myelin and of mid-brain. Enlarged cortex
(5) Warm-blooded animals (350×10^6 years)	As for (4) with temperature controls. Enlarged cortex and states resembling emotions and consciousness

Taken in part from J.M. Allman, *Evolving Brains*, Scientific American Library, New York, 2000.

TABLE 9.4

SOME MAJOR FUNCTIONS OF BRAIN ZONES[a]

Amygdala, thalamus (limbic system)	Emotional and behavioural responses
Hippocampus	Spatial resolution, learning, memory autobiographical events
Cerebellum	Movement coordination
Tectum (pineal)	Light and auditory responses
Hypothalamus	Controls autonomic nervous and endocrine systems: internal milia
Red nucleus and substantia nigra	Motor system connections
Locus ceruleus	Vigilance, sleep/wake cycle
Medulla	Control of heart and skeletal muscle

Taken in part from J.D. Fix, *High-Yield Neuroanatomy*, 2nd ed., Lippincott, Williams and Wilkins, Philadelphia, PA, 2000.
[a] Zones are interconnected but the very fact that they differ in chemical composition, physical appearance and that damage to them causes differential disturbance indicates that they have major, separate functional roles in addition to general cooperative ones.

of several glands in relation to the brain, the nerves which form an effective continuation down the spinal chord and the final substantial size of the cerebellum in humans. (Note again that in addition to nerve cells, the brain has an even larger number of non-conducting *glial cells*, which are probably cells for looking after the nerves as house-keepers (see Pellerin and Magistretti in Further Reading). The earliest parts of the brain as we noted were those connected to the automatic reflexes, now much associated with the spinal chord contrast the nematode. These automatic responses are traceable to the genetic structure, but it is unlikely that much of the other sophisticated responses can be so directly traced no matter that they have an underlying initial genetic origin. The reason for this doubt lies in the nature of the development (growth) of the brain. Before describing it we return to the glands, some of which are in the brain. These were seen as part of the development of the *chemical message* system in multi-cellular eukaryotes (Section 8.13), but in higher animals rather than being solely under direct chemical transfer in the external fluids the glands are activated, with chemical release, and coordinated with the nervous system and the brain. Consequently, the release of hormones (as opposed to transmitters) is in part controlled by images formed in the brain from data sensed in the environment. What was just an internal physiological control system has become linked to external circumstances and as a consequence the psychological and the physiological became intertwined. The fifth unusual feature quoted above of the brain is the huge relatively very recent increase in the cerebellum, which we see in mankind (see Fig. 9.4). Is it this that has allowed the peculiar evolution of mankind? How otherwise has "random" mutation produced such quick local change? Now as in all chapters of this book we must ask about the general chemical content of the brain in the developing physical constructions. Observe one other physical factor first: later animals came to have a fixed internal temperature (37°C for humans) and this thermodynamic control over kinetics has greatly helped to refine the use of the brain chemicals.

9.6. The Chemical Element Composition of the Brain

So far we have treated the brain mainly as one large chemical compartment. In fact, it has a great variety of compartments and we have shown in Fig. 9.4 and Table 9.4 that they have different chemical functions. The compartments are surrounded by the CSF, the content of which, as shown in Table 9.5, is quite different chemically from the other external fluids in the bodies of organisms. Now each of the other compartments has its own cellular functions, and its own complement of enzymes associated with a particular complement of metal ions and transmitters (Table 9.6). Outstanding are the different sites of iron, copper and zinc concentrations (see Williams (2003) in Further Reading), and as a result some zones even have different colours. We have no understanding of why different transmitters are synthesised and/or used to a greater or lesser extent in different brain regions, necessitating different metalloenzymes for their syntheses. In fact, the basic nature of the relationship of this chemistry to brain function is not known. The possible times of the stages of evolution of different chemical systems, are given in Table 9.7 but they are largely speculative. Clearly, much analytical work is needed to supplement the Tables in this section. Great interest also surrounds the genetic

TABLE 9.5

ELEMENTS AND COMPOUNDS IN RAT BRAIN CSF AND ORGANS (meq kg^{-1} H$_2$O)

	Plasma	CSF	Cytoplasm
Na	148	152	20
K	5.3	3.36	140
Ca	6.14	2.2	10^{-4}
Mg	1.44	1.77	8.0
Cl	106	130	40.5
Glucose	7.2	5.4	
Pyruvate	0.17	0.18	
Lactate	0.7	2.0	
Proteins (mg per 100 ml)	6500	25	10,000

From R.J.P. Williams and J.J.R. Fraústo da Silva, (1997). *Bringing Chemistry to Life*, Chapter 15. Oxford University Press, Oxford

Note: For recent data on transition metals in human CSF, blood serum and whole blood (TXRF analysis) (see Boruchowska *et al.* in Further Reading).

TABLE 9.6

ELEMENT CONTENT OF BRAIN ZONES (μg g^{-1} DRY WEIGHT)

Zone	Zinc	Copper	Non-haem iron
Hippocampus	> 65	6.6	
Cerebellum	30–50	10.4	30
Cortex			10
Thalamus	< 25		50
Locus ceruleus	—	62.0	
Substantia nigra	—	18.8	

basis of brain development (see Pennisi in Further Reading.) The peculiarity of the chemistry of the brain of animals with nerves again sets these organisms aside as one novel chemotype.

9.7. The Brain Development as an Information Store: The Human Phenotype

The brain is the strangest of all organs as it is divided into so many highly linked parts (Fig. 9.5). The most interesting feature from the point of view of this book is that it is very large, containing in mankind over one million billion intercommunicating nerve contacts as well as many more glial cells of several types and functions. It is very malleable after initial formation, much more so than any other organ, and its late development allows the individual phenotype to be strongly delineated in

TABLE 9.7

POSSIBLE EVOLUTIONARY STAGES OF CHEMICAL SYSTEMS IN BRAIN

Animal	Innovation
Nematode neurons (Pre-cambrian)	Na^+/K^+, Ca^{2+}, acetylcholine? glycine? Butyrate, GABA Recovery by re-entering synapse
Chordate neurons (early Cambrian)	As above plus first hydroxylations giving serotonin and dopamine; iron/pterin chemistry in cytoplasm; vesicle filled in centre of cell Recovery by amine oxidation (flavoenzymes)
Jawless fish (Cambrian)	As above second hydroxylation giving norepinephrine and amidated peptides; copper chemistry in vesicles Recovery additionally by hydrolysis Zinc enzymes
Complete vertebrates (late Cambrian)	As above plus myelinated neurons ?use of zinc enzymes in glial cells Free zinc in nerve messages NO/haem chemistry? in glial cells?

Note: Many transmitters are synthesised in special zones.

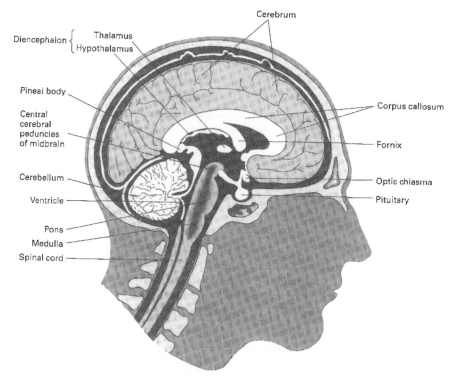

FIG. 9.5 A midsagittal section through the human brain. Note that in this type of section half of the brain is cut away so that structures normally covered by the cerebrum are exposed.

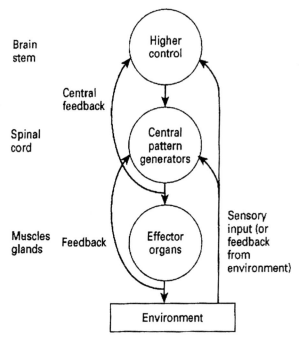

FIG. 9.6   Hierarchical organisation of neuronal circuits governing the control of movement. Note the nesting of feedback loops. (After a figure in Shepherd, G.M. (1988). *Neurobiology*, Oxford University Press, Oxford.)

contrast with phenotypic differences of earlier chemotypes which we have deliberately put to one side. Notice that at birth of a human and some other mammals, such as the mouse, the senses seem to be extremely poorly connected and even the organisation of the nerve connections to the muscles seem of little use in self feeding or movement. It is almost as if the embryo had deliberately insufficient development for self-preservation, so as to allow the brains growth to become environmentally dependent. We only need to contrast a baby fish or a tadpole, which can manage the outside world relatively well immediately after birth, with a baby human which can manage almost nothing, to see the possible advantages and disadvantages of delayed management. The increasing importance of the environment as a source of information in organisms has become strongly related to brain development in humans (see Fig. 9.6). While DNA is fixed and time-independent, the connectivity of nerves yields a time-dependent information store due to a chemistry dependent on environmental input vastly greater than is possible in DNA.

9.8.   A Note on Animal Genes and Morphology

Towards the end of each chapter, we have put a section on genetic changes in order to relate the account of evolution we are giving to that of more traditional references to biopolymer sequences, especially DNA, and their most obvious relationship to morphology. The difficulty for this chapter is that the genes responsible for the development of the nervous system and the brain are, as yet, ill-defined. It will be necessary to have a large number of sequences before a pattern of DNA changes relating to the fast development of the brain becomes clear. An obvious difficulty is in finding how one part of the animal, its brain, has evolved so quickly. The introduction of the Na/K ATP-ase only occurs late in embryonic development, so that even functioning nerves do not exist very early in the *in utero* life of animals. The use of genetic manipulation in efforts to understand the DNA sequences related to nerve and brain may be the only way forward (see Chapter 11). It is very clear that the later developments of the brain are not connected to large changes in morphology. The brain is clearly increasing in functional strength, though not as fast in other evolutionary families as that leading to mankind.

9.9.   The Biological Chemotypes of the Ecosystem: A Summary

We have now completed the account of biological evolution except for one special case, that of *cognitive Homo sapiens*, a single species which we must put to one side for separate analysis much though we believe this species is but another step in a systematic evolution. Before we discuss further, the most recent stage of evolution of organisms, that of *Homo sapiens* in Chapter 10, we pause to summarise the classification of non-cognitive organisms into chemotypes remembering all the while their strong connection both in time of appearance and chemical composition with the changes of the environment, and their mutual dependencies. We shall do this in sections separating especially plants from animals and then in a final section we give an overview of all biological chemotypes in the single ecosystem, (Section 9.13).

The features which distinguish a chemotype as described in Chapter 4 are: (1) the chemical element content and the use of chemicals; (2) the use of energy; (3) the spatial divisions within organisms and exterior to them; and (4) the organisation of their chemistry. We do not separate the external and internal organisation as they are linked and became more and more so as the organisms became able to organise as well as able to sense and use their environment. From Chapter 5 to this chapter we have identified in these terms the evolution of the chemotypes, see Table 4.2 with the final type in Table 9.8. In particular, under (1) we have observed notable chemical changes in a sequence from the initial system of the earliest prokaryote in the order: (a) increased complexity in the use of iron, magnesium, cobalt, nickel, vanadium and molybdenum, especially including multiplying their functions using low-spin states in porphyrins; (b) the introduction of major functions for manganese and calcium; (c) the more gradual increasing use of selenium, zinc and then copper; (d) the employment of oxidised sulfur and halogens, such as iodine, and even of carbon and nitrogen; (e) the introduction of extensive uses of sodium, potassium and chloride. As a background there is the ever-increasing special use of kinetic management of phosphate compounds. Throughout, of course, organic chemicals have evolved, but we stress particularly here the use of their more oxidised frameworks and side chains.

TABLE 9.8

SUMMARY OF CHEMICAL CHARACTERISTICS OF MULTICELLULAR ORGANISATION

Characteristic	Novel Elements or Element Uses Involved
Growth from single cell	Many including Zn, Cu
Differentiated cells	Many in different concentrations
Extra-cellular structures	Especially Zn, Cu, Ca and sulfated sugars (cross-linked celluloses and proteins)
Extra-cellular messengers	Organic compounds synthesised using Cu and haem Fe
Intra-cellular communication	Zinc fingers
Extra-/Intra-cell communication	Many novel Ca^{2+} trigger proteins (see Table 9.9)
Extra-cellular hydrolysis	Zinc proteases
Defence mechanisms	Selenium peroxidases, oxidised halogens
Circulating fluids	Controlled element composition
Mineralisation	Especially internal bone containing oxidised protein side chains
Nerves, brain	Electrostatic currents of mainly Na^+, K^+ and Cl^- (see Table 9.10)

TABLE 9.9

INCREASE IN INFORMATION TRANSFER IN EVOLUTION

Organism	Information Transfer with Sensors
Prokaryotes	Food, light gradients, magnetic/gravitational fields
(a) Aerobes only	Oxygen gradient, CO/NO gradients
Eukaryotes	As above plus the following
(a) Single cells	Contact – Ca^{2+} gradient
(b) Multi-cell	Cell–cell connection by organic chemicals
(c) Nervous systems	Na^+/K^+ gradient connection

Under (2) we observed the change from the use of energised inorganic minerals to that of light, and then that of the oxidised environment, mainly rejected oxygen, with excess reduced internal and debris chemicals and the possible use of gradients of rejected ions. The developing changes in spatial divisions (3) is open to clear confirmation even in the titles of the chapters. Development of (4) coincides with the changes in uses of elements and oxidation states of elements mentioned and seen in communication networks (see Table 9.9). Here a quite remarkable feature is the development of the functional value of Ca^{2+} (see Table 9.10), then of small oxidised organic molecules and finally of Na^+, K^+ and Cl^-. We linked all the changes to those in the environment and then the sensing of them. The next sections indicate that the whole of the development of chemotypes was linked to mutual organism dependencies and the reasons for this.

TABLE 9.10

CALCIUM IN EVOLUTION

Cellular organization	Calcium function

A. Prokaryotes
  single cells
	Externally – wall crosslinks – random
	  Mineralisation
	Internally – very slow signal: swimming
	No message to genetic code except for
	  sporulation under starvation

B. Early eukaryote
  single cells
  (no organelles)
	As for prokaryotes (A) but external events
	  relayed to internal structures
	(a) Shape response: contractile devices
	(b) Metabolic and energy response
	(c) Controlled external mineralisation
	(e) Control over cell death

C. Late eukaryotes
  single cells + organelles,
  e.g. see *Acetabularia*
	As for (B) but
	(a) Extra response and energy release
	  from mitochondria and chloroplasts
	(b) Necessary connection to genetic code
	  through phosphate compounds to give
	  growth, see C(a)
	(c) Growth pattern linked to constant
	  Circulating calcium current around cell
	  edges: differentiation

D. Early multi-cellular
  animals (nematodes)
  plants are similar but without
  nerve structures
	As for (C) except
	(a) Link to nerve response as well as
	  chemical response to the environment
	(b) Control of circulating fluids inside
	  organism by organs, releasing hormones
	(c) Connective tissue allows build-up of
	  internal mineralisation
	(d) Protective and immune system
	  organisations

E. Late multi-cellular
  animals with brains
	As for D except
	(a) the brain is a new plastic organ
	  (unlike other organs) and calcium
	  currents around nerve cell tips of the
	  brain are now linked to local growth to
	  give memory
	Apoptosis

9.10. The Relationship between Plants, Fungi and Bacteria: A Summary

We have explained that the difficulty with any highly organised single organism is that the more complicated it becomes the more difficult it is to manage its chemistry within one central control unit. Bacteria developed in many forms – anaerobes, aerobes, sulfobacteria, nitrobacteria, photosynthesisers, and so on – their debris forming nutrients for other species, groups of chemotypes, managing energy and element capture differently. Though to some degree they exchanged products, even genes, this is hardly an organised system of cells, but it is believed that these prokaryotes can be attracted by sources of food to form very loose "colonies", that is by sensing the presence of other cells or of debris of other prokaryotes they can become cooperative in the ecosystem. Unicellular eukaryotes, split into fungi and plant-like chemotypes which appeared next (we discuss animals in Section 9.11), can manage more activities within single cells than prokaryotes, but many of these organisms rely upon some bacteria for basic food, limiting the need for synthesis. They are able to detect food, environmental features, other organisms in their environment and they sense the environment in a more advanced way but note that they rely on incorporated bacteria, organelles, for energy, and even for iron metabolism. Multi-cellular organisms which came next, and here we deal with plants not fungi, became very successful in utilising much space to absorb light and take in CO_2 + H_2O to make saccharides. They developed large organs connected by internal fluid flow. The cost of size was the need for organs of differentiated cells operating in different parts of space to obtain different chemicals and hence there needed to be an increase of organisation between cells during each stage of development (contrast here the mixed colonies of bacteria). It is easier for such a complex organism to survive if some nutritional requirements are met by association with a population of successful lower organisms which provide nutrients, since this reduces their genetic load. We find, in fact, that plants do not even fix nitrogen to make ammonia but have symbiotic bacteria that carry out this task. In exchange, the bacteria take excess carbon compounds from the plants. Again fungi in the soil are very good at taking in mineral elements but poor at direct energy capture. Their scavenging and mineral collecting depends upon extensive hyphae entering deep into soils, though they seem less organised than plants. Plants rely then on fungi for mineral supplies (note that multi-cellular fungi work and scavenge in the dark) while plants supply energised carbon/hydrogen compounds to the fungi. In conclusion, higher multi-cellular plants, lower unicellular plants, multi-cellular and unicellular fungi and prokaryotes include several different chemotypes which cooperate in an organised ecosystem. The development of modern plants was very unlikely (impossible?) without this mutualism or symbiosis. Our claim is that the plant coverage of Earth today (see Table 8.2) with assistance of fungi and bacteria could be close to the optimum possible for the production of carbohydrates given the present environment as a source of material and energy (light) and this organisation of chemotypes. Within each of these large groups, such as plants, we can distinguish chemotypical variants, e.g. in the use of silica in certain plants, but the present day lack of chemical analysis prevents complete listing as yet. We shall have to refer later to the intervention of humans, asking what is the long-term future of plant life under the stress of agriculture and generally exploitation of land.

We shall now turn to animals to show how they, with fungi, that is the total living non-photosynthetic multi-cellular eukaryote system, have developed in step with and dependent upon the evolution of plants much as we stressed the evolutionary development of non-photosynthetic single cells based on the increase in population of single photosynthetic cells in Chapters 5–7. It will then be possible to see human organisms in a systematic relationship to that of plants for gathering energy and animals for scavenging the chemicals produced by plants.

9.11.   The Relationship between Plants and Animals

We must go back to the beginning of life to appreciate the value of the combination of plants and animals in the general development of an ecosystem.

There is only one source of required materials – the geosphere – constantly developing from a more reduced to a more enforced oxidised condition. This is well described today in chemical terms (see Chapters 1–4). There are two sources of energy: first that from already energised chemicals of the Earth and of organic debris from this life, or any system pre-dating life, and second from the sun. Maximum energy retention in chemicals demands coexistence of the two synthesising systems using the two energy sources, photosynthesising and non-photosynthesising, since it is difficult to conceive of a highly mobile plant, a light harvester. The photosynthesising plant system with attendant bacteria and fungi is most effective as these, almost immobile, organisms readily capture C/H/O directly from air and water solutions but less readily elements from scattered debris. The effective reuse of scattered debris demands cooperative action with mobile organisms, animals. It is the combination of large, specialised organs, internal circulating fluids and nerves connected with brains which has produced higher animals very efficient in scavenging. The ability to use light by plants is also limited to time of day and temperature zones while the animals on Earth can use all zones rich enough in initially energised material no matter what the light intensity. As mobile organisms the animals assist plant growth (as well as they eat plants) in that they aid recovery and degradation of the synthetic kinetically stable compounds in a cycle in the ecosystem. They also help the distribution of many elements from Earth (see Table 9.11). The following scheme of materials is illustrative:

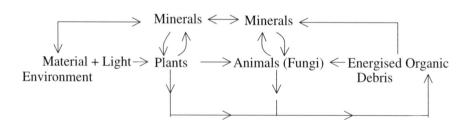

Animal life is as inevitable as is plant life, and always was so, if energy retention and utilisation is to be optimal since it removes energised debris all of which would otherwise finish as coal, oil or gas and/or burnt off very slowly as $CO_2 + H_2O$ and heat. The overall steady final state involves all forms of life to take advantage of the chemical change of environment. *Note that the whole is just a maintained but somewhat incomplete energised cycling of 20 elements between the two parts of the ecosystem, organisms and the environment while light is converted into heat.*

The overall result is that plants, animals, fungi, of all levels of eukaryote complexity, and prokaryotes together with the environment form a cooperative whole in which effectiveness of energy storage in chemicals is moving towards an optimum value by using each chemotype as a system of value to the whole (Fig. 8.7). The parallel stepwise chemical and organisational development of very different chemotypes, which became separated very early in evolution, and the fact that the time of these developments is not linked by any genetic relationship, is highly suggestive that for all organisms evolution is environmentally driven since environmental change came first and is common to all. It is in this light and in that provided by the next section that we shall view the rise and activity of human beings in Chapter 10 before we look at the whole ecosystem development in Chapter 11.

9.12. Energised Inorganic Elements and their Uses by the End of Biological Evolution

It is worthwhile pausing again at the end of this chapter to appreciate the reason for the stress we are placing on the (inorganic) environment, its changes with time in evolution and the ability to utilise its chemicals in organisms. This is in marked contrast with all previous approaches to evolution. We have noted from the beginning of the description of energised chemical systems of any kind in Chapter 4 and through all chapters, Chapters 5–9, on organisms, leaving *Homo sapiens* to one side, the changing involvement of the available 20 elements in inorganic chemical processes as well as the uses of most of them in organic chemical constructs in organisms (Table 9.11). This approach underlines the strong connection of the environment to the ecological system. It is important to reflect too on the changing flow of energy in organisms which created evolution and made possible the use of different elements in time. The flow of energy and material is through more and larger complex machines doing work, physical and chemical. Taking the elements first we observed that the "organic" chemical elements plus certain, then available, inorganic elements dominated activity in cells, that is H, C, N, O, P, S, Se with K, Mg, Fe, Mo(W), while others were rejected to considerable degrees, notably Na, Ca, Mn, Cl and a further group, V, Co, Ni, Zn and (Mo) were of very low availability, or virtually not available, e.g. Cu, and were then used sparingly if at all. The evolution from reducing to oxidising conditions as oxygen was rejected from organisms increased the environmental concentrations of Co, Ni, Cu, Zn and Mo but decreased markedly that of Fe while managing to preserve the use of the essential first dozen essential elements mentioned above despite the oxidised state of C,

TABLE 9.11

EXPLOITATION OF AVAILABLE ELEMENTS

Elements	Use	Time ago to today
C/H/N/O/P/S	Biological synthesis	4×10^9 years increasing
Mg, Fe[a]	Catalysis and control in cells	4×10^9 years increasing
Ni, Co	Catalysis	4×10^9 years decreasing
Mn[a]	Catalysis (O_2 production)	3.5×10^9 years to today
Mo[a]	Catalysis	4×10^9 years (increasing)
Ca	External hydrolysis	4×10^9 years to today
	Messenger	2×10^9 years to today
Zn	Catalysis	4×10^9 years increasing
	Control messenger	
Na/K	Osmotic/electrolytic balance	4×10^9 years increasing
	Messages	1×10^9 years increasing
Cl	Osmotic/electrolytic balance	4×10^9 years increasing
	Messages	1×10^9 years increasing
Cu	Catalysis	2×10^9 years increasing
Se[a]	Catalysis	4×10^9 years increasing

[a] Changed function 2×10^9 years ago.
Note: This table should be matched with the changing chemotypes

N, S and Se which came about. Hence later in evolution, under oxidising conditions, all the free elements Na, Ca, Mn, Cl, Co, Ni, Cu, Zn and Mo had to be lowered in concentration in the cytoplasm relative to the environment, as they are poisonous in high free concentration. However, considerable levels of them were used, mostly in catalysts but restricted in free concentration in the cytoplasm by pumping and by the thermodynamic equilibria internally employing selectivity of reaction with organic chemicals, for example as in Irving–Williams series, and hence maintained in bound forms. These mainly catalytic elements then contributed to increased flow and therefore the synthesis and degradation of organic material with an overall increase in energy retention (temporary) in useful materials as a cyclic steady state was approached with a simultaneous increase in energy degradation rate. In these direct catalytic respects however Na, Cl and Ca are not valuable but because they are bulk elements, in considerable free ionic state concentration externally and are rejected from cells to a large degree, they created a significant energised concentration and charge gradient across cell membranes. While such physical gradients were of little use to very small single prokaryote cells except in exchange, they became of increasing value especially in nerves in animal evolution. Now the overwhelming needs of all these organisms are sources of nutrients and the great advantage of single cells of eukaryotes over prokaryotes, which the calcium gradient also assisted, is recognition of many more physical and chemical environmental circumstances from which they can take advantage. Later, together with calcium ion as a messenger, the use of particularly zinc and copper enzymes engendered the creation of multi-cellular organisms with novel organic messengers and connective tissues in large assemblies of cells. As collectors of light (plants) and in sensing and scavenging (animals) these large eukaryotes gained further access to the environment. Now, as mentioned, just because Na and Cl ions, although in a parallel energised condition to Ca^{2+}, cannot activate as their binding is too weak, they can act as carriers of information to a target *by bulk physical transfer*. They then become useful even through their *inability to bind* in a quite novel way, that is, in very fast transmission over a long distance, which is only valuable for multi-compartmented organisms. No other ions, except the complementary K ions, opposed in gradient across membranes from Na ions, are of equally low binding strength. These four ions, Na^+, K^+, Cl^- and Ca^{2+}, with organic messengers became the long-range messengers of multi-cellular animals and allowed the development of the nerves and brain. By so doing in animals they advanced scavenging, the capture, trapping and effective use of energy and material and hence gave a novel twist to the evolution of the ecosystem. In the multi-cellular organism the special circulating extra-cellular fluids gave the additional advantages of internal management of distribution and avoidance of energy and material loss, even of Ca^{2+}, Na^+, K^+ and Cl^-, to the environment. We wish to draw attention again here to the fact that *the energised ecological system has used the elements of the Periodic Table in a remarkable efficient way*. Note how the major functions are separated in precisely the manner expected from the chemical properties which allowed the table to be devised. The availability limits the way they can be employed but observe the separate action of Na^+, K^+, Cl^- in bulk properties and exchange, Mg^{2+}, Ca^{2+} and Zn^{2+} in acid functions, V to Cu with Mo(W) in oxidation/reduction and the non-metals, H, C, N, O, S, P and Se, in organic biopolymers, all put to appropriate functional use. There is no better illustration of the optimalisation of the chemical function in organisms noting also its changes with the oxidising condition of the environment. Observe that it was somewhat more difficult to make this argument in Chapter 4 with regard to organic compounds much though we presented the same case. It is the close to equilibrium states of inorganic elements which allows this visualisation of an optimal chemical system.

9.13.   The Direction of Biological Evolution

Now it is against this background of environmental element and energy uptake and combination in materials and then their close to cyclic return to the environment that we must see the changing and increasingly effective development of life employing ever more intensively the possible energy sources. The initial energy for the working molecular machinery of organisms was probably from energised inorganic materials resulting from the kinetics of Earth's formation. Very shortly there was a change to the use of light, and it produced oxygen and oxidised non-metals, e.g. NO_3^- and SO_4^{2-} and liberated many metal ions which together and with organic chemical debris gave further sources of energy to plants in the first place and then to animals. As cell organisation grew so there has been a shift to macro-machines using these energy sources through combinations of molecular machines.

The stress throughout our account up to and in Section 9.12 is therefore on the developing optimalisation of the interaction of the energy and elements from the environment with their use in organisms together with an awareness by organisms of the physical nature and chemical content of the environment. It is this awareness of the environment which eventually became the basis of the evolution of a special species – mankind. This points to the fact that there may well be a logical progression to an endpoint of biological evolution on Earth which can only come about through the fullest interaction possible of external chemical elements and energy supply with organisms. This endpoint could arise when each individual in a species and each species has a final controlled form and behaviour based on DNA overwhelmingly. There are then only cyclic chemical products in organisms and in the environment energised by processes connected to organisms. It is this final step which is approached, not reached, before the introduction of the latest chemotype, mankind. As a chemotype mankind is not just a product of biological evolution in the sense we have used this phrase so far in that it largely involved internal metabolic events. As we know, while all previous developments were led by environmental change generated biologically, mankind is proactive, cognitive, and forces "deliberate" change on both the environment and other organisms but while the changes can only be in the same general sense of all previous evolution – use of more elements, more space and more organisation to increase energy flux – they may fail to enter the cycle. The risks of speed of change are obvious, note the conservative nature of DNA, and we shall discuss them in Chapter 10, but remember that without cognition organisms produced oxygen and this gave rise to the whole of biological evolution in an earlier inevitable progression with risks. While this development was adaptation to the changes of the environment, even an environment in part (far from totally) created by organisms themselves, it was always at first a change, which was damaging to life, e.g. oxygen and metal ion increases (contrast the Gaia hypothesis, see Appendix 11A). All change of environmental chemicals is of the same character – it is damaging, but evoling biological cycles "aim" to become element neutral, non-polluting, in contrast to the activity of mankind. We shall also see that human beings introduce a further problem, very probably not associated directly with DNA, that of individual phenotypes.

Further reading

Books

Allman, J.M. (2000). *Evolving Brains*. Scientific American Library, New York

Berry, M., Butt, A.M., Wilking, G. and Perry, H. (2002). Structure and function of glia in the central nervous system. In D.I. Graham and P.L. Plantos (eds.), *'Greenfield's Neuropathology'* (7th ed., Vol. 1). Arnold, London, pp. 75–122, Chap. 2

Cairns-Smith, A.G. (1996). *Evolving the Brain*. Cambridge University Press, Cambridge

Calvin, W.H. (1996). *How Brains Think*. Weidenfeld and Nicholson, London

Clarkson, N.R. (1998). *Physiology of Behaviour.* Allyn and Bacon, Needham Heights, M.A.

Cox, P.A. (1995). *The Elements of Earth.* Oxford University Press, Oxford

Crick, F. (1994). *The Astonishing Hypothesis.* Simon and Schuster Ltd., London

Pinker, S. (1997). *How the Mind Works.* Weidenfeld and Nicholson, London

D. Randall, Burggren, W. and Trench, K. (2002). *Eckert Animal Physiology – Mechanisms and Adaptations* (5th ed.). W.H. Freeman and Company, New York

Rang, H.P., Dale, M.M., Ritter, J.M. and Gardner, P. (1999). *Pharmacology.* (4th ed.). Churchill-Livingstone, Edinburgh

Selinus, O., Alloway, B., Centeno, J.A., Finkelman, R.B., Fuge, R., Lindh, U. and Smedley, P. (eds.) (2004). *Essentials of Medical Geology.* Elsevier, Amsterdam

Williams, R.J.P. and Fraústo da Silva, J.J.R. (1993) *The Natural Selection of the Chemical Elements – the Environment and Life's Chemistry.* Oxford University Press, Oxford

Williams, R.J.P. and Fraústo da Silva, J.J.R. (1997). *Bringing Chemistry to Life – from Matter to Man.* Oxford University Press, Oxford

PAPERS

Barres, B.A. and Smith, S.J. (2001). Cholesterol – making or breaking the synapse. *Science, 294,* 1296–1297

Bohening, D. and Snyder, S.H. (2003). Novel neural modulators. *Ann. Rev. Neurosci., 26,* 105–131

Boruchowska, M., Lankosz, M., Adamek, D., Ostachowicz, B., Ostachowicz, J. and Tomik, B. (2002). X-ray fluorescence analysis of human brain tissue and body fluids. *Polish J. Med. Phys. Eng., 8*(3), 173–181

Bush, A.I. (2000). Metals and neuroscience. *Curr. Opin. Chem. Biol., 4,* 184–191

Chen, J. and Berry, M.J. (2003). Selenium and selenoproteins in the brain and brain diseases. *J. Neurochem., 86,* 1–12

Fields, R.D. (2004). Glial cells – The other half of the brain. *Sci. Am., April,* 27–33

Fields, R.D. and Steven-Graham, B. (2002). New insights into neuron-glia communication. *Science, 298,* 556–562

Frederickson, C.J., Suh, S.W., Silva, D. and Thompson, R.B. (2000). Importance of zinc in the central nervous system: the zinc-containing neuron. *J. Nutr, 130,* 1471S–1483S

Holt, M. and Jahn, R. (2004). Synaptic vesicles in the fast lane. *Science, 303,* 1986–1987

Kandel, E.R. (2001). The molecular biology of memory storage: a dialogue between genes and synapses. *Science, 294,* 1030–1038

Kimura, Y. and Kimura, H. (2004). Hydrogen sulfide protects neurons from oxidation shocks. *FASEB J.,* May 20 (e-pub ahead of print).

Magistratti, P.S., Pellerin, L., Rothman, D.L. and Schuman, R.G. (1999). Energy on demand. *Science, 283,* 496–497

Meyerowitz, E.M. (2002). Plants compared to animals: the broadest comparative study of development. *Science, 295,* 1482–1485

Miller, G. (2005). The dark side of Glia. *Science, 308,* 778–781

Moos, T. (2002). Brain iron homeostasis. *Danish Med. Bull. 49,* 279–301

Pellerin, L. and Magistretti, P.J. (2004). Let there be (NADH) light. *Science, 305,* 50–52, and see Kasischke, K.A. *et al.* in pages 99–103

Pennisi, E. (2003). Genome comparisons hold clues to human evolution. *Science, 302,* 1876–1877

Sandstead, H.H., Frederickson, C.J. and Penland, J.G. (2000). History of zinc as related to brain function. *J. Nutr., 130,* 496S–502S

Takeda, A. (2001). Zinc homeostasis and functions of zinc in the brain. *Biometals, 14*(3–4), 343–351

Thompson, S.M. (2005). Matching at the synapse. *Science, 308,* 800–801

Ullian E.M., Sapperstein, S.K., Christopherson, K.S. and Barres, B.A. (2001). Control of synapse number by glia. *Science*, *291*, 657–661

Verkhratsky, A., Orkand, R.K. and Kettermann, H. (1998). Glial calcium homeostasis and signalling function. *Physiol. Rev.*, *78*, 99–141

Williams, R.J.P. (2003). The biological chemistry of the brain and its possible evolution. *Inorg. Chim. Acta*, *356*, 27–40

Williams, R.J.P. (2004). Uptake of elements from a chemical point of view. In O. Selinus, B. Alloway, J.A. Centeno, R.B. Finkelman, R. Fuge, U. Lindh and P. Smedley (eds.), *Essentials of Medical Geology*. Elsevier, Amsterdam, pp. 61–85, Chap. 4

CHAPTER 10

Evolution due to Mankind: A Completely Novel Chemotype (Less than One Hundred Thousand Years Ago)

10.1. Introduction

The genetic apparatus, as described in the earlier chapters, was largely the source of inheritance of information from between 4.0 and 3.0 billion years, the origin of life, until some 300 million years ago when animals began to develop tubular cells, nerves and then brains (Table 10.1). As a result of this development, especially in those animals making use of tools, that is those species with larger brains, they came to have a communication means and a novel information centre that could learn and teach i.e. transfer information independently from genetic structures. The difficulty that arises now lies in the appreciation of the nature of information since its transfer changes. Previous information has been concerned with messages obliging an activity – they were instructions (see Sections 3.8–3.13), which described biological cellular controls. All messages had only a chemical code to and from DNA and later the environment. We have described these codes and that necessary binding of the message material to the receptors meant that messages had concentration and temperature dependence. The whole of the information transfer led to automatic responses. The brain carries a different code, related to information, in the novel form of gradients and stores of ions and organic chemicals in nerves interconnected by current – carrying circuits. This combination of

TABLE 10.1

THE DEVELOPMENT OF CODES AND INFORMATION

1.	The original code: RNA/DNA/proteins
2.	Storage of chemical effectors in zones in single cells, e.g. Ca^{2+} and organic molecules which respond to the environment in selected ways independent of (1). Different effects at receptors giving a crudely coded definition of the environment
3.	Storage of electrolytic charge and chemicals in zones of multi-cellular organisms: nerves and brain. Faster chemical response than in (1) or (2) due to environment change plus a memory of external events
4.	Sign codes of animals
5.	Sound-related codes: languages also in written codes
6.	Electronic and radiation codes: radio
7.	Electronic and radiation codes: computers

Note: (1) Observe how coding spreads out from the DNA in a cell to greater and greater connection to distances outside a cell.
(2) We do not include information between cells.

novel use of space and elements made for a new chemotype, i.e. animals with brains, as described in Chapter 9. As stated in the previous chapter, the information store itself is not connected to an all or nothing fixed computer or DNA form but has extensive thermodynamic features related to unstable chemical/physical concentrations and potentials. Moreover, novel inputs increase information storage by causing growth of nerves using synaptic stored materials including proteins not requiring immediate synthesis at the nucleus, which provides a back-up supply for them. The code is then adjustable and the information relates very largely to knowledge of the outside world and how to act quickly on it, allowing action externally, unlike the information in the DNA/RNA machinery that relates largely to internal organisation and use. Information transfer to the brain is now quite different in that much of it is by sound or electromagnetic radiation, light, resulting in electrical and chemical memory storage at synapses without an effect at the level of DNA. The storage is of events outside the organism and of experience. Response to a message is not obligatory or automatic but depends upon the nature of previous experience that has an individual character. Thus, new words are needed to describe the information that can be an advice, a request, a demand or mere gossip, all of which have no meaning for earlier species. Because of the peculiarity of the human brain we shall put aside until Section 10.10 any consideration of this major feature of its development – the need to be concerned with individuals. We consider first the corporate management of the use of human brains in developing chemicals in space and devising management of industry and society, as well as controlling other forms of life. The end result to date, the rise of one species, mankind with an exceptional brain, which probably evolved 50,000 years ago, is the subject of this chapter (see Goudie in Further Reading). From their beginnings humans developed the capability of exploiting new external organisation, including *the application of all chemical elements* in the creation of new objects and machinery, and well-defined organisation and they therefore became the latest, perhaps the last possible, and an extremely powerful, chemotype. As we have stated repeatedly, while the development of this nervous system will be seen to increase the use of energy and of chemicals, that is increased efficiency by organisms in the general expected direction, there is no way the drive in this direction could have predicted the time of appearance of this particular chemotype, here a single species, *Homo sapiens*, among one division of animals but with a brain having a huge qualitative difference from that of all other animals. However, we can see with hindsight that this is a move towards an efficient final product of the drive to increased interaction between energy and the environment in the activity of organisms. The arrival of such an organism should in principle lead in the fullness of time to the ultimate logical conclusion of an overall optimal utilisation of energy, space and chemical elements in the environment together with that in organisms through an understanding of the ecosystem as a whole, but we shall note that this outcome is far from certain.

It is readily seen that mankind became able to plan, on a large scale the energised chemical activities on Earth's surface, but now of necessity *in space outside organisms* as well as within them to some degree (see Tables 10.1 and 10.2). The tables deliberately link mankind's endeavours to those of all previous organisms – a stance we stress throughout this book, that is the continuous directed thermodynamic nature of evolution. At the same time human beings are today quickly bringing all other forms of life under their control and they may even turn one day to the better use of energy and material inside organisms, including the human body, by genetic modification, but this is still a doubtful endeavour. We turn to a description of the general nature of this change of use of the environment, which is becoming undoubtedly a case of directed evolution of the ecosystem, and to some extent, of organisms by mankind (see Goudie in Further Reading).

In this chapter, then, although there appears to be some mystery surrounding the nature of the evolution of human beings, we wish to show that the appearance of this organism and its activities is just a more or less final step in the way in which energy can be put into all materials in an effort to create in space an optimal final cyclic steady state, given and utilising fully the present environmental conditions. As in other cases we concentrate attention on new uses of elements and of existing energy sources, the discovery of new energy stores in external space and their use in the organisation of new chemical materials by machinery before we consider genetic alterations. The final steps in the progression cannot be achieved without a further evolution in organisation with all that this implies – for example in advances in structures, messages and transfer apparatus and centralised global command and in fact we see this developing in our very society. At the same time we wish to stress that the dependence for survival of humankind is increasingly on the whole of the previously evolved ecosystem. Mankind cannot act independently from the rest of the ecosystem except in a damaging way. Thus, we must be very careful as to how we use our power for we are capable of reducing the possibility of reaching an efficient final cyclic condition and even of damaging the biological development as it is happening today.

10.2.   The Nature of Homo Sapiens

We must see *homo sapiens* in this physical/chemical perspective as a very advanced animal, an animal with by far the greatest organisational capacity of any species that had developed previously and with quite new capabilities. The activities of human beings are, in a sense, only half controlled by DNA/RNA information since they are also half controlled by the unique development of the brain. Parts of the brain circuits are certainly inherited and may be coded, but the actual growth of the brain is environmentally determined to a novel large degree (Section 9.7). The human brain is in fact highly formed by education, passing on acquired knowledge in all forms of learning to the next generation which is, in one sense, an inheritance but it is also formed by novel direct experience. It might be thought that this vast development in the ability to handle the environment would remove our dependence on other organisms. We must remember however that initially primitive life was chemically dependent on the environment for all chemicals and energy while each cell was independent of all other separate cells and was poorly interactive with external events and conditions. Certainly by the time of the eukaryotes, a new extra dependence of separate organisms upon one another, mutualism or symbiosis, had evolved due to the difficulty of retaining all novel activities in just one cell or organism (see Chapters 7 and 8). All animals which developed subsequently depended upon plants, and this dependence has increased, as described in earlier chapters, with the complexity of animals and is now seen to the greatest degree in humans (Table 10.3). Human beings require many essential chemicals, apart from bulk food, some of which are called vitamins, also many small molecules, saccharides, lipids, amino acids, etc., and the chemical elements themselves, some 20, that are essential for our life (Table 10.2). It is likely from DNA studies that even the code for selenium amino acids has been dropped. All elements must be in balance in the diet largely from plants directly or indirectly through eating animals. Moreover, we have no source

TABLE 10.2

DEVELOPMENT OF PHYSICALLY DEFINED COMPARTMENTS AND SPACE OF CHEMOTYPES

Prokaryotes	Single internal cytoplasm
	Periplasm as second compartment
Eukaryotes (unicellular)	As for prokaryote plus internal vesicles and organelles
Eukaryotes (multicellular)	As for unicellular eukaryote plus cell/cell differentiation
	Organs and extracellular body space
Later Animals	Animals with a nervous system senses and a brain able to search external space
Mankind	As for later animals plus control of external space
	External storage of information and of activities
	Knowledge of chemistry and physics and their applications

TABLE 10.3

DEPENDENCIES OF HUMAN BEINGS

Dependence	Source
Minerals	Plants and other animals
Amino acids	Mainly plants and other animals
Saccharides	
Lipids	
Coenzymes (vitamins)	Other organisms generally
Digestion and protection	Lower organisms

of energy except from plants. Thus, we are not unlike other animals except for the extent of this dependence, and we have probably become the species most dependent on external chemicals from other organisms. We are then the poorest organisms as a chemical factory, and with the activity of our brains the fastest heat producers. While there is this dependence in humans on energy, elements and compounds from external organisms there is also a direct dependence upon "internal" organisms. It is said that the human body has at least as many cells coded with non-human DNA as with human DNA. Many of the internal organisms living symbiotically in the human body are needed for digestion and protection: that is essential bacteria and unicellular eukaryotes. The "wrong foreign" organisms (and viruses) internally are the causes of many diseases, but it is clear that every human being is internally an "ecosystem" of required internal organisms and is also dependent upon a vast external ecosystem of organisms, plants and animals, which in turn depend upon other organisms down to prokaryotes: anaerobic and aerobic. At the base of this pyramid of "organic life" is the "inorganic" or geochemical (environmental) set of chemical elements in compounds and the sources of energy, see cover diagram. This entire ecosystem is evolving interactively. The evolution of human life is just another piece of this whole global ecosystem in which novelty, including increasing interaction with the environment, is introduced only with dependence on simplicity in earlier life forms. (We must also ask: how did human beings as an animal come to lose so many useful genes?) Thus, the brain was introduced only at the cost of biological weakness, and despite the huge dependence on other organisms as described in this chapter, it is the powerful consequences of having such a brain that concern us now. To indicate the extent of this power to look after ourselves, cognitively not genetically, we have invented the enormous and widespread activities of agriculture and farming (control over plants and animals), erected massive areas of novel structures (buildings), created a huge industry and produced new chemicals for internal use (medicines) all within a very extensive organisation using and integrated by *external* machines of various kinds. All this has demanded a network of novel transport and communications facilities covering the whole Earth. We are making our lives as secure and comfortable as possible but we must be aware of the cost side of the achievements knowledge has brought and will be bringing. We must see ourselves as part of a very large system and are therefore not exempt from its rules. Perhaps our history reveals our nature most clearly.

10.3. The Evolution of Human Beings from 100,000 Years Ago

History shows that mankind as a chemotype, putting aside later individuals, is part of a general trend in animal evolution and until about 40,000 years ago human beings were not in fact very different animals from the apes and gorillas (Table 10.4). They were hunter-gatherers of greater skill but were hardly better organised than other animals. Undoubtedly, they had the beginnings of the understanding of how they might gain control over the natural world outside themselves (see Table 10.5 and Dawkins in Further Reading).

The first big step was the switch from a wandering lonely hunter/gatherer style, common to all animal life, to farming in a locality (Table 10.5). Farming utilises a control of certain plant and animal life by selective breeding. It led to the development of human communities, increased organisation, with the further development of new tools. The activity demanded mutual dependence on one another in a group not obviously required by genetic structures. A remarkable probably earlier development seen first to a small degree in many animals was a new external message system – the production and reception of organised sound, its transmission and reception, which once developed proved extremely useful in this new organisation.

TABLE 10.4

HUMAN EVOLUTION

Years ago	Stages of Evolution
4 million	*Australopithecus afarensis*
3.2 million	'Lucy' (*Australopithecus afarensis*)
2.5 million	Several *Australopithecus* species
2 million	*Homo habilis*
1.6 million	Homo erectus
1.4 million	Australopithecines become extinct
1 million	*Homo erectus* settles in Asia
400,000	*Homo erectus* settles in Europe
	Homo sapiens begins to evolve
250,000	Archaic forms of *Homo sapiens*
	Homo erectus becomes extinct
125,000	*Homo neanderthalensis*
100,000	*Homo sapiens* fully evolved in Africa and Asia
40,000	*Homo sapiens* (Cro-Magnon) fully evolved in Europe
35,000	Neanderthals become extinct; *Homo sapiens* remains the single surviving human species

TABLE 10.5

A RECENT HISTORY OF HUMAN EVOLUTION

Years ago	Period	Human population size	Technology or event	Effect on the biosphere
1,000,000	Lower Paleolithic	125,000	Fire use (Africa) Tool use	Increased incidence of fire
300,000	Middle Paleolithic	1,000,000	Migrations to Europe and Asia	Pleistocene over-kills in Europe and Asia
25,000	Upper Paleolithic	3,300,000	Migrations to North and South America Domestication of the dog	Pleistocene over-kills in North and South America
10,000	Mesolithic	5,000,000	First farming in the 'Fertile Crescent' Domestication of sheep, goats and cattle	Beginnings of soil erosion in the Middle East which spread with farming to North Africa and Europe
8,000	Neolithic	86,500,000	Early villages and small communities, beginnings of many crafts Use of gold	Accelerated soil erosion
5,000	'Civilisation' Bronze Age		Use of copper, first city states in Mesopotamia Beginnings of trade	Transfer of plants, animals and disease about the Eurasian world
3,000 1,971 (0 AD)	Iron Age	133,000,000	Use of iron Roman, Indian and Chinese civilisations	
1,000	Middle Ages	200,000,000	Maori settlement of New Zealand	Pleistocene over-kil in New Zealand
600	Enlightenment (Science)	300,000,000	Black Death sweeps Europe	Extensive deforestation in Europe
300		550,000,000	Colonisation of 'New World' by Europeans	
150		900,000,000	Beginnings of modern medicine	

It is connected to the control of muscle movement by the nerves, so that images of internal and external conditions are relayed in sound patterns and interpreted by the receiving brain, but it requires a large memory. While it is clearly an evolving attribute it is this that becomes the most remarkable human tool, language, and is the first fully sophisticated use of transmission by coded waves much though other animals had some forms of such communication earlier. Notice that it is based on a *coding in sound frequencies*. Many believe this is as important as any step in the evolution of mankind for it gave us what is often called culture as opposed to basic organism drives, but once again there is no known connection to genes. Observation led to some empirical understanding of biological growth patterns, of the weather and of the soil conditions. The use of the empirical method later became the basis of science but it was aided earlier by a second "language", that of mathematics in a separate code, which has become very strongly developed more recently (see below). Mankind settled down in constructed ever-improving buildings in strengthened communities, and though many lower animals also make rough external structures, such as nests, and act in communities, there is no real comparison after a short period of innovation with mankind's works. None of these first steps of external organisation and construction in evolution could be inherited in the genetic apparatus, but they were easily handed down using the spoken word and memory, enabling transfer through generations. Later, language developed into extensive writing, a novel, more permanent and reliable form of coded information store and transfer based on sophisticated creation and interpretation of written sounds. Other advances of the same period can be seen in the early use of minerals and energy (fire) to produce new materials. Apart from changed external activity, use of space, the new chemotype began thereby to change the use of chemical elements, a very basic drive of evolution as we have portrayed it. The beginning of this skill in handling the mathematics, physics and chemistry of the Earth had to wait for substantial advance until 1500 AD, about 15,000 years later than the beginning of organised dwellings and this has led to the revolution in application of energy and use of chemicals, which we see all around us today (Table 10.6). Notice that much though individual self-consciousness grew earlier and was expressed through classical times in language and writing (even in philosophical ideas) knowledge of the external material world, science, which is not subjective, was relatively slow to be kick-started but once it began it caused an equally remarkable fast development of understanding together with an extraordinary change in our whole environment greater even than that brought about by language. Whether it was the Greeks, Chinese or others who first began to use this scientific empirical method is of little consequence compared to the use that has been made of it, and also theoretical mathematical constructions concerning it, in recent years, say since 1500 AD. It is the understanding of the physics of energy, the chemistry of the elements and mathematics generally which has caused the development – the full realisation of a new chemotypical evolution using empirical methods, and judgements based on probabilities expressed numerically. No other chemotype has shown any development approaching the sophistication and application of these new codes.

TABLE 10.6

HISTORY OF ELEMENT USAGE: THE USE OF NATURAL MATERIALS (E.G. WOOD, STONE) WITHOUT
PROCESSING IS EXCLUDED

Application	Era of first major usage of element or its compounds			
	Prehistoric (before 2500 BC)	Pre-Industrial (2500 BC– AD 1750)	Industrial (AD 1750–1940)	High-technology (after AD 1940)
Metals: vessels, tools, coins, weapons, etc.	Cu, Au	Fe, Zn, Ag, Sn, Pb	Al, Ni, Mo, W	Zr, Nb
Metals: construction, transport	—	—	Al, Cr, Mn, Fe	Be, Mg, Ti
Fuels and explosives	C	N, S	H	U, Pu, Th, H
Glass, ceramics, refractories	Na, Al, Si, Ca (clays)	Pb	Mg, Zr, Th	Li, B, La–Lu
Pigments and dyeing	—	Al, Fe, Co, Cu, Cd, Hg, Pb	Ni, Zn, As, Se	Ti
Pharmaceutical	C, N, O, H (plants)	S, As, Sb, Hg	Bi, Br, I, Ra	Li, Pt
Fertilisers and pesticides			N, K, P, Cl, As, S, B, Br, Hg, Cu	Sn
Industrial chemicals and catalysts	—	—	C, N, F, Na, S, Cl, K, Hg, Pt	Ar, Rh, Ba, La–Lu, Re
Electrical and electronics	—	—	Fe, Cu, In, Pb, W	Si, Ga, Ge, As, Li, Cd, Cd, Ni, Se, Ta, Ir
Household goods and chemicals	—	—	C, N, Na, Cl	B, P, Br, Sn, C, H, N, O, F

Note: Each column adds to the previous ones.
Source: After Cox (1995) in Further reading.

10.4. The Coming of Science

In the previous chapters we had to delay the treatment of chemical content and
energy associated with a given chemotype until we had described changed use of
compartments, space, and an outline of organised activity. For mankind it is

science that has revolutionised both uses. The outstanding feature of mankind's physical/chemical activity has developed only over the last 400–500 years that is based eventually upon the full knowledge of the chemical elements of the Periodic Table and the compounds they can form (see Chapter 2), and then using an understanding of their properties in physics, chemistry, and recently in biology and of energy generally, especially to develop machines (see Chapter 3). Unlike any other part of organised life human beings can employ not just some 20 available elements common to all life but all the total number of the 92 elements of the Periodic Table plus even some further unstable ones (see Fig. 2.2). (Note again the astonishing insight this coded table brings to appreciating, and assisting advance, of chemical evolution at least potentially.) This has been achieved by a vast extension of chemistry not just in space completely outside organisms, that is in the environment itself, but also in physical controls over chemical methods involving temperature and pressure and in their management using energy deliberately and externally in synthesis and construction (Table 10.7 and Chapter 2). New chemical processes also use a variety of oxidation and reduction reactions in controlled atmospheres. Of course, knowledge of improved physical technology is also involved, for example in power generation, transport, communication, agriculture, building and mining. Many thousands of chemicals never before seen on the Earth over its preceding 4.5 billion years of history have been produced and crafted into physical forms using old and new sources of energy, all external to organic life. We describe some of these separately in succeeding paragraphs but we advise the reader to look back through Chapters 5–9 to see the parallels with changes in chemistry and material form not over 500 but over 3 or 4 billion years internally within organisms. Moreover, it has so developed that initiated by curiosity and the wish to use the environment we have found ourselves recognising that our brains cannot produce images of the fundamentals of the material universe except by mathematical abstraction (see Chapters 1 and 2), where spoken language is quite

TABLE 10.7

EVOLUTION OF CONSTRUCTS

Organism	Construction
Single-Cell Eukaryote	Multiple internal vesicles
	Filaments, shells
Multi-cell Eukaryote	Extracellular matrices
	Organs, shells
Plants	Roots, trunk (stem), leaves (flowers)
	Silicated leaves
Animals	Head, arms, legs, body
	Bones and shells
Mankind	External buildings, machines, roads,
	railways, canals, etc.

inadequate. Yet this abstraction itself has produced great extra ability to transform the world around us.

In addition to these physical/chemical features of our daily lives we have altered our geological and our biological surroundings in different ways (see Goudie in Further Reading). Much of the soils of the Earth have been used for agriculture and many forests have been destroyed or created. In agriculture man has selected and bred plants and animals, so that the yields of food from organisms have changed. The scale of this change is not yet obviously damaging, but we must be aware of such factors as the reduction in the coverage of the Earth by trees and the increase in grass-related plants. Mankind's bias in the type of animals that are allowed to exist is also apparent. While these switches do not yet dominate there is the accidental effect of eliminating directly or through side-effects large numbers of mankind friendly parts of the total ecosystem of yesterday without taking into account the effect that this may have upon certain "troublesome" organisms, especially bacteria which adapt quickly. Mankind through breeding is directing the course of evolution even using genetic engineering. Again in order to manage favourable changes synthetic chemicals of many kinds are applied to soils so that previous biological balances are altered, all in the name of the yield of mankind's need for "food". This together with protective practices has led to a huge increase in human population, which could lead to a loss of balance of organisms in the ecosystem which has been built up over billions of years. Note that human beings are causing change so rapidly that it leaves little scope for the ecosystem to adjust by strictly biological means. Mankind itself is the only manager now. The question as to whether or not these activities can alter favourably the optimum energy retained on Earth's surface in biological cycles has to be faced sooner or later, since we have seen the total dependence of the whole ecosystems on all the biota and the environment (Chapter 9).

All of the above has required a completely different type of organisation but we leave this theme until after we have given a more detailed view of the changes in the use of elements and of energy.

10.5. Mankind and the Detailed Use of Chemical Elements

Much of what can be said under this heading differs from that in previous chapters in that the application of chemistry by humankind is based overwhelmingly on the use of the brain, cognitive activity and not related to the information content of DNA. The exploitation of the chemical elements outside ourselves is all around us in our daily lives in our cities, in the countryside, and dominates our civilisation. It has been greatly aided by the discovery of the underlying limitations of the fixed number of elements which are placed in systematic relationships in the Periodic Table, one of, if not the, greatest discovery made by mankind (see Chapter 2). There are no more elements to be discovered and not so many more useful ways of combining them. There is hardly an element among the 92 of the Periodic Table that has no employment within compounds in structures, communication, energy distribution, transport devices, and so on. Today we can also add direct interference internally in biological organisms in the introduction of applied chemicals of medicine and agriculture and also in genetics. There are many books and a huge literature covering all these subjects (see Further Reading), and here we mention just one or two facts to give the flavour of the new chemistry on the Earth since 1500 AD.

Outstanding is the quite novel use of *inorganic* chemical elements. In structures we have, for example, the huge involvement of iron alloys, steels, everywhere (see Table 10.6). Together with minerals made into bricks, mortar, plaster and concrete, there are mined minerals including marble and granite. Here the elements calcium, carbon, sulfur and some ionic metals are dominant in salts. In tonnage the scale is vastly greater than that of human turnover of synthetic organic chemicals though not yet greater than that of the organic chemistry of the whole of the biological system. The metals or semiconductors of inorganic chemistry are used in communication devices of all kinds, for example electrical and radiation transmission and reception. New vast information storage from these materials is now possible in computers, an externally *coded* further extension of human memory and information handling, but with no input from genetic information. *Personal computers*, for example, can be considered to be part even of phenotypes, but not of genotypes. Added inorganic chemicals are slowly accidentally transforming the surface of the Earth and some are even introduced deliberately to create fertile soils for the growth of foodstuffs. (Note especially chemicals based on B, N and P but also trace elements such as Se, Cu and Mo.) By way of contrast, the organic chemicals introduced by man are sophisticated synthetic products based on carbon chemistry, including medicines, herbicides and pesticides and, the most used, plastics. Only reference to chemistry and technology textbooks can give a feeling for these developments and their scale but we note that they all require considerable energy for their production. Some of the energy is retained in the new, kinetically stable but thermodynamically unstable chemical products but much is lost as heat. Thus, human activity is in line in part with the proposition that evolution concerns energy degradation, but to make it fully compatible with long-term sustainability all our chemical products must cycle. The objective should be to make all elements non-polluting, i.e. the cycles are complete and element neutral, see carbon for example. Mankind needs to be aware of this need for otherwise these chemicals can become useless, even inhibitory, waste. Like a sudden volcanic eruption or pollution from a meteorite the result could send evolution backwards for a while but then it will recover to some degree as waste is absorbed. While we, mankind, operate on short-term basis, at best 5–10 years, biological or rather, to include the environment, ecological evolution is based on hundreds of millions of years. Unfortunately, the time scales of the two are so disparate that the two systems may well come into conflict rather than into cooperative activity.

It is interesting to note that industry extracts many metals from sulfide ores by oxidation/reduction coupling (compare living processes)

$$MS + O_2 \quad \rightarrow \quad M + SO_2 \uparrow + heat$$

followed by later decay

$$2M + O_2 \quad \rightarrow \quad 2MO$$

and solubilisation in the waters of the Earth. This is the same thermodynamic reaction that caused metal pollution by aerobes and later new uses of the few extra released metals in the activities of organisms (Zn, Cu) but it has been held in check previously by barriers to uptake or by rejection introduced in the coded form in DNA. Humans are increasing metal pollution insofar that their use of the metals obtained from sulfides does not trap them totally with the result that by erosion (corrosion) they become available in surface waters as waste. The metals in question are especially Al, Co, Ni, Cu, Zn, Cd and As. Note again that all *free* divalent metals apart from Mg^{2+} and Fe^{2+} may be inherently poisonous in the cytoplasm at above a very low level. Metal pollution is a risk both short term and very long term though it may be reduced by precipitation of oxides and carbonates. Genetic response is slow and genetic constructs need great protection from unusual high concentrations of such elements. We repeat there is no chance of fast compensating changes in the DNA and its expression. (We see this risk is parallel to that produced by material interaction with the deep crust and mantle of the Earth.)

A further peculiarity of humans is that they have come to understand much of their own and other animals' physiology and biochemistry. Apart from deliberately handling diet, that is the bulk input of required chemical elements in "desirable" compounds in general animal, including mankind's own husbandry, and treatment of plants to increase yields, they have sought for remedies to internal chemical maladjustment, illness, by introducing special medicines. This is not the place to go into detail but we note that very many elements, known to be non-essential for life, usually in compounds, are recommended as valuable medicines. The list of such elements includes Li, F, Al, Cl, As, Sb, Bi, Hg, Pt and Au. Often these medicines are used to destroy invading organisms as well as to attack inborn errors. Of course, dose is a very important control in all applications. Is it wise to be evermore protective of life and to increase the population? Is it wise to be so protective of old age?

In this book, at the end of chapters 5–11, following an outline of the nature of organisms and chemical changes involved in evolution, we have been concerned with the problems of further environmental changes produced by organisms. Here we see that chemical operations of mankind, which are now on a significant scale and kind, are beginning to introduce problems for the whole ecosystem, and it is obvious that it is the whole geochemical/biological chemical system that is becoming of concern. For example, we use reduction as well as oxidation to produce many chemicals such as metals by high-temperature processes. In some way this takes us back to the initiating events that started life since the start was reduction of geochemicals driven by available energy followed by sequential and inevitable change of the environment due to waste, i.e. oxidative pollution. However, the elements we introduce have higher risks than possible value in future

organisms. In discussions of evolution in conventional biology texts this coordinated interaction is usually overlooked. The changes caused by mankind today are
due to the use of scientific knowledge and should be and can be purposeful but
must be controlled to the advantage of the global ecosystem, not just to mankind.
We shall review next in somewhat more detail the new ways of energy, transport
and information handling in human organisations and return to look at the problems in the future in Chapter 11.

10.6.   Mankind, Energy and External Machines

Repeatedly we have stressed that our concern is with the ecosystem development stimulated by energy input to material (see Table 3.8) for sources of energy.
Thus, we are concerned with energy input as well as with material input by man.

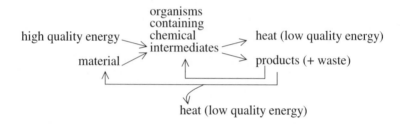

The energy input from the Sun did create material waste such as oxygen but that
has come to be used in cycles. When we describe the new energy inputs devised by
humans we must ask if they can devise methods of using waste *in the cycle* to generate energy. Also note that many new applications of energy are dependent not just
on understanding the nature of energy, but on the development of chemicals capable of being used in appropriate constructions (machines) needed in energy generation and application. Put simply, mankind is developing energy sources and
element uses alongside those in organisms and in the previous environment but
although the drive generates (new) harmless energy degradation the waste is common and possibly harmful.

One available waste from earlier life is fossil material such as natural gas, oil and
coal, which is an obvious energy source on reaction with oxygen. This material is
being put back in the cycle today by mankind as CO_2 and H_2O helping to increase
the intermediate system above (organisms), while increasing the energy and material in the cycle contributing to an increase in heat production. Here the skill is in
the construction of power stations. The increase in CO_2 in the environment is a
cause for concern *for mankind* since it aids heat retention, causing a small temperature rise. Other partly wasted resources of the Sun's radiation and of possible use
in valuable energy generation are the movement of air, sea and water. These are
continuous physical energy sources related to temperature and gravity gradients

and hardly affect chemistry directly, but by conversion to electric power they are beginning to be more used to increase high quality available energy input without material waste. This energy will increase the chemical production if captured in a sufficiently smart way (that is directed to cyclical (biological) growth) but inevitably will also increase heat output. (Note that organisms have always used wind and water currents to circulate nutrient elements, even seed, to and from them.) Finally, nuclear energy stored from the consequences of the Big Bang is an almost infinite supply and any use of it again increases total heat output, but we must take into account its dangers in that it does produce harmful waste. In each of these cases complex machinery had to be invented and employed. (Remember the complex molecular machinery of chloroplasts and mitochondria.)

The greater the increase in more effective energy input by mankind the greater will be the increase in the heat and useless (poisonous) chemical output. Hence, we shall have to manage the results–some kind of global warming as well as chemical pollution, as the scale of activity increases. A difficult problem arises if mankind uses nuclear power as a source of energy, which, as mentioned above, may generate quite new radioactive, often permanent, waste that cannot enter the cycle. If such uncyclable products are produced then we risk acting in a destructive sense much though we may improve our style of living temporally. Here a benefit/risk analysis is all we can manage. Of course, the ecosystem is not concerned with what is called the "quality" of life which is just a selfish concern of mankind to increase energy use often for "trivial" purposes, comfort and pleasure, where trivial is in relation to the underlying processes of the ecosystem. Thermodynamics has no sympathy and, therefore, much thought about our future is needed.

Turning to the application of energy flow we are concerned with machinery and its efficiency in doing work. Therefore, along with the changes in the materials and energy sources, it is essential to see that much of human lifestyle depends upon macro-machines devised by man and capable of working on a massive scale. Here we wish to put this machinery in the context of evolution. In Chapters 5–7, we described the molecular machines of single cells with short-range electrical and molecular flows in devices such as those for pumping, energy transduction, synthesis in guided, coded paths and so on. In chapter 8 we saw a change to multi-cellular machines such as muscles with increasingly long-range cell/cell cooperation ensured by large-scale flow of messenger molecules and ions over larger and larger distance. Workings, on the scale of dams built by beavers, large nests by birds and the early buildings of mankind, appeared before external machines were developed by mankind. The external machines we see around us have grown in size and in ability to handle large amounts of energy, electrical and mechanical, in all manner of constructions. The industrial revolution is one result of this scientific ability to increase work in an unparalleled way. It has also proved possible to develop mechanical machines for transport of goods and electronic machines for message transmission and reception to sustain a very comfortable living style and purely for pleasure. Attempts by man at making miniature machines are now approaching the small size of molecular machines but all our efforts are to increase the use of energy and resources. We must ask again and again: is this wise in a non-cyclic material system?

Now that we have given a quick summary of chemicals, energies and machinery introduced by mankind we must see that like just as in all other developments of new chemotypes it is necessary to devise new transport and communication systems to increase survival strength.

10.7. Transport

In all biological development we have stressed the need for directed transport of material in an organisation. The original mode in primitive life was that after arriving at a membrane receptor a useful chemical was pumped inward and waste was pumped out by both applied energy. Refinement in simple eukaryote cells allowed for guided transport in vesicles or along field gradients into or out of the cell, i.e. on molecular routes. In multi-cellular organisms this transport across a cell membrane increased in capacity, for example between cells in the circulating intake and rejection systems, that is by physically controlled circulation in extracellular fluids. Recognition was again by receptors in receiver cells. Animals with brains came later and were able to transport natural materials for their "buildings". Humans have learnt to increase transport of material through external space first using animals, including themselves, and then by sea, road, rail and air where navigation is controlled along routes by themselves or by robots (Table 10.8). These are developments resulting from increased knowledge of both physics and chemistry but this transport network too has to be controlled by a new central organisation. The organisation clearly requires messages between activities over long distances as the transport now extends through a huge space (see Table 10.9). Here computers are taking over much of the effort in the collection and use of *coded* (in digital numbers) information. (Note the parallel with biological evolution: as variety of activity increases so also transport with communication must increase. Animals transport quite large quantities of materials for

TABLE 10.8

THE EVOLUTION OF TRANSPORT

Organism	Material carriers	Energy carriers
Prokaryotes	Substrates	e/H^+
	Coenzymes	Ion gradients
Single-Cell Eukaryotes	Vesicles	O_2, SO_4^{2-}, NO_3^-
	Tubulin systems	Fe^{3+}
Multi-cell Eukaryotes	Circulating fluids outside cells	Sugars outside cells
Modern animals	External limbs	
Mankind	Boat, road, rail,	Fossil fuels
	air traffic	Water
		Batteries

their own constructions. Human activity is then just a novel extension of previous biological chemotype activity but now based on using macro-machines.)

10.8. Human Message Systems

We have observed in each chapter that evolution of organisational complexity has to be accompanied by increase in communication systems. All the early messengers were ions or small molecules (see Chapters 5–9). The change in the message systems through external space began with signs and sounds and developed as the language of humans. It was extended in distance by smoke signals and instruments such as drums, and through carriers – animals (pigeons) and mechanical devices. The major change today is the use of electron transfer in wires and radiative transfer, guided or not. We do not need to stress the huge advances that have come in organisation through the use of new materials and new devices suitable for transfer of these messages, i.e. of information all in novel "coded" forms (Table 10.9), only some of which, e.g. computer language, are in all or none, digital form like DNA. We cannot go into detail here but it is left as an exercise to draw up a comparison

TABLE 10.9

THE EVOLUTION OF MESSAGES

Type of organism	Medium[a]	Carrier[a]	Speed[b]
Prokaryotes	Water (Cytoplasm)	Coenzymes Substrates, Mg^{2+} Controls Fe^{2+} c-AMP, ATP	milliseconds
Single cell Eukaryote	+ Across membranes	As above Ca^{2+}	milliseconds
Multi-cell Eukaryote	+ Extracellular space	As above Ca^{2+} increasingly Hormones Zn^{2+}	As above seconds to years
Modern animals (Nerves)	+ Electrolytic charges in water	As above Na^+, K^+, Cl^-, (Ca^{2+}) Transmitters	As above ≪milliseconds
Mankind	+ Extra-organism space (a) Wires (b) All space	Sound Electrons Radiation	As above + ≪10^{-10} ≪10^{-10}

[a] Going from top to bottom the systems add on to the previous systems so complexity increases, and message media and carriers were extended. The available chemicals also change from top to bottom systematically.
[b] Time from initiation to action.

between the message circuitry of ions and organic molecules inside organisms and those of electrons and radiative waves outside organisms. The demand for efficiency of organisation has many parallels with early life. (Note that we do not understand the coding in the brain by a variety of inorganic and organic transmitters.)

All that we have to see is that these activities of mankind cannot be other than an extension of the evolution we have seen in biological systems, both linked to an environment. The whole including mankind itself is an unavoidable consequence of the efficient use of more powerful energy forms put into environmental chemical elements for ordered or organised constructions and it rests intrinsically upon an ecosystem driven by more effective energy degradation, that is increase of thermal entropy. The process is associated with survival fitness of the ecosystem but there is always risk in providing protection for a given species since it can be overwhelmed by hidden activity.

10.9. Organisation and Mankind

Organisation of human life is the only organisation in animals inherited overwhelmingly through learning, not just from the DNA, while much of organisation of other animals, for example those of ants and bees, even dolphins, has little cognitive intellectual content. A puzzling feature, however, is that much of the limited external organisation in these lower animals *appears to owe itself to learning*, that is constructions within the brain after birth. For example, navigation by birds or fish in organised patterns may be in large part inherited but there may also be a *learning* of the ability to navigate while in possession of a rough "map". Birds do have a certain learning ability without doubt but how any basic "map" provided by genes is inherited is difficult to appreciate, since it implies an inherited construct in the individual's brain of an individual map. Detailed maps have a huge data content and surely genes cannot cope with the storage of them. Even so each animal goes long distances to its own "home", and knowledge of migration paths appear to be inherited. How? It is also difficult to see how human beings appear to have an inherited basic grammatical structure of language as indicated by Chomsky and also by Pinker (see Further Reading), but this possibility is under intense study. We have to consider too that the brain is only a very basic product of DNA inheritance since it is malleable opposite experience and we have to puzzle and enquire about inheritance in this context. Is the degree of malleability inherited and what are its genes? It is development by use, learning and constant reinforcement in the brain, which is obviously extremely dominant in mankind's organisation much though we trace its beginnings in earlier animals. Slowly we are learning global management of our various capacities, an unavoidable necessity, but at the same time we struggle to allow individual free action (to which we turn shortly). Note that education systems often stress the way structures of and attitudes to the development of society developed in the Ancient World, 5,000 to 0 BC, under the heading of "culture", but neglect the even more extensive changes in the last 500 years due to advance based on empirical scientific approaches. Empirical judgements are of just as much value in action but often actions lack the near certainty of cause and effect we have met previously through inheritance of DNA. We now stress the peculiarity of this uncertainty.

10.10.   The Development of Self-Consciousness

In this section, we return in a small way to the problems of "information" in a human context mentioned in the first paragraph of this chapter before we return to them in Chapter 11. There is another feature of human beings, which seems to distinguish them from all animals, self-consciousness, which is related to the separation of a species into individuals. We all perceive that human brain is different, not just in capacity, and we are inclined to think that it differs in kind from that of all other animals (see Figures 9.4 and 9.5). For reasons that greatly trouble philosophers and scientists alike this difference is seen in self-consciousness, an ability to explore even ourselves mentally on a scale and in a way almost unknown apparently in other animals. If we seek a reason for its appearance, and we believe that it could not happen by an extraordinary jump in evolution, then we need to find a continuous link of it within other animals of the monkey variety and of the then earlier ones. (In physical chemistry an attribute of a system may arise with but an extremely small change in parameters.) It is likely that the advances came slowly at first, but more quickly with the development of tools, the use of the hands, sign and then spoken language. We have to consider further if it was the very fact that mankind could distinguish clearly internal from external events, which led to self-knowledge. The development of self-consciousness is then a final expansion of animal nature in each individual of one species bringing space outside the body with its sources of material and energy under control for useful effective constructs by that individual and at the same time evolving sensations and "feelings", which have led to self-awareness. What has this to do with the general thesis of this book that all life is an evolving chemical system with a coded reproduction via genes? The problem lies in how does this property, self-awareness of individuals, relate to the physical chemistry of the living system. Let us look again at genes in a search for a solution before we look at the individual human in Chapter 11, where we shall stress again how information transfer, communication, based on DNA, common to all individuals in previous organisms as an unavoidable instruction and so making all individuals very closely alike, differs from information transfer from and to individual brains in humans, which has no such common instructive power.

10.11.   Human Genes

Traditional views would have led to the expectation that considerable differences between humans and other animals would have been found in a construction linked to the brain and directly connected to novel genes. It was a considerable surprise in the 17th and 18th centuries to find the brain so clearly linked to that of earlier species (see Chapter 9), but it was a greater shock to find that human genes are not greater in number than those in some higher plants and animals and are not even different in kind from those of a chimpanzee, with whom we share 99.0% of our genes. The human genome has 3×10^9 bases, which express 35,000 proteins, neither of which is unusual for a modern animal or plant. There is no known metabolic pattern different from other animals, nor is there a known significant difference in *internal* element content or use. The difference between humans and chimpanzees lies as far as we can tell in the functional relationship between observation and brain cell memory based on brain growth and connectivity to the external environment. Although human brains have more cells – there are some one hundred billion, 10^{11}, neurons at least and many more so-called glial cells of various types – it is the way in which they retain transferable usable knowledge that is much more developed than in other animals. The advance is in systems not in DNA coding. DNA has never been the only source of information. How has this evolved in perhaps 100,000 years *involving only 5,000 generations*, compare the billions of bacterial and even of some eukaryote generations which took many million years to evolve new chemotypes? Surely it is too fast for random genetic mutation, which is only apparent in humans in a relatively minor way. Against the 3×10^9 bases of the human DNA there are more than 10^{13} synapses of nerves. It may well be that the development is the result of a simple "trick" involving differential growth of the skull leaving space for a larger more flexible brain in which growth opposite experience dominates and its information store is individually different; the phenotype then greatly extends the genotype. This results since phenotypic experience, not DNA, dominates growth of cell organisation in the brain after birth. It would mean that a small adjustment in controls over genes has come about making a huge difference in a contingent (chemical) capability – realisation of a large change of capacity to act not related easily to genetic alteration. The capacity develops thus by teaching and learning not by genetic change and all is lost if education is lost. The secret certainly lies in the ability to absorb, store and utilise information not contained in the DNA. The chemotypic system seen in phenotypes advanced far faster than the genotypic could and may well arise since the brain construction is initially ill-organised in the DNA-controlled foetal state but develops quickly after birth. We return to this discussion in Section 11.13. Here, we see the advantage in a classification of organisms not based on classical morphological or more modern biomolecular sequences of RNA/DNA or proteins but on chemical thermodynamic properties included in uses of components, space and organisation. The particular properties rest in fair part in the brain. Humans are just the consequence of different, during individual lives, (late) onsets of chemical organisation in a large brain, that is internal system development by experience. We turn in the final chapter to a very brief account of how in some 30,000 years, but more especially in the last 500, the peculiarity of mankind has completely altered the ultimate possibilities of evolution in the foreseeable future. There is no way this development can be linked to genetic change but it is part and parcel of the evolution of chemotypes and we look at it in this way more closely in Chapter 11.

10.12.   Summary

Our approach to the ecological evolution leads us to consider mankind as a very advanced, probably a final chemotype in the physical/chemical evolution of an open ecological system of chemicals exposed to a high quality energy source, the Sun, and degrading its energy to heat. It is then largely a thermodynamic approach. The problem with this latest development has only become apparent in the last 500 years since man has come to understand and utilise chemistry and physics to manipulate powerfully the whole ecosystem using macro-machines. In the last chapter, we summarise the changes over 4.5×10^9 years before analysing the effects of mankind's nature. This will lead us to a very uncertain outlook for the future success of the system as a whole in its ability to be able to reach an optimised cyclic steady state of energy uptake, use and degradation. If it does reach it, ecological evolution would then be complete but it could be that the very activity of a large number of individual phenotypes (not of a cooperative genotype), which condition has also arisen through the nature of the brain, will take us further and further from this endpoint. Will mankind increasingly poison the environment causing reduced biological activity for many, many years? In the next chapter, we shall summarise all the first 10 chapters and then turn to the peculiar twist, which the appearance of individuals (phenotypes) has had within the one particular species, *Homo sapiens*, a most unusual chemotype. Sooner or later we shall have to ask what population can Earth sustain with what life style.

10.13.   Note on Creation and Intelligent Design: Mankind's Inventions

Given the intensity of discussion of both creation and intelligent design, both of which propose an outside influence on evolution which has no simple scientific explanation, we must make our position clear. As far as we can see the directional character of evolution of our ecosystem, illustrated by the cone on the cover of this book, requires only one act for which we can see no explanation. We know of no cause of the Big Bang and the limitations it imposed on the cosmos observed in the laws of Nature. It is these laws alone which we use in our analysis. The laws contain possibilities both of systematic development which is the centre of our discussion and of random events. The first we relate, in the evolution of life, to chemotypes and the second to the appearance of species within chemotypes, see the cover of this book. At no time in this chapter, or in any other chapter do we invoke any other kind of activity.

10.14.   A Note on General Culture

It might be argued that we have paid too little attention to mankind's total cultural achievements and are over-stressing those derived just from scientific knowledge and its exploitation. Our response is that all the other cultural values and achievements are well recognised as having an individual character as well as a general one and so differ from scientific advances. They do not affect to any large degree our discussion of evolution with one major exception. If cultural achievement outside science leads to the belief that *Homo sapiens* is *totally different in kind* from all other animals, these cultural achievements as well as the advance of the use of science could blind us to the risks we take. If we set ourselves outside the ecological system and look only at how we need to behave towards one another within a framework of the best possible way for us to live as selfish individuals, as selfish communities and as a selfish species then we shall surely eventually put much at risk. We shall clarify these problems at the end of the last chapter.

Further Reading

BOOKS

Calvin, W.H. (1996). *How Brains Think*. Weidenfeld and Nicholson, London
Chomsky, N. (1975). *Reflections on Language*. Pantheon, New York
Cox, P.A. (1995). *The Elements of Earth*. Oxford University Press, Oxford
Damasio, A.R. (1995). *Descartes' Error*. Picador, London
Dawkins, R. (1998). *Unweaving the Rainbow*, Penguin, London
Durham, W. (1991). *Coevolution: Genes, Culture and Human Diversity*. Stanford University Press, Stanford, USA
Goldsmith, T.H. (1991). *The Biological Roots of Human Nature*. Oxford University Press Inc., New York
Goudie, A. (2000). *The Human Impact on the Natural Environment*. Blackwell Science, Oxford
Pinker, S. (1994). *The Language Instinct*. Morrow, New York
Pinker, S. (1998). *How the Minds Work*. Allen Lane – The Penguin Press, London
Southwood, R. (2003). *The Story of Life*. Oxford University Press, Oxford.
Wilson, E.O. (1978). *On Human Nature*. Harvard University Press, Cambridge, MA

PAPERS

Balter, M. (2005). Are humans still evolving. *Science*, *309*, 234–237
Dolan, R.J. (2002). Emotion, cognition, and behaviour. *Science*, *298*, 1191–1194
Fields, R.D. (2005). Making memories stick. *Sci. Am.*, *291*, 59–65
Hauser, M.D., Chomsky, N., and Fitah, W.T. (2002). The faculty of language; what is it, who has it and how did it evolve. *Science*, *298*, 1569–1579
Lisman, J.E. and Fallon, J.R. (1999). What maintains memories? *Science*, *283*, 339–340
McGaugh, J.L. (2000). Memory – a century of consolidation. *Science*, *287*, 248–251
Miyashita, Y. (2004). Cognitive memory: cellular and network machineries and their top-down control. *Science*, *306*, 435–440
Pennisi, E. (2003). Genome comparisons hold clues to human evolution. *Science*, *302*, 1876–1877
Premack, D. (2004). Is language the key to human intelligence? *Science*, *303*, 318–320
Ridderinkhof, K.R., Ullsperger, M., Crone, E.A., and Nieuwenhuis, S. (2004). The role of the medial frontal cortex in cognitive control. *Science*, *306*, 443–447
Rugg, M.D. (1998). Memories are made of this. *Science*, *281*, 1151–1152
Tononi, G. and Edelman, G.M. (1998). Consciousness and complexity. *Science*, *282*, 1486–1451

CHAPTER 11

Conclusion: The Inevitable Factors in Evolution

11.1. Introduction

In this final chapter before we look at the peculiarities of human beings and examine future ecological problems we wish to pull together the various parts of earlier chapters, while summarising our overriding view that the broad features of evolution in chemotypes were inevitable, though their timetable was uncertain and there are a large variety of species, randomly selected within a chemotype. This conclusion cannot be derived from considerations of the observable biological nature of species such as their morphology nor from any analysis of coded molecules or their structures, studied under molecular biology. Our conclusion, no matter what its final standing in detail, has to be based on the general principles of chemical thermodynamics of energised ecological systems. We set out the geochemistry and general chemistry needed for an approach to such systems in Chapters 1 and 2 with the necessary outlook upon their energisation in Chapter 3. Several features stand out. First, the geochemistry of the early Earth, given in Chapter 1, is an outcome of events related to the Big Bang and the consequential kinetic limitations imposed on the abundance of the possible

415

elements in giant stars. *There are 90 kinetically stable elements, no more and no less.* The Sun, derived from a giant star, was a precursor of the planetary system, and among the planets formed from it, by an accident, the cool Earth had a particular distribution of elements, largely in reduced or semi-reduced oxidation states. The Earth is peculiar not so much in the element distribution, though the nature of the surface was a somewhat unpredictable outcome, but in its physical characteristics (Table 1.5), including its controlled surface temperature of close to 300 K over the whole of its lifetime of 4.5×10^9 years, a result of unavoidable balancing of unusual physical/chemical circumstances (see Table 11.1 and Section 1.7). The temperature ensured the constant presence of liquid water and hence the special but limited possibilities of aqueous solution chemistry. The circulation of water and the production of ozone described in Chapter 3 are an essential part of the present day ecosystem and man interferes with them with a risk to all organisms on land (see Table 11.2).

Knowledge of the 90 chemical elements and their properties in compounds led to the construction, by man, of a *unique table of elements, the Periodic Table*, of 18 Groups in six periods in a pattern fully explained by quantum theory, described in Chapter 2. There is then a huge variety of chemical combinations possible on the Earth and limitations on what is observable are related to element position in this Table. It also relates to the thermodynamic and/or kinetic stability of particular combinations of them in given physical circumstances (Table 11.3). The initial state of the surface of the Earth with which we are concerned was a dynamic water layer, the sea, covering a crust mainly of oxides and some sulfides and with an atmosphere of NH_3, HCN, N_2, CO_2(CO, CH_4), H_2O, with some H_2 but no O_2. This combination of phases and their contents then produced an aqueous solution layer of particular components in which there were many concentration restrictions between it and the components of the other two layers due to thermodynamic stability, equilibria, or kinetic stability of the chemicals trapped in the phases. It is the case that equilibrium

TABLE 11.1

EVIDENCE OF SINGULAR NATURE OF EARTH

1.	The relative sizes and positions of the Sun, Earth and Moon
2.	General physical features of Earth as listed in Table 1.5.
3.	Temperature at surface ~300 K for 4.5×10^9 years
4.	The separate chemical zones of the Earth including core, mantle, crust, aqueous and atmospheric zones
5.	The relative amount of sea on the surface leading to interaction between land and water, weathering, and the production of fertile land
6.	The graded temperature from the poles to the equator. The resultant flows of the seas to and from the ice of the poles to the equator generate a graded steady state. There have been hot and cold periods of virtually no ice and an ice-covered Earth but correction has always set in. Global warming today is not open to serious doubt. Ecologically it is a minor problem.

TABLE 11.2

INTERACTIVE INORGANIC FLOWS IN EARTH'S ECOSYSTEM

Material Flow	System
Water	The flows are in the circulation from the sea to the land (see Figures 1.6 and 1.13). The sea itself flows in vast circles
Air	The air flows help to establish the average climate temperature
Ozone	The ozone layer reduces the effect of UV light on Earth's surface
Molten Rock	(a) The magma is a constant flow carrying sediments downwards only for them to emerge in transformed states (b) Upwelling of minerals in the black smokers (also contributes to temperature) (c) Continental drift causes changes in environment
Carbon Dioxide	The CO_2 of the sea and its reserves in some carbonates equilibrates with the atmosphere and is one source of green house gases
Oxygen	The O_2 exchanges with some soluble inorganic materials especially in waters

TABLE 11.3

BASIC CHEMICALS ON EARTH

Chemicals	Examples
Inorganic close to Equilibrium	Oxide minerals, e.g. Fe^{3+} ions and carbonate and seawater contents
Inorganic far from Equilibrium	Deep zones of metals and very reduced minerals in the mantle Some parts of the atmosphere O_3, N_2/O_2
Organic C/H/O compounds far from Equilibria	All such chemicals are thermodynamically unstable in the absence of air relative to CO_2, CH_4 and H_2O and in air all are unstable relative to CO_2 and H_2O but all have considerable kinetic stability

in water is quickly reached for certain inorganic metal ion/non-metal anion reactions while kinetic stability is very high for many C/N/O/H compounds. The compounds of the elements P and S with lighter non-metals are intermediate in character. The solubilities and the vapour pressures and kinetic stabilities of chemicals are strict limitations *on availability for any processing in water*. We wish to impress on the reader that the origin of organisms is strictly an outcome of the basic abundance and of the chemical properties of elements in compounds as grouped in the Periodic Table, that is especially of thermodynamic or kinetic stability at the temperature of 300 K of the sea, which limited availability. Later, water became present on land due to its circulation via sea, clouds, and rivers, which also create sediments (see Chapter 3). It is upon the very limited *available* chemical elements in water that the energy of sunlight fell and has driven the chemistry, which is what we see as life and its evolution over a long time period. We cannot give, as yet, any realistic account of the pre-biotic chemistry or of the way life's basic components arose (see Westall in Further Reading), and so we made general statements of what we believe had to happen before life got fully underway. On exposure to energy, mainly radiation from the Sun, the light non-metals became readily *energised* in *reduced chemical* kinetic traps, often large organic compounds, including biopolymers and lipids, simply because of the closeness of the Sun, the energy supply, and the controlled temperature of the Earth. This is a consequence of charge separation leading to separation of the reduced compounds from the oxidised ones. The formation of membranes from lipids then gave rise to the earliest cells with physically trapped ions and organic molecules inside them. It is this energisation and the subsequent energisation of the flows in and out of the cell, which, in ways unknown, led to the selected organic compounds seen in all cells that is mainly one set of biopolymers (proteins, nucleotides, saccharides and lipids) and their precursors, which may well be a unique set for living processes (see Chapters 3 and 4). The energised system in cells can be looked upon as a set of flows driven by molecular machines. Activation of many required steps during and subsequent to the initial energisation needed among other things the presence of selected catalysts – very frequently that of the readily available metal ions which were bound selectively to organic ligands at equilibrium which also became trapped. Control over salt content of Na^+, K^+, Cl^- and Ca^{2+} ions was also essential for cell stability, and other poisonous elements had to be rejected. The resulting mixed inorganic and organic cytoplasmic chemistries in cells may be the only general composition able to result in living systems, which, as we know, require 15–20 elements, metals and non-metals. In an unknown manner the membrane-limited energised chemistry became coded and thence reproductive, and it was managed by feedback within the reaction systems, linked with the environment and to the coded molecule, as described in a general manner in Chapter 4 (see Table 11.4). We stressed in Chapters 3–9 that the code provides a set of *unavoidable* products and instructions (information transfer) and is conservative. Through cooperativity, both molecular and cellular, the system gained greatly in survival strength, which we describe more thoroughly in Section 11.4.

Now, most metal ion/organic molecule chemical reactions inside cells also come to equilibrium rapidly. The organic products, made irreversibly available by synthesis under feedback control, contain a broad set of possible binding sites for selected metal ions mainly in soluble proteins (enzymes) and in the pumps for uptake or rejection managed at the cell membrane, as well as in the factors, transcription factors, necessary for controlled production of those organic products under the direction of the coded system. These ion-selective binding sites are common to all cells so that while all cells are based on similar major organic reactions and similar but specific biopolymer products, they also have in common a set of

TABLE 11.4

THE ESSENCE OF CELLS

1.	Energy capture from energised minerals (first?) but overwhelmingly from light giving charge separation and then reduced compounds separated from oxidised ones
2.	Compartments isolated from environment
3.	Interior chemicals reduced relative to environment
4.	Rejection to environment of certain oxidised chemicals, mainly oxygen
5.	Syntheses of a particular set of polymers
6.	Self-synthesis of a particular set of catalysts, pumps and so on
7.	Inclusion of a wide diversity of available metal ions and exclusion of others
8.	Reproductive through DNA code
9.	Progressive change of cell metabolism and organisation in compartments to manage an oxidised environment
10.	Cells do work in converting light energy to synthesised chemicals and can then be treated as containing molecular machines
11.	The fundamental instability of cells leads to death and cycles of chemicals plus waste. Consequently, the overall process of a cell is the accelerated conversion of light into heat, increased thermal entropy production

similar but specific metal/organic complexes at equilibrium with free ion levels in their cytoplasm. The special set of organic compounds, proteins, nucleotides, saccharides and lipids are then complemented by a common set of free, equilibrating, metal ions (and of some small organic anions including coenzymes) at almost fixed concentrations in all cell cytoplasm. In large part these concentrations of certain small organic compounds, particularly phosphates, and of the metal ions Na^+, K^+, Ca^{2+}, Mg^{2+}, Fe^{2+} originally, perhaps later Mn^{2+}, Co^{2+}, Ni^{2+}, and again later still Zn^{2+} and Cu^+, are connected by feedback and feed-forward controls both within metabolism and at the level of expression of the code and to the degradation and membrane pumping processes. The whole activity of cells is then homeostatically managed in what can be thought of as an autocatalytic feedback controlled cyclic system (see Section 3.9). The essential feature of an autocatalytic set of reactions such as this is that the formation from the reduced metabolic units of all large molecules including the catalysts themselves and the coded molecules is catalysed under control of both the metabolites linked to the environment, and the coded molecule, the expression of which is also under this control. The metabolites here include metal ions. In this way, the cell produces a singular small set of pathways for efficient growth and survival (see Table 11.5).

The combination of the material of the first four chapters led us then to give an outline of the thermodynamic systems chemistry common to all cells. We offered little explanation of the origins of cellular life, as such an explanation could well remain beyond our insight. We then concerned ourselves with the way in which the

TABLE 11.5

THE FITNESS OF LIFE'S CHEMICAL SYSTEM

1.	Modes of energy capture in a confined space, a cell or an organism (but see Section 11.14)
2.	Major metabolic chemistry leading to long-life but unstable polymers using the available non-metal elements H, C, N, O, P and S in covalent attachments but not halogens. The polymers may be the fittest for the purpose they perform
3.	Reproduction using code based on the same covalent chemistry as in (1) above
4.	Control features of exchange of covalent fragments from P, S, and aromatic ring N- or O-centres
5.	Catalytic activity by available metal ions including (V), Mn, Fe, Co, Ni, Cu, Zn, Mo(W), selected for chemical/biological fitness
6.	Osmotic/electrical potential control using K, Na, Cl
7.	The development of highly efficient energy transduction systems later using chloroplasts and mitochondria
8.	Ability to recognise the environment and to consume large particles
9.	Ability to adapt to environmental change
10.	Control of temperature
11.	(Most recent). General control of much of the environment's energy and materials.
12.	Rapid recycling of debris through the catalysed reactions with the oxidised environment. Unlike the faster oxidation of basic inorganic elements in the environment (see Fig. 1.14), the rate of reduction and oxidation of organic materials is faster utilising uptake and catalysts

total ecological system of life and the environment could evolve. Evolution of chemotypes is open to chemical investigation as there are now adequate data concerning the composition, concentrations and structures of all major groups of organisms and the geochemical environment at different times. We set out to enquire first about the essential use of 15–20 elements, non-metals and metals, in the first cells, prokaryotes. Put in a somewhat different way, life as we know it had to start in some space-limited proto-cell with the energised uptake of very basic environmental inorganic chemicals, to reduce many of them, and then to synthesise a special set of molecules, including machines, catalysts and coded molecules, and for it to prosper it had to become tightly organised and reproductive. The system always decays. It could have been just produced again and again but it became reproductive and developmental. Now our contention is that, once these conditions arose, life could evolve in but one way, that is, at first using the available chemical elements and energy sources in a somewhat reduced and then in a slowly changing more oxidised environment. The environmental change was inflicted by organisms through release of oxygen from them, necessitated by their own reductive activity. To manage the new environment in a singular sequence, organisms adjusted their chemistry and increased organisation of necessity in compartments, that is, increased use of space, together with an ever-increased capture and use of energy and of some elements from newly soluble salts; while generally the elements were used both as previously and in novel ways, as described in Chapters 5–10. Organisms thus evolved to face and utilise environmental change in a systematic direction. While all types of organisms advanced, the ecosystem as a whole gained in its ability to capture energy and material through the creation of diversity. The increase in chemical organisation with major diversity is reflected in a succession of new chemotypes with new uses of space but which also required new organisation and new messengers (see Table 11.6). The term chemotype is not just analytically descriptive but includes concentrations, energy content, space limitations and organisation, and is therefore a comprehensive thermodynamic description. We have shown that evolution is not constrained by the changing information in coded molecules, which had to follow rather than lead change, but depends upon an ever wider ability of organisms in the ecosystem to sense, obtain information about, and then exploit both changing environmental materials and energy sources not just internally but, finally, also externally. The whole system is an inevitable, not a random, development and is a cooperative ecosystem of energy stored in chemicals both in cells and in the environment (Figure 11.1 and see cover). The different chemotypes became cooperative rather than competitive again because of necessity. Thus, it is found that the whole system developed extra physical/chemical increased "fitness", initially in small local spaces, the prokaryotes, by combining their functions while optimalising chemicals, energy, space and organisation in greater and greater compartmental combination, the eukaryotes. The final biological consequence was recognition of physical and nutritional value of the environment and lies in the use of external physical chemistry by organisms, as we see it today, using the development of the brain in mankind. The full use of chemicals and energy by this species has required an "understanding" of the physics and chemistry of the environment and then application of this knowledge to build macro-machines and employ them to do work (Chapters 9 and 10). The major novel evolutionary feature introduced by mankind was and is that there is no need to wait, as in the past, for geochemical or genetic change before employing chemicals

TABLE 11.6

OUTLINE OF CHEMOTYPES

1.	Anaerobic prokaryotes including archaea and bacteria: chemolithotropes
2.	Photosynthesising prokaryotes
3.	Proto-aerobic prokaryotes utilising sulfate, nitrate or ferric ions as energy sources
4.	Aerobic prokaryotes
5.	Various unicellular eukaryotes in plant, fungal and animal groups
6.	Various multicellular eukaryotes in the same general groups as (5)
7.	Animals with nervous systems and brains
8.	*Homo sapiens*

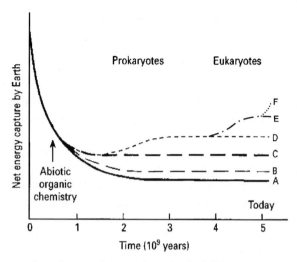

FIG. 11.1   The upgrading of energy by organisms. A, the fall in energy content of the original
Earth; B, the capture of energy in abiotic times mainly organic chemicals; C, the capture of
energy by anaerobic life from chemicals (A + B); D, the capture of energy by unicellular aerobic life with capture plus C; E, the capture of energy in multicellular organisms plus C + D; F,
the capture of energy by man, plus C + D + E.

and energy in new ways, held back by the need to modify a conservative code.
Mankind deliberately accelerates and initiates change but this activity is a part of
the general increasing pattern of energisation in the environment, a source of materials and energy, in kinetically (not thermodynamically at equilibrium) stable constructs. Mankind is then a final chemotype (today), and we do not distinguish
cognitive and metabolic activities as is conventional in discussions of evolution. In
all the discussion of the ever more efficient absorption and degradation of energy
the ideal endpoint is for all materials in the system to cycle, as shown in Chapters
3 and 10, but we doubt whether such a condition can be reached. Note that by
breeding, mankind has altered the populations of other organisms and by chemical
methods mankind has begun to alter the rate of change of the code.

    Before we turn to some of the details of our argument we need to place the con-
ventional Darwinian view of evolution, based on species and genes in the context of
this view of the inevitable *thermodynamics* of an energised ecosystem. Note that our
view sees the advance of evolution in the major branches of the evolutionary tree
within an ecological cone, see Fig 11.7. The Darwinian approach applies, of course,
to biological change but not to the activities of mankind and conventionally does not
relate to a rational, continuous geochemical environmental change or to increased use
of energy. It is only a correlational approach, not an explanation. There have been, in
fact, two such approaches: first the earlier study of the development of the nature,
morphology and behaviour of cellular systems in a variety of environments and later
the examination of coded molecules of cells, both in a historical perspective.

11.2.   The Darwinian Approach to Evolution

The following is the merest sketch of the early Darwinian approach to evolution so that its ideas can be seen in the context of this book (see Further Reading for more extensive descriptions). It is based on observations of form and function of species of organisms, not on chemistry, and in it there is recognition of the huge diversity of living organisms in species. There is no doubt either in this approach that life has evolved in part in organisation while simpler cellular organisms developed and still exist. It is also true that some individual species have themselves increased in numbers and sophistication and that many have been lost. The study by Darwin of the form and behaviour of the organisms in species classified earlier (see Table 4.1) led to a proposal for their geographical and physical diversity, that is that the evolution of organisms took advantage of the variety of the environment (see Burrow in Further Reading). Hence a special environmental feature gave rise to a special species, e.g. among Darwin's finches, and isolation led to differences such as marsupials in Australia. Survival with reproduction is the theme and into this theme more and more varieties of form and activity can be introduced by *chance within natural selection*, where an increasing organisational degree of awareness of the locality is just a way of utilising the environment. Competition then shows itself in the survival of the fittest, best adapted and reproductive species locally (see Maynard-Smith and Szathmáry in Further Reading. (Note that "survival of the fittest" has taken on a very sophisticated meaning but still refers to species or individuals.) The increase in size and complexity of organisms is readily built into a branching tree of diverse types and both early and later branches evolved separately in time. There is nothing said about systematic advance of chemistry or energy use in the environment or in organisms in this picture, about which Darwin himself could not have known. The whole of the Darwinian description of evolution is then a selection of possibilities open to mishap or gain if sources of nutrients or conditions of temperature etc. fluctuate. A species may be lost now and then and a new related one may appear but there is no recognisable progression, only inherited relationship. If we look at, say four-legged animals, we see a huge variety of sheep, horses, cattle, cats, dogs, kangaroos and so on, amongst a greater animal variety and there is an equally puzzling set of very similar plant species. Following the above line of argument, there is no possible reason for the existence of all the species and the multitude of variants except those linkages which can be connected in a tree of variations (Figure 11.2). We are in fact in complete agreement with this view of *similar species* and with the accepted idea that, given a set of environmental circumstances (variable) and a possible way for a system to divide amongst them, bifurcate, then it is entirely possible that an initial single type of organism will give rise to a variety of species, all of almost equal survival strength and each looking for some minor advantage by moving to a territory most suitable for it. Moreover, these similar species are likely to compete especially if sources of nutrients diminish. Further evolution follows from further bifurcations in a spreading tree (see Fig. 11.2), with small advantages in different niches. There is with time a sub-branching of one

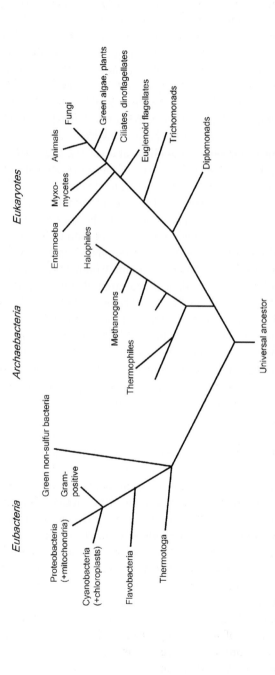

FIG. 11.2   The 'Universal tree' based on 16S and 18S rRNA, which is not very different from that generated earlier based on the observable characteristics of organisms.

or more early domains of life and following certain relationships they can be placed in a number of kingdoms, orders etc. down to species (see Table 4.1), and thus geographical separation is explicable. We accept the general connectivity in this tree which has required two centuries of analysis from Linnaeus to the present time to complete (see Gould in Further Reading). In such approaches major developments in organisation, sudden branches of the tree, such as the appearance of a new domain, of say eukaryotes, simply reflected a gain in survival value, in this case for the chance major reason that they can consume larger particles of food including bacteria which pre-dated them (see Maynard-Smith and Szathmáry in Further Reading and see references in Chapters 8 and 9). Initially, the argument could not be related to chemistry, but more recently new organisms are seen as arising in part as variations in ways of living off very basic chemicals, largely the elements C, H, N, O, S and P and their environmental compounds, concerned in the synthesis of biologically important organic molecules, and then off one another. Rarely are inorganic chemicals mentioned. There is no suggestion of systematic improvement, but uniformity of development of all branches of the tree is advocated while there is, however, a clear overlying pattern of development in distinct broad groups: anaerobic prokaryotes, photosynthetic bacteria, aerobic prokaryotes, single-cell eukaryotes, multi-cell eukaryotes, animals with nervous systems and brains, and mankind. (It is this major splitting which we are re-examining.) It is then generally agreed that there is an obvious line of increase in organisation, and a connection to morphological and organic chemical variety. At this stage of analysis all we ask our reader to do is to notice that "continuity" of a new major branch from the general pattern is more to be likened to a deviation from the pattern than a simple continuity, for example, in the use of light (Chapter 5), in the appearance of eukaryotes (Chapter 7), and in the development of *Homo sapiens* (Chapter 10). It is with the stress on the role of simple chance in the appearance of these major branches that we disagree.

11.3. Genes and Darwin's Proposals

The knowledge, subsequent to Darwin's proposals, that there was Mendelian-type inheritance of genes in organisms and then the realisation some 50 years ago that the genes could be linked to coded molecules, DNA, RNA and proteins, made for the need to connect the evolution of species with that of such molecules. The generally accepted approach is that chance random modification, mutation, of the genes gave rise to a new genotype matching a particular feature of the environment so as to gain advantage for that organism. (An explanation of this view is often covered by Dawkin's phrase "The Selfish Gene".) There is no doubt that to a large degree analysis of coded molecules DNA, RNA and proteins has given a well-defined picture of the link between changes of form in organisms and the codes of these molecules. This is still not seen as "progress" since in this picture all organisms from bacteria advance upwards in time by random variation. The picture remains of a connected set of branches of differences in a tree-like pattern (see Fig. 11.2). There is no endpoint to the variations as there is no apparent logic to what is observed. There is therefore no hidden causative factor in this treatment of sequences but it does relate closely to the evolutionary trees previously devised based on physical and behavioural features of organisms. Today there is renewed interest in the possible influence of the environment on genes often discussed under the heading of epigenetics, and here much is required to be learnt. It may well support the picture of evolution we have developed in this book to which we return below.

It is important to stress the nature of this combined Darwinian/genetic approach to the evolution of organisms. It was devised originally by biologists concerned mainly with static morphological characteristics of species and augmented by biochemists concerned with sequences of polymeric molecules and their properties, especially comparisons of the sequences and structures of large molecules from diverse species, often called molecular biology. The weakness in the analysis is that it does not examine the nature of the physics and chemistry of open systems which must include energy and their environment, yet organism activity has to be part of systematic thermodynamics. Darwin himself knew little physics or chemistry, as already remarked, and was completely unaware of the thermodynamics of systems. In his footsteps few scientists have done other than to look for parallel descriptive connections as in the trees (such as Fig. 11.2) and the possibility of causality in terms of physics and chemistry is rarely analysed. It must be the case, however, that any causal explanation meets the knowledge of evolutionary connections in the above tree and must then explain general changes in morphology and of the code. If possible an explanation should lead to the origin of mankind.

11.4.   The General Thermodynamic View of Ecosystem Evolution in this Book

We return now to our physical/chemical view of evolution. In the above molecular descriptive correlation between the organisms and genetic information seven important physical/chemical changes with time are missing: (i) the exposure to the observed gradual switch to an increasing presence in the environment of oxidised chemicals; (ii) the sequential increase and use of the chemical and energy stores in chemicals in organisms; (iii) the way increasing energy is put selectively into changing patterns of cellular reactions against a fixed background of biopolymer synthesis in different organisms; (iv) the gradual introduction of new compartments, the extension of cellular organisation in space with time (see Tables 11.9 and 10.8 and Sections 3.9 and 4.2); (v) the increasing complexity of the management of the changing flow of chemicals in the organisms (see Table 11.8); (vi) the increasing total uptake and degradation of energy in the ecosystem as organisms increasingly synthesise chemicals and establish chemical gradients which are then degraded, all approaching a total cyclic steady state, and (vii) the increasing ability to do work as seen in the switch from single molecular machines in the simplest cells to the cooperative activities of many such machines in more complicated single cells, to the synchronised activities of macro-machines in multi-cellular organisms, and finally to large macro-machines devised by mankind and operating external to cells in and on the environment. Their inclusion requires a switch in thinking from sequences of particular molecules and from morphology of organisms to more general chemical considerations and their spatial energies, i.e. to thermodynamics, while keeping in mind the great advances as seen in the alternative approaches and evolutionary trees. The whole is driven by the effect of energy (light) in causing separation of charges and then of reduced (organisms) from oxidised (the environment) materials and then by the inevitable decay partly due to back reaction resulting in a cycle of material but a continuous conversion of light into heat. Our approach therefore has been to look for a chemical thermodynamic reason for the relationship between these developing features of a total ecosystem, biological and geological. It is consistent with the way we have described the evolution of chemotypes. In doing so we have related these changes to a general rule for chemical systems exposed to energy. While in equilibrium thermodynamics there is a law, the second law, postulating that all systems have a direction of change toward a state of equilibrium and maximum thermal entropy, we consider that in an open system the absorption of effective energy of high frequency sets in motion flow and that this flow adjusts materials aiming as far as possible to generate thermal low-frequency energy as it moves towards an optimal cyclic material steady state. This final cyclic steady state is one which retains a fixed amount of energy in the flow of material while generating optimum energy throughput and degradation (thermal entropy production) (see Corning in Further Reading and references therein). It thereby generates a randomisation more rapidly than would otherwise have been the case. We believe we have demonstrated that this is a general rule for systems which absorb energy: they optimalise the rate of thermal entropy production. The rule is in accord with the second law of equilibrium thermodynamics but relates to kinetic, not equilibrium thermodynamic, factors. It follows that evolution must have an inevitable direction toward a cyclic steady-state condition which simultaneously optimalises use of the materials and degradation of the energy of the environment in this process. We have therefore referred frequently to the following cycle:

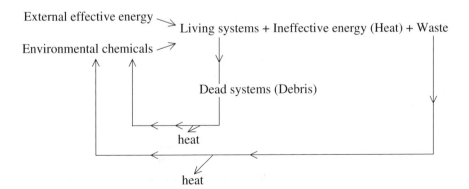

Within this system each dynamic chemotype with its environment is a part of an identifiable chemical thermodynamic system, or can be separately recognised by differences in any one of the seven features of evolution given above. The chemo-types can still be characterised by sets of their genes or morphology (forms) but not explained by either of them. Now fitness becomes described by the total effec-tiveness in the creation of an all-embracing ecology system (environment + organ-isms) in which organisms together give optimal survival, in life/death cycles of material, while generating optimal energy degradation, i.e. optimal rate of thermal entropy production.

At the beginning of this chapter we described the origin of the initial environ-ment and its maintained, approximately fixed energy sources (see also Chapter 1) and the possible initial uses of the elements in life were given in Tables 5.8–5.10. Note that by deliberately concentrating attention on the roles of some 20 elements in these chemotypes, established by analysis as those essential for life, we remove the limitations which have arisen by previous reference to only organic chemicals in evolution and have automatically connected organisms and the environment. Because there is reductive synthesis in life and oxidised chemical waste the origi-nal environment was in danger of being slowly stripped of certain of its chemicals and the above diagram would then have become a flow of diminishing rate since there was the possibility that the waste would have just accumulated and would have diminished the capacity or even poisoned the living system. For example, all carbon could have finished as trapped wasted coal, while oxygen could have built up as a poisonous gas. In such a case, there is nothing that can be said by way of a prediction of the evolution of the living system which requires carbon and oxygen as essential elements except that it would begin to fail relatively quickly. It would fail to collect more energy. Similarly if individual organisms lived forever there would be no return of elements to starting material, no cycle, and any evolution would be frustrated. In fact they cannot do so as they are a system of unstable chemicals, doomed to die. In this book we show instead that the system slowly adapted itself to incorporate much of the waste, composed of oxidised and rejected elements and the debris from death, and therefore bit by bit, in an unavoidable way, the system, environment plus organisms, attempts to reach a fully cyclic condition (see Chapter 3). (This is in overall agreement with the general features of many previous partial ecological analyses, for example, the cycles of C, N, P and S as they exist today.) Life is then seen as a catalyst of energy degradation, thermal entropy production. While stressing the involvement of the 20 elements we have directed attention to the way the changes in them and their concentrations in com-partments made possible the systematic development of the organic chemicals in their materially dominant roles. One further feature needs to be commented. We offer no explanation on the diversity of species within chemotypes and hence of the appearance of species related in chemotype in restricted (geographical) loca-tions; both are accepted as random. However, no matter what the geographical location, we state that the environment and organisms develop together in the same progression of chemotypes. The major types of flora and fauna of Australia or of the deep ocean trenches have evolved in the same pattern as that on the major linked geographical oceans and continents. We are not concerned to explain phenotypes, variation within species, that is within genotypes, with but one exception – mankind (see below).

Here we must stress that this thermodynamic approach has a considerable history. Boltzmann, a physical chemist and leading figure in statistical thermodynamics, wrote in 1886: "The struggle for life is not a struggle for *basic* elements nor energy on Earth .... It is a struggle for the less probable forms of energy which are available in its transfer from the hot Sun to the cold Earth. Utilising these forms with maximum efficiency through yet unexplored ways plants force the energy of the Sun to make chemicals." Since that time a variety of physicists and chemists quoted in Chapters 3 and 4 have put forward more detailed considerations as to how energy drives flows but on the whole biologists and physical scientists have ignored this approach. We have added to previous views a description of the nature of these flows in life by reference to chemical analysis and chemical interactions, especially those of the basic elements now known to be essential for life.

We turn now to a more detailed account of the four major directional thermodynamic characteristics of the evolution of the ecosystem which incorporate the seven features of change which we noted above. They are:

(1) the chemical composition changes in the environment and in organisms;
(2) the increase in energy utilised by the organisms;
(3) the increasing use of space by organisms including eventually the space outside the bodies of the organisms;
(4) the changing pattern of organisation required to manage (1)–(3).

Once we have completed this survey we shall turn to a speculative look at the problems facing an ecosystem open to mankind's nature and activities.

11.5. The Chemical Sequence of the Environment

As we have described in Section 1.11 the environmental sequence of states of elements in evolution followed their aqueous oxidation/reduction potentials as the concentration of oxygen increased (see Fig. 1.14). The rise was very slow at first, for up to two billion years, since change was buffered by the presence of large quantities of Fe^{2+} and HS^-, and then was quite rapid. Critical to the process was the slow rise at first in the resultant environmental redox potential close to an equilibrium with the O_2 partial pressure, for example, the rise of the ratios, $[Fe^{3+}]/[Fe^{2+}]$ and of $[S_n]/[HS^-]$ or $[SO_4^{2-}]/[HS^-]$ included in the generalised ratio of [oxidised]/[reduced] chemicals in the environment. A change, an increase in potential with the [oxidised]/[reduced] ratio, is then imposed on the whole environment where change is relative to the starting potential of the environment, say -0.2 V on the H^+/H_2 scale at pH 7.0 (see Section 1.14). The critical changes in the equilibrated inorganic chemistry of the surface of the Earth occurred as the potential increased towards that of the ambient oxygen of today, that is $+ 0.8$ V on the same scale. Among non-metals the changes were:

1. All available carbon became CO_2 not CO or CH_4.
2. All available nitrogen became N_2, later some NO_3^-, not NH_3 or HCN.
3. All available sulfur became SO_4^{2-}, not HS^-.
4. All available selenium became SeO_4^{2-}, not HSe^-.
5. All available hydrogen became H_2O not H_2S.
6. Oxygen became available as O_2 and the ozone layer was created.
7. Iodide could be oxidised to iodine.

Note that other halogens, phosphorus, boron, and silicon were of unchanged oxidation state and therefore their availability remained very largely constant. No other non-metals are of significance.

The changes among metal ions were equally striking.

8. All available iron became Fe^{3+}, not Fe^{2+}, that is, iron became of low availability since Fe^{3+} forms a virtually insoluble hydroxide at pH 7.
9. All heavy metals in the Periodic Table Groups 12, 13, 14, and 15 became more available, as sulfide was removed. Note especially zinc, the most abundant.
10. In the first transition metal series, ions of vanadium, cobalt, nickel and copper became more available later as most sulfide was removed.
11. Manganese became slightly less available as it formed some MnO_2.
12. Molybdenum became available, again due to the removal of sulfide, while tungsten availability did not change, perhaps some going from WS_4^{2-} to WO_4^{2-}.

Note that the states and availability of sodium, potassium, magnesium and calcium did not change, nor did that of metals of Group 3 or 4.

Most importantly, these changes occurred in time in the order of their equilibrium redox potentials including those between precipitated sulfides, oxygen and soluble sulfates, so that, for example, 3, 7 and 12 occurred before 4, 8 (Zn) and 10 and 11 above in that order. We need also to be aware of other small inputs and outputs to Earth's surface chemicals but we wish it to be noted that *it is these major inorganic non-metal and metal chemical changes which, we consider, directed evolution*. The changes can be followed using geochemical markers and they are all due to the intervention of organisms' chemistry through release of oxygen. We need to note also some physical changes of the environment which, due to erosion, generated soils and the run-off of especially, sodium, potassium, chlorine, magnesium, calcium and silicon to the sea.

During the long period since the Earth was formed the ecosystem has lost available quantities of carbon and some nitrogen (see Section 1.3), but much of both elements has been retained inside life and in the case of carbon much is also locked in coal, oil and gas as well as in carbonates. There is then a compensation in that these stores and life itself have prevented greater loss of CO_2 (and N_2) from the atmosphere. Insofar as man is bringing back stored carbon (fuels) into circulation as CO_2, he is in fact restoring the cycling of this element in line with the drive of evolution. That this produces an upward fluctuation in temperature may be a disaster for human population, a problem for this one species, but may not be so for the slow advance of the ecosystem as an energy-capturing system. In the past there have been many ages of low and high temperature, variable CO_2 and O_2 levels and losses and gains of groups of organisms but each time the difficulties have been overcome and evolution slowed, not stopped, and this is likely to be the case again. This process is not without end since eventually too much carbon and nitrogen may be lost from the atmosphere and the Sun will certainly die. The risk of the loss of the other light elements, hydrogen and oxygen, is much reduced by their storage as a condensed liquid, water, and oxygen is of great abundance in rocks as oxides.

Chemicals previously not in contact with organisms including radioactive products from nuclear power stations are the new waste due to the introduction of new physical and chemical processes by mankind. Some of this waste is damaging but some could possibly become a new drive for local evolution although it can be incorporated only exceedingly slowly in higher organisms because of the hindrance to change due to the basic nature of the cytoplasm and general genetic, conservative DNA, control. There are some bacteria which are strongly resistant to radioactivity (see Gross in Further Reading), and others which adapt very quickly.

We turn to the involvement of the elements in the chemical components of organisms.

11.6.   Chemicals and their Changes in Organisms: Chemotypes

We turn now to the way in which the element content of organisms, chemotypes, has changed as we analyse in the sequence anaerobic prokaryotes, aerobic prokaryotes, single-cell eukaryotes, multi-cellular organisms, animals with nerves, mankind. Note that this is not a conventional division of organisms, especially the inclusion of the two last chemotypes, but we justify the division by reference to the points we made in Sections 4.3 and 11.4. We have shown in detail in Chapters 5–10 that between these groups and strongly between some of them there is a sequence of use of elements, energy, space and/or of organisation in keeping with the change of the environment and/or the ability to use it.

In summary we see these chemotypes as starting (see Table 11.7) and then developing, with energy intake, *the direct employment of chemical elements in components,* as follows:

(1) Anaerobic prokaryote cells based initially on pre-existing chemical and energy stores, within which conventional reductive synthesis was established, some 3.5 billion years ago. The major non-metal elements used were H, C, N, O, P, S and Se, but not B, Si, halides, or heavier non-metals.

TABLE 11.7

REJECTION AND RE-USE

Rejected Elements	Use
Sulfur from H_2S	Deposits of S_n (not easily used)
Oxygen from H_2O	With C/H/O compounds: energy source
Protons (gradient)	Energy transduction
	Acid-catalyst of hydrolysis
	Pumping of other ions and compounds
Calcium ions (gradient)	Signalling, chemical and physical
Sodium, Chloride ions (gradient)	Signalling, physical
Many heavy metal ions	Use at low levels or rejected as poisons
e.g. Zn^{2+}, Cu^{2+}, Pb^{2+}, Hg^{2+}	
Ca^{2+}, SO_4^{2-}, CO_3^{2-}, SiO_2	Mineral protection

The major metal ions used were of K, Mg, Fe, somewhat assisted by Co, Ni and Mo(W) and possibly some V, Mn and Zn. Ca, Cl and Na were largely rejected (see Table 11.7), and Al, Ti and all other metals were not used.

(2) Use of light in anaerobes produced oxygen, utilising novel incorporation of Mn, which generated O_2. $CaCO_3$ was formed on the outside of cells. Porphyrin-bound Mg, Fe, Co, Ni and coenzyme-bound Mo appeared (three billion years ago).

(3) Use of the first oxidised products by proto-aerobic organisms – sulfate, ferric ions, and probably nitrogen oxides. Molybdenum became more available and was generally required for N and S metabolism (three to two billion years ago).

(4) Aerobic prokaryotes utilising (later) copper released from its sulfide by oxygen. Increasing use of oxygen and its environmental products was seen in oxidative pathways and synthesis (around two or so billion years ago).

(5) Single-cell eukaryotes, large cells with separate compartments for digestion, glycolysis, sulfation, mineral formation, etc. and organelles for energy generation. Minerals now included silica, $CaCO_3$, $SrSO_4$, even $BaSO_4$ and some phosphates. Calcium was used in signalling, and then some oxidised organic molecules from vesicles. Zinc and copper were used somewhat more extensively (about two to one billion years ago).

(6) Multi-cellular organisms with considerably increased development of the use of zinc, copper and selenium. New uses of minerals and reactions in controlled *extracellular* fluids; lesser use of Co and Ni. Changed use of Se, halogens and vanadium; more extensive use of Ca and oxidised organic molecule signalling (one billion years or so ago).

(7) Animals with nerves as in (6) but with new uses of Na, K, Cl and Ca ions in currents carrying messages. Central control of nerves led to brains and greater "reach" to use the environment in external devices by animals (some three hundred million years ago).

(8) Mankind began exploring use of all possible sources of energy, space and many elements (some 25,000 years ago).

(9) Increased exploration of uses of all elements, energy and space by humans. If the uses become fully effective further evolution of chemotypes has to be very limited if not impossible (some 500 years ago).

We stress that the use of virtually all available elements of the Periodic Table in life is much as expected from their chemical value, but added to this value is selective uptake, controlled binding, activation, and compartmental siting in cells. This is consistent with optimum fitness of the ecosystem for energy degradation. Thus, use is energised in synthesis and in gradients in a thermodynamic system. The parallel element changes in the environment and uses of elements by organisms are shown in Fig. 11.3, and this is seen to be the major directing force of evolution where abiotic systems generated the initial stages of change. (Note that the continuity does not separate the cognitive activity of mankind from metabolic activity by our treatment of use of elements in a total ecosystem.) It is very important to

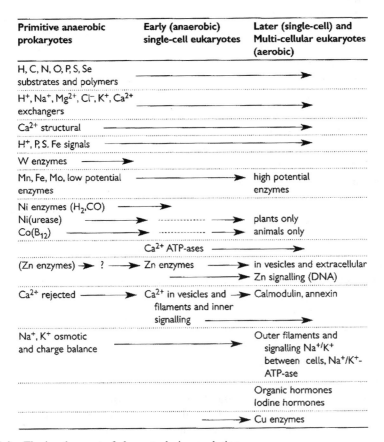

FIG. 11.3   The involvement of elements during evolution.

observe that oxidation of the inorganic environment came before oxidation of organic material by cells but that the second oxidation became catalysed by organisms at a rate faster than that in the environment. Both drove evolution in one direction. There were three linked needs to allow this increasing involvement of novel chemical reactions to be incorporated: (1) greater energy intake in single organisms, (2) more space and a larger number of compartments, and (3) an increase in sophistication of information systems. Finally, all the changes have to be related to some degree to genetic apparatus adjustments since reproduction is required.

11.7.   The Continuous Gain in Use of Energy and its Degradation

In this concluding chapter we have now dealt with material change but we have not so far considered explicitly the input of energy. Energy in its many forms was described in Chapter 3 and applications to biological systems were examined in Chapter 4. In Chapter 5 we described the initial energisation of organisms, the chemolithotrophic prokaryotes by energised minerals, and introduced the capture of light which continued to increase until the present day plant capture system had evolved. Effectively, the environmental inputs of energy are constant and it is incorporation in systems which has increased. A second great advance in useful energy recovery in organisms lay in the reactions of waste dioxygen and excess C/H/O compound production, oxidative phosphorylation, on top of oxidation using Fe^{3+}, sulfate and nitrate in cells. All these chemicals are stores of energy, created by light, when seen together with the reduced chemicals of organisms. All added successively to the very early secondary internal use of energy from degradative disproportionation of sugars, i.e. to pyruvate in glycolysis. It is clear that energy, like material, flows from outside the *surface* ecological system, from the Sun, into both organisms and the surface environment, via a tortuous path. We see that all the biological chemical advances came together in the combined increasing employment of perhaps all possible energy sources in one ecosystem some half a billion years ago. Generally speaking, the several different cooperative uptake systems and uses of energy are still to be found today in different chemotypes which evolved in a rational sequence in time. Not until the coming of mankind have further completely new sources of energy been pressed into service and historically new energy storage has been created in the environment both deliberately and accidentally. Table 11.8 and Section 10.6 outlined all these energy sources.

   In all these considerations we have to remember that all the energies entering the ecological system are reflected or finally degraded to heat. It is the change toward optimal rate of the energy absorption, use and degradation simultaneously which is the directing influence of evolution. Now this energy includes that of the Earth's interior and it is its degradation that in large part maintains the possibility for life on the Earth through temperature control. It also includes therefore the constant radioactive decay and the chemical reactions of the deep Earth, which continue quite independently from the processes in organisms. Life is locked into a huge ecosystem.

TABLE 11.8

CHEMICAL THERMODYNAMICS OF FLOWING SYSTEMS

1.	Outward flow of energy and material: Big Bang
2.	Condensation of material in local zones: Earth, Sun
3.	Local energy flow from inner Earth and from the Sun to Earth's surface
4.	Energised flow of energised mineral materials near Earth's surface
5.	Fixed flowing phases at fixed temperature on Earth's surface
6.	Local flows inside confined (protocell) volumes: pumps
7.	Preferential local cyclic synthesis/degradation of kinetically stable components in protocells
8.	Continued degradation of energy in (1), (3)–(7)
9.	Autocatalysis of preferred system from (7)
10.	Initial organisation of material and energy flow using a code (DNA) and feedback: prokaryotes
11.	Self-induced modification of the Earth by (3), (4) and (10)
12.	Further chemical and physical limitations on flow: evolution of directed flow; single and multi-cell eukaryotes
13.	New coded messages using signalling to direct flow: hormones, morphogens, nerves: animals with a brain
14.	New codes with huge memory storage controlling flow in brains: language, mathematics, radio, television: *Homo sapiens*
15.	New code using computers to control flow: automatons

Note: All flow runs with thermal entropy increase toward a final system.

11.8.   The Changing Use of Space

Table 11.9 describes the way space has been utilised by organisms describable by compartments. *The new compartments* themselves relied on *syntheses of specialised membranes* and filaments to secure their positions and special new chemistry to allow growth that is in eukaryotes (see (5) above), all requiring oxidising conditions. The new flexible membranes were in fact made by the introduction of steroids, mainly cholesterol produced by oxidation. *New filaments*, found only inside cells at first, required new proteins, actino-myosin and tubulin, not due to oxidation, but later outside cells other filaments needing oxidative reactions for cross-linking supported multi-cellular organisation. Here new changes in space external to cells necessarily required copper and zinc which had become available (Section 8.7). Frequently copper provided the active sites of novel enzymes, activating oxygen to cross-link connective filaments and so held cells together while zinc became the essential element for controlled external degradation of the external filaments so as to allow growth. Additionally external calcium was bound to *oxidised* surface amino acids of the filaments, to strengthen interaction between them, and sulfation of external saccharides made possible the creation of an open matrix through which diffusion could occur. Overall, there is a general increase in structures allowing cell extension in space, on the increase in oxidising environmental conditions.

TABLE 11.9

USE OF SPACE BY ORGANISMS

Organism	Space
Prokaryote	Simple single compartment, the cytoplasm, with at most two surrounding membranes and an intermembrane space, the periplasm. Now and then species may have a vesicle. Small cells
Single-cell eukaryote	The cytoplasm of a much larger cell has a nucleus and many vesicles including organelles. Complex outer structure often mineralised
Multi-cell eukaryote	Much larger organisms with cells grouped into organs. They occupy space considerable distances from a surface and can connect different parts of this space
Animals with brains	Multi-cellular organisms with a new type of compartment, a brain, connecting to all organs and activities
Homo sapiens	Animals with an exceptional brain able to make use of space outside themselves as well as to manipulate space inside organisms

11.9. The Changes in Organisation

Changes in organisation required greater detection by senses of events inside and outside cells, the relaying of the information via message systems to the different but connected activities of the organisms, and a controlled response to utilise the events or to mitigate against them. The controls needed to act at different speeds and can be assisted by memory devices in both molecular codes, DNA, and a variety of physical/chemical stores especially in the brain. We have described the successive advances in detection by senses from prokaryotes up to mankind and there is a clear direction towards a final ability to sense sound and electric-magnetic information subsequent to chemical and physical contact (Table 11.10). Dependence on complex apparatus was required in multi-cellular systems requiring oxidising conditions.

We have shown that *the changes of signalling with senses* have also been in a systematic sequence with oxidising conditions. The overall pattern of evolution of messengers was of a remarkable change in external carriers of information with a multiplication of varieties of a limited number of internal carriers:

Single cell, internal:	Mg^{2+}, Fe^{2+}, substrates, free phosphates (NTP, NDP, NMP, C-AMP, -SH) (Chapters 5 and 6).
Single cell, external:	Ca^{2+}; released new phosphates from membranes (Chapter 7).
Cell:cell by diffusion, (External:)	(Oxidised) organic molecules; oxidation using iron but most notably copper; zinc in transcription factors and hydrolytic enzymes (Chapter 8).

TABLE 11.10

CHEMOTYPES, THEIR MESSENGER CHEMICALS AND THEIR ORGANISATION

Organisms (Chemotypes)	Messengers	Organisation
Anaerobic species	Fe, Mg, coenzyme substrates	Internal network
	c-AMP	Sense of sources of nutrients
Aerobic bacteria	As above plus O_2, NO, CO, SO_4^{2-}, NO_3^{2-}	Senses New sources of environment content
Single-cell eukaryotes	As above plus Ca^{2+}	Source of environmental knowledge
Multi-cell eukaryotes	As above plus Zn^{2+} Organic chemicals Cu^+	Internal network External network New synthesis agent
Multi-cell eukaryotes with nerves and brains	As above plus Na^+, K^+, Cl^- potentials	New memory and responses to environment
Mankind	As above plus e, $h\nu$, sound	Vast communication network Control of the environment

Cell:cell by bulk electrolytics, external: Na^+, K^+, Cl^- (Ca^{2+}) (Chapter 9)

Mankind's devices; external: A variety of sound and electromagnetic signals.

The whole pattern also involved a gradual systematic increase and use of information storage initially in vesicles (Table 11.11), following changes in the environment together with changes in organisation in and out of cells, the opportunity for which was created by the environment. Later storage is in physical apparatus made by man.

It is easily seen that this changing pattern inside the organisms arises as oxidation forced the need for the production of more and more compartments together with and the development of suitable membranes, the use of oxidised ions and molecules and of ions released by oxidation (Mo, Cu^{2+}, Zn^{2+}). As a consequence of the increasing complexity an obvious need is greater central command. Our treatment cannot be in conflict with the known changes in gene structure since the genetic structure reflects the basic features of the thermodynamic system changes (see Section 11.11). Information is stored in and relayed from DNA in bacterial cells so as to control the inner workings of the cell in relatively slow responses and these workings have been kept throughout evolution. Later multi-cell eukaryotes evolved communication between adjacent cells, as described in Section 8.13, which utilised considerably zinc finger transcription factors, and animal organisms developed

TABLE 11.11

EVOLUTION OF INFORMATION IN SYSTEMS

Organism	Code	External/Internal Gradients
Bacteria	Circular DNA Satellite DNA (RNA)	External gradients of food Magnetic and gravity fields (Oxygen, CO, NO) Internal flows of substrates, NTP, Fe and Mg
Single-cell eukaryotes	Linear DNA Organelle DNA (RNA)	As above Ca^{2+} gradients (external store)

Multi-cellular
  organisms
Plants and animals
	As above
	Chemical hormones in
	  vesicle stores
		As above
		Extra cytoplasmic organic
		  chemical gradients (internal
		  vesicle stores)
		Sound and light external

Animals with brains	As above Sign language Electrolytic storage	As above Electrolytic gradients of Na^+, K^+ and Cl^- (internal to organism, external to cells)

Mankind	As above
	Language spoken
	Electronic storage
	Digital computer
	  storage, etc.
		As above
		Electron and radiation due to
		  external energy store

nerve networks to collect and use information more quickly through their senses and brains in a responsive mode unrelated to DNA information. The nerves co-ordinated cells separated by large distances, especially those of senses and muscles. This led to the development of larger brains with a novel coded system and response, based on gradients of ions and molecules, independent of DNA, and which in turn led to mankind. The continuity of later chemotypes is more closely related to rapid activity, not to the genotype, and to use of external, not internal, chemistry and space. The development required the change in central command.

Mankind, to whom we turn below, evolved into a thinking animal developing new external communication codes (languages) and command structures. *Probably advances in communication are the dominant leading novel feature.* Understanding of physics and chemistry, needed to develop increased external organisation, led to the very fast development of new chemistry, new compartmental structures, new energy sources and uses, new transport and message systems and new information storage by mankind, e.g. in computers, but the basic nature of this development is no more than a remarkably fast and large addition to all previous steps in evolution.

We must also stress that the control over *the dynamics of individual organisms* increased, which needed further increased organisation. Initially molecular machinery, particularly pumps were devised, but subsequently mobility developed in *bulk* organs, muscles, used for capture and digestion of food, for handling objects and then for motion in chosen directions. Very recently, natural molecular machines have been replaced by manufactured equipment (see Gibbs in Further Reading). The increased physical scale of operations is clear but there is no clear-cut category distinction in the activities. The aim must be globalisation of the ecological system.

11.10. Symbiosis: A Form of Compartmental Collaboration

We have described in Chapters 4–10 and summarised in this chapter (Section 11.6), different ways of improving effective use of energy and materials in internal compartments linked to individual cellular command centres. Here we turn to collaboration between command centres. A first step in collaboration between organisms was the utilisation of oxidised chemicals, for example, sulfate, nitrate and ferric iron, produced by the actions of some but used by other, different bacteria. Here new organisation results from almost random transfer of debris with some sense of corporate signalling. A more striking example is the appearance in unicellular eukaryotes of energy-generating activity in separate compartments with separate DNA, mitochondria and chloroplasts, as oxygen partial pressure increased. They were created from a symbiotic relationship between bacteria and the very early precursors of the large eukaryote cell. The centralised control is the genetic apparatus of each organism. Now eukaryotes as they evolved in higher organisation to multi-cellular plants, fungi and animals selectively lost certain genes – in fact genes for many essential syntheses of components of all life as we know it – and came to rely more considerably on symbiotic relationships with (lower) organisms. Many plants, for example, do not fix gaseous nitrogen but rely on bacteria to supply this essential element. Again many plants use fungi to deliver minerals, while animals depend on bacteria for digestion and on scavenging debris of plants for basic energy, chemical element and compound supplies. At the same time fungi feed on chemicals from plants. As we have stressed, as organisation in cell communities evolved; earlier chemotypes did not disappear but the old and the new assisted each another in an ecosystem. Hence, evolution of different biological genotypes changed by chance variation in time while competing, but organisms moved forward within different chemotypes (of multiple species) increasingly together and interactively (Fig. 11.4). All the chemotypes developed with the environment, each depending on another. Even anaerobic bacteria help the life of the most advanced animals. They all continue to evolve with time. (The ecosystem that formed is now being transformed by mankind as this species dominates, cultivates and herds all others in an effort to reach apparently a selfish life of comfort: see Section 11.12.)

Looking at evolution of organisation of cells from its beginning to the present in what is considered to be separate organisms, at first of single cells, we see that

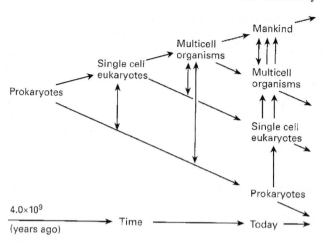

FIG. 11.4   The dependencies between chemotypes.

from the earliest times to the present day organisation and specialisation within a whole, internal cells and cells in different organisms, have increased. The organisation needs structures in the two senses of stationary frameworks and dynamic potential energy constraints. For maintenance of flow, and more so for growth with communication in the organisations, more and more energy is trapped in devices (made eventually from elementary chemicals) which are useful to the whole.

11.11.   Different Environmental Possibilities

Now we must return to difficulties, inherent in our description of the approach of the ecosystem to a cyclic state, which we have mentioned several times but glossed over in order to show the major way the *surface ecosystem* has evolved. The difficulties arise from the character of two inputs to this system: (1) light, which does not penetrate deeply into water so that photosynthesis and oxygen production cease below about 50 m in the sea which can be more than 10,000 m deep; and (2) from the bottom of the sea there is considerable constant input of reduced sulfide minerals from the black smokers of the crust. These two factors produce gradients of chemicals and redox conditions made uncertain by mixing so that the degree of aerobic character differs at different sea depths and in different places (see Section 1.12). However, the real difficulty for the approach to a cyclic state is that the flow from the oxidised surface is down towards the reduced crust of an enormous capacity so that *a full cycle cannot be achieved in a measurable time*. Fortunately, the activity is so slow that we have been able to ignore it while describing the reactions in the surface waters and on land.

We must also mention that in a very different environment life may well be able to use the differently available chemicals in a somewhat different way. However, we consider that the development of form with internal compartments and utilising symbiosis must result in a parallel evolving chemical system. An example is in the deep sea trenches, where there is a source of H_2S and to which oxygen is carried from the sea's circulation. Here novel, but worm- and clam-like, organisms together with prokaryotes have been discovered. They take up nutrients through their skins. Their source of energy is from the metabolism of H_2S in captured (symbiotic) anaerobic bacteria (compare mitochondria). However, the construction of their bodies depends on oxidation – they are overall aerobic organisms of a novel chemotype limited by certain novel element availabilities and not yet fully characterised though they employ at least 15 of the same elements in the same ways functionally as do all other organisms. We see them as at the stage of development parallel to that on the surface of the Earth at least half a billion years ago. This system is out of balance with the sea as a whole in a different way from surface organisms and it is known that their environment is relatively tungsten rich and molybdenum poor as expected from the presence of H_2S. They have tungsten and not molybdenum enzymes. Here is a clear-cut example of a switch of elements for use in a given process, O-atom transfer, dependent on element availability and functional value (see Table 11.12). It is not certain which was used first, but molybdenum is the element present in all life in oxidising conditions. It is the fitness of an ecosystem in energy degradation that is dominant.

Another major hidden problem for the ecosystem's systematic development is a possible local physical disaster, as mentioned earlier. Events such as meteorite strikes and massive volcanic eruptions could cause and have caused considerable disruption in what could be called steady progress, but the general trend on the surface towards oxidation has and will resume after such set-backs due to the very nature of life's reductive chemistry. We turn away from all these considerations of the difficulties we face in any attempt to predict the future to make the statement

TABLE 11.12

PROPERTIES OF TUNGSTEN AND MOLYBDENUM

Element	W	Mo
Abundance in Earth crust (ppm)	1.55	1.00
Soils (ppm)	0.1–3.0	0.3–3.9
Seawater today (ppb)	0.0002	8.9–13.5
Fresh water today (ppb)	<0.1	0.1–13.5
Earth hot springs (ppb)	15–300	3–69
Deep-sea vents (top) (ppm)	180–585	47
Oxidation state (more abundant)	VI (WO_4^{2-})	IV (MoS_2)
Function	(a) O-atom and electron transfer	O-atom and electron transfer
	(b) S-atom transfer	S- and N-atom transfer

After Kletzin, A., and Adams, M.W.W. (1996), Tungsten in biological systems. *FEMS Microbiol. Rev. 18*, 5–63.

that the description of the past evolution of *the surface ecosystem*, geochemistry plus biochemistry, as struggling to reach a cyclic state is not affected by these difficulties (fluctuations) in the long term but the Earth may be some long distance from a final condition, and mankind's contribution has added greatly to that uncertainty.

In conclusion, to a somewhat unknown degree the environment cannot be cycling and cannot then enter into a fully balanced state with life. We are better advised to think of a steady state of the whole environment with inputs, for example of sulfides, and outward losses, for example of N_2, CO_2, which cannot be compensated in a complete balanced flow but within which there is much cyclic activity.

We must not miss noting that the release of energy from unstable material in itself has nothing to do with life but is, like the Sun's radiation, another example of energy degradation which continues in the universe and here on Earth provides life (see Chapter 5).

11.12. Summary of Thermodynamic Chemical Approach to Evolution

In Sections 11.5–11.10 we believe we have demonstrated that three of the thermodynamic characteristics of chemotypes (components with their concentrations, the space they use, and their organisation) have evolved systematically and inevitably following the equally inevitable changes of the environment. The other possible variables, external energy input, temperature and pressure, which characterise a dynamic flow system, have remained approximately fixed. The basic idea is the drive to an effective ("economically efficient") ecosystem (see Appendix 4C for our use of the term "efficient"). We treat organisms as a machine performing work in an incomplete cycle and seeking to utilise all possible sources of energy and materials (see Table 10.6) in the degradation of energy. We need to remember that all the systems use specific elements in particular oxidation states, not just selectively held but optimally functional in organisms as they absorb and degrade energy. This is not just a feature of mankind's aim in chemistry but is an obvious requirement throughout evolution for "economic" effectiveness or fitness.

We conclude these sections on the chemistry of organisms therefore by stating that we do not consider that "evolution strives for diversity, not improvement" (see Maynard Smith and Szathmáry in Further Reading). Evolution may be blind in its diversification of similar organisms (species) but it expands within a directed time cone of physical and chemical opportunity in an ecosystem, increasing and improving the retention and use of elements and energy – a sophisticated way of increasing the rate of approach to equilibrium in a continuous increase of fitness within several large groups of species, in chemotypes (see cover and Fig 11.7). The randomness of individual species is then not very different from the randomness of molecules in any dynamic non-living system while selected *ordered* phases also occur, all of which are fundamental features of all equilibrium chemical systems, and can be rationalised. However, energy absorption allows ensemble structures within systems in flow to be *organised*, not ordered, and they will evolve advantageously where the advantage lies in a thermodynamic efficiency of energy degradation. We now attempt to show that the idea of the directional change of chemotypes is not in conflict with our knowledge of the changes of genetic structures.

11.13. Chemotypes and Genotypes

Before proceeding with the discussion of our systems approach to evolution as we see it developing now under mankind's influence we stress again that we are not concerned with species, except mankind itself, but with the general development of chemotypes which are described as those broad groups of organisms that have major distinctive chemical, energetic, spatial or organisational features. *Species* within chemotypes may well be a chance selection but new chemotypes only arise to take advantage of major new ways of using and adapting to the changing environment of elements, space and energy. Now the problem as to how such persistent chemotypes arose must be in some manner related to the way new species occur since both changes parallel changing genes: chemotypes must be related to genotypes. How can the inevitable change of the environment be linked to gene change so as to allow the deeper evolution of response to and use of chemicals, energy and space? Let us go back to the beginning of cellular life.

We all accept that in order for a living system to have evolved from its initial state of non-reproductive being it had to discover a mode of coded reproduction coupled with an efficient capture and cooperative use of energy and materials. The code is very well established experimentally, but the genesis of it remains a mystery, though it is the only way to ensure substantial energy retention in the system since the organic chemical system of life by itself is thermodynamically unstable in the presence of oxygen and cannot be of great kinetic stability. (Autopoeisis itself is not as effective a style of energy capture.) This coded reproductive activity was then the only way forward from abiotic to the biotic systems and which we recognise as living. However, a code-based reproduction system, once attained, is conservative and there then must be a way of adjusting what is coded if the system is to evolve. It is this problem which has reinforced the well-established idea that the whole development of life's species was due to random mutation of the code in a variable environment which, on this view, had no systematic change imposed upon it. On this view therefore there can be no direction or causation since it is not clear what could cause anything else but such a random change of genes in species that is by nucleotide base change or shuffling of base sequences. As stated above, these suggestions are backed by correlations between the evolving form and the DNA/RNA sequences of many species but it appears to conflict with our approach. We shall try to resolve this conflict by suggesting that minor variations in species are of this random kind but that they are different in selection from the development of major ways of evolving in a *directional, changing environment*, either in energy or element content, as we see them in chemotypes (see Sections 7.14 and 11.6).

There are some features of the genetic apparatus which are often lost from sight while concentration is engaged on one essential feature, the linear DNA that codes the proteins of the cell. The other features are:

(1) The large percentage of sequences which are not translated but are transcribed to RNA from introns. The value of these RNA_i is now becoming recognised. Their properties are concentration and environment dependent not just sequence dependent.
(2) The non-linear connections between DNA regions brought together by coiling and which are subject to control by transcription factors. These are then concentration-dependent cooperative features internal to the code.
(3) The expression in transcription and translation is not just qualitatively that of DNA sequences but arises from quantity controls: expression is a concentration-, temperature- and environment-dependent feature both of RNA and proteins.
(4) There are ways of modifying all the four bases of DNA chemically and of modifying the proteins bound to them, such as histones, which affect expression. Modification is environment dependent, e.g. availability of modifying agents.
(5) The DNA of advanced organisms is broken up into chromosomes so that there has to be cooperative connections between stretches of DNA well separated in space.
(6) The DNA is not reproduced exactly as the telomeres of later eukaryotes decrease in length with the cell cycle. Telomere reduction is environment dependent and does not happen in cancer cells.
(7) In evolution DNA is adaptive, some say by *random* mutation only, but we have given reason to think that the adaptation is selective in parts of DNA (see Section 7.14).
(8) In evolution parts of the code, even parts essential for independent functioning, have been lost so that higher organisms are dependent on vitamins, for example for coenzymes, and on amino acids, lipids and saccharides, or even on selenium incorporation, from other forms of life. This is selective loss in an environment of supportive organisms. Can this be random?

The features together with the demonstration, in this book, that evolution in the broadest sense is systematic with changes in the environment leads us to consider that the changes in DNA are linked to the environment changes not by simple random selection over the whole DNA. Especially the last feature shows that survival is of an ecosystem of organisms containing DNA and is not selfish in one organism but that a DNA in an organism can generate products for other distant species. Survival is of general fitness of all organisms, not of the fittest ones though fitness of species does provide local survival strength, in an environment. Symbiosis is of the essence of fitness of a total system.

Maintaining our starting point that a DNA code is essential we wish to show how we think a new chemotype could arise. Let us look more closely at the genetic apparatus. We shall next describe again our view of evolution (see Section 7.14), using as an example the effect of light which we consider created a new chemotype from the original chemoautotrophic one existing in the dark. It is likely that life did not start from the use of light as an energy source since it requires some sophisticated equipment to convert this energy into a useful chemical form: it is clearly easier to use energised chemicals at first to drive chemistry. In all probability light was a menace for these original organisms since it produced arbitrary damaging free radical chemistry. Remember that there was no ozone layer to protect from UV energy. Let us set aside damage to DNA itself by light which can produce random mutations and consider the possible damaging of chemicals including proteins and RNA in its cells. The very nature of the cells, as judged by knowledge of today's anaerobes and later cells, is that damage can be overcome. We see this in terms of replacement of damaged proteins (or RNA) by accelerated synthesis, and in the context of very early life it would have been those proteins (or pieces of RNA) most sensitive to light which had to be replaced. Such replacement involves activating corresponding genes to the damaged proteins by feedback initiation of specific protein synthesis (see Chapter 4 and Section 7.14). We have stressed that undoubtedly during such extra synthesis of a specific group of proteins (DNA $\rightarrow$ RNA $\rightarrow$ proteins), there is a greater chance of mutation of the region of the DNA which is activated, since it becomes single stranded, hence more exposed. In the case of radiation damage by light the genes related to the light-sensitive protein/RNA system are likely to be preferentially mutated. There is then the further possibility that mutations which direct the light-generated free radical reactions away from causing damage will result in the energy of light being diverted from damage and becoming used by the cell. The general change which we have taken as a trend in evolution is the adaptive sequence:

New environment (poison) $\rightarrow$ protection $\rightarrow$ use

which we now detail here as

new environment (light) $\rightarrow$ damage to proteins,

protein damage requires new protein production,

new protein production involves local loss of DNA protection,

loss of local DNA protection $\rightarrow$ localised random mutation,

localised random mutations $\rightarrow$ proteins protective against light,

further mutations of local region $\rightarrow$ proteins making use of light.

Support for such a progression is provided by the knowledge that the protein which carries the O_2-producing Mn site (see Section 5.8) is extremely sensitive to radiation and must be replaced once every half an hour. The very incorporation of Mn in this evolution can be given an explanation consistent with the above sequence of events. Protection from offensive agents such as radiation has been much discussed (see Mulkidjanian in Further Reading). The progress to the use of a new environment (a poison) clearly requires gene duplication as a consequence of gene exposure as well as gene mutation so that the good of the old is conserved while the new is introduced. Certain novel "species" would then arise but they are created not by a bifurcation leading to a slightly different cell but to a cell different in kind (Fig. 11.5), which here does not depend on energised minerals of the Earth. There is then no direct competition with the earlier organisms but selection of a separate chemotype able to obtain energy from a novel source – here light – in an environmental light-bathed zone to which the new chemotype is adapted, while it retains much of the rest of its metabolism intact, and it can even reject the unwanted genes of early forms of energy capture. We consider the general advantage to the ecosystem of the evolution of such a new chemotype albeit associated with a genotype change in Section 11.10. As well as exposure to new energy sources, new chemicals such as oxygen and copper could be effective in such major changes. Once the development occurred two (and more) chemotypes could exist separately but cooperatively since debris from the light or other novel zone would help existence in the earlier zone, and *vice versa*.

The suggested principle to explain "directed" evolution is then that mutation is not random over the whole genome but that its intensity is related to the harmful effect of a new environmental energy source or any new damaging substance. Several such substances were released in turn in time due to the oxygen increase in the atmosphere and so new chemotypes of organisms evolved in a sequence as new groups of genotypes were better able to handle the damaging environment.

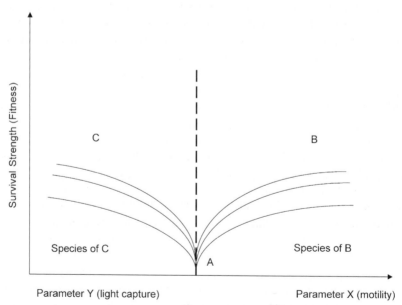

FIG. 11.5   A plot of fitness, survival strength, of two types of organism (chemotypes), one of which adapts to light and the other which does not. Both chemotypes improve in fitness with time shown by increase in use of parameters dependent on light or upon a quite other parameter which engenders motility. Intermediate cases can also occur.

There remains an orthodox Darwinian approach to arrival of new species within the chemotypes, that is by the competition for survival between similar species which could result from the random mutation of DNA. Our discussion suggests that there is an inevitable development of chemotypes with quite different capabilities as the environment changes in major ways but what we see around us is similar species in a confusingly haphazard number and to a degree competing. Evolution is then blind in similar species production but it feels its way along a cone-like constraint of increasing variety with the direction fixed by major environmental features while the diversity of its ever-increasing small variations is not directed (see cover and Fig. 11.7). All the time the efficiency of the whole ecosystem in the degradation of energy increases; while there is considerable cooperation between chemotypes and limited competition between species and individuals.

While it is possible that this explanation is sufficient for us to see how novelty in chemotypes can arise, it leaves the puzzle of the rejection of genes which are no longer of value. This rejection is more and more obvious in evolution as symbiosis increases especially as it is selective. Genes are lost for the most important processes such as energy capture and transduction found in mitochondria and chloroplasts but not fully coded in the central cytoplasmic DNA. There is also the inability to obtain essential elements and to synthesise coenzymes, and even later to synthesise "essential" amino acids, sugars and fats. Genes for steps of synthesis such as those of selenium compounds, Fe/S and porphyrins are also lost to the bacterial symbionts, i.e. organelles, and are not all programmed in the main DNA of eukaryotes. Notice the contrasting rapid increase in genes for control in the main DNA in Mg kinases, calcium proteins and zinc fingers in response to the changed possibility of sensing the environment requiring selected organisation of messengers often based on oxidation.

While the above discussion is general to the problem of mutation in organisms ranging from the original prokaryotes to the most advanced eukaryotes there was in fact a change in the nature of genes when especially in eukaryotes the DNA became wrapped in nucleosomes and higher order structures, chromosomes. These structures contain proteins, especially histones, and removal of histones during transcription again makes local DNA relatively more vulnerable to mutation and hence to the above mechanism for systematic evolution. However, these DNA-binding proteins introduce also a quite novel complication.

It is now well known (see Caporale and also Turner in Further Reading) that the histones, apart from the DNA, are often chemically modified, methylated, acetylated and phosphorylated, which affects expression. In addition to enzymes for these modifications, there are enzymes for removing them and some of both of these types of enzymes depend on exchangeable Mg^{2+} or Fe^{2+}. These modifications turn out to be systematic so that in addition to the DNA code, some scientists consider that there is a "histone code". Unfortunately, for simple description of genes and their reading the histone code is different in kind from that of the DNA code. The DNA is an all or none intensive code, though reading of it is an extensive, not an intensive, thermodynamic activity. The histones are produced in controlled concentrations and their modifications are directly dependent upon energy and both enzyme, metal ion, and substrate concentrations, that is they again have a variable probability and are not of an all or none character. In thermodynamic language they too are *extensive*, not *intensive* factors. Protein molecules are synthesised and modifications are made in a way related to the cell contents as well as to the coded units. As the chance of change is now dependent on the internal concentrations of components we also note that these extensive factors are dependent on the environment and such dependence can be inherited. Notice that there are temperature-sensitive organisms as far as protein expression is concerned. A second example is the growth of a tumour in which the tumour growth consists of cells which have inherited histone modifications. The correlation between environment evolution and organism development can then be causative. The upshot would be that the more these probabilistic factors are dependent upon the environment, the less an individual organism's inherited character would depend upon its DNA code alone. There is much discussion of this possibility today and we cannot give a definite answer yet concerning the several so-called epigenetic mechanisms.

There are other ways of evolution of organisms than by simple mutation of individual bases or changes in bound proteins. As explained in Section 7.14, it may become advantageous to move genes around before or, especially, after gene doubling so that they fall under different promoter controls at the level of transcription/translation. We know that gene shuffling is possible and is again due to built-in enzyme activity but it may well be difficult to see the effects – almost impossible from the analysis of DNA without the *quantitative* examination of RNAs, the proteome, the metabolome and of the metallome. This gene shuffling may also be activated by environmental insult (see Caporale in Further Reading). We do not know of clear-cut examples but we do know that evolution in a relatively short period of time can produce a huge variety of species, and at the same time some within a very different chemotype. An example is the explosion of species and shortly thereafter of chemotypes at the beginning of the Cambrian Period. A singular development too with little change in DNA is the separation of *Homo sapiens* from chimpanzees which has developed very quickly.

Now recently and increasingly it has been observed that there are also feedbacks to DNA mutation not just from DNA exposure or chemical modifications of the DNA or the histones associated with it but by small interfering RNA molecules derived from *intron transcription*, i.e. RNA_i (see Couzin in Further Reading). (Remember that the productions of RNA_i are not in a one-to-one relationship to a part of a code, they are extensive thermodynamic properties and unlike DNA itself are variable in expression with cell conditions, both chemical and physical.) To quote Matzke and Matzke (see Further Reading) "We know that RNA_i, after being transcribed from DNA, can feedback to direct modifications of the genome. These modifications can be inherited through cell divisions and influence development". Of course, the RNA_i binding will also affect local mutational rates as they then act to protect introns. Intron DNA sequences have a low level of mutation. It therefore appears that DNA changes, through multiple circuitous routes, can sense and respond to the environment of a cell.

TABLE 11.13

STRATEGIC MUTATIONS[a]

Organisms	Nature of Mutation
Snail	Variation of poisons
Lyme spyrochetes	Variation of protection
Many bacteria	Development of resistance in plasmids
Human immune system	Variable region of antibodies, histocompatibility genes
Human serine protease	Variable region of receptor site genes
Hot spots	For example in DNA within local exons (hot spots) recombination[b]
Especially plants	Jumping genes (Transposon sites)

[a] See L.H. Caporale, *Darwin in the Genome*, McGraw-Hill, New York (2002).
[b] Recombination, the cross-over between male and female genes on fertilisation occurs at local regions of the DNA within exons not within introns.

In Table 11.13 we tabulate other indications of what has been called strategic mutations, that is local mutations or other changes in genes which are stimulated by environmental stresses (see Caporale in Further Reading). It is clear that we need to consider mechanisms of connecting the environment and genes much more directly than we have analysed so far.

11.14. Mankind's Industrialised Society

We turn now to the final steps of evolution to the present time as seen in developments of and due to mankind. These changes, we propose, are but a sophisticated extension of the pattern of evolution of chemotypes though they are much more dramatic in nature and time. Using our approach we cannot allow separation of cognitive from metabolic activity as an excuse for removing human kind from general evolution as some biologists appear to wish to do since we are analysing by continuous chemical thermodynamics. Human beings present as great a change in evolution of chemotypes in 10,000 years as in the preceding four billion and in one sense they were predictable with hindsight. They seem to represent the last steps of the expansion of cooperation from energised chemistry inside organisms to that outside them.

We have described the development of mankind's society in Chapter 10. Here we draw together some of its threads in order to show how the advances interact with biological evolution. Our major point is that development is based on access to new chemicals and new energy sources together with new space and organisation and the necessary communication systems, that is in exactly the same way as all previous biological evolution, to aid survival while generating heat. However, it is not due to the unavoidable introduction of a product such as oxygen as in the case of earlier evolution, an environmental change which was at first countered and then used, but to the *deliberate* introduction of change based on the understanding of chemistry and physics, including the thermodynamics of the environment. For example, especially recently mankind has learnt to use greater energy and higher temperatures to enable access to a diversity of elements and then to use them in new materials. These ways of handling the environment are deliberate, that is cognitive. Thus, historically the ages of the history of mankind are labelled Stone (physical, not chemical, but note bricks and mortar), Bronze (Cu, Sn), Iron, and finally Industrial or should it be called The Age of all the Elements (see Table 10.5), when evolution has entered a final stage. (Mankind could be considered as a very rapid succession of chemotypes and note that the order of elements used is that of availability as in all other stages of evolution.) The availability of the appropriate elements has followed the rise in working temperatures (i.e. intense use of energy) from 300 K (primitive) via 1,000 K (Iron Age) (coal) to 3,000 K (Al) in element production. (Note that early access to new elements by primitive organisms also required the use of a higher energy source, light and then oxygen, but observe too how long that change took.) Mankind's sequence is systematic as in biological evolution, although in a quite different way and much faster, but there is again only one direction to change, exploitation of new made-available elements in new spatial constructs. Now the use of new materials requires energy from sources other than directly from the Sun to achieve high temperatures and extra skills. History shows how mankind has increased energy sources systematically, such as fossil fuel and water power and uses in novel furnaces (Section 10.6). We must be aware that the parallel with the introduction of a new element, oxygen, carried a threat to the anaerobes which produced this "waste". What are today's threats?

At the same time mankind had to engage in building an increased organisation as the uses of chemicals and energy increased much like that in all other chemotype evolutions. The build up must be toward a complete global activity in biological/environmental ecology as organisation increases in efficiency. Command, transport and messenger activities corresponding to those in organisms have been introduced (see Sections 10.9 and 10.10) around a structure based on localised industries delivering products under central management for the purposes of overall organisation. The necessary increasing change of mode of information transfer by faster messages over longer distances, electro-magnetic transmission, also follows that seen in biological evolution. The parallel with the ecosystems of organisms in all these features, energy, chemicals, space, organisation and communication is not exact but close. Organisation, biological or cognitive, assists effectiveness.

The changes need to be seen against the general proposition that as far as possible they must be made compatible with the need to optimalise in an ecosystem the absorption of energy and its degradation in a cyclic system. Now change always brings with it the risk of biological poisoning by elements or of excess heat (global warming), which in practice may be very difficult to avoid if change is rapid. We have to see evolution and the part played by ourselves in a rapid continuum towards novelty in the context of increasing risk to life, which can be circumvented only if we act rationally (Table 11.14). In the past ecological systems reduced risks slowly in well-defined ways (see Table 11.7). If it is the case that we can understand evolution in thermodynamic terms then it is time to ask if we are behaving to our own long-term

TABLE 11.14

SOME ECOSYSTEM SIDE-EFFECTS OF HUMAN ACTIVITY

Habitat and species loss (including conservation areas)
Truncated ecological gradients
Loss of soil fauna
Side effects of fertilisers, pesticides, insecticides, and herbicides
Proliferation of resistant strains of organism
Genetic loss of wild and domestic species
Over harvesting of renewable natural resources
High soil surface exposure and elevated albedo
Accelerated erosion
Nutrient leaching and eutrophication
Pollution from domestic and commercial wastes
Atmospheric and water pollution
Global changes in lithosphere, hydrosphere, atmosphere, and climate

After D. Western, see Further Reading.

advantage or are we in danger of damaging the purely "biological" system but mainly ourselves quickly, while the ecosystem searches slowly for a different way forward.

11.15.   Mankind's Interference with the Ecosystem

We have stated that the geological/biological ecosystem evolved so far along certain inevitable lines of necessity, but this drive cannot allow for rapid intervention by knowledgeable animals, using new chemicals, new energy and space in a novel way (Table 11.15). We now have to look at mankind's approach against the background of what was previously pre-determined in our opinion by physical and chemical environmental and energy constraints. Let us return to the following general diagram:

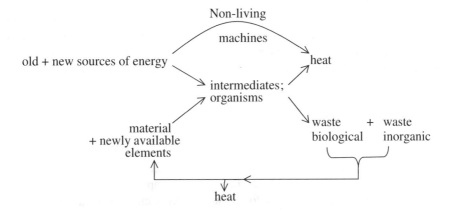

The additional input by mankind is the directed use of energy, even in novel forms, to newly available as well as previously available materials. The new inorganic

TABLE 11.15

ECOLOGICAL PRINCIPLES FOR CONSERVING ECOSYSTEM PROCESSES

Maintain or replicate
   Species richness
   Internal regulatory processes (e.g. predator–prey interactions)
   Large habitat areas and spatial linkages between ecosystem
   Ecological gradients

Minimise
   Erosion, nutrient leaching and pollution emission
   Landscape homogenisation

Mimic
   Natural process in production cycle

After D. Western, see Further Reading.

materials are largely metals, especially iron, and building materials, such as bricks, concrete, glass, etc., and organic plastics. All these generated chemicals, which cannot assist evolution of the organisms in the above diagram, could be called just polluting waste. Again much of the new energy with materials goes into products for comfort and pleasure, e.g. heating, lighting, travel, etc. which are activities unrelated to the cycle but they produce side-products such as CO_2 and cause excess temperature change, all of which can again be looked upon as waste. The tendency by mankind is also to increase production of animals rather than plants, that is cattle not forests. The plants are often seen as mainly food for animals including mankind. The fact that plants are not seen as the major source of retention of energy (even fossil fuels) will clearly lead to a reduction in activity unless new energy sources are introduced. The time scale of the problem is not clear since other factors such as climate change, introduced quite probably by mankind's activities, have an unknowable consequence. Within the next 100 years there may be a crisis for mankind, not particularly for the total ecosystem which has always fluctuated, if the production of materials not usable by organisms is not monitored and managed and new large energy sources are found. The only suitable energy known at present is nuclear with its risks though a desirable alternative would be to increase the cycle

$$H_2O \xrightarrow{\text{(light)}} H_2 \rightarrow H_2O + \text{useful energy} + \text{heat}$$
$$\downarrow O_2 \nearrow$$

In this cycle the reaction of $H_2 + CO_2$ to give reduced carbon-based fuels such as sugar or *oil with oxidation back to H_2O and CO_2 could be useful. It is then precisely the same as the biological cycle*. Note that at present mankind and organisms utilise only a small fraction of the Sun's energy, which could be used as the source of hydrogen from water if waste is to be avoided. The insight is essential for us all to understand; in effect we should be copying plants and animals within modern industry thereby preserving the ecosystem. The drawbacks are the ever-increasing heat production and solving the energy problem may not solve that of material waste. While undoubtedly we are increasing the rate of energy degradation we must also be aware that rejected chemicals (waste) can affect the cycle so as to reduce the amount of "structure" in it (see the ozone layer in Section 3.4). One may say additionally that much of the changes produced by man are not an advance of life as they are not self-reproducible. Let us then examine reproduction.

11.16. Reproducibility: Human Inheritance of Information

Reproducibility depends on transfer of a code in a space containing equipment for gathering energy and material. DNA (RNA) alone are inadequate and produce dead viruses on their own. The reproduction of organisms is in fact by a code with a reading machine which is interactive with the ability to collect energy and material. An alternative could be to develop (chemical or physical) systems of a different kind which can reproduce or be reproduced not internally in what we call organisms but externally. In principle, they could use quite different types of code and hence of reproduction. Their activity must be coded and must be transferable. The obvious examples of codes we use already are symbols contained in "language", whether in letters or numbers (mathematics) that are in books or computers. They provide an inheritance, a memory, but do not change the organisms, humans which pass them down the generations, nor can they reproduce such inheritance via genes. We are aware that this form of transfer looks vulnerable and has lasted for less than 10,000 years or so. The understanding of biological reproduction leads to the need for reproduction to sustain one or other of these activities, i.e. in machines which make machines. There is no inherent impossibility for otherwise there would have been no life. It is for mankind to think if he wishes this to happen and then to try to do it. Notice the new system will have to use "available" materials and energy from the environment and it cannot escape the necessary consequence of general energetics treated by thermodynamics. All we wish to do here is to remove the idea that mankind as a chemotype cannot possibly devise ways of creating a new chemotype, which is a product of cognitive, not metabolic, change until it becomes independent and is capable of producing itself. Maybe the first steps will be to change the present genetic system and use it to produce new animals able to inherit different types of "genes", e.g. for knowledge of basic chemistry and physics. After all a bird inherits knowledge of how to build a nest, how to travel and how to communicate and there are many who believe we inherit the rules of language so that one form of code may be transferred to another one. While we realise that all of this is presently science fiction it does indicate that selection within existing "chance" schemes is not the final limit of evolution.

It is then up to society to educate itself while being cautious before rushing into new advances. Our present procedure of educating individuals so as to continue our inheritance is quite different from genetic inheritance. Previous organisms were largely restricted to patterns of behaviour by genes in closely knit populations, but today we face a quite different problem, that of the individual, which introduces a hazard of a quite novel kind, which only education can rectify.

11.17.   The Individual as a Problem

We now turn to a peculiar twist in evolution only reached clearly in the case of mankind. We have defined chemotypes not just in terms of chemical content but also as possessing an organisation dependent on inherited sets of genes. While each chemotype included many similar species with very similar chemistry and organisation, we have avoided analysis of species considering them to have arisen by chance. The variation within a species, that is, of the phenotype, has been put totally to one side as being beyond fruitful physical/chemical analysis. Additionally, we have had no occasion to refer to the individual in any previous population. By contrast we have seen that in the evolution of *Homo sapiens* we have had to consider one single species as a chemotype. It so happens that this difference with all other chemotypes, which arises from the strong involvement of a new interaction with the environment, also forces us to re-appraise our interest in individuals. While a certain amount of a human individual's behaviour in the external world is dominated by genetic inheritance, much of it depends upon the information gained from experience. While the genetic information in each individual in a species, including humans, as seen in the new born, is very similar, not identical for reasons connected to sexual reproduction and to small gene modifications during the life of the parents, external influences including education, have contributed greatly to the nature of the mature individual. We can see this trend to increased individual development in higher animals while it is unremarked in insects and lower organisms, though they may work in multiples of units, flocks or herds. In other words, the human phenotype becomes a strongly observed feature which we cannot predict directly from the genes of an individual but is forseeable from later animal behaviour. Moreover, for a reason which it is not for us to examine here, the evolution of the brain has taken a peculiar twist – strong individual awareness or self-consciousness. Jointly the variation in capability and awareness produced in individuals the ability to act independently, which has meant that organisation of individuals has to be different from anything achieved earlier – that is by consent. It is this self-interest which creates a problem for the ecosystem. (Some would like to distinguish the human species as a "nootype"* (see Morowitz in Further Reading; a word used by Teilhard de Chardin.))

---

* From the Greek *nous* meaning thinking and compare the English "nous".

A peculiar difficulty has then arisen for group behaviour with human development as adult individuals, now as much a product of internal phenotypic as of
chemotypic or genotypic characteristics. Decision making, through the advance of
mankind's brain, consciousness, technology and culture, has become part of the
"right" of a single human individual to act alone though living within a huge organisation. This is part of the nature of a society of individual organisms under the title
of "freedom within a democracy". The ability to choose or create a personalised
lifestyle has the great satisfaction of boosting the ego but is difficult to keep under
the full control of this kind of central command. It is easy to see that the individual
presents problems which have never been of consequence before in evolution. The
dichotomy is now in the *design* of society, organisation, by mankind itself to allow
individual action within necessary central control of the whole society where the
limits must be introduced of necessity. It is quite peculiar that a single organism has
to understand and accept constraints *rationally*, some of which cannot be forced
upon it, for the good of the whole system. The social system has to set up external
coded rules (not to be found in genes) to govern while allowing for the individual
freedom. It is not just that people must manage and accept management of themselves but they have to accept that certain restrictions of their social environment are
necessary for them if mankind is to manage the whole geological/biological ecosystem in a sustainable way. There are limits to exploitation of the environment
which are a direct consequence of the very nature of this evolving ecosystem (see
Tables 11.14 and 11.15) with the mutual dependence on all life and the environment.
This is a matter of great chemical, physical and social concern for the ecosystem.

Now there is another risk for the ecosystem which has arisen from mankind's
knowledge and increasing power, and which is unfortunately readily passed on by
teaching. The clear-cut difference between humans and all other forms of life has
led to the idea that the species is special in the so-called "design" of the whole
ecosystem (even in the universe). This is not just expressed in religions but in a
more general view of human superiority leading to the idea that selected people,
mostly groups of people, have rights based on individual freedom which are not to
be curtailed. Such a view can be held in either democratic or dictatorial societies.
This belief of superiority is often at the root of education and contains a fundamental error. In many ways as a chemotype, mankind is not superior but is just different. The difference has allowed him to become the major controller of much of the
ecosystem on Earth's surface but he faces the same problems as any other novel
chemotype in that he is a product of and a part of the ecosystem and totally dependent upon it. This problem is the same as that of the anaerobes which took advantage
of getting hydrogen from water but "overlooked" the waste oxygen. No part can be
as great as the whole and it is the whole ecosystem which has to be preserved carefully. This means that we have to think globally in the widest sense whether we like
it or not. Scientists have a special responsibility in bringing attention to the problems based on scientific ideas, which are universal. In essence our explanation of
how a system of energised chemicals can evolve and not decline is that it must not
store up waste which could be a threat to it or could diminish resources. Waste must

be used or completely hidden or eventually it becomes a threat. Individual activity, unrestricted, is extremely unlikely to be restrained in this way. Yet our future depends on understanding by individuals of features of our present and past and responding responsively. It relies on individual and corporate human understanding not just of new scientific advances, modifications of DNA, or novel reproduction schemes but of societal needs. Further advance, scientific and otherwise, has to be toward management of the whole ecosystem, which now rests in mankind's hands, for purposes to be decided by mankind, desirably as a consequence of developing a well-advised population, which is not the same as decisions made by popular vote for the creation of wealth. Wealth creation by itself is dangerous.

11.18. Summary and the Possible Future of the Ecosystem

The unavoidable sequence of events on the Earth from the beginning of life which has created our ecosystem (see Fig. 11.6), we consider are the following:

1. Life started as a single compartment, a protocell or cell, of selected reduced chemicals and a restricted number of internal ions using external energy, probably from mineral sources (see Chapter 1 to 4).
2. Genes stored information about the internal system and allowed cell reproduction. Probably all organisms belonged to one chemotype swapping genes before more restricted chemotypes evolved.
3. The necessary communication network in all chemotypes was based originally on exchangeable reduced available internal ions such as Mg^{2+} and Fe^{2+} and reduced organic chemicals, substrates and coenzymes, often linked to phosphate, and feedback and forth of information about cell conditions to and from the genetic central control.
4. There was an improvement in the use of material and energy capture due to the introduction of coenzymes and such chemicals as metal porphyrins. Very distinctive chemotypes evolved to optimalise the use of particular environmental circumstances. The environment presented very few hazards if we disregard light.
5. The major energy source from the Sun (the hazard, light) was introduced and oxygen generated. A very different chemotype evolved.
6. Evolution of the environment was inevitable but gradual due to oxidation of the environment by the oxygen since the internal cytoplasmic constructs of cells had to remain reduced. Change was initially faster in the inorganic environment. Together organisms in chemotypes and the environment formed an evolving ecosystem moving in one chemical direction, the separation of reduced and oxidised chemicals. The environment was changed at close to equilibrium but change was slow due to the immense reserves (buffers) of environment reducing materials, Fe/S.

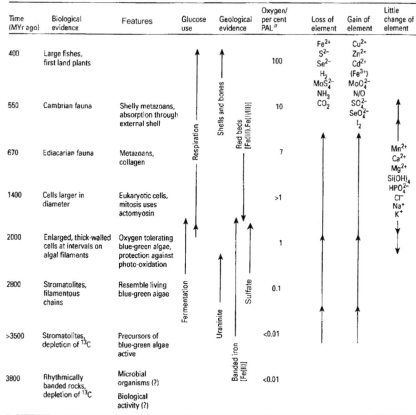

^{a}PAL = present atmospheric level.

FIG. 11.6 A diagram correlating the rise of O_2 partial pressure with the ecological evolution indicating element availability.

7. To make use of the increases in oxygen and oxidised chemicals, an inevitable change, organisms were forced to develop means of utilising them and reduced materials at catalysed rates now faster than those of the environment. Compartments in single cells, to keep oxidative and reductive chemistry apart, increased effectiveness of these reactions but at a price. Novel messengers linking the environment and cell compartments more closely were introduced, note especially Ca^{2+} messengers, to secure longer life. The number of chemotypes co-existing and supportive increased. Genes became more complex and symbiosis became general. The price was slower reproduction rate and adaptability but greater complexity.
8. Cells became differentiated and linked, by oxidised cross-links, to obtain and conserve energy and control space more effectively in multi-cellular organisms. Again this change had its origin in further oxidation of the environment with

which cells interacted more strongly; note the increases of Cu and Zn released into the environment. The environment and organisms became more closely linked but adverse factors also increased.

9. A communication network of organic messengers advanced further as compartments multiplied to form organs. Cells and organs informed one another in an enclosed total body with internal "environmental" fluids. Step by step, organisms became more aware of the external environment and responded to it independently of direct genetic involvement. The multi-cellular organisms advanced into new chemotypes.

10. New extended filamentous cells extended in cross-linked networks to organs and gave faster communication in animals using expelled Na^+ and Cl^- and internal K^+ flows in nerves and using other messengers from (6) and (8) at synapses.

11. Throughout these stages of environment and organism evolution, reliance of chemotypes on one another, symbiosis, increased to increase energy and material capture in a more economical way. Complexity was relieved by dependence on lower organisms. Fitness in the uptake and degradation of incoming energy increased.

12. To maintain organisation in mobile animals the brain evolved from nerve cell packages. The brain stored information about the environment so that increased interaction between environment and organisms evolved again.

13. The brain allowed greater and greater use of the environment since it, unlike the genes, had information from the environment, made useful in a rapid operational way and hence made possible environmental control by organisms. The brain developed its own coded information.

14. Mankind evolved and came to use all possible chemical combinations of elements, energy sources, and communication means to make every type of device in external space and to control all life forms consistent with mankind's desires and advancement. In effect one species became a new chemotype. In principle mankind is the final chemotype but it is the most dependent of all on basic chemicals of all other chemotypes.

15. Concomitantly individuals evolved and with this "advance" the need to act by consensus with all that this implies for society. Simultaneously mankind came to consider that it had a special function (of selfish interest) in the ecosystem.

16. Throughout the whole of this evolution it is the ecosystem which evolves with all kinds of chemotypes. The whole of biological evolution is but advancing the effect of high-frequency energy applied to material and released as low-frequency heat – an ever-increasing rate of thermal entropy production.

17. Mankind is introducing quite a new approach, as he seeks material and energy sources not available previously. What effect does this have?

We can now present the treatment of the evolution of our ecological system as a diagram in the form of a cone with a cut-away sector (see Fig. 11.7 and the cover of this book). The cone space encloses all the chemistry of the environment and organisms from the beginning of life on Earth to today in a continuous manner. Its central

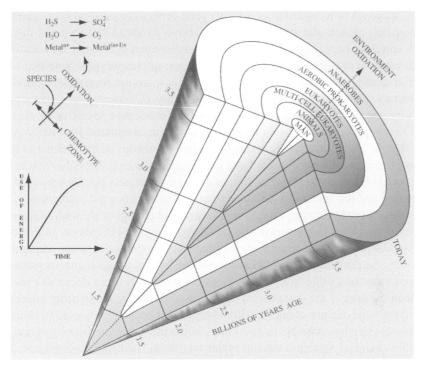

FIG. 11.7 A diagram representing the development of our ecosystem. Time is along the axis of the cone with separation of oxidised chemicals in the environment and reduced chemicals in increasing numbers of chemotypes, see text. The Darwinian tree of species evolution fits into the cone and has linear connectivity while the ecological cone is continuously filled. The upper side-figure indicates the extent of each zone and the species in it. The lower side-figure shows the increase of use of energy, the rate of entropy production, with time.

axis is time, but with time has come an increasing separation of oxidised (mainly environmental) and reduced (mainly organism) chemicals. We can represent this change also by a qualitative increase of oxygen in the environment along the cone axis. During time new major chemotypes have appeared as shown as zones within the cone and those present at any time are in a cross-sectional plane of the cone, while species of any chemotype are seen in the width of the chemotype zone, see upper side-figure of Fig. 11.7. Proceeding up the centre of the cone we observe that the number of compartments and organisation increase as the more recent chemotypes arise. There is of necessity a systematic increase in messengers and in interaction with the environment. The increasing width of the cone also depicts the increasing conversion of energy from the sun into heat, see lower side-figure of Fig. 11.7, the increasing rate of thermal entropy production, and with it the increasing diversity of chemicals in the ecosystem. A three-dimensional, Darwinian, branched line tree of species evolution (see Fig. 11.2) can be fitted neatly into this continuous three-dimensional cone of ecological evolution, in which organisms and the environment are increasingly becoming one system though this end point has not been reached yet, contrast Gaia.

The approach to a condition of a fully cyclic ecosystem must now be examined by mankind, against the premises of (15) above, to avoid self-inflicted disaster which could set back, not stop, the positive development of biological evolution to which we have drawn attention in our description of chemotypes. Note that some material changes of today, e.g. in nuclear reactions, are not reversible. Here some consideration must be given also to the possibility of evolving new chemotypes either by introducing changes in the present day genetic equipment, or by creating quite new reproductive systems using very different chemistry. It must be made clear that the underlying thesis in this book, that organisms adapt to make as full a use as possible of the environment and its changes, makes it inevitable that evolution would eventually generate cognition and self-awareness but the timing of this happening and the species in which they would occur is beyond prediction.

Finally, instead of stating evolution in terms of chemotypes we wish to return to the description of it just in terms of total energy flow, see (16) above. In this book we have shown that the general development of evolution including mankind is driven by the over-riding thermodynamic necessity that all spontaneous processes proceed towards equilibrium. This includes the possibility of an interactive process in which the drive is the degradation of high-quality energy, but in order to degrade it efficiently the capture of high-quality energy is, of necessity, coupled to chemical synthesis as much as possible to give kinetically semi-stable but thermodynamically unstable chemical systems, leaving organisms in an oxidising environment. This coupling increases inevitably towards optimal trapping in a cycle of capture of high quality and loss of low-quality energy (higher frequency energy goes to lower frequency) since this increases the rate at which the whole ecosystem of external energy plus chemicals goes towards equilibrium, while generating an increasing rate of thermal entropy production. The increase of energy capture in useful material for energy retention is a slow process since new mate-rials are produced by side reactions and they have to be incorporated. The incorporation requires systematic increase in the use of space and in control, that is, in organisation, and in genetic structure. In other words, the ecosystem advanced towards optimal energy fitness. Until mankind arose, the steps of change were very slow due to the environmental buffering and the conservative nature of DNA. The question arises as to whether mankind, an organism capable of understanding the whole of this ecosystem and its evolution, can manage it appropriately (see Goudie and Western in Further Reading). In our last section we have shown that this is dependent upon mankind's own understanding of the nature of itself as a chemotype within an energised ecosystem but also as a special organism able to extend influence to the whole ecosystem. The ability to reach this endpoint due to cognitive activity and from the ability to store and use information almost outside any genetic control gave rise to the phenotypic individual. Only if individuals come to accept through education that we are locked into the whole ecosystem can our activities be consistent with any certainty for our future. Darwinian ideas may well have been used to over-stress the competition between species as opposed to the wider cooperativity in the total system of organism and this has had unfortunate consequences, e.g. in eugenics.

The application of science is an intensely political problem for scientists who should engage in politics for only they can give appropriate information. A very extraordinary development would be if mankind corporately, yet voluntarily, recognised that through inheritance of mastery of external events it must accept responsibility for the whole ecosystem and act appropriately. That is no small act of responsibility with very far-reaching educational and poli-tical implications.

Appendix 11A. A Note on Gaia

An original suggestion by Lovelock that the organisms on the Earth manage the environment to their advantage led to the suggestion that changes in the environment and of organisms were tightly coupled. Despite the very many modifications of it the original idea of tight coupling persists so that it is considered that the Earth acts as one "living" system on its surface. The problem we face with this description of coupling has been voiced by many authors who point out that though organisms disturb the environment and they adjust to these changes there is really only adaptation. We subscribe to this view. The activity of organisms is not very well integrated with the environment though it attempts to follow its changes, irrespective of the source of environmental change. In no sense is there any kind of a "deliberate effort of organisms" to improve the environment for the good of organisms. Rather the changes are accidentally caused by organisms which have to adapt so as to increase survival. Of course, the two are linked as life has to use available material and energy from the environment. We must observe that environmental change has also features independent of life. Consider the original sea with the desirable Fe^{2+} and HS^- concentrations for anaerobic life. The production of oxygen as waste by these very organisms led to a stripping of the ocean of both Fe^{2+} and HS^- as they were converted to Fe^{3+} and SO_4^{2-}. This led to much heat production. Much of the iron is buried in the Earth as iron oxides. There is no way this was other than a disaster for anaerobic life which helped to generate it. The biological release of O_2 generated consequences quite outside its control. By adaptation (the "enforced" slow change of genes) organisms worked hard to obtain Fe^{3+} and SO_4^{2-} and then to remake Fe^{2+} and HS^-, but there was faster energy degradation due to this considerable energy cost. The organisms cannot react so as to improve the environment for themselves, but they can evolve so as to become able to survive efficiently. Note that finally the oxidation of the environment led to a more effective total rate of energy degradation within the ecosystem as a whole. The original forms of life were banished from all aerobic environments. Another example of non-biological development is erosion. The sea is saturated with $CaCO_3$ and SiO_2 and subsequent precipitation, formation of shells, is connected with life but this precipitation of carbon and silica would occur without life. Thus, the run-off of water from land carries Ca^{2+} to the sea and there they precipitate aided or not by biological activity. These are consequences of inorganic cycles and equilibria and all steps release heat. In fact, much CO_2 is locked up permanently as dolomite due to subsequent transformations of $CaCO_3$ to $Ca_xMg_{1-x}CO_3$ and hence the carbon is lost from the carbon cycle. Another loss of carbon is as methane, oil and coal, about which organisms can do very little as it is stored deep below the surface in the absence of O_2. Furthermore erosion creates the basic condition for the deve-lopment of soils and then of land organisms. Finally, the deeper reduced zones of the Earth which mop up O_2 slowly are of huge capacity making it impossible to achieve balances. The chemical and energy fluxes of the Earth quite free from life are on a huge scale, for instance, keeping the temperature of the Earth suitable for life. These are natural inorganic processes and though some can become coupled to organisms by adaptation, others cannot. Our approach to systems change is that indeed it is directed but not just to organisms advantage. It is directed by energy flux and, because of the way pre-existing materials are of necessity reduced by energy uptake to create organisms, waste is generated. Organisms are a part of this energy flux and intermingle with that due to inorganic processes dependent on the Sun, e.g. erosion, or on the energy of the deep hot levels of the Earth. It is the total degradation of this energy that drives the system and though it all may strive to reach a final steady state of energy flux and a state of cyclic material flow (no waste) which could be an ultimate form of Gaia this state has never been achieved and is not obviously achievable.

References to Gaia

Kirschner, J.W. (2002). The Gaia Hypothesis: fact, theory and wishful thinking. *Climate Change*, 52, 391–408

Lenton, T.M. and Wilkinson, D.M. (2003). Developing the Gaia theory. *Climate Change*, 58, 1–12

Lovelock, J. (1979). *Gaia – New Look at life on Earth*, Oxford University Press, Oxford

Volk, T. (1998). *Gaia's Body*. Springer-Verlag, Copernicus, New York

Further Reading

BOOKS

Barrow, J.D., Freeland, S.J., Harper, C.L. and Morris, S.C. (eds.) (2006). *Fitness of the Cosmos for life: Biochemistry and Fine-Tuning* (promoted by Templeton Foundation). Cambridge University Press, Cambridge, to be published, (promoted by Templeton Foundation)

Burrow, J.W. (1968). *Introduction to Charles Darwin's The Origin of Species by Means of Natural Selection*. Penguin Books, London (1st ed. 1859).

Butcher, S.S., Charlson, R.J., Orians, G.H. and Wolfe, G.V. (eds.) (1992). *Global Biogeochemical Cycles*. Academic Press, London.

Caporale, L.H. (2003). *Darwin in the Genome*. McGraw-Hill, New York.

Corning, P.A. (2003). *Nature's Magic: Synergy in Evolution and the Fate of Humankind*. Cambridge University Press, Cambridge.

Dawkins, R. (1986). *The Blind Watchmaker*. Oxford University Press, Oxford.

Dawkins, R. (2003) *The Ancestor's Tale – a Pilgrimage to the Dawn of Life*. Weidenfeld & Nicholson, London.

Goudie, A. (2000) *The Human Impact on the Natural Environment*, Blackwell Science, Oxford

Gould, S.J. (1989). *Wonderful life*. Norton, New York.

Gross, M. (1998). *Life on the Edge*. Plenum Press, New York.

Lovelock, J. (1988). *The Ages of Gaia*. W.W. Norton and Co., New York.

Maynard Smith, J. and Szathmáry, E. (1998). *The Origins of Life*. Oxford University Press, Oxford.

Morowitz, H. (2002). *The Emergence of Everything*. Oxford University Press Inc., New York.

Mulkidjanian, A.Y., and Junge, W. (1999). Primordial UV-protectors as ancestors of the photosynthetic pigment protein. In G.A. Peshek, *et al.* (eds). *The Phototropic Prokaryotes*. Kluwer Academic Press Pub., New York, pp. 805–812

Selinus, O. Alloway, B., Centeno, J.A., Finkelman, R.B., Fuge, R., Lindh, U. and Smedley, P. (eds.) (2005). *Essentials of Medical Geology*. Elsevier, Amsterdam.

Southwood, R. (2003). *The Story of Life*. Oxford University Press, Oxford.

Turner, B.M. (2001). *Chromatin and Gene Regulation – Molecular Mechanisms in Epigenetics*. Blackwell Science, Oxford.

Ulanowicz, R.E. (1997). *Ecology – the Ascendant Perspective*. Columbia University Press, New York.

Wilson, E.O. (1999). *Consilience – The Unity of Knowledge*. Little, Brown and Co., London.

Wilson, E.O. (2003). *The Future of Life*. Abacus, London.

PAPERS

Bitbol, M. and Luisi, P.L. (2005). Autopoiesis with or without cognition: defining life at its edge. *J. Royal Soc. (London). Interface*, *1*, 99–107 and references therein.

Colaço, A., Dehairs, F. and Desbruyères (2002). Nutritional relations of deep-sea hydrothermal fields at the mid-Atlantic ridge: a stable isotope approach. *Deep-Sea Res. I, 49*, 395–412.

Couzin, J. (2002). Small RNAs make big splash. *Science, 298*, 2296–2297.

DeLong, E.F. (2004). Microbial life breathes deep. *Science, 306*, 2198–2200.

Gibbs, W.W. (2004). Synthetic Life. *Sci. Am., 290*, 74–81.

Kirschner, J.W. (2002). The Gaia hypothesis: fact, theory, and wishful thinking. *Climatic Change, 52*, 391–408.

Lenton, T.M. and Oijen, M. (2002). Gaia as a complex adaptive system. *Phil. Trans. R. Soc. London*, B, *357*, 683–695.

Lenton, T.M. and Wilkinson, D.M. (2003). Developing the Gaia theory. *Climatic Change, 58*, 1–12.

Llewellyn-Smith, C. and Ward, D. (2005). Fusion Power. *European Review, 13*, 337–360.

Mattick, J.S. (2004). The hidden genetic program of complex organisms. *Sci. Am.*, October, pp. 31–37.

Matzke, M. and Matzke, A.J.M. (2003). RNA extends its reach. *Science, 301*, 1060–1061.

Neuberger, M.S., Harris, R.S., Di Noia, J. and Petersen-Mahrt, D. (2003). Immunity through DNA deamination, *Trends in Biochem. Sciences, 28*, 305–312.

Mulkidjanian, A.Y., Cherepanov, D.A. and Galperin, M.Y. (2003). Survival of the fittest before the beginning of life: selection of the first oligonucleotide-like polymers by UV-light. *BMC Evolut. Biol., 3*, 12.

Westall, F. (2005). Life on early Earth: a sedimentary view. *Science, 308*, 366–367.

Western, D. (2001). Human modified ecosystems and future evolution. *Proc. Natl. Acad. Sci. USA, 98*, 5458–5465.

Note Added to Proof

A book entitled "Fitness of the Cosmos for Life", see Barrow et. al., in Further Reading will give a view of "Fitness" from the viewpoint of many different scientists.

A scientific discussion meeting was held (2005) by The Royal Society (London) on "Major steps in cell evolution" and will be published *Phil. Trans. Roy. Soc. B 361* in 2006, eds. Cavalier-Smith T., Embley, T.M., and Brasier, M.D.

Index